SOIL MECHANICS:
Principles and Practice

SECOND EDITION

Graham Barnes

First published 2000 by

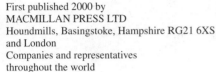

MACMILLAN PRESS LTD
Houndmills, Basingstoke, Hampshire RG21 6XS
and London
Companies and representatives
throughout the world

ISBN 0–333–77776–X

A catalogue record for this book is available from the British Library.

This book is printed on paper suitable for recycling and made from fully managed
and sustained forest sources.

10	9	8	7	6	5	4	3	2	1
09	08	07	06	05	04	03	02	01	00

Printed and bound in Great Britain by
Antony Rowe Ltd
Chippenham, Wiltshire

Contents

Preface

Aims of the book

The main aims of this book remain the same as for the first edition. The book is intended to provide:

- an understanding of the nature of soil;
- an appreciation of soil behaviour; and
- a concise and clear presentation of the basic principles of soil mechanics.

Who is *Soil Mechanics* for?

The book is intended to support the main courses for undergraduate civil and ground engineering students. It provides the basic principles of the subject and illustrates how, why and with what limitations these principles can be applied in practice.

It is also intended that the book will be retained by these students when they become practitioners and used by professionals already in practice as a reference source offering guidance and information for the solution of real geotechnical problems.

Changes to the Second Edition

For the second edition all chapters have been reviewed to update the existing material and to ensure a thorough explanation of each topic. Additional topics have been incorporated into most of the chapters with more reference material and further worked examples.

In addition, the following features have been included:

- A list of Objectives at the beginning of each chapter. These should guide the student through the required key elements before embarking on detailed reading.
- Worked Examples have been retained at the end of each chapter to ensure a consistent flow of the subject material, but markers have been inserted in the text to identify where the

student can pursue the relevant numerical example.

- Summaries have been included at the end of each chapter to allow the student to review the main points before moving to another chapter.
- A Glossary for the more unusual terms that are not explained in the text.

Case Studies

A major addition to the book is the introduction of case studies. These are included in most chapters to give an insight into the problems that engineers did not foresee but how, in most instances, they have learnt from these studies and adapted their design philosophies. They are also intended as supportive reading, with the aim of exciting the committed geotechnical engineer but also to stimulate those less committed civil engineers. At the very least they should be read as a warning that things can go wrong and that too much reliance on theories will never be a substitute for experience.

Limit state design

In chapters 8, 10, 11 and 12 the design philosophy of limit states and partial factors is introduced. This is incorporated into the European geotechnical codes, the Eurocodes and in the more recent British Standards.

Lecturer's manual

A manual containing the worked solutions to all of the end-of-chapter problems is available free to lecturers using *Soil Mechanics* on their courses. For more information, contact your Macmillan Press academic sales representative, or the Sales Support Department of Macmillan Press in Basingstoke.

Acknowledgements

I wish to record my thanks to all those researchers, writers and practising engineers who have investigated the subject and collected information over the last 70 years, without whom no standard textbook could be written. My thanks also go to those publishers, organisations and individuals who have granted permission to use material from their publications. In particular, I wish to thank Professor Burland of Imperial College for his assistance in the preparation of the case study on the Leaning Tower of Pisa.

The photograph of the Tower of Pisa on the front cover of the book was provided by Max Soudain of Ground Engineering magazine. The other photograph on the front cover is of the quick clay slide at Trogstad, Norway in 1968 and was kindly provided by Kjell Hauge of the Norwegian Geotechnical Institute.

Graham Barnes

List of symbols

A	Activity
A	Area
A'	Effective area
A	Pore pressure parameter
A_c	Ash content
A_b	Pile base area
A_f	Pore pressure parameter at failure
A_r	Area ratio
A_s	Pile shaft area
A_v	Air voids content
a	Slope stability coefficient
B	% of particles passing maximum size
B	Width of foundation
B	Pore pressure parameter
$\bar{B}$	Pore pressure parameter
B'	Effective width
b	Slope stability coefficient
C_E	Energy ratio correction
C_N	Correction for overburden pressure
C_R	Correction for rod length
C_W	Correction for water table
C_c	Compression index
C_c	Coefficient of curvature
C_α	Coefficient of secondary compression
C_s	Soil skeleton compressibility
C_s	Swelling index
C_w	Compressibility of pore water
CD	Consolidated undrained
CU	Consolidated undrained
CI	Consistency index
CSL	Critical state line
c_a	Adhesion
c_b	Adhesion at underside of foundation
c_r	Remoulded undrained cohesion
c_u	Undrained cohesion
c_v	Coefficient of consolidation, vertical direction
c_H	Coefficient of consolidation, horiz. direction

c_w	Adhesion between soil and wall or pile
c'	Cohesion in effective stress terms
D	Depth of foundation
D	Depth factor of slip circle
d	Diameter, depth of penetration, particle size
d	Length of drainage path
d_0	Initial depth of embedment
D_r	Relative density
E'	Young's modulus in terms of effective stress (drained condition)
E_p	Pressuremeter modulus
E_u	Young's modulus in terms of total stress (undrained condition)
E	Lateral force on side of slice
ESP	Effective stress path
e	Eccentricity
e	Void ratio
e_0	Initial void ratio
e_f	Final void ratio
e_{max}	Void ratio at loosest state
e_{min}	Void ratio at densest state
F	Factor of safety, length factor
F	Force
F_d	Enlargement factor
F_B	Correction for roughness
F_D	Correction for depth of embedment
f	Shape factor or intake factor
f_0	Slope stability correction factor
f_s	Skin friction, sleeve friction
f_s	Shape factor
f_t	Correction for time
f_t	Permissible tensile strength of reinforcement
f_y	Yield factor
f_1	Thickness factor
G	Shear modulus
G_s	Specific gravity of particles
g	Gravitational acceleration (9.81 m/s^2)
g	Soil constant for the Hvorslev surface

H	Height, thickness, horizontal force	N_q	Bearing capacity factor
H_c	Constant head above the water table	N_γ	Bearing capacity factor
H_0	Initial head above the water table	NC	Normally consolidated
H_t	Head at time t	N_0	Specific volume at $p' = 1.0$ kN/m^2 on K_0CL
h	Head difference	N_s	Stability number
h_c	Capillary rise	n	Porosity, number of piles
h_m	Mean head	n	Slope stability coefficient
h_p	Pressure head	n	Ratio R/r_d
h_s	Fully saturated capillary zone	n_d	Number of equipotential drops
h_w	Depth to water table	n_f	Number of flow paths
h_z	Elevation or position head	O_c	Organic content
I	Influence value or factor	OC	Overconsolidated
ICL	Isotropic normal consolidation line	OCR	Overconsolidation ratio
I_c	Compressibility index	PL	Plastic limit
I_p	Plasticity index (or PI)	PI	Plasticity index (or I_p)
I_z	Strain influence factor	P	Force
i	Hydraulic gradient	P_a	Resultant active thrust or force
i_c	Critical hydraulic gradient	P_p	Resultant passive thrust or force
i_e	Exit hydraulic gradient	P_{an}	Normal component of active thrust
i_m	Mean hydraulic gradient	P_{pn}	Normal component of passive thrust
J	Seepage force	P_w	Horizontal water thrust
K	Absolute or specific permeability	p	Pressure, contact pressure
K_0	Coefficient of earth pressure at rest	p	Stress path parameter (Total stress)
K_0CL	K_0 normal consolidation line	p'	Stress path parameter (Effective stress)
K_a	Coefficient of active earth pressure	p_c'	Preconsolidation pressure
K_{ac}	Earth pressure coefficient	p_e'	Initial isotropic stress
K_p	Coefficient of passive pressure	p_0'	Present overburden pressure (Effective stress)
K_{pc}	Earth pressure coefficient	p_0	Total overburden pressure
K_s	Coefficient of horizontal pressure	Q	Steady state quantity of flow
k	Coefficient of permeability	Q_{ult}	Ultimate load
k	Coefficient for modulus increasing with depth	Q_s	Ultimate shaft load
		Q_b	Ultimate base load
L, l	Length, lever arm	Q	Line load surcharge
L'	Effective length	q	Flow rate
LL	Liquid limit	q	Uniform surcharge
LI	Liquidity index	q	Stress path parameter (Total stress)
M	Moment	q'	Stress path parameter (Effective stress)
M	Gradient of the critical state line on $p'-q'$ plot	q_a	Allowable bearing pressure
MCV	Moisture condition value	q_{app}	Applied pressure (or q)
m	Mass	q_b	End-bearing resistance
m	Slope stability coefficient	q_c	Cone penetration resistance
m_v	Coefficient of volume compressibility	q_{max}	Maximum bearing pressure
N	Normal total force	q_s	Safe bearing capacity
N	Stability number	q_{ult}	Ultimate bearing capacity
N	Specific volume at $p' = 1.0$ kN/m^2 on ICL	R	Resultant force, distance
N	Standard penetration test result, No. of blows	R	Dial gauge reading
N'	Corrected SPT value	R	Radius of influence of drain
N'	Normal effective force	R_f	Friction ratio
N_c	Bearing capacity factor		

R_s	Pile group settlement ratio		W_t	Total weight
R_3, R_t	Time correction factors		W_w	Weight of water
R_T	Correction for temperature		X	Shear force on side of slice
r	Radial distance, or radius		x, y, z	Coordinate axes
r_d	Radius of well or drain		Z	Dimensionless depth
r_u	Pore pressure ratio		Z_I	Depth of influence
S_r	Degree of saturation		z	Depth
SL	Shrinkage limit		z_a	Height above the water table
s	Spacing of drains, spacing of piles, anchors		z_c	Critical depth
s	Stress path parameter (Total stress)		z_c	Depth of tension crack
s'	Stress path parameter (Effective stress)		z_0	Depth of negative active earth pressure
T	Shear force, surface tension force, torque		α	Angle, angular strain
T	Tensile force in reinforcement		α	Shaft adhesion factor
TSP	Total stress path		α_F	Settlement interaction factor
T_v	Time factor for one-dimensional consolidation		α_p	Peak adhesion factor
T_R	Time factor for radial consolidation		β	Angle, relative rotation
t	Time		β	Skin friction factor
t	Stress path parameter (Total stress)		χ	Proportion of cross-section occupied by water
t'	Stress path parameter (Effective stress)		δ	Angle of wall friction, base sliding, piles
U	Water force		Δ	Delta, change in, increment of
U, U_c	Uniformity coefficent		Δ	Relative deflection
U_c	Combined or overall degree of consolidation		$\delta\rho$	Differential settlement
U_R	Degree of radial consolidation		$\delta\rho_h$	Differential heave
U_v	Degree of one-dimensional consolidation		ε_a	Coefficient of secondary compression
$\bar{U}$	Average degree of consolidation		Φ	Potential function
UU	Unconsolidated undrained		ϕ	Friction angle
u	Horizontal displacement		ϕ_1	ϕ before pile installation
u_a	Pore air pressure		ϕ_u	Angle of failure envelope, undrained condition
u_w	Pore water pressure			
V	Volume		ϕ_{cv}	ϕ at constant volume
V_a	Volume of air		ϕ_μ	Particle–particle friction angle
V_0	Initial volume		ϕ_m	Mobilised friction angle
V_s	Volume of solids		ϕ_r	ϕ at residual strength
V_T	Total vertical laod		Γ	Specific volume at $p' = 1.0$ kN/m^2 on *CSL*
V_v	Volume of voids		γ	Unit weight
V_w	Volume of water		γ_b	Bulk unit weight
v	Velocity		γ_d	Dry unit weight
v	Specific volume		γ_{min}	Dry unit weight in loosest state
v_κ	Specific volume on isotropic swelling line at $p' = 1.0$ kN/m^2		γ_{max}	Dry unit weight in densest state
			γ_{sat}	Saturated unit weight
$v_{\kappa0}$	Specific volume on anisotropic swelling line at $p' = 1.0$ kN/m^2		γ_{sub}	Submerged unit weight
			γ_w	Unit weight of water
v_s	Seepage velocity		η	Efficiency, viscosity of fluid
w	Water content or moisture content		κ	Slope of overconsolidation line
w_c	Saturation moisture content of particles		λ	Slope of normal consolidation line
w_e	Equivalent moisture content		λ	Pile adhesion coefficient
W	Weight		θ	Rotation, inclination of a plane
W_p	Weight of pile		μ	Interparticle friction

μ	Vane correction factor	ρ_T	Total settlement
μ	Correction for consolidation settlement	ρ_y	Immediate settlement including yield
μ_1	Influence factor	σ	Total stress
μ_0	Correction for depth	σ_N	Normal total stress
μ_r	Correction for rigidity	σ'	Effective stress
v	Poisson's ratio	σ_N'	Normal effective stress
ρ	Mass density	σ_m	Mean stress
ρ_b	Bulk density	$\sigma_1,\sigma_2,\sigma_3$	Major, intermediate and minor principal total stresses
ρ_d	Dry density		
ρ_f	Fluid density	$\sigma_1',\sigma_2',\sigma_3'$	Major, intermediate and minor principal effective stresses
ρ_s	Particle density		
ρ_w	Density of water	σ_H,σ_H'	Total and effective horizontal stresses
ρ_{all}	Allowable settlement	σ_v,σ_v'	Total and effective vertical stresses
ρ_i	Immediate settlement	τ	Shear stress
ρ_c	Consolidation settlement	τ_y	Yield stress
ρ_h	Heave	ω	Tilt, correction for strength of fissured clays
ρ_s	Secondary settlement	Ψ	Flow function
ρ_t	Consolidation settlement at time t		

Note on units

SI Units

The International System of units (SI) has been used throughout in this book. A complete guide to the system appears in ASTM E 380 published by the American Society for Testing and Materials. The following is a brief summary of the main units.

The base units used in soil mechanics are

Quantity	Unit	Symbol
length	metre	m
mass	kilogram	kg
time	second	s

Other commonly used units are:
for length:

micronmetre (μm)
millimetre (mm)

for mass:

gram (g)
megagram (Mg)
1 Mg = 1000 kg = 1 tonne or 1 metric ton

for time:

minutes (min)
hours, days, weeks, years

Mass, force and weight

Mass represents the quantity of matter in a body and this is independent of the gravitational force. Weight represents the gravitational force acting on a mass. Unit force (1 N) imparts unit acceleration (1 m/s^2) to unit mass (1 kg). Newton's law gives

weight = mass × gravitational constant

The acceleration due to gravity on the earth's surface (g) is usually taken as 9.81 m/s^2 so on the earth's surface 1 kg mass gives a force of 9.81 N.

The unit of force is the newton (N) with multiples of

kiloNewton (kN) = 1000 N
megaNewton (MN) = 10^6 N

Measuring scales or balances in a laboratory respond to force but give a measurement in grams or kg, in other words, in mass terms. Confused?

Stress and pressure

These have units of force per unit area (N/m^2). The SI unit is the pascal (Pa).

1 N/m^2 = 1 Pa
1 kN/m^2 = 1 kPa
(kilopascal or kilonewton per square metre)
1 MN/m^2 = 1 MPa

Density and unit weight

Density is the amount of mass in a given volume and is best described as mass density (ρ). The SI unit is kilogram per cubic metre (kg/m^3). Other units are megagram per cubic metre (Mg/m^3).

Density is commonly used in soil mechanics because laboratory balances give a measure of mass.

Unit weight is the force within a unit volume (γ) where

$$\gamma = \rho g$$

The common unit for unit weight γ is kiloNewton per cubic metre (kN/m^3) or sometimes MN/m^3.

Unit weight is a useful term in soil mechanics since it gives stress directly when multiplied by the depth.

List of case studies

CHAPTER 1

Soil formation and nature

OBJECTIVES

- To appreciate the wide variety of soil types that exist in nature and the importance of geological processes in their individual depositional and post-depositional environments.
- To understand how the nature of the soil particles, their combinations and their structural arrangements determine the engineering behaviour of the soil.

This chapter has been divided into three sections: soil formation; soil particles and soil structure.

Soil formation

Soils in the engineering sense are either naturally occurring or man-made. They are distinguished from rocks because the individual particles are not sufficiently bonded together.

Man-made soils

These are described as made ground or fill (BRE Digest 274:1991). The main types of made ground are:

- waste materials
- selected materials.

Waste materials

These include the surplus and residues from construction processes such as excavation spoil and demolition rubble, from industrial processes such as ashes, slag, PFA, mining spoil, quarry waste, industrial by-products and from domestic waste in landfill sites. They can be detrimental to new works through being soluble, chemically reactive, contaminated, hazardous, toxic, polluting, combustible, gas gener-ating, swelling, compressible, collapsible or degradable.

All made ground should be treated as suspect because of the likelihood of extreme variability and compressibility (BS 8004:1986). These deposits have usually been randomly dumped and any structures placed on them will suffer differential settlements. There is also increasing concern about the health and environmental hazards posed by these materials.

Selected materials

These are materials that have none or very few of the detrimental properties mentioned above.

They are used to form a range of earth structures such as highway embankments and earth dams, backfilling around foundations and behind retaining walls. They are spread in thin layers and are well compacted. This gives a high shear strength and low compressibility, to provide adequate stability and ensure that subsequent volume changes and settlements are small.

Contaminated and polluted soils

Due to past industrial activities many sites comprising naturally occurring soils have been contaminated (there are potential hazards) or polluted (there are recognisable hazards) by the careless or intentional introduction of chemical substances. These contaminants could comprise metals (arsenic, cadmium, chromium, copper, lead, mercury, nickel, zinc), organics (oils, tars, phenols, PCB, cyanide) or dusts, gases, acids, alkalis, sulphates, chlorides and many more compounds.

Several of these may cause harm to the health of people, animals or plants working in or occupying the site and some may cause degradation of building materials such as concrete, metals, plastics or timber buried in the ground.

Naturally occurring soils

The two groups of naturally occurring soils are those formed *in situ* and those transported to their present location. There are two very different types of soils formed *in situ*: weathered rocks and peat. Transported soils are moved by the principal agents of water, wind and ice although they can also be formed by volcanic activity and gravity.

In situ soils – weathered rocks

Weathering produces the decomposition and disintegration of rocks. Disintegration is produced largely by mechanical weathering, which is most intense in cold climates and results in fragmentation or fracture of the rock and its mineral grains. Chemical alter-

ation results in decomposition of the hard rock minerals to softer clay minerals and is most intense in a hot, wet climate such as in the tropics.

A scale of weathering grades for rock masses is given in BS 5930:1981 and summarised in Table 1.1. It can be seen that several of the grades could behave as a soil rather than a rock but the soil/rock interface could be placed at different levels, as shown in Table 1.2, according to the engineering application. Thus, a 'moderately weathered' rock may behave as a soil!

In situ soils – peat

Peats are almost entirely organic matter, they are referred to as cumulose soils and may occur as high or moor peats comprising mostly mosses, raised bogs consisting of sphagnum peat and low or fen peat composed of reed and sedge peat. Moor and bog peats tend to be brown or dark brown in colour, fibrous and lightly decomposed while fen peats are darker, less fibrous and more highly decomposed.

Landva and Pheeney (1980) suggested a suitable classification for peats based on genera, degree of humification, water content and the content of fine fibres, coarse fibres and wood and shrub remnants. This is summarised in Table 2.9 in Chapter 2.

Water-borne soils (Figure 1.1)

For soils to be deposited they have to be first

Table 1.1   *Scale of weathering grades (From Anon, 1977)*

Grade	Symbol	Term	Comment
VI	RS	residual soil	100% soil, mass structure destroyed
V	CW	completely weathered	100% soil, mass structure intact
IV	HW	highly weathered	> 50% soil, rock present as discontinuous framework or as corestones
III	MW	moderately weathered	< 50% soil, rock present as continuous framework or as corestones
II	SW	slightly weathered	rock, discoloured only, weaker than fresh
I	F	fresh	rock, no visible sign of weathering

TABLE 1.2 *Soil/rock interface*

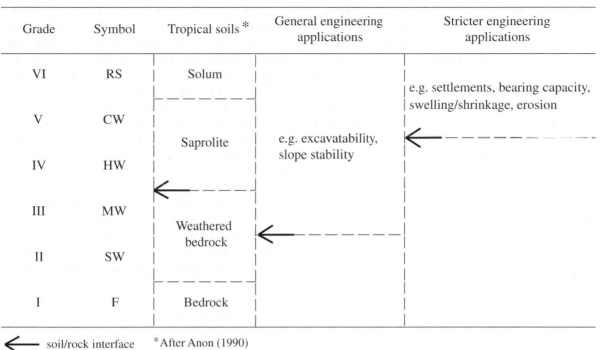

Grade	Symbol	Tropical soils *	General engineering applications	Stricter engineering applications
VI	RS	Solum		e.g. settlements, bearing capacity, swelling/shrinkage, erosion
V	CW	Saprolite	e.g. excavatability, slope stability	
IV	HW			
III	MW	Weathered bedrock		
II	SW			
I	F	Bedrock		

⟵ soil/rock interface   *After Anon (1990)

removed from their original locations or eroded and then transported. During these processes the particles are also broken down or abraded into smaller particles. The most erosive locations are in the highland or mountainous regions and upper reaches of rivers, especially during flood conditions and along the coastline, particularly at high tides and during storms. Cliff erosion can produce a wide variety of

particles sorted into beach materials (sands, gravels) and finer materials which are carried out to sea.

Soils deposited by water tend to be named according to the deposition environment as shown in Figure 1.1, e.g. marine clays. Whether a particle can be lifted into suspension depends on the size of the particle and the water velocity so that further downstream in rivers where the velocity decreases certain

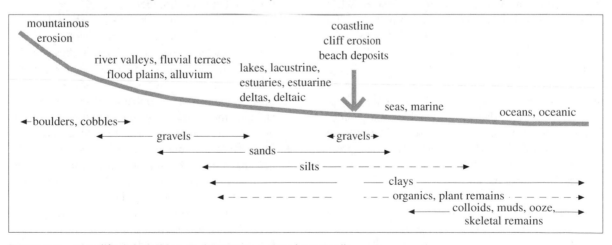

FIGURE 1.1 *Simplified deposition environment – water-borne soils*

particles tend to be deposited out of suspension progressively. However, various geological processes such as meandering, land emergence, sea level changes and flooding tend to produce a complex mixture of different soil types.

Glacial deposits

During the Pleistocene era which ended about 10,000 years ago the polar ice caps extended over a much greater area than at present with ice sheets up to several hundred metres thick and glaciers moving slowly over the earth's surface eroding the rocks, transporting rock debris and depositing soils of wide variety over northern Europe, the United States, Canada and Asia.

The deposits are generally referred to as glacial drift but can be separated into:

● soils deposited directly by ice
● soils deposited by melt-waters.

Soils deposited by ice

These are referred to as till. Lodgement till was formed at the base of the glaciers and is often described as boulder clay. Unless the underlying rock was an argillaceous shale or mudstone the fine fraction consists of mostly rock flour or finely ground-up debris with the proportion of clay minerals being low. Gravel, cobble and boulder-size lumps of rock are embedded in this finer matrix. These deposits have been compressed or consolidated beneath the thickness of ice to a much greater stress than at present and are overconsolidated which makes them stiff and relatively incompressible. The deposits have been left as various landforms: oval-shaped mounds of boulder clay called drumlins are a common variety.

Ablation till was formed as debris on the ice surface and then lowered as the ice melted and typically consists of sands, gravels, cobbles and boulders with little fines present but because of their mode of formation they are less dense and more compressible. Melt-out till was formed in the same way but from debris within the ice.

Soils deposited by melt-waters

These may be referred to as outwash deposits, stratified deposits or tillite. Close to the glaciers the coarser particles (boulders, cobbles, gravels) will have been deposited as ice contact deposits but streams will have provided for transport, sorting and rounding of the particles and subsequently producing various stratified or layered deposits of sands and gravels including outwash and fluvio-glacial deposits.

As the glaciers melted and retreated leaving many large lakes the finer particles of clays and silts were deposited (glacio-lacustrine deposits) producing laminated clays and varved clays. In deeper waters or seas where more saline conditions existed glacio-marine deposits are found.

From the soil mechanics point of view the term 'till' is of little use since it can describe soils of any permeability (very low to very high), any plasticity (non-plastic to highly plastic) with cohesive or granular behaviour. Although glacial soils are often considered to be varied and mixed it may still be possible at least within a small site to identify a series of layers or beds of different soil types and it is important to attempt a reasonable geological interpretation during the site investigation.

Post-depositional changes

These have altered glacial and many other soils in the following ways:

● Freezing/thawing – this tends to destroy the structure of the soil so that the upper few metres of say, laminated or varved clays have been made more homogeneous. Associated with desiccation vertical prismatic columnar jointing has been produced in many boulder clays.
● Fissures – these may be produced in boulder clays due to stress relief on removal of ice. Where they have opened sufficiently they may be filled with other clay minerals making them much weaker or with silt and sand particles making them more permeable. Due to chemical changes they may be gleyed (reduced to a light grey or blue colour and of softer consistency).
● Shear surfaces – due to moving ice shear stresses may have produced slip surfaces in the clay soil which can be grooved, especially if gravel particles are present, slickensided or polished.
● Weathering – oxidation will change the colours in the upper few metres, especially of clays and leaching of carbonates is likely.

- Leaching – where this has been extensive in post-glacial marine clays such as in Norway, Sweden and Canada the removal of some of the dissolved salts in the original pore water by the movement of fresh water has resulted in a particle structure which is potentially unstable. This structure can support a high void content or high moisture content (usually greater than the liquid limit) and can be fairly strong (soft, firm or stiff) when undisturbed but when it is disturbed the soil structure collapses and with the excess of water present the soil liquefies. The reduction in strength is called sensitivity and when the reduction or sensitivity is high the soils are referred to as quick clays.

Wind-blown soils

Wind action is most severe in dry areas where there is little moisture to hold the particles together and where there is little vegetation with no roots to bind the soil together. Wind-blown or aeolian soils are mostly sands and occur near or originate from desert areas, coastlines and periglacial regions at the margins of previously glaciated areas.

There are basically two forms of wind-blown soils:

- *Dunes* – these are mounds of sand having different shapes and sizes. They have been classified with a variety of terms such as ripples, barchan dunes, seif dunes and draas which are found in desert areas and sand-hills which are found in temperate coastal regions. They are not stationary mounds but will move according to wind speed and direction with some desert dunes moving at over 10 m/year and coastal sand-hills moving at a slower rate. Sand sizes are typically in the medium sand range (0.2–0.6 mm) with coarser sand particles forming the smaller mounds (ripples). In coastal regions vegetation such as marram grass is promoted to bind the sand together and stabilise the dunes for coastal protection.
- *Loess* – silt mixed with some sand and clay particles is stirred up by the wind to form dust-clouds which can be large and travel several thousand kilometres. For example, the loess found in Russia and central Europe is believed to

have originated from the deserts of North Africa. Loess deposits cover large areas of the earth's surface especially the United States, Asia and China.

During deposition a loose structure is formed but loess has reasonable shear strength and stability (standing vertically in cuts) due to clay particles binding the silt particles and to a lesser extent secondary carbonate cementation. Loess is typically buff or light brown in colour but inclusion of organic matter gives it a dark colour as in the 'black earth' deposits of the Russian Steppes. Fossil root-holes provide greater permeability especially in the vertical direction than would be expected of a silt so making the soil drain easier. However, when loess is wetted the clay binder may weaken causing collapse of the meta-stable structure and deterioration of the soil into a slurry. Therefore, loess is very prone to erosion on shallow slopes.

Soil particles

Nature of particles

The nature of each individual particle in a soil is derived from the minerals it contains, its size and its shape. These are affected by the original rock from which the particle was eroded, the degree of abrasion and comminution during erosion and transportation and decomposition and disintegration due to chemical and mechanical weathering. A discussion on particle size, shape and density and the tests required to identify these properties is given in Chapter 2.

The mineralogy of a soil particle is determined by the original rock mineralogy and the degree of alteration or weathering. Particles can be classed as:

- Hard granular – grains of hard rock minerals especially silicates, from silt to boulder sizes.
- Soft granular – coral, shell, skeletal fragments, volcanic ash, crushed soft rocks, mining spoil, quarry waste, also from silt to boulder sizes.
- Clay minerals – see below.
- Plant residues – peat, vegetation, organic matter. These are discussed in the section on soil formation, above.

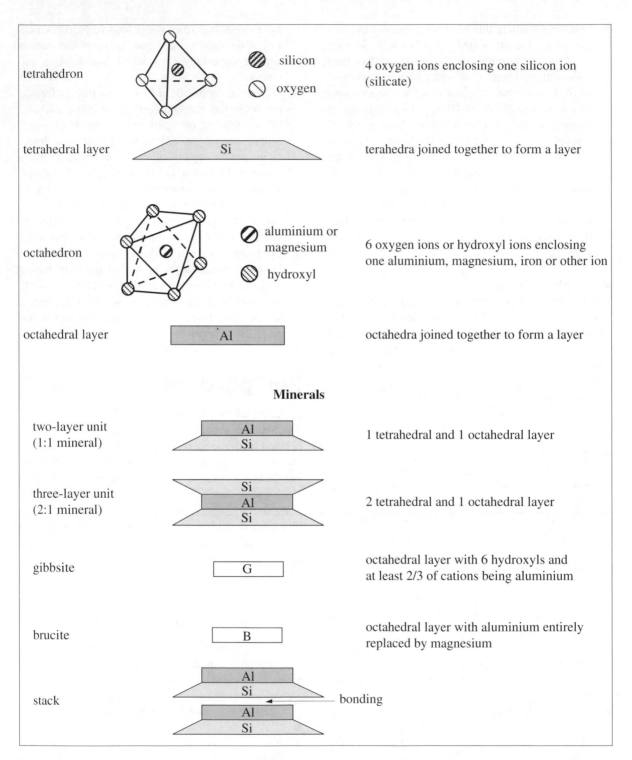

FIGURE 1.2 *Clay minerals*

Most granular particles are easy to identify with the naked eye or with the aid of a low magnification microscope after washing off any clay particles present. The hard granular particles consist mostly of quartz and feldspars and are roughly equidimensional. The quartz particles, in particular, have stable chemical structures and are very resistant to weathering and abrasion so these minerals comprise the bulk of silt and sand deposits. Gravel, cobble and boulder particles are usually worn-down fragments of the original rock.

Soft granular particles will produce a more compressible soil since the particles can be easily crushed and they are more likely to be loosely packed.

Clay minerals (Figure 1.2)

The term 'clay' can have several meanings:

1. *Clay soil* – the soil behaves as a 'clay' because of its cohesiveness and plasticity even though the clay mineral content may be small.
2. *Clay size* – most classification systems describe particles less than 2 μm as 'clay' which is a reasonably convenient size. However, some clay minerals may be greater than 2 μm and some soils less than 2 μm such as rock flour may not contain many clay minerals at all.
3. *Clay minerals* – these are small crystalline substances with a distinctive sheet-like structure producing plate-shaped particles.

Clay minerals are complex mineral structures but they can be visualised and classified by considering the basic 'building blocks' which they comprise, as shown in Figure 1.2. The octahedral sheet and the tetrahedral sheets combine to form layered units, either two-layer (1:1) or three layer (2:1) units. The octahedral sheets are not electrically neutral and therefore do not exist alone in nature. However, the minerals gibbsite and brucite are stable.

The oxygen and hydroxyl ions dominate the mineral structure because of their numbers and their size; they are about 2.3 times larger than an aluminium ion and about 3.4 times larger than a silicate ion. Even if their negative charges are satisfied, because the O^{2-} and OH^- ions exist on the surface of the sheets they will impart a slightly negative character.

If substitution of the cations has occurred, for example, Al^{3+} for Si^{4+} or Mg^{2+} for Al^{3+} because these ions were more available at the time of formation then there will be a greater net negative charge which is transmitted to the particle surface. Isomorphous substitution refers to the situation when the ions substituted are approximately of the same size. Base-exchange or cation-exchange capacity is the ability of a clay mineral to exchange the cations within its structure for other cations and is measured in milli-equivalents per 100 grams of dry soil.

The resulting negative charges are neutralised by adsorption on the mineral surfaces of positive ions (cations) and polar water molecules (H_2O) so that with various combinations of substituted cations, exchangeable cations, interlayer water and structural layers or stacking a wide variety of clay mineral structures is possible.

The structure of the more common types is illustrated in Table 1.3 and the nature of the clay particles is described in Table 1.4. Most clays formed by sedimentation are mixtures of kaolinite and illite with a variable amount of montmorillonite whereas clays formed by chemical weathering of rocks may also contain chlorites and halloysites.

Cation exchange is important in that the nature and behaviour of the clay minerals is altered. This can occur as a result of the depositional environment, weathering after deposition and leaching due to sustained groundwater flow. It can also be promoted by chemical stabilisation methods for engineering purposes such as the addition of lime (calcium hydroxide) to strengthen a soil for road building.

Soil structure

The way in which individual particles arrange themselves in a soil is referred to as soil structure. This structure is sometimes referred to as a soil skeleton.

Granular soils (Figure 1.3)

Granular soils comprise coarse silts, sands, gravels and other coarser particles. Their soil structure will depend on:

TABLE 1.3 *Structure of clay minerals*

Mineral	Layer structure	Stack structure	Bonding between layers	Base exchange capacity me/100 g
kaolinite	1:1	Al or G / Si / Al or G / Si	hydrogen bonds (strong)	3 – 15
halloysite	1:1	Al or G / Si / O O O O / Al or G / Si	hydrated with water molecules	6 – 12
illite	2:1	Si / Al or G / Si / K K K / Si / Al or G / Si	potassium ion (weaker than hydrogen bonds)	10 – 15
montmorillonite	2:1	Si / Al or G / Si / Si / Al or G / Si	van der Waal's forces exchangeable ions water molecules (weak)	80 – 140
chlorite	2:1:1	Si / Al or G / Si / B / Si / Al or G / Si	brucite sheet	20

- the size, shape and surface roughness of the individual particles
- the range of particle sizes (well graded or uniformly graded)
- the mode of deposition (sedimented, glacial)
- the stresses to which the soil has been subjected (increasing effective stresses with depth, whether the soil is normally consolidated or overconsolidated)
- the degree of cementation, presence of fines, organic matter, state of weathering.

The state of packing of a granular soil whether loose or dense can be visualised in a number of ways as shown in Table 1.5. The limits of packing can be illustrated by considering the void spaces within an ideal soil consisting of equal sized spherical particles as illustrated in Figure 1.3. In practice, typical values of porosity lie within the limits of about 35–50% with the lowest porosity and greatest density produced with:

- larger particle sizes
- a greater range of particle sizes
- more equidimensional particles
- smoother particles.

TABLE **1.4** *Nature of clay mineral particles*

Mineral	Diameter: thickness ratio	Surface area m^2/gram	Nature
kaolinite	10 – 20	10 – 70	Hydrogen bond prevents hydration and produces stacks of many layers (up to 100 per particle). Particle size up to 3 µm diameter, low shrinkage/swelling
halloysite	–	40	Two water layers between stacks when fully hydrated ($4H_2O$) distort structure to a tubular shape. Low unit weight. Water (in crystal) irreversibly driven off at 60 – 75°C affecting moisture content, classification and compaction test results
illite	20 – 50	80 – 100	Common mineral but varies in chemical composition. Particles flaky, small, diameter similar to montmorillonite but thicker. Moderate susceptibility to shrinkage/swelling
montmorillonite (smectite)	200 – 400	800	High surface area due to small (<1 µm) and thin (<0.01 µm) particles produced by water molecules and exchangeable ions entering between layered units and separating them. A good lubricant. Water readily attracted to mineral causing very high susceptibility to expansion, swelling and shrinkage

The difference in porosity between the densest state and the loosest state of a granular soil is typically about 9–10%.

Relative density

Because of the variables discussed above, porosity or void ratio are not good indicators of the state of packing so relative density, D_r is often used. This compares the *in situ* or as placed state with the loosest and densest possible states:

$$D_r = \frac{e_{max} - e_{in\ situ}}{e_{max} - e_{min}}$$

(1.1)

Name	Plan view	Elevation	Points of contact per particle	Porosity %	Void ratio
Cubic	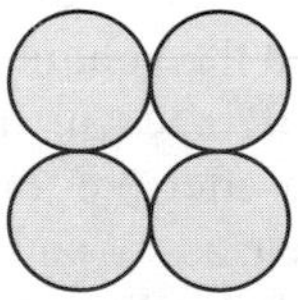	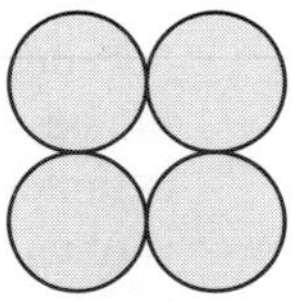	6	47.6	0.91
Rhombic		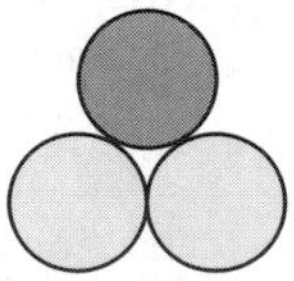	12	26	0.35

FIGURE **1.3** *Particle packing*

TABLE **1.5**   *Packing of granular particles*

Effect of packing	Loose	Dense
distance between particles	furthest apart	closest together
void space	maximum (maximum *e* and *n*)	minimum (minimum *e* and *n*)
density	minimum	maximum
particle contacts	least	most
freedom of movement of particles	most	least

where:

e_{max} = maximum void ratio
e_{min} = minimum void ratio
$e_{in\ situ}$ = *in situ* void ratio

or

$$D_r = \frac{\gamma_{in\ situ} - \gamma_{min}}{\gamma_{max} - \gamma_{min}} \frac{\gamma_{max}}{\gamma_{in\ situ}}$$ (1.2)

where:

γ_{max} = maximum dry unit weight
γ_{min} = minimum dry unit weight
$\gamma_{in\ situ}$ = *in situ* dry unit weight

Values of densities can be used instead of unit weights. Descriptive terms for the state of packing are given in Table 1.6.

The minimum and maximum densities (BS 1377:Part 4:1990) are determined for oven-dried soils so the *dry* density of the soil *in situ* must be determined.

TABLE **1.6**   *Terms for state of packing*

Term	Relative density %	SPT *'N'*
very loose	0 – 15	0 – 4
loose	15 – 35	4 – 10
medium dense	35 – 65	10 – 30
dense	65 – 85	30 – 50
very dense	85 – 100	> 50

The minimum density test consists of measuring the maximum volume that a 1 kg sample of clean, dry sand can occupy. The sand is placed in a measuring cylinder and sealed with a rubber bung. The cylinder is then shaken and inverted to loosen the sand and the volume measured when placed the right way up. The maximum volume is assessed by repeating this procedure at least 10 times.

The maximum density is determined by placing a water-soaked sample in three layers in a 1 litre mould with water in the mould above the top of the layers. Each layer is compacted for at least 2 minutes with a vibrating hammer until there is no further decrease in volume. The dry mass of soil in the mould is then determined by oven-drying and the maximum (dry) density calculated.

The *in situ* density can be obtained from the sand replacement test (BS 1377:Part 4:1990). A cylindrical hole approximately 100 mm diameter and 150 mm deep is excavated on a levelled ground surface without disturbing the soil around the sides of the hole and the mass of soil removed is determined. A clean, dry uniform sand of known or calibrated bulk density is poured into the hole from a pouring cylinder and the volume of the hole is determined from the difference in mass of the pouring cylinder before and after pouring. The bulk density is obtained from the mass of soil removed divided by the volume of the hole. The dry density is derived from the moisture content or determined from the dry mass of soil.

Relative density is not a widely used parameter apart from laboratory research and perhaps compaction control since:

1. The maximum and minimum density tests should

Elementary particle arrangements

Face to face groups of particles arranged as:

dispersed – mostly face-face arrangement of groups

flocculated – mostly edge-face and edge-edge

partly discernible – no strong structural
tendency, difficult to distinguish

Particle assemblages

clay coating
on silt and sand particles

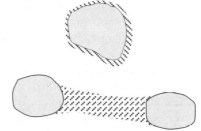

connectors
'bridges' of mostly clay particles between
silt and sand particles

aggregations
silt to fine sand size mixtures of elementary
particle arrangements

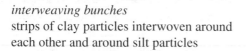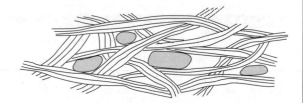

interweaving bunches
strips of clay particles interwoven around
each other and around silt particles

particle matrix
present where clay content is high binding
other assemblages together

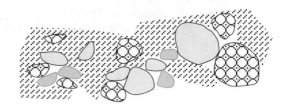

FIGURE **1.4**   *Clay particle arrangements (From Collins and McGown, 1974)*

not be carried out on slightly cohesive sands or with more than about 10% silt fines, or on particles which are crushable.

2. The range of density values between minimum and maximum can be quite small and the accuracy of the minimum, maximum and *in situ* density tests can be poor so the errors in the calculation for relative density are compounded.

3. The sand replacement test can only be carried out at or near to ground surface. Sampling of sand at depth from pits or boreholes is prone to disturbance affecting any *in situ* density determination. Relative density at depth is normally assessed by relation to the standard penetration test, SPT 'N' value, in Table 1.6.

Cohesive soils (Figure 1.4)

Clay mineral particles are too small to be seen by the naked eye so their arrangements are referred to as microstructure or microfabric and our knowledge of particle structure comes largely from electron microscope studies.

Clay mineral particles (see above) have electrically charged surfaces (faces and edges) which will dominate any particle arrangement. The microstructure of clay soils is very complex but appears to be affected mostly by the amount and type of clay mineral present, the proportion of silt and sand present, the deposition environment and chemical nature of the pore water (Collins and McGown, 1974).

These authors have observed within a number of natural, normally or lightly overconsolidated soils a wide variety of structural forms as illustrated in Figure 1.4. The engineering behaviour of clay soils (shear strength, compressibility, consolidation, permeability, shrinkage, swelling, collapse, sensitivity, etc.) will be better understood if the nature of the microstructure of the soil is appreciated.

The macrostructure of a clay soil comprises the structure which can be seen with the naked eye and generally consists of features produced during deposition such as inclusions, partings, laminations, varves and features produced after deposition such as fissures, joints, shrinkage cracks, root holes.

Moisture content is a commonly used parameter to represent the structural nature of a clay soil since it is related to the openness of the microstructure, provided the soil is fully saturated. If the soil is partially saturated then void ratio, porosity or specific volume are also used.

The state of packing of a clay soil cannot be represented by a relative density approach since maximum and minimum densities cannot be sensibly defined. The liquidity index and consistency index are parameters sometimes used to represent the structural state of a clay soil since they compare the natural moisture content with two limits, the plastic limit and the liquid limit. This is described further in Chapter 2.

SUMMARY

There is a wide variety of soil particles, combinations of particles and structural arrangements produced by geological processes in nature. Each depositional environment has its own characteristics and post-depositional processes can modify the soils dramatically. When soils are affected by engineering structures or reworked during construction the particles and their structure may be irretrievably altered.

By appreciating the importance of the undisturbed nature of soils and the effects of remoulding a better understanding of the properties of soil such as shear strength, compressibility, consolidation, permeability, shrinkage, swelling, collapse and sensitivity will be gained.

CHAPTER 2

Soil description and classification

OBJECTIVES

- To be able to provide a systematic soil description for both a hand specimen and a stratum within a soil deposit in order to convey an unambiguous statement of the nature of the soil.
- To determine values of soil classification parameters from laboratory tests including particle density, grading, bulk density, moisture content and consistency limits.
- To understand the purpose and uses of the classification tests.
- To apply the soil model to the determination of a range of parameters used in soil mechanics to denote the state or condition of a soil.

This chapter has been divided into two sections: soil description and soil classification.

Soil description

Designers of geotechnical works and the contractors constructing the works will often have seen little, if any, of the soils they intend to build on or work with. Soil description must convey sufficient information to enable the designers and constructors to appreciate the nature and properties of the soils and to anticipate the likely behaviour and potential problems. It is then essential that the site investigation company's soils engineers and geologists provide accurate and sufficient detail on the borehole or trial pit records and that the designers and constructors understand these descriptions.

Considering its importance, this topic should be given more coverage in university courses.

Classification of soils

In the UK soils can be classified according to the British Soil Classification System (BSCS) and in the United States the Unified Soil Classification System (USCS) is adopted. In the sections below the British system for soil description has been followed primarily, using the scheme given in BS 5930:1981 and as suggested by Norbury et al. (1986). Various simple tests can be carried out to help in classifying soils and these are described in the second part of this chapter, soil classification.

The major soil types are given in Table 2.1, classified according to their particle nature or size.

Scope of description

Both granular and cohesive soils can behave in an undrained and a drained manner. To avoid confusion between these terms, soils are described as:

- *very coarse soils* – particles greater than 60 mm, e.g. cobbles and boulders
- *coarse soils* – more than 65% of the soil is sand and/or gravel
- *fine soils* – more than 35% of the soil is clay and/or silt
- *organic soils* – these comprise either peat or fine, coarse or very coarse soils with an organic content.

13

TABLE **2.1** *BSCS Group symbols (From BS 5930:1981)*

Soil type	BSCS Group symbol
MADE GROUND or FILL	None
TOPSOIL	None
CLAY	C
SILT	M
SAND	S
GRAVEL	G
COBBLES	Cb
BOULDERS	B
ORGANIC SOILS	O
PEAT	Pt

There are two levels of description for soils where the material characteristics are considered distinct from the mass characteristics. Material characteristics would apply to a hand specimen which may be disturbed and refers to the nature of the soil grains, its colour, grading, shape, composition, the soil name and in the case of a clay soil, a measure of its intact or intrinsic strength and degree of plasticity.

Mass characteristics would apply to the small and large-scale structures or fabric of the soil such as bedding, joints, layering or fissures and an assessment of how this structure could affect the overall properties of the soil. These features could only be observed in larger undisturbed samples and field exposures.

When the soils have been described it is necessary to carry out a geological interpretation by considering their stratification across a site and variations within each deposit. There are two ways in which this could be achieved:

1. By applying the geological formation or type to each deposit identified, e.g. London Clay, alluvial sand, etc. These could be subdivided to give more detail such as weathering zones or grades, soil type, strength, structure, etc. Do not be reluctant to name a deposit yourself, one day it may become famous!
2. By applying a classification symbol or title to each soil type. This is useful when the soils are to be used for fill or construction materials so that soils of the same nature and properties can be easily identified, their volumes quantified, and extracted for the relevant purposes. An example of this could be where a gravel bed within other deposits is to be extracted for use as an aggregate.

Made ground

This is a man-made layer of material deposited or dumped over the natural ground. It can be either:

1. a carefully controlled construction made of suitable material spread and compacted in layers to form an engineered fill; or
2. a random, variable material formed from dumping of a variety of waste materials such as excavation spoil, demolition rubble, domestic refuse and industrial by-products.

It is best described by listing its major constituents with some estimate of relative proportions, followed by its minor constituents. A description of its mass structure and degree of compactness will provide information on its compressibility. The major constituents such as soil, ashes, rubble, refuse, degradable materials together with the age of deposition will also give some indication of its compressibility, both under its own weight and superimposed loads, and its suitability for improvement.

The minor constituents can be of equal if not greater importance since these could be contaminated, degradable, combustible, hazardous and toxic. They could produce harmful by-products such as methane gas which can explode and leachates which can pollute watercourses or aquifers and permeate into natural soils producing contaminated ground.

Topsoil

Humus is formed from the microbial breakdown of plant and animal tissues in the soil. Topsoil comprises an accumulation of humus-rich soil covering the natural inorganic soils or rocks and provides

TABLE 2.2   *Strength terms*

Term	Undrained shear strength c_u kN/m^2	Field identification
very soft	< 20	exudes between fingers when squeezed in hand
soft	20–40	moulded easily by finger pressure
soft to firm	40–50	
firm	50–75	can be moulded by strong finger pressure
firm to stiff	75–100	
stiff	100–150	cannot be moulded by fingers but can be indented with thumb
very stiff	150 – 300	can be indented by thumb nail
hard	> 300	broken with difficulty

support to plant life. It is usually no more than 100–300 mm thick although in the tropics, where intense vegetation occurs and erosion is limited, topsoil thicknesses over 1 m can exist.

Its major use in civil engineering works is for landscaping and supporting erosion protection.

Fines

These are particles smaller than 60 µm and comprise silt and clay particles. They must not be confused with the term 'fine soil' used in the BSCS which may contain up to 65% of sand or coarser particles.

Clay

According to the BSCS this is a soil comprising 35–100% fines where the clay particles predominate to produce cohesion, plasticity and low permeability. They usually plot above the 'A' line on the plasticity chart, see below.

The description of a clay soil is commonly given in the following order:

strength/mass structure/colour/soil NAME/ of – plasticity /with other structure

followed by the name of the geological formation or the type of deposit, e.g. London Clay, Estuarine clay. Some examples of soil descriptions are given in Table 2.10.

Strength terms according to laboratory tests and hand identification are given in Table 2.2. Colour terms can be obtained from Table 2.4.

The soil NAME is given in capitals. Examples are:

CLAY
Silty CLAY
Sandy CLAY   (sand 35–65%)
Very Silty CLAY
Very Sandy CLAY   (sand 65%+)

It is, however, possible for a soil to contain less than 35% clay particles but to have sufficient cohesion, plasticity or a low permeability so that it still behaves as a clay soil. This soil should be described as a clay since it is the engineering behaviour which is paramount, not the particle size distribution.

Plasticity terms are given in Table 2.5.

Mass structure
This can consist of the following.

Bedding
Terms for bedding spacing are given in Table 2.3.

Interstratified deposits
These could be partings where bedding surfaces separate easily, such as silt dusting in laminated clay, or interlaminated or interbedded silt and clay, e.g. varved clay.

Discontinuities
Bedding is treated separately, as above. Shear planes and faults are best described individually. Joints and fissures can be described in a number of ways:

TABLE 2.3 *Discontinuity terms (From BS 5930: 1981)*

Mean spacing mm	Term	
	Bedding	Other discontinuities
over 2000	very thickly bedded	very widely spaced
600 to 2000	thickly bedded	widely spaced
200 to 600	medium bedded	medium spaced
60 to 200	thinly bedded	closely spaced
20 to 60	very thinly bedded	very closely spaced
6 to 20	thickly laminated	extremely closely spaced
less than 6	thinly laminated	

- intensity – this could be described as highly fissured, fissured or poorly developed fissuring
- spacing – use a term (taken from Table 2.3) or give the range of spacings in mm
- block size – relate the size of soil lump to the particle size, such as cobble size blocks.

Further factors should be reported including:

- tightness or aperture – use a term (slightly, moderately, very open) or give the size of aperture
- orientation – give the direction and angle of dip and whether these are planar, curved, undulating or wavy
- surface texture – use terms such as rough, stepped, ridged, grooved, striated, smooth, slicken-sided, polished, or state the condition if weaker than the mass
- infilling – this could be peat, silt or sand-filled fissures.

Weathering

This is less important for soils than for rocks but it can be appropriate for tropical and residual soils and the older more heavily overconsolidated soils. Weathering grade schemes have been devised for Keuper Marl (Chandler and Davis, 1973), Lias Clay (Chandler, 1972), London Clay (Chandler and Apted, 1988), Glacial Clay (Eyles and Sladen, 1981). These are generally based on the proportion of matrix (homogeneous clay) to the remnant core-stones or lithorelics (original rock), fissure intensity, degree of oxidation and Atterberg limits.

Other structure could consist of:

- Discrete pockets, lenses or layers of peat, clay, silt, sand, gravel.
- Inclusions of gravel or cobbles in a matrix of the clay. Table 2.6 gives terms which could be used to describe the proportions present.
- Root, plant, organic, peat inclusions.

Silt

This is a soil comprising 35–100% fines where the silt particles predominate producing marked dilatancy and fairly low permeability but little cohesion or plasticity. Two types of silt can be distinguished: non-plastic and plastic silt.

With non-plastic silt the plastic limit test, see later, would tend to be difficult to perform satisfactorily and the liquid limit test would meet with varied success. This type of silt would display marked dilatancy which can be demonstrated simply by placing a small pat of wet silt (or fine sand) in the hand and squeezing it. As the soil structure expands the water is drawn into the soil giving it a matt surface and a greater strength. If the soil pat is then jolted or shaken the soil structure contracts and water oozes out of the soil giving it a shiny surface and a weak consistency.

This soil could be described in the following order:

TABLE **2.4** *Colour terms (From BS 5930: 1981)*

Tone		Shade		Colour
light	|	pinkish	|	pink
dark	|	reddish	|	red
mottled	|	orangeish	|	orange
variegated	|	yellowish	|	yellow
	|	greenish	|	green
	|	brownish	|	brown
	|	olive	|	olive
	|	blueish	|	blue
	|	purplish	|	purple
	|		|	white
	|	greyish	|	grey
	|		|	black

density/mass structure/colour/grain size/non-plastic/soil NAME/with other structure

Density terms based on the standard penetration test (SPT) are given in Table 1.6, in Chapter 1.

The mass structure and any other structure may not be as distinct as for clays but any present may be better observed by allowing a sample of the silt to partially dry. Colour terms can be obtained from Table 2.4.

Grain size can be reported as fine, medium or coarse based on the results of a sedimentation test.

The soil NAME is given in capitals as:

SILT   (sand 0–35%)
sandy SILT   (sand 35–65%)

with any gravel or cobbles described according to Table 2.6.

Plastic silt will display some cohesion and plasticity with inhibited dilatancy and can be described in the same way as a clay:

strength/mass structure/colour/soil NAME/ of – plasticity/with other structure

The soil name should reflect the degree of cohesion or plasticity with terms such as:

TABLE **2.5** *Plasticity terms (From BS 5930: 1981)*

Liquid limit	Plasticity term	Group symbol	
		Clay	Silt
< 20 (or PI < 6)	non-plastic	--	M
20 – 35	low	CL	ML
35 – 50	intermediate	CI	MI
50 – 70	high	CH	MH
70 – 90	very high	CV	MV
> 90	extremely high	CE	ME

clayey SILT – moderate cohesion, plasticity, inhibited dilatancy
SILT – (sand 0–35%)
sandy SILT – (sand 35–65%) little cohesion, greater dilatancy

Silt soils will tend to lie below the A-line on the Casagrande plasticity chart, see Figure 2.7 and can be distinguished from clay soils in this way.

TABLE **2.6** *Gravel and cobble inclusions (From BS 5930: 1981)*

Term	Approx. % of inclusions
with a little gravel or *occasional* cobbles	5
with gravel or *with* cobbles	5 – 20
with much gravel or *with many* cobbles	20 – 40
and gravel or *and* cobbles	50 – 65

Gravel should be described according to size and shape e.g. with fine to medium subangular gravel. Cobbles could be small or large with shape described

TABLE 2.7   *Soil NAME – Sand and Gravel (From BS 5930: 1981)*

Term	Fines content %	Term	Sand content %
slightly silty* or clayey*	up to 5	slightly sandy GRAVEL	up to 5
silty* or clayey*	5 – 15	sandy GRAVEL	5 – 20
very silty* or clayey*	15 – 35	very sandy GRAVEL	20 – 40
*describe according to the prominent soil type		SAND AND GRAVEL	40 – 60
		very gravelly SAND	60 – 80
		gravelly SAND	80 – 95
		slightly gravelly SAND	> 95

However, other soil types such as organic, micaceous, diatomaceous, volcanic soils and halloysite rich soils may also lie below the A-line.

Sand and gravel

According to BS 5930:1981 this is a soil containing up to 35% fines. However, if these fines are clayey or silty then they will probably dictate the behaviour of the soil so that it may still be better described as a clay or a silt. Care and judgement are required in this situation.

The order of description for a sand and/or gravel could be:

density/mass structure/colour/minor constituent/ grain size/grain shape/soil NAME/with other structure

followed by the name of the geological formation or the type of deposit such as Terrace Gravel or Glacial Sand. Some examples are given in Table 2.10.

Colour terms are given in Table 2.4 and density terms in Table 1.6.

The mass structure and any other structure may not be as distinct as for clays and will only be preserved in field exposures but terms given for the clay descriptions may be appropriate.

Grain size can be assessed by visual inspection supplemented by sieving tests and described as fine, medium and coarse sand or gravel. Grain shape can also be assessed visually. Grain size and shape should refer to the major constituent so it is prefer-

able to split the soil name, e.g. sandy fine to medium subangular GRAVEL.

Soil names as given in BS 5930:1981 are summarised in Table 2.7. The percentages are based on the whole sample less any cobbles and boulders. The terms used are satisfactory but strict adherence to the proportions can lead to conflict, especially for very clayey SAND or GRAVEL whose mass behaviour could be more like a clay when this soil would be best described as a clay.

If gravel is present conflict is avoided by treating the soil as a clay matrix, since this largely dictates its properties, with the gravel acting as inclusions, see Table 2.6. The gravel will only tend to dictate the mass behaviour when it is in sufficient proportion so that the gravel particles interfere with each other. The name claybound gravel has been used to describe this sort of material.

Cobbles and boulders

These are particles greater than 200 and 600 mm, respectively and are classed as very coarse soils. They are not usually included in a particle size distribution test by sieving. A minimum mass of about 50 kg is required for a representative sample of even small cobbles. This is an impractical amount from most site investigation methods. Instead the particles are removed from a sample by hand and their proportion estimated.

Soils containing very coarse particles and finer material can be described using the terms given in

Table 2.8 with the finer material described as a separate soil, e.g. gravelly SAND with occasional cobbles; small to medium COBBLES with some finer material (silty very sandy gravel).

Peat and organic soils

Peats are easily distinguished by their dark brown to black colour, high organic content, high moisture content and lightweight nature, especially when dried.

When significant inorganic particles such as extraneous sediments are present the soils are referred to as organic soils rather than peats. A useful classification of these soils is given by Landva *et al.* (1983) and this is summarised in Table 2.9. This is based largely on the organic content ($O_c\%$) which is obtained from an ash content ($A_c\%$) test where

$$O_c\% = 100 - A_c\% \qquad (2.1)$$

These authors recommend that the moisture content be determined by oven-drying at a temperature no greater than 90°C to expel water but also avoid charring or oxidation of the organic matter. The ash content can then be determined by igniting the oven-dried soil at a temperature of 440°C for about 5 hours to remove the combustible organic matter.

Organic soils can be distinguished from inorganic soils by their grey, dark grey or black colour and their distinctive odour which can be enhanced by gentle heating. Another approach is to determine whether there is a significant difference in liquid limit values of specimens tested from the natural state and from the oven-dried state.

Peats and peaty organic soils (see Table 2.9) could be described with the same order as for clays whereas soils with organic content should be described according to the type of inorganic soil present. The intermediate organic soils group would probably be best described according to their prominent engineering behaviour.

The organic content is determined from the masses of organic matter present. Considering the relatively low specific gravity of the organic matter compared to the mineral content their volumetric proportions will be much greater. Assuming the specific gravity of the organics to be at least half that of the mineral content then a soil with an organic

TABLE 2.8 *Cobbles and boulders (From BS 5930: 1981)*

Term	Cobble or boulder content %
COBBLES or BOULDERS with a *little* finer material	> 95
COBBLES or BOULDERS with *some* finer material	80 – 95
COBBLES or BOULDERS with *much* finer material	50 – 80
FINER MATERIAL with *many* cobbles or boulders	20 – 50
FINER MATERIAL with *some* cobbles or boulders	5 – 20
FINER MATERIAL with *occasional* cobbles or boulders	up to 5

content of 5% by mass will have an organic content by volume of over 10%. These organics will be able to support much higher moisture contents so even small organic contents can have a significant effect on the engineering behaviour.

Types of description

There are basically two types of description. One is of a piece or discrete unit of the soil such as a hand specimen when the material characteristics can be described and the other is of a thick stratum, layer or deposit when the mass characteristics can be assessed. The former should depict the classification and basic intrinsic material properties of the soil such as permeability, compressibility and shear strength while the latter should demonstrate the variability of these properties within a particular stratum or with depth (in a borehole) or in plan or section (in a trial pit or exposure).

Some examples of typical soil descriptions used on borehole or trial pit records are given in Table 2.10.

TABLE 2.9   *Classification of peats and organic soils (After Landva et al., 1983)*

Soil type	Peats	Peaty Organic soils	Organic soils	Soils with organic content
Group symbol	Pt	PtO	O	MO or CO
Ash content %	< 20	20 – 40	40 – 95	95 – 99
Organic content %	> 80	60 – 80	5 – 60	1 – 5
Particle density	< 1.7	1.6 – 1.9	> 1.7	> 2.4
Moisture content %	200 – 3000	150 – 800	100 – 500	< 100
Liquid limit %	difficult test to perform		> 50	< 50
Fibre content %	> 50	< 50	insignificant	—
Degree of decomposition (von Post)	H1 – H8	H8 – H10	H10	

von Post Classification based on degree of humification of peat

H1	no decomposition	entirely unconverted mud-free peat
H5	moderate decomposition	fairly converted but plant structure still evident
H8	very strong decomposition	well converted, plant structure very indistinct
H10	complete decomposition	completely converted

Soil classification

Various laboratory tests are available to classify soils. The main tests relate to the nature of the particles (particle density, shape, size distribution, packing or structure) and the relationship with water (moisture content, shrinkage, plastic and liquid limit).

Particle density (Figure 2.1)

Soil particles consist of a range of mineral grains of differing molecular weights with the minerals present in any soil determined by geological processes. Particle density is the mass per unit volume of the mineral grains, so for soils containing a wide range of minerals only an average value will be obtained.

Particle density can be useful in identifying the presence of certain minerals but since for most soils it has a small range of values (2.6 to 2.8) it is of limited use for this purpose. Its main use is in the determination of other derived soil properties such as the void ratio and degree of saturation, and for tests such as sedimentation tests and the air voids lines on a compaction test result.

Particle density should be quoted in units of Mg/m^3 when it will be numerically equal to the specific gravity. Specific gravity is the ratio of the unit weight of soil particles to the unit weight of distilled water and as it is a ratio it has no units.

TABLE **2.10** *Examples of soil descriptions*

Soil type	Soil description (on borehole or trial pit record)
MADE GROUND	Brick rubble, concrete pieces and lumps of clay with some slate, timber, plaster, plastic Loose state of packing with some small voids present (Demolition and excavation waste)
PEAT	Soft medium brown fibrous PEAT High fibre content, some wood and other plant remains, moderate decomposition (von Post H5) (Moor Peat)
CLAY	Very soft grey organic silty CLAY of intermediate plasticity with fine root inclusions and small pockets of peat Weathered in upper horizon with oxidised root holes and orangeish brown patches (Estuarine Clay) Firm reddish brown sandy CLAY of low plasticity with much fine to medium subangular to subrounded gravel and occasional cobbles Thin lenses (< 100 mm thick) of fine to medium sand between 5.50 and 8.50 m (Glacial Clay) Stiff fissured greyish brown silty CLAY of high plasticity with occasional bands (< 10 mm) of silt and fine sand near base of stratum Fissures planar, smooth, generally tight, mostly 50 – 100 mm apart (London Clay)
SILT	Medium dense very thinly bedded light brown non-plastic sandy medium to coarse SILT (Alluvial Silt) Firm pinkish brown clayey SILT of low plasticity with many thin (< 2 mm) bands of silty clay of high plasticity (Laminated Silt/Clay)
SAND AND GRAVEL	Medium dense brown slightly silty very sandy fine to coarse subrounded to subangular GRAVEL Sand mostly fine to medium (Flood Plain Gravel) Loose yellowish brown very silty gravelly fine to medium SAND Gravel fine to coarse subrounded Structure not distinguished (Glacial Sand)

The test requires the determination of the volume of a mass of dry soil particles. This is obtained by placing the soil particles in a glass bottle filled completely with de-aired distilled water. To remove all of the air trapped between the soil particles the bottle and its contents are either shaken vigorously for coarser-grained soils or placed under vacuum for finer-grained soils. This procedure is the most critical part of the test. The volume of the soil particles is determined from differences in mass, see Figure 2.1,

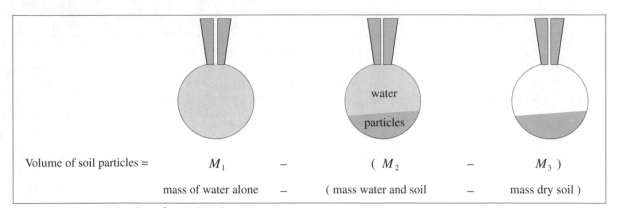

Volume of soil particles = M_1 – (M_2 – M_3)

mass of water alone – (mass water and soil – mass dry soil)

FIGURE 2.1 *Determination of particle volume*

assuming the specific gravity of water to be unity so masses in grams or Mg are equivalent to volumes in ccs or m^3, respectively.

SEE WORKED EXAMPLE 2.1

Some typical values of particle density are:

Peat	< 1 to 1.2
Feldspars	2.55 to 2.70
Quartz	2.65
Calcite	2.75 to 2.9
Micas	2.7 to 3.1
Haematite	5.0 to 5.2

For light coloured sand a particle density of 2.65 is usually adopted, with slightly higher values for darker coloured sands. For most clays a value of 2.70 to 2.72 is usually adopted in the absence of test results.

Particle shape

The shape of clay minerals is typically platy with the thickness of the grains being an order of magnitude less than its other dimensions. Clay minerals are described in more detail in Chapter 1.

For coarser grained soils, silts to gravels, the particle shape can affect the properties of the soil. The more the particle shape deviates from that of a perfect sphere the greater the angle of shearing resistance, and the greater the range between the maximum and minimum densities.

Various methods are available to quantify particle shape using the laws of granulometry but for most practical purposes it is sufficient to use general terms based on:

- origin – such as crushed rock which produces angular, rough surfaces or natural sands and gravels which are more rounded and smoother
- closest geometrical form – terms such as spherical, cubic, elliptic, prismatic, disc, cylindrical, plate, blade, needle-shaped could be used
- surface asperity or particle edges – these could be angular, subangular, subrounded, rounded or irregular.

Particle size distribution

Particle sizes vary considerably from those measured in microns (clays) to those measured in metres (boulders) – a difference of one million! Most natural soils are composite soils; a mixture of different particle sizes and the distribution of these sizes gives very useful information about the engineering behaviour of the soil.

The distribution is determined by separating the particles using two processes – sieving and sedimentation.

Sieving (Figure 2.2)

Sieves separate particles in the range between 75 mm and 60 μm (gravel and sand) and sedimentation separates particles less than 60 μm (silt and clay). Particles greater than 75 mm are cobble sized and not usually included in the tests. They are removed before testing and an estimate made of their proportion.

SEE WORKED EXAMPLE 2.2

The test can be a lengthy and involved procedure as shown in Figure 2.2, for three reasons:

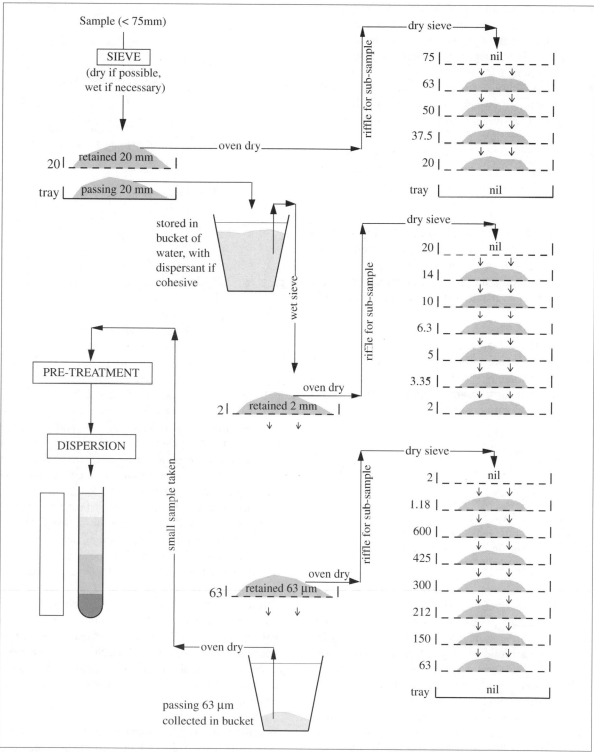

FIGURE 2.2 *Particle size distribution test*

- To ensure separation of the particles wet sieving is preferred, where the soil particles are soaked and washed through the sieves. If the soil fines are cohesive a dispersant should be added to the wash water.
- It is necessary to commence sieving and sedimentation with a known mass of dry soil. Oven-drying after wet sieving at several stages will then be necessary.
- The sieves should not be overloaded. Each sieve can support only a limited weight so smaller but representative subsamples should be obtained by riffling at several stages.

Fortunately, not all soils contain the full range of particle sizes so the test can be simplified. Clean, non-cohesive soils may only require dry sieving. It is usually considered that the sedimentation procedure is not necessary if the soil contains less than 10% fines.

Sedimentation (Figure 2.3)

This test procedure is based on Stokes' Law which states that a smooth spherical particle suspended in a fluid (water and dispersant solution) will settle under gravity at a velocity given by:

$$v = \frac{d^2}{18\eta}(\rho_s - \rho_f)g \tag{2.2}$$

where:

d = diameter of particle
ρ_s = particle density
g = gravitational constant, 9.81 m/sec^2
ρ_f = density of fluid (water and dispersant)
η = viscosity of fluid

It is essential that the soil particles are separated from each other before sedimentation so a pre-treatment procedure is carried out where any small amounts of organic matter are removed with hydrogen peroxide followed by dispersion comprising mechanical shaking for several hours in a dispersant such as a sodium hexametaphosphate solution.

Two test methods are available, the pipette and hydrometer methods. The pipette method requires taking a small sample of the soil suspension (about 10 ml) from 100 mm below the fluid surface at a particular time as illustrated in Figure 2.3. From the

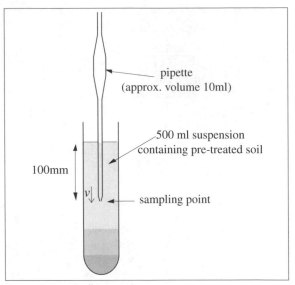

FIGURE 2.3 *Sedimentation test*

above expression, using velocity = 100/time (t), it can be seen that time $t \propto 1/d^2$ so that if the samples are taken at fixed times these will relate to the standard particle sizes. The times chosen will also depend on the particle density but for typical silts and clays, ρ_s lies between 2.65 and 2.75.

If a sample is taken by the pipette at a time of about 4 minutes then the coarse silt particles (>20 µm) will not be present and the proportion of this size range can be obtained by difference. The other times are chosen to represent the medium silt, fine silt and clay sizes but remember that these are only the equivalent spherical diameters. For silts this test may produce an adequate guide for engineering purposes but where a high proportion of flaky clay minerals is present the result will not be as accurate.

The alternative hydrometer method is based on measuring the density of the fluid and soil suspension as it reduces with time as the particles settle around the hydrometer. More readings can be taken with this test than the pipette method but it can be more prone to errors since difficulties can be experienced in reading the meniscus around the hydrometer and calibration of the hydrometer scales is required.

Grading characteristics (Figures 2.4 and 2.5)

Clay particles are distinct in their small size, flaky shape and mineralogy and give a soil its important

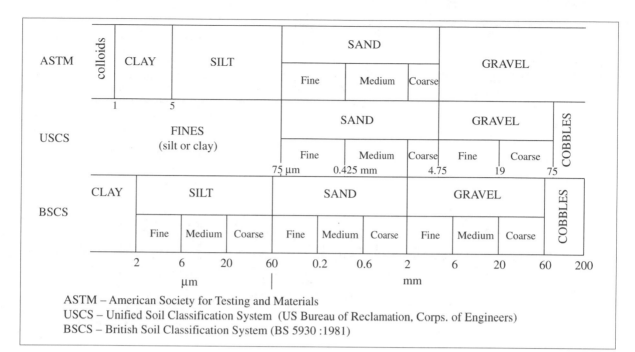

FIGURE 2.4 *Soil classification systems*

property of plasticity. However, there is no internationally agreed size for these particles, see Figure 2.4, so clay contents will differ according to the scheme adopted. Of perhaps less significance, but equally confusing, the grades of sands and gravels do not agree.

It is generally recognised that the properties of a composite soil containing a wide range of particle sizes are dictated by the finer particles, the coarser particles often simply acting as a 'filler' in a finer matrix. The British Soil Classification System (BSCS) gives more recognition to this phenomenon:

Soil group	USCS	BSCS
fine soil	> 50% fines	> 35% fines
coarse soil	< 50% fines	< 35% fines

Some 'strategic' particle sizes are defined in Figure 2.5 from which the uniformity coefficient and the coefficient of curvature are determined. These parameters represent the shape of the grading curve and denote whether the soil is well graded or poorly graded.

SEE WORKED EXAMPLE 2.2

Density

It is necessary to distinguish between density and unit weight. Density (or mass density) is the mass of a material in a unit volume. It is denoted by ρ and has units of kg/m^3, Mg/m^3 or g/cm^3. The latter two are numerically equal. Bulk density ρ_b is the total mass of soil (solid particles, water, air) in a given volume. Dry density ρ_d is the mass of just the solid particles in a given volume.

Unit weight (or weight density) is the weight of soil in a unit volume. It is denoted by γ and has units of kN/m^3. Weight is the force exerted by a mass due to gravity:

$$\text{weight} = \text{mass} \times g$$

where g is the acceleration due to gravity (m/s^2). Unit weight is related to mass density, therefore, by:

$$\gamma = \rho \times g$$
$$(kN/m^3) = (Mg/m^3) \times (9.81)$$

Unit weight is a useful parameter since it gives the vertical stress in the ground directly from:

$$\sigma_v = \gamma_b \times z$$

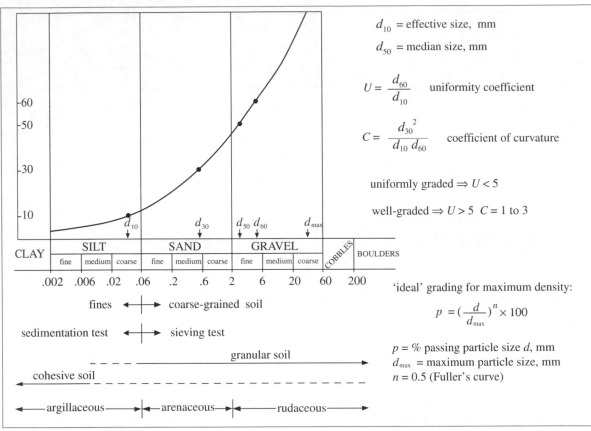

d_{10} = effective size, mm

d_{50} = median size, mm

$$U = \frac{d_{60}}{d_{10}} \quad \text{uniformity coefficient}$$

$$C = \frac{d_{30}^{\,2}}{d_{10}\,d_{60}} \quad \text{coefficient of curvature}$$

uniformly graded $\Rightarrow U < 5$

well-graded $\Rightarrow U > 5 \quad C = 1$ to 3

'ideal' grading for maximum density:

$$p = (\frac{d}{d_{max}})^{n} \times 100$$

p = % passing particle size d, mm
d_{max} = maximum particle size, mm
$n = 0.5$ (Fuller's curve)

FIGURE 2.5 *Particle size distribution parameters*

where:

σ_v = total vertical stress
γ_b = bulk unit weight
z = depth

This is described in more detail in Chapter 4.

To determine the density of soil it is necessary to measure the volume of a sample (as well as its mass). The simplest procedure is to use volumes of regular shapes such as the right cylinder used for a triaxial specimen whose volume can be obtained from linear measurements. This approach is only suitable for soils of a cohesive nature and which are little affected by sample preparation.

SEE WORKED EXAMPLE 2.3

For soils which cannot be easily formed into suitable shapes without breaking or structural disturbance then the volume of a lump of soil can be found by first coating it with a layer of molten paraffin wax

and then using either:

1. *Immersion in water*
   The coated lump is placed on a cradle. It is weighed in air and then weighed fully immersed in water. The volume of the lump of soil is obtained from:

$$V = m_w - m_g - \frac{m_w - m_s}{\rho_p} \qquad (2.3)$$

$$\frac{\text{volume}}{\text{of soil}} = \frac{\text{mass (= volume)}}{\text{of water displaced}} - \frac{\text{volume}}{\text{of wax}}$$

where:

m_s = mass of soil lump before coating with wax (g)

m_w = mass of soil lump and wax coating (g)

m_g = mass of soil lump and coating when suspended in water (g)

ρ_p = density of wax (g/cm^3)

SEE WORKED EXAMPLE 2.4

or

2. *Water displacement*

A metal container with an overflow or siphon tube is required, filled with water up to the overflow level. The wax-coated sample is then fully immersed below the water level so that the water overflowing equals the volume of the coated sample. The volume overflowing V_0 could be measured in a measuring cylinder or more accurately by weighing in grams to give volume in cm^3. The volume of the lump of soil V is obtained from:

$$V = V_0 - \frac{m_w - m_s}{\rho_p} \qquad (2.4)$$

The bulk density of the soil is then given by:

$$\rho_b = \frac{m_s}{V} \qquad (2.5)$$

The bulk density of a soil will depend on the particle density, the distances between the particles or state of packing and the degree of saturation or the amount of water and air in the void spaces. Some typical values for saturated soils are given in Table 2.11.

Moisture content or water content

Apart from soils in dry desert areas, the voids within all natural soils contain water. Some soils may be fully saturated with the voids full of water, some only partially saturated with a proportion of the voids containing air as well as water. Moisture content (or water content) is simply the ratio of the mass of water to the mass of solid particles and is an invaluable indicator of the state of the soil and its behaviour.

SEE WORKED EXAMPLE 2.3

If a sample of clay is taken from the ground the water in the voids is held in the sample by suction stresses so a true 'natural' moisture content is determined.

If a sample of sand (or silt, gravel, some fibrous peats) is taken from the ground below the water table then the voids will contain some 'free' water

TABLE 2.11 *Typical densities*

Soil type	Bulk density Mg/m³	Unit weight kN/m³
sand and gravel	1.6 – 2.2	16 – 22
silt	1.6 – 2.0	16 – 20
soft clay	1.7 – 2.0	17 – 20
stiff clay	1.9 – 2.3	19 – 23
peat	1.0 – 1.4	10 – 14
weak intact rock (mudstone, shale)	2.0 – 2.3	20 – 23
weak rock crushed, compacted	1.8 – 2.1	18 – 21
hard intact rock (granite, limestone)	2.4 – 2.7	24 – 27
hard rock crushed, compacted	1.9 – 2.2	19 – 22
brick rubble	1.6 – 2.0	16 – 20
ash fill	1.3 – 1.6	13 – 16
PFA	1.6 – 1.8	16 – 18

which will drain or flow under gravity out of the sample. The remaining capillary-held water or films of water on the particle surfaces will then be driven off in a moisture content test but the result will represent a partially drained sample not the fully saturated soil. Caution must therefore be exercised with 'free-draining' soils sampled below the water table.

The test is carried out by oven-drying a sample at a temperature between 105 and 110°C until all of the 'free' moisture is expelled. This usually means leaving the sample overnight so a result would not be available on the same day as sampling. Microwave ovens may be faster but are not recommended due to the lack of temperature control and the possibility of alteration of the solid particles, especially clays and peats.

Some soils contain minerals which on heating lose their water of crystallisation, producing a moisture content higher than the 'free' moisture content.

Examples are soils containing gypsum and certain tropical soils for which maximum temperatures of 80°C and 50°C, respectively, are suggested.

The 'natural' moisture content result is often compared with the liquid and plastic limit test results (the Atterberg limits are both moisture contents) and the optimum moisture content from a compaction test. For soils containing coarse particles (sandy or gravelly clays) a correction to the natural moisture content should be made since the latter tests are carried out on a selected or 'scalped' portion of the total sample thus:

Atterberg Limits – on particles less than 425 μm
Compaction tests – on particles less than 20 mm

If a clay soil contains sand (>425 μm) or gravel (>20 mm) particles these are likely to have little, if any, moisture associated with them.

The natural moisture content w_n% of the whole soil sample should be corrected to give the equivalent moisture content w_e% of that portion of the sample without the oversize particles using:

$$w_e = \frac{w_n}{P} \times 100$$

(2.6)

where:

$P = \%$ passing 425 μm for comparison with the Atterberg limits
$P = \%$ passing 20 mm for comparison with the compaction test result

SEE WORKED EXAMPLE 2.5

The equivalent moisture content should be used in determining the liquidity index and consistency index values, see later.

If the sand or gravel particles contain moisture due to their porous nature then the saturation moisture content w_c% of these particles should be determined and the equivalent moisture content can then be obtained from:

$$w_e = \frac{100}{P} \left[w_n - w_c \left(\frac{100 - P}{100} \right) \right]$$

(2.7)

Moisture contents of natural soils will depend on their degree of saturation but can vary considerably from less than 5% for a 'dry' sand to several hundred % for a montmorillonitic clay or a peat. As a guide

some typical values of natural moisture content are given in Table 2.12.

Consistency and Atterberg limits
(Figure 2.6)

Soils containing fines (silt and clay) display the properties of plasticity and cohesiveness where a lump of soil can have its shape changed or remoulded without the soil changing in volume or breaking up. This property depends on the amount and mineralogy of the fines and the amount of water present, or moisture content.

As the moisture content increases a clayey or silty soil will become softer and stickier until it cannot retain its shape when it is described as being in a liquid state. If the moisture content is increased further then there is less and less interaction between the soil particles and a slurry, and a suspension is formed.

If the moisture content is decreased the soil becomes stiffer as shown in Figure 2.6 until there is insufficient moisture to provide cohesiveness when

TABLE 2.12 *Typical moisture contents*

Soil type	Moisture content %
moist sand	5 – 15
'wet' sand	15 – 25
moist silt	10 – 20
'wet' silt	20 – 30
NC clay *low plasticity*	20 – 40
NC clay *high plasticity*	50 – 90
OC clay *low plasticity*	10 – 20
OC clay *high plasticity*	20 – 40
organic clay	50 – 200
extremely high plasticity clay	100 – 200
peats	100 – > 1000

NC – normally consolidated
OC – overconsolidated

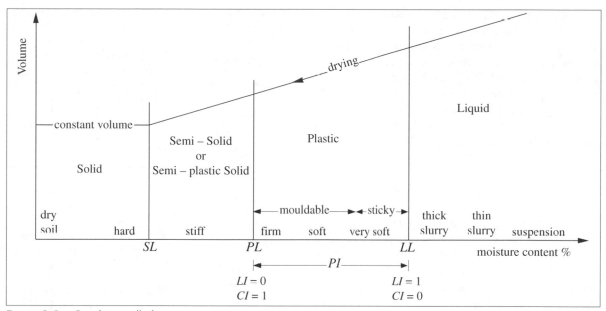

FIGURE 2.6 *Consistency limits*

the soil becomes friable and cracks or breaks up easily if remoulded. This state is described as semi-plastic solid or semi-solid. If the moisture content is decreased further there is a stage when the physico-chemical forces between the soil particles will not permit them to move any closer together and the soil is then described as a solid.

The transitions between these states are gradual rather than abrupt but it is very useful to define moisture contents at these changes as:

liquid limit – LL
plastic limit – PL
shrinkage limit – SL

even though these moisture contents may be somewhat arbitrary.

Tests for the liquid and plastic limits were devised by a Swedish soil scientist named Atterberg and they are now known together as the Atterberg limits. Both tests are carried out on the portion of a soil finer than 425 μm.

Liquidity index (LI)
This is defined as:

$$LI = \frac{w - PL}{LL - PL}$$
(2.8)

and is a good indicator of where the natural moisture content w lies in relation to the Atterberg limits. If the soil contains particles greater than 425 μm then the equivalent moisture content w_e obtained from Equation 2.6 or 2.7 should be used for the liquidity index.

SEE WORKED EXAMPLE 2.7

Consistency index (CI)
This is defined as:

$$CI = \frac{LL - w}{LL - PL}$$
(2.9)

but is used less often than the liquidity index.

SEE WORKED EXAMPLE 2.7

Liquid limit
The most reliable method for the determination of the liquid limit is the definitive cone penetrometer method (BS 1377:1990). This test consists of a 30° cylindrical cone with a sharp point and a smooth, polished surface and a total mass of 80 g which is allowed to fall freely into a cup of very moist soil which is near to and just below its liquid limit.

The penetration of the cone into the soil and the moisture content of the soil are measured. As the moisture content is increased by small amounts the penetration increases so the test is repeated with the same soil but with further additions of distilled water and a plot of cone penetration versus moisture content is obtained. The liquid limit of the soil is taken as the moisture content at a penetration of 20 mm. The test is easier to perform, less prone to operator error and gives more reproducible results than the alternative Casagrande method.

SEE WORKED EXAMPLE 2.6

The alternative method was developed many years ago by A. Casagrande where a pat of soil is placed in a circular brass cup and a groove 2 mm wide is made in the soil to separate it into two halves. The cup is then lifted and allowed to drop onto a hard rubber base. The number of such blows to cause the two soil halves to come together over a distance of 13 mm is recorded and the moisture content of the soil is determined. The test is repeated and a graph of moisture content versus number of blows is plotted. The flow of the two halves towards each other is related to the moisture content of the soil and the liquid limit is defined as the moisture content when this condition is achieved after 25 blows.

However, this method can provide poor reproducibility since it is more sensitive to operator error, requires judgement concerning the closing of the gap and it has been found that the results are affected by the hardness of the rubber base on which the cup is dropped. Comparative tests have shown that for liquid limits less than about 100 the two methods give similar results.

It has been found that the liquid limit can be affected by the amount of drying a sample of soil undergoes before the test is carried out so the current British Standard recommends that the test is carried out on soil wetted up from its natural state with any particles greater than 425 μm picked out by hand or removed by wet sieving.

Table 2.13 gives some indication of the effects of drying on index properties. These effects are most marked for less plastic soils, organic soils and tropical soils where the clay minerals aggregate to form larger particles.

Plastic limit

For reasons given above it is preferable to carry out

TABLE 2.13 *Effect of drying on index properties, Tropical soils (From Anon, 1990)*

Soil type	Test	Natural state	Air-dried	Oven-dried
Red clay Kenya	LL	101	77	65
	PL	70	61	47
	CC	79		47
Weathered shale Malaysia	LL	56	48	47
	PL	24	24	23
	CC	25	36	34
Weathered granite Malaysia	LL	77	71	68
	PL	42	42	37
	CC	20	17	18
Weathered basalt Malaysia	LL	115	91	69
	PL	50	49	49
	CC	80	82	63
Volcanic ash Vanuata	LL	261	192	N-P
	PL	184	121	
	CC	92	57	6

CC = clay content, % < 2 μm

this test on material prepared from the natural state. The soil is dried to near its plastic limit by air drying, moulding it into a ball and rolling it between the palms of the hands. When the soil is near its plastic limit a thread about 6 mm diameter and about 50 mm long is rolled over the surface of a smooth, glass plate beneath the fingers of one hand with a backward and forward movement and just enough rolling pressure is applied to reduce the thread to a diameter of 3 mm. If the soil remains intact and does not shear at this stage then it is dried further by rolling it into a ball and the test is repeated until the thread crumbles or shears both longitudinally and transversely at the 3 mm diameter. The soil is now considered to be at its plastic limit and its moisture content is determined.

It will be appreciated that the test is prone to variability since it is dependent on the individual operator's fingers, hand pressure and judgement concerning the achievement of the crumbling condition at the required diameter.

Soils also behave differently during the test. Higher plasticity (heavy) clays may not crumble easily, becoming quite tough at this low moisture content whereas for lower plasticity soils it may be difficult to produce a 3 mm thread without premature crumbling.

Plasticity index (PI)

This is defined as:

$$PI = LL - PL \qquad (2.10)$$

SEE WORKED EXAMPLE 2.7

Plasticity chart (Figure 2.7)

The Atterberg limits are useful in identifying the type of clay mineral present using the plasticity index and the liquid limit. The British Soil Classification System gives a plasticity chart (Figure 2.7) distinguishing fine-grained soils on the basis of predominantly clays (C) or silts (M) lying above or below the A-line and varying degrees of plasticity from low (LL < 35%) to extremely high (LL > 90%) with symbols for each type of soil.

SEE WORKED EXAMPLE 2.7

Organic soils usually lie below the A-line and are given the symbol O or Pt for peat. Most soils are found below the B- or U-line.

Activity

The moisture content of a clay soil is affected not only by its particle sizes and mineral composition but also by the amount of clay present. Silt and sand particles will be present in a clay soil when carrying

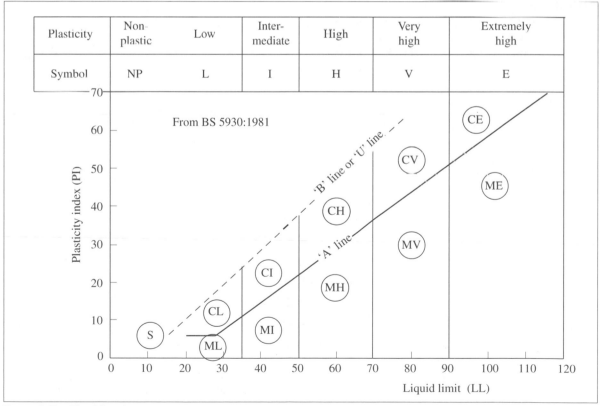

Plasticity	Non-plastic	Low	Inter-mediate	High	Very high	Extremely high
Symbol	NP	L	I	H	V	E

From BS 5930:1981

FIGURE 2.7   *Plasticity chart*

out the plasticity tests and will affect the moisture content value (see moisture content, above) but will have little effect, if any, on the plasticity properties of the soil since the clay particles dominate. Activity was defined by Skempton (1953) as:

$$\text{Activity} = \frac{PI}{C} \tag{2.11}$$

where:

PI = plasticity index %
C = clay content %

and from Equation 2.6 it can be seen that activity represents the plasticity index of the clay fraction alone.

SEE WORKED EXAMPLE 2.7

Four groups of activity are defined from inactive to highly active and some typical values for different soils are given in Table 2.14.

Shrinkage limit

Shrinkage refers to the reduction of volume as the moisture content decreases. It is represented by the gradient of the drying line or shrinkage curve, in Figure 2.6. Silts and sands are not particularly sus-ceptible to shrinkage and will produce a fairly flat shrinkage curve. A clay soil will reduce significantly in volume if its moisture content is reduced, denoted by a steep shrinkage curve.

The amount of shrinkage will depend on the clay content and its mineralogy but it will also be significantly affected by the structural arrangement of the particles and the forces between them as produced in the soil's natural state. For this reason, this test is carried out on an undisturbed specimen.

The definitive test in BS 1377:1990 consists of measuring the volume of a cylindrical specimen of soil as its moisture content decreases. The shrinkage limit is determined as the moisture content below which the volume ceases to decrease. The cylindrical specimen which may be 38–50 mm diameter and 1 to 2 diameters high is placed in a cage and lowered into a tank containing mercury until it is fully submerged.

Volume measurements are carried out by observing the rise in level of the mercury with a micrometer which is adjusted until its tip is in contact with the surface when a completed electrical circuit causes a lamp to light. The mass of the specimen, as its wet weight is determined for each volume reading. The volume measurements are determined as unit volume, U given by:

TABLE 2.14   *Activity of clays*

Groups		Typical values	
Description	Activity	Soil/mineral	Activity
inactive	< 0.75	kaolinite	0.4
		Lias Clay	0.4 – 0.6
		Glacial clays	0.5 – 0.7
		illite	0.9
normal	0.75 – 1.25	Weald Clay	0.6 – 0.8
		Oxford, London Clay	0.8 – 1.0
		Gault Clay	0.8 – 1.25
active	1.25 – 2.0	calcium montmorillonite	1.5
		organic alluvial clay	1.2 – 1.7
highly active	> 2	sodium montmorillonite (bentonite)	7

$$U = \frac{V}{m_{\mathrm{d}}} \times 100 \qquad\qquad (2.12)$$

where:

V = volume of the specimen in cm³
m_{d} = oven-dried mass of the specimen in g

The units of U are cm³ per 100 g of dry soil.

The moisture content for each reading can then be obtained from the dry weight, after the volume measurements are completed. The test is continued until three sets of readings show no volume change even though the moisture content is decreasing and a shrinkage curve similar to Figure 2.6 is plotted.

The shrinkage limit of clay soils from temperate climates typically lies within the small range of 10–15% so it is difficult to distinguish soils on the basis of this test. If the natural moisture content lies above the shrinkage limit then the soil will shrink on drying and the gradient of the volume–moisture content plot will indicate the amount of shrinkage.

Soil model (Figure 2.8 and Table 2.15)

The soil model, illustrated in Figure 2.8 considers the masses and volumes of the three constituents of a soil – solids, water and air – as separate entities. This is a useful way to determine relationships between the basic soil properties and to derive other parameters.

Table 2.15 lists the main properties used in soil mechanics to denote for example:

- how much mass is present in a given volume – density
- the weight or force applied by a given volume – unit weight
- how much void space is present – void ratio, porosity, specific volume
- how much water is present – moisture content
- how much air is present – degree of saturation, air voids content

SEE WORKED EXAMPLE 2.8

It can be seen that many of the properties are ratios. These are dimensionless and, therefore, useful in comparing results and plotting data.

Another purpose of the classification tests is to derive values for these properties, using the expressions given in Table 2.15. For example, in order to determine the degree of saturation of a sample of soil, three classification tests are required:

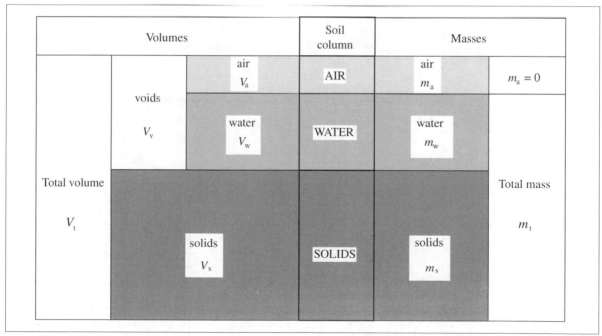

FIGURE 2.8   *Soil model*

TABLE 2.15 *Soil model expressions*

Term	Symbol	Units	Expression		Formulae
moisture content	w	%	$\dfrac{\text{mass water}}{\text{mass solids}}$	$\dfrac{m_w}{m_s}$	w is a fraction in formulae below
void ratio (partially saturated)	e	ratio	$\dfrac{\text{volume voids}}{\text{volume solids}}$	$\dfrac{V_a + V_w}{V_s}$	$e = \dfrac{n}{1-n} = \dfrac{wG_s}{S_r}$
void ratio (fully saturated)	e	ratio	$\dfrac{\text{volume water}}{\text{volume solids}}$	$\dfrac{V_w}{V_s}$	$e = wG_s$
porosity	n	ratio	$\dfrac{\text{volume voids}}{\text{total volume}}$	$\dfrac{V_a + V_w}{V_t}$	$n = \dfrac{e}{1+e} = \dfrac{wG_s}{S_r + wG_s}$
specific volume	v	ratio	$\dfrac{\text{total volume}}{\text{volume solids}}$	$\dfrac{V_a + V_w + V_s}{V_s}$	$v = 1 + e$
degree of saturation	S_r	%	$\dfrac{\text{volume water}}{\text{volume voids}}$	$\dfrac{V_w}{V_a + V_w}$	$S_r = \dfrac{\rho_b wG_s}{\rho_w G_s (1+w) - \rho_b} \times 100$
air voids content	A_v	%	$\dfrac{\text{volume air}}{\text{total volume}}$	$\dfrac{V_a}{V_t}$	$A_v = n(1 - S_r)$
particle density	ρ_s	Mg/m^3	$\dfrac{\text{mass solids}}{\text{volume solids}}$	$\dfrac{m_s}{V_s}$	$G_s \, \rho_w$
specific gravity	G_s	ratio	$\dfrac{\text{solids density}}{\text{water density}}$	$\dfrac{m_s}{V_s}\dfrac{1}{\rho_w}$	$\dfrac{\rho_s}{\rho_w}$
water density	ρ_w	Mg/m^3	$\dfrac{\text{mass water}}{\text{volume water}}$	$\dfrac{m_w}{V_w}$	$\rho_w = 1.0 \ \text{Mg/m}^3$
bulk density (partially saturated)	ρ_b	Mg/m^3	$\dfrac{\text{total mass}}{\text{total volume}}$	$\dfrac{m_s + m_w}{V_a + V_w + V_s}$	$\rho_b = \dfrac{G_s(1+w)\rho_w}{1+e}$
bulk density (fully saturated)	ρ_{sat}	Mg/m^3	$\dfrac{\text{total mass}}{\text{total volume}}$	$\dfrac{m_s + m_w}{V_s + V_w}$	$\rho_{sat} = \dfrac{(G_s + e)\rho_w}{1+e}$
dry density	ρ_d	Mg/m^3	$\dfrac{\text{mass solids}}{\text{total volume}}$	$\dfrac{m_s}{V_t}$	$\rho_d = \dfrac{\rho_b}{1+w}$
bulk unit weight (partially saturated)	γ_b	kN/m^3	$\dfrac{\text{total weight}}{\text{total volume}}$	$\dfrac{m_t g}{V_t}$	$\gamma_b = \dfrac{G_s(1+w)\gamma_w}{1+e}$
bulk unit weight (fully saturated)	γ_{sat}	kN/m^3	$\dfrac{\text{total weight}}{\text{total volume}}$	$\dfrac{m_t g}{V_s + V_w}$	$\gamma_{sat} = \dfrac{(G_s + e)\gamma_w}{1+e}$
dry unit weight	γ_d	kN/m^3	$\dfrac{\text{weight solids}}{\text{total volume}}$	$\dfrac{m_s g}{V_t}$	$\gamma_d = \dfrac{\gamma_b}{1+w}$
unit weight water	γ_w	kN/m^3	$\dfrac{\text{weight water}}{\text{volume water}}$	$\dfrac{m_w g}{V_w}$	$\gamma_w = \rho_w g = 9.81 \ \text{kN/m}^3$

- moisture content
- bulk density
- particle density

To illustrate the use of the soil model the derivation of the formula for the degree of saturation is given below. This is for a partially saturated soil so the bulk density is used:

$$\rho_b = \frac{m_s + m_w}{V_a + V_w + V_s} \qquad (2.13)$$

Divide by m_s to give:

$$\rho_b = \frac{1 + \dfrac{m_w}{m_s}}{\dfrac{V_a + V_w}{m_s} + \dfrac{V_s}{m_s}} \qquad (2.14)$$

Now:

$$\frac{m_w}{m_s} = w \qquad \frac{m_s}{V_s} = G_s\rho_w \qquad S_r = \frac{V_w}{V_s + V_w} \qquad \rho_w = \frac{m_w}{V_w}$$

so:

$$\frac{V_a + V_w}{m_s} = \frac{V_a + V_w}{m_w} \times \frac{m_w}{m_s} = \frac{V_a + V_w}{V_w}\frac{1}{\rho_w}\frac{m_w}{m_s} = \frac{1}{S_r}\frac{1}{\rho_w}w$$

Putting this into equation 2.14 gives:

$$\rho_b = \frac{1 + w}{\dfrac{1}{S_r\rho_w}w + \dfrac{1}{G_s\rho_w}} \qquad (2.15)$$

which is rearranged to:

$$S_r = \frac{\rho_b w G_s}{G_s\rho_w(1 + w) - \rho_b} \qquad (2.16)$$

In this formula the moisture content w would be a ratio not a percentage, the specific gravity G_s is dimensionless and the bulk density of the soil ρ_b and of water ρ_w must have the same units. This will give the degree of saturation S_r as a ratio. Multiply by 100 for the percentage value.

SUMMARY

It is essential that if you examine a soil and prepare a description you must convey sufficient information to those who have not seen the soil. It is often these persons who must make geological interpretations and assess the soil parameters and behaviour.

A systematic approach to soil description is discussed based on the British Soil Classification System.

Classification tests for particle density, particle size distribution, bulk density, moisture content, liquid limit, plastic limit and shrinkage limit are described and some typical values are given.

The classification tests are used to indicate the nature of the soil including parameters such as the liquidity index and activity. They have a major contribution to the classification of soils according to the Casagrande plasticity chart.

The applications of the soil model in soil mechanics are described by providing a summary table of expressions for the most common terms used and by deriving the formula for the degree of saturation.

Worked Example 2.1   Particle density of sand

From the following masses determine the particle density of a sample of sand.

mass of bottle (with stopper) $m_1 = 27.464$ grams
mass of bottle and sand $m_2 = 33.660$ grams
mass of bottle, sand and water $m_3 = 84.000$ grams
mass of bottle and water $m_4 = 80.135$ grams

$$\text{particle density} = \frac{\text{mass of sand particles}}{\text{volume of sand particles}}$$

mass of sand particles $= m_2 - m_1 = 33.660 - 27.464 = 6.196$ g
volume of sand particles $= M_1 - (M_2 - M_3)$ (From Figure 2.1)
$$M_1 = m_4 - m_1$$
$$M_2 = m_3 - m_1$$
$$M_3 = m_2 - m_1$$
volume of sand $= (m_4 - m_1) - (m_3 - m_2)$
$\qquad = (80.135 - 27.464) - (84.000 - 33.660) = 2.331$ g $(= 2.331$ cm$^3)$ (for water, $\rho_w = 1$ g/cm^3)

particle density $= \rho_s = \dfrac{6.196}{2.331} = 2.658$, say 2.66 g/cm^3 or Mg/m^3

Specific gravity $G_s = 2.66$

Worked Example 2.2   Particle size distribution

The total mass of a sample of dry soil is 187.2 grams before being shaken through a series of sieves. From the masses retained on each of the following sieves determine the percentages passing and give a description of the soil tested.

Sieve size (mm or μm)	6.3	2	1.18	600	425	300	212	150	63	tray
	0	6.8	4.7	7.2	27.8	31.3	33.1	28.7	26.5	21.1

sieve size	mass retained	mass passing	% passing
6.3	0	187.2	100
2	6.8	180.4	96.4
1.18	4.7	175.7	93.9
600	7.2	168.5	90.0
425	27.8	140.7	75.2
300	31.3	109.4	58.4
212	33.1	76.3	40.8
150	28.7	47.6	25.4
63	26.5	21.1	11.3
tray	21.1	–	

The soil could be described as a silty (or clayey) fine and medium SAND with a little fine gravel.
Alternatively, it could be described as a silty (or clayey) slightly gravelly fine and medium SAND.
The particle size distribution (or grading) curve is obtained by plotting the % passing versus the sieve size.
From this curve the following values can be obtained, see Figure 2.5.

effective size $d_{10} = 60$ μm or 0.06 mm

$d_{30} = 0.17$ mm

median size $= d_{50} = 0.27$ mm

$d_{60} = 0.31$ mm

uniformity coefficient $= \dfrac{0.31}{0.06} = 5.2$

coefficient of curvature $= \dfrac{0.17^2}{0.06 \times 0.31} = 1.55$

The soil could be described as a moderately well graded sand.

Worked Example 2.3   Density, unit weight and moisture content

A triaxial specimen of moist clay has a diameter of 38 mm, length of 76 mm and a mass of 183.4 g. After oven drying the mass is reduced to 157.7 g. Determine for this soil:

i) bulk density and bulk unit weight

ii) dry density and dry unit weight

iii) moisture content

Volume of specimen $= \dfrac{\pi \times 38^2}{4} \times \dfrac{76}{10^9} = 8.62 \times 10^{-5}$ m^3

i)   Bulk density $= \dfrac{183.4 \times 10^5}{8.62 \times 10^3} = 2127.6$ kg/m^3 or 2.13 Mg/m^3

Bulk unit weight $= 2127.6 \times \dfrac{9.81}{1000} = 20.87$ kN/m^3

ii) Dry density $= \dfrac{157.7 \times 10^5}{8.62 \times 10^3} = 1829.5$ kg/m^3 or 1.83 Mg/m^3

Dry unit weight $= 1829.5 \times \dfrac{9.81}{1000} = 17.95$ kN/m^3

iii) Moisture content $= \dfrac{183.4 - 157.7}{157.7} \times 100 = 16.3\%$

Alternatively, dry density can be obtained from the soil model, see Table 2.15

$\rho_d = \dfrac{2127.6}{1 + 0.163} = 1829.4$ kg/m^3

The difference in values is a result of the degree of accuracy adopted in the calculations.

Worked Example 2.4   Irregular lump test for density

An irregular lump of soil has a mass of 96.7 g. It is then completely coated in wax and weighed again in air to give a mass of 109.5 g. Its mass when fully immersed in water is 48.2 g. Determine the volume of the soil lump and its bulk density. The density of the hardened wax is 0.92 g/cm³.

From Equation 2.3:

Volume = $109.5 - 48.2 - \dfrac{109.5 - 96.7}{0.92} = 47.4$ cm³

Bulk density = $\dfrac{96.7}{47.4} = 2.04$ Mg/m³ (or g/cm³)

Worked Example 2.5   Equivalent moisture content

In an earthworks contract it is stated that the natural moisture content of a gravelly clay soil must not exceed the optimum moisture content (OMC) obtained from a standard compaction test plus 2% (OMC + 2%). Otherwise, it is unacceptable and cannot be used. Determine whether the following soil should be classified as acceptable or not.

natural moisture content = 18.5%   (including particles > 20mm)
optimum moisture content = 19%   (for particles < 20mm)
% passing 20mm = 83%

Equivalent moisture content = $18.5 \times \dfrac{100}{83}$ = 22.3%

At first glance, it would appear that the clay soil is acceptable, its natural moisture content being numerically less than the optimum moisture content. However, the equivalent moisture content of the soil, excluding the coarse gravel (>20 mm) is much greater than the OMC so the soil should be treated as unacceptable. The compaction test is described in Chapter 13.

Worked Example 2.6   Liquid limit

The following are the results of a cone penetrometer test on a sample of clay soil. Determine its liquid limit.

Mass of wet soil, g	9.29	9.05	8.83	9.17
Mass of dry soil, g	5.78	5.51	5.27	5.34
Cone penetration, mm	15.6	18.8	21.0	24.7
Moisture content, %	60.7	64.3	67.6	71.7

From a plot of moisture content versus cone penetration the liquid limit is the moisture content at a cone penetration of 20 mm, 66%. The plastic limit of the soil is determined as 29%. From the plasticity chart the soil would be described as a clay of high plasticity.

Worked Example 2.7   Classification tests

The following test results were obtained for a fine-grained soil.
LL = 48%   PL = 26%
Clay content = 25%
Silt content = 36%
Sand content = 39%
Natural moisture content = 29%
Determine appropriate parameters to classify the soil.

Plasticity index = $LL - PL = 48 - 26 = 22\%$

According to the Plasticity chart, Figure 2.7, the soil would classify as CI, clay of intermediate plasticity. The large proportion of silt and sand (75% in total) will have little effect on the engineering behaviour, particularly the shear strength.

Activity $= \dfrac{22}{25} = 0.88$

Liquidity index $= \dfrac{29 - 26}{22} = 0.14$

Consistency index $= \dfrac{48 - 29}{22} = 0.86$

The clay is of normal activity and should be of a firm to stiff consistency.

Worked Example 2.8   Void ratio, degree of saturation, air voids content

For the triaxial specimen in Worked Example 2.3 determine:
i) void ratio and porosity
ii) degree of saturation
iii) air voids content
given the particle density of 2.72 Mg/m³.

From Table 2.15 use:

$$\rho_b = \rho_s \left(\frac{1 + w}{1 + e} \right)$$

i)   Void ratio $e = \dfrac{\rho_s}{\rho_b}(1 + w) - 1 = \dfrac{2.72}{2.128}(1 + 0.163) - 1 = 0.487$

Porosity $n = \dfrac{0.487}{1.487} = 0.328$

ii) Degree of saturation $S_r = \dfrac{w.G_s}{e} = \dfrac{0.163 \times 2.72}{0.487} \times 100 = 91.0\%$

iii) Air voids content $A_v = n(1 - S_r) = 0.328(1 - 0.91) \times 100 = 3\%$

EXERCISES

2.1 The results of a particle density test on a sample of gravel were as follows:
mass of empty gas jar = 845.2 g
mass of gas jar full of water = 1870.6 g
mass of gas jar with soil sample = 1608.7 g
mass of gas jar with soil sample and full of water = 2346.0 g
Determine the particle density of the gravel. All of the above masses include the mass of the cover lid. Assume the density of water to be 1.0 Mg/m^3 or 1.0 g/cm^3.

2.2 The results of a dry sieving test are given below. The total mass of the sample of dry soil was 2105.4 g. Determine the percentage passing, plot a particle size distribution curve and give a description for the soil.

sieve sizes (mm or μm)	20	14	10	6.3	3.35	2	1.18	600
mass retained (g)	0	18.9	67.4	44.2	75.8	122.1	193.7	240.0
	425	300	212	150	63	tray		
	282.2	242.1	233.7	265.3	240.0	80.0		

Determine the effective size, median size, uniformity coefficient and coefficient of curvature.

2.3 An irregular lump of moist clay with a mass of 537.5 g was coated with paraffin wax. The total mass of the coated lump (in air) was 544.4 g. The volume of the coated lump was found to be 250 ml by displacement in water. After carefully removing the paraffin wax the lump of clay was oven-dried to a dry mass of 479.2 g. The specific gravity of the hardened wax is 0.90. Determine the water or moisture content w, the bulk density ρ_b, the bulk unit weight γ_b, dry density ρ_d and the dry unit weight γ_d of the clay.

2.4 For the sample of clay in Exercise 2.3 determine the void ratio, porosity, degree of saturation and air voids content. The specific gravity of the (dry) clay particles was 2.72.

2.5 For the sample of clay in Exercise 2.3 what would be the density and moisture content if the soil was fully saturated, assuming that the void ratio remained the same.

2.6 It is required to prepare a specimen of soil at a moisture content of 13.5% and an air voids content of no more than 5%. To what density must it be compacted? The particle density is 2.68 Mg/m^3.

2.7 The results of a liquid limit test and a plastic limit test on a sample of soil are given below:
liquid limit

cone penetration (mm)	14.8	16.9	19.1	21.2	23.2	24.7
water content (%)	50.8	52.9	54.2	56.0	57.3	58.7

plastic limit
water content of threads (%)   26.6 and 27.3
Determine the liquid limit, plastic limit, plasticity index and soil classification.

2.8 The natural water content of the clay in Exercise 2.7 is 35%. Determine the liquidity index and the consistency index.

CHAPTER 3

Permeability and seepage

OBJECTIVES

- To appreciate that groundwater flow causes the most severe problems when dealing with construction works below the water table and in earth structures retaining water.
- To understand the fundamental Darcy law of permeability and to be able to determine values of the coefficient of permeability from laboratory and field tests.
- To understand the theory of seepage and to be able to construct a flow net by sketching.
- To determine seepage quantities and pore pressures and to assess the stability of a soil mass subjected to seepage forces.
- To appreciate the factors governing the performance of soil filters.

Permeability

Introduction (Figure 3.1)

As a result of the hydrological cycle (rainfall, infiltration) it is inevitable that the voids between soil particles will fill with water until they become fully saturated. There then exists a zone of saturation below ground level the upper surface of which is called the water table.

The water table generally follows the shape of the ground surface topography but in a subdued manner. A sloping water table surface is an indication of the flow of groundwater or seepage in the direction of the fall. Water tables change with varying rates of infiltration so that in winter they can be expected at high levels and at lower levels in summer.

Groundwater (Figure 3.2)

The voids of permeable deposits such as sands will fill up easily and also allow this water to flow out easily. Such permeable deposits are called aquifers (bearing water). The void spaces in a clay will also contain water but these void spaces are so small that flow of water is significantly impeded making a clay impermeable. Clay deposits will then act as aquicludes (confining water). The location and state of groundwater in soil deposits is often determined by the stratification of sand–clay or permeable–impermeable sequences. Some commonly used terms are described in Figure 3.2.

Flow problems

In nature groundwater may be flowing through the ground but this flow will not normally be large enough to cause instability so the ground is stable or in equilibrium. Ground engineering works, particularly excavations, will disturb this equilibrium and alter the pattern of flow. It is the responsibility of the geotechnical engineer to identify where a problem

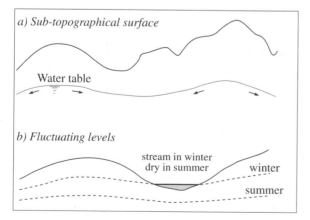

a) Sub-topographical surface

Water table

b) Fluctuating levels

stream in winter
dry in summer

winter

summer

FIGURE 3.1   *Fluctuating water tables*

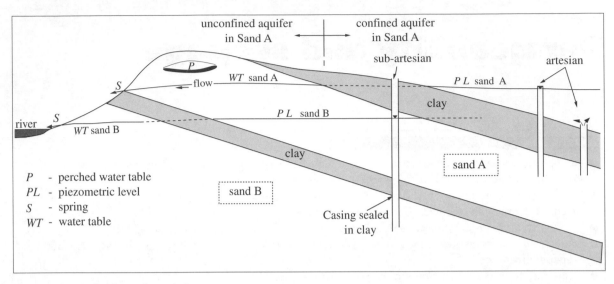

FIGURE 3.2 *Definitions of groundwater terms*

may be encountered and how to ensure that stability of the works is maintained.

Flow into excavations (Figure 3.3)

An estimate of the quantity of flow must be obtained so that adequate numbers and capacities of pumps can be provided to remove inflows. This technique is referred to as sump pumping and relies on the groundwater emanating from the soil, leaving the soil particles behind. This is only feasible for open, clean gravel deposits under small heads of water, otherwise stability problems arise, see below.

Flow around cofferdams (Figure 3.4)

The required pumping capacity must be assessed but also the quantity of flow can be altered by changing the length of sheet piling penetrating below excavation level. This increases the length of seepage path and reduces the quantity of inflow.

SEE WORKED EXAMPLE 3.9

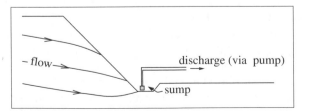

FIGURE 3.3 *Flow into excavations*

Dewatering (Figure 3.5)

Where the risk of sump pumping cannot be accepted (which is the case for most permeable soils) then it is far more suitable to lower the water table temporarily before excavating so that excavation can be carried out in the 'dry'. This is achieved by using dewatering techniques such as pumping from well-points or shallow wells inserted below the water table. A knowledge of the rate of flow of water through the soil is then required to determine the numbers, depths and spacing of these well-points and the required pumping capacity.

SEE WORKED EXAMPLE 3.7

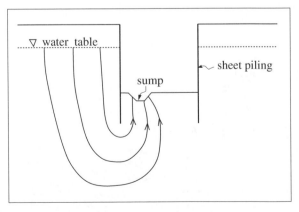

FIGURE 3.4 *Flow around cofferdams*

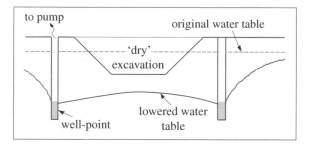

FIGURE 3.5   *Dewatering excavations*

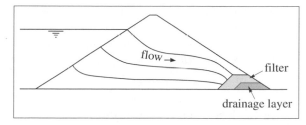

FIGURE 3.6   *Flow through earth structures*

Flow through earth structures (Figure 3.6)

An earth structure impounding water will allow water to flow through it or beneath it, if it is permeable. Measures to minimise these flows through the dam such as a central clay core and beneath it such as a grout curtain may have to be incorporated. Alternatively the earth structure may only be required to contain water for a short period such as for flood banks alongside rivers and such measures would then be uneconomical. In these cases some flows may be permitted but they must be controlled using drainage measures including permeable layers and filters. These drains must allow water to enter and pass through them unimpeded so they require protection by filters which prevent soil particles being carried into the drainage layers and blocking them.

Stability problems – 'running sand'
(Figure 3.7)

It is incorrect to state that 'running sand' exists naturally on a site. Sand and groundwater exist naturally but 'running sand' is a man-made condition caused by allowing water, carrying soil particles, to flow out of the sides of an excavation. This results in the excavation sides slumping, soil above being undermined and soon after collapse of the excavation sides.

Boiling or heaving in cofferdams
(Figure 3.8)

When the upward seepage forces (in the water) are greater than the downward gravity forces (in the soil particles) localised boiling will occur. This is often observed as small 'volcanoes' of soil. When the average upward seepage forces are not sufficiently

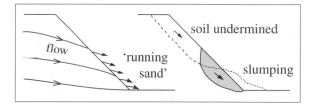

FIGURE 3.7   *The 'running sand' condition*

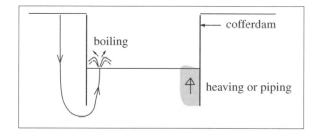

FIGURE 3.8   *'Boiling' or heaving in cofferdams*

balanced by the downward gravity forces of the soil mass below excavation level, a more severe heaving condition can occur. This results in separation of the particles, increased permeability and increased seepage resulting in progressive and rapid loss of passive resistance in front of the sheet piling. This is soon followed by complete collapse of the cofferdam.

SEE WORKED EXAMPLE 3.8

Piping (Figure 3.9)

The erosive force produced by water passing out of an earth structure can be large, resulting in the formation of 'pipes' which increase in size and flow capacity leading to progressive erosion, undermining and eventual instability. Certain soils, mainly non-cohesive and fine soils have a low resistance to erosion and should not be used in earth dams, flood banks or canal linings unless they are protected by other resistant materials.

SEE WORKED EXAMPLE 3.9

Heaving beneath a clay layer (Figure 3.10)

As shown in Figure 3.2 the groundwater and the pressure within it can be confined beneath a relatively impermeable stratum (clay or silt) which produces an uplift pressure. If the downward pressure from the clay below the base of an excavation is insufficient to balance the uplift pressure beneath the clay then the clay will be lifted, causing it to crack and allowing water to flow through. This will carry soil particles upwards leading to erosion of the sand, undermining of the clay and a severely disturbed excavation.

SEE WORKED EXAMPLE 3.10

Uplift pressures (Figure 3.11)

Permitting flow through a permeable stratum beneath a structure will reduce the hydrostatic pressure in the water due to energy losses. However, the pore water pressures remaining will produce uplift beneath any impermeable structure. There must be sufficient deadweight (or anchors, if necessary) to balance these remaining uplift pressures.

Soil voids (Figure 3.12)

All soils contain voids or pores and can be described as porous. However, to allow water to flow at least some of the voids must be continuous and then the soil can be described as permeable. Permeability is not dependent on the amount of voids present. The porosity of a clay is usually greater than that of a sand but the clay could be a million times less permeable than the sand. Some typical values are given below:

Soil	Porosity	Typical coefficient of permeability k m/s
clay	0.3–0.5	10^{-9}
sand	0.2–0.4	10^{-3}

This difference is due to the nature and size of the voids. In sands, particularly clean sands without

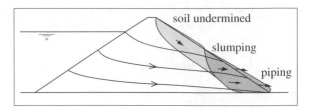

FIGURE 3.9 *Piping through an earth dam*

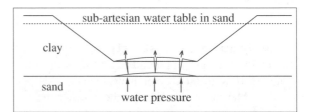

FIGURE 3.10 *Heaving of clay layer at the base of an excavation*

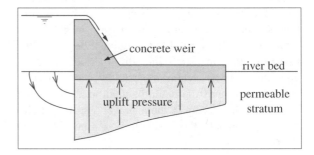

FIGURE 3.11 *Uplift pressure beneath structures*

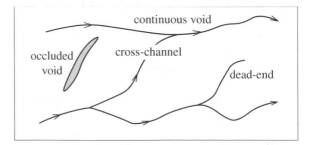

FIGURE 3.12 *Soil voids*

fines most of the voids are continuous, more direct and relatively large, while in clays the voids are smaller, more tortuous and with several stagnant voids. Three types of void are described in Table 3.1

TABLE 3.1 *Influence of voids on flow*

Effect on flow of:		Void name	Typical soil
particle surface	void size		
negligible	negligible	macropores	gravel
negligible	moderate	macropores and capillaries	sand
small	significant	capillaries	silt
significant	very significant	micropores	clay (intact)

TABLE 3.2 *Influence of voids in clays*

Clay type	Voids present
fissured clays	macropores and micropores
desiccated clays	
compacted clays (dry of optimum)	
intact clays	micropores
compacted clays (wet of optimum)	

(macropores, capillaries and micropores) showing the influence on flow of their size (as in flow through pipes) but also the retarding effects of the particles' surfaces (their mineralogy, water adsorption and particle orientation) as water flows over and around them.

The types of voids present in different clay soils will determine their permeability, as shown in Table 3.2.

Pressure and head (Figure 3.13)

Pressure is the force per unit area (kN/m²) acting at a point in the water. Head is a measure of pressure but in terms of a height (metres of water). The total head at a point such as point B in Figure 3.13 is given by Bernoulli's equation:

$$H = h_Z + h_B + v^2/2g \qquad (3.1)$$

| total head (at B) | = | position head (at B) | + | piezometric head (at B) | + | velocity head (negligible) |

The velocity head can be ignored since the velocity of flow of water through soil is quite small. If the water levels in the piezometric tubes *a* and *b* are the same then no flow will occur. Flow will only occur if there is a total head difference between *a* and *b*. This

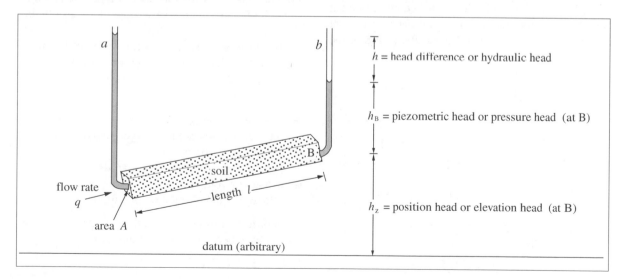

h = head difference or hydraulic head

h_B = piezometric head or pressure head (at B)

h_z = position head or elevation head (at B)

datum (arbitrary)

FIGURE 3.13 *Darcy's law*

head difference is known as head loss or hydraulic head.

SEE WORKED EXAMPLE 3.8

Darcy's Law (Figure 3.13)
This states that the discharge velocity, v of water is proportional to the hydraulic gradient, i.

$$\frac{q}{A} = v = k\,i \qquad (3.2)$$

where:

k = Darcy coefficient of permeability, m/s

The hydraulic gradient i is the ratio of the head loss h over a distance l.

The discharge velocity v is defined as the quantity of water, q percolating through a cross-sectional area A in unit time. This is not the same as the velocity of the water percolating through the voids of the soil which is known as the seepage velocity.

Since porosity $n = \dfrac{\text{volume of voids}}{\text{total volume}}$

'area' of voids $A_s = nA$

seepage velocity $= v_s = \dfrac{q}{A_s} = \dfrac{v}{n}$ $\qquad (3.3)$

The flow of water through most soils is laminar and Darcy's law applies to most soils. However, with open, large-void gravels flow can become turbulent and Darcy's Law may not be valid. The coefficient of permeability, k, is dependent on the nature of the

voids (see above) and the properties of the fluid, particularly its viscosity. It is given by:

$$k = K\,\frac{\gamma_w}{\eta} \qquad (3.4)$$

where:

γ_w = unit weight of fluid
η = viscosity of fluid
K = absolute or specific permeability (m^2)

SEE WORKED EXAMPLES 3.1 AND 3.8

Effect of temperature (Figure 3.14)
Unit weight, γ_w is affected very little by changes in temperature but viscosity is affected. It is conventional to report k at a temperature of 20°C so a correction factor, R_T, should be applied if the test is carried out at a temperature of T°C.

$$k_{20} = R_T k_T \qquad R_T = \frac{\eta_T}{\eta_{20}} \qquad (3.5)$$

where:

k_{20} = permeability at 20°C
k_T = permeability at T°C
R_T = correction factor, see Figure 3.14

Empirical correlations for k (Figure 3.15)
Typical values of k for various soil types are given in Table 3.3 illustrating the vast range of values obtainable.

The absolute or specific permeability K has been found to be related to the macropore sizes between the soil particles (for sands and gravels) which are in turn related to:

● particle size – d_{10}
 (larger particles produce larger voids)
● grading of particles – U_c
 (smaller particles clog larger voids)
● density or void ratio – e
 (closeness and orientation of particles)

Empirical correlations relating all three of these factors have been suggested in the form:

$$k = C\;d_{10}{}^{a}\;U_c{}^{b}\;e^{c} \qquad (3.6)$$

where C, a, b and c are constants.

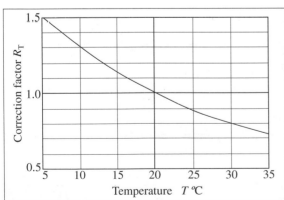

FIGURE 3.14 *Temperature correction*

TABLE 3.3 *Typical values of k*

k m/s	Soil type		Drainage characterisitics
10	Coarse gravel, cobbles, boulders flow may become turbulent ∴ Darcy's Law may not be valid		Very good
1			
10^{-1}	Clean gravels		
10^{-2}			
10^{-3}	Clean sands		Good
10^{-4}	Clean sand-gravel mixtures		
10^{-5}		Impervious soils modified by the effects of:	
10^{-6}	Very fine sands		
10^{-7}	Silty sands	fissuring	Poor
10^{-8}	Silts	desiccation	
	Stratified clay/silt deposits	weathering	
10^{-9}	Unweathered, unfissured,homogeneous clays (Clay content > 20%)		Practically Impervious

None of these correlations is particularly reliable especially compared to *in situ* test results. Determination of the effective size d_{10} can be very inaccurate for gap-graded materials, samples of stratified soils, or where finer particles are lost during sampling as is often the case. The most commonly used correlation is Hazen's relationship for clean (no fines) filter (fairly uniform) sands in a loose condition (near the maximum void ratio):

$$k = C d_{10}^2 \text{ m/s} \qquad (3.7)$$

where:

d_{10} = effective size, mm
C = coefficient, 0.01 to 0.015

SEE WORKED EXAMPLE 3.1

Compared to some *in situ* permeability tests this relationship appears reasonable, see Figure 3.15. Laboratory tests also show (Cedergren, 1989) that k can vary by as much as one order of magnitude between the loosest and densest states of a soil. For example, a sandy gravel gave:

● loose state – $k \approx 10^{-3}$ m/s
● dense state – $k \approx 10^{-4}$ m/s

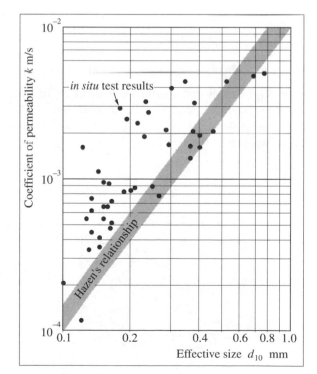

FIGURE 3.15 *Empirical correlations (From Leonards, 1962)*

Layered soils (Figure 3.16)

These are quite common especially where they have been deposited in lakes, estuaries, deltas, flood plains or rivers where layers of clay, silt and sand are deposited either alternately as with laminated clays and varved clays or a thin layer of one soil type may exist within a thicker deposit of another soil type.

The permeabilities of these soil types are several orders of magnitude different so that a layer of sand in a thick clay deposit will greatly increase overall horizontal permeability while a thin clay layer in a thick sand deposit will greatly decrease overall vertical permeability.

Horizontal flow

Consider layers 1 and 2 in Figure 3.16 with thicknesses L_1 and L_2 and with different permeabilities k_1 and k_2. The equipotentials are vertical and equidistant so the hydraulic gradient, i is the same in both layers:

$$i_1 = i_2 = i$$

Applying Darcy's law:

$$q = A\,k\,i$$

Total horizontal flow:

$$q_H = 1\,(L_1 + L_2)\,k_H\,i \text{ per unit width (1 m)}$$

where k_H is the overall coefficient of permeability in the horizontal direction for the layered soil. The overall total flow q_H is equal to the sum of the flows in the individual layers, i.e.

$$q_H = q_1 + q_2$$

where $q_1 = 1\,L_1\,k_1\,i_1$ and $q_2 = 1\,L_2\,k_2\,i_2$

$$\therefore (L_1 + L_2)\,k_H\,i = (L_1\,k_1 + L_2\,k_2)i$$

and

$$k_H = \frac{L_1 k_1 + L_2 k_2}{L_1 + L_2} \tag{3.8}$$

or

$$k_H = \frac{\sum_1^n L_i k_i}{\sum_1^n L_i} \tag{3.9}$$

which is a general expression for several layers.

For horizontal flow $i_1 = i_2$ $\therefore$

$$\frac{q_1}{q_2} = \frac{L_1 k_1}{L_1 k_2} \tag{3.10}$$

Vertical flow

For continuity the rate of flow must be the same in each layer, i.e.

$$q_1 = q_2 = q_v$$

For unit horizontal area ($A = 1\ \text{m}^2$)

$$q_v = 1\,k_v\,i_v$$

where k_v is the overall coefficient of permeability in the vertical direction. The loss in total head over length $L_1 + L_2$ will be equal to the sum of losses in total head in the individual layers, i.e.

$$i_v(L_1 + L_2) = i_1 L_1 + i_2 L_2$$

$$i_v = \frac{i_1 L_1 + i_2 L_2}{L_1 + L_2}$$

$$i_1 = \frac{q_1}{k_1} \text{ and } i_2 = \frac{q_1}{k_2}$$

$$k_v = \frac{q_v}{i_v} = \frac{q_v\,(L_1 + L_2)}{q_1 \dfrac{L_1}{k_1} + q_2 \dfrac{L_2}{k_2}} = \frac{L_1 + L_2}{\dfrac{L_1}{k_1} + \dfrac{L_2}{k_2}} \tag{3.11}$$

or

$$k_H = \frac{\sum_1^n L_i}{\sum_1^n \dfrac{L_i}{k_i}} \tag{3.12}$$

See Worked Examples 3.3 and 3.4

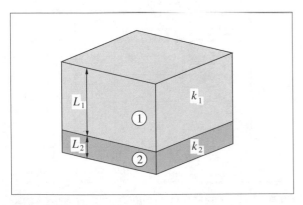

Figure 3.16 *Layered soils*

Laboratory test – constant head permeameter (Figure 3.17)

The procedure for carrying out this test using a constant head permeameter is described in BS 1377:1990. The test is only suitable for soils having a coefficient of permeability in the range 10^{-2} to 10^{-5} m/s which applies to clean sand and sand–gravel mixtures with less than 10% fines (silt or clay).

The test is carried out by adjusting the control valve and waiting until the flow q through the sample and the hydraulic head loss h between the manometer points have reached a steady state. The flow rate q and head loss h are measured and the coefficient of permeability is calculated. The test is repeated at different hydraulic gradients by adjusting the control valve to obtain a number of q and h values. An intermediate manometer point is recommended to ensure that the hydraulic gradient through the sample is uniform.

SEE WORKED EXAMPLE 3.1

The permeability values should be corrected for the effect of temperature using the correction factor R_T (see Figure 3.14) and the average value is reported. The dry mass of the test sample should be obtained and the dry density of the soil determined. The void ratio can be calculated providing the particle density is known. It is preferable to carry out the test on samples compacted at a range of densities and to plot k against density or void ratio. From this an estimate of the *in situ* value can be interpolated for the field density.

The test is rarely used for natural soil samples since:

- The samples are unrepresentative of *in situ* conditions. The samples are small (75 or 100 mm diameter) and are unlikely to contain sufficient macrostructure (bedding, fissures, laminations, root holes etc.) to give the appropriate k value.
- One-dimensional (vertical) flow of water *in situ* is an unlikely event on its own whereas the test is constrained to this direction.
- Installation and sealing of 'undisturbed' specimens in the permeameter is difficult and further disturbance is inevitable during sample

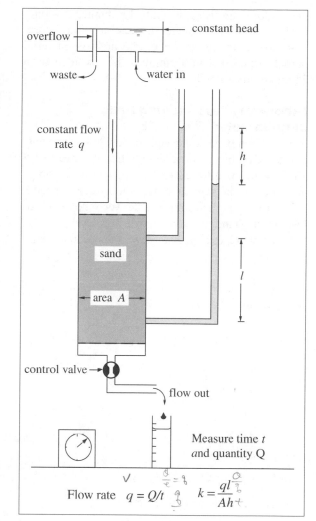

FIGURE 3.17 *Constant head permeameter*

preparation. This changes the microstructure and macrostructure and tends to reduce flow.
- Leakages around the sides of the sample may give erroneously high results and air bubbles trapped would produce low results.

The test can be suitable for soils when used in their completely disturbed or remoulded states such as for drainage materials and filters to confirm that their performance will be adequate. Again, a range of tests at different densities and placement moisture contents would be required to relate k to the compaction characteristics of the materials. The maximum particle size must be limited to one-twelfth of the permeameter diameter to ensure representative sampling, so for coarse drainage material particles greater than 6 mm must be discarded for a 75 mm diameter cell.

Laboratory test – falling head permeameter (Figure 3.18)

This test can be used for soils of low to intermediate permeability but its procedure is not included in BS 1377. Water is allowed to flow down the standpipe and through the sample so that the hydraulic head h at any time t is the difference between the meniscus level in the standpipe and the overflow level.

In the standpipe the water level falls dh in dt and:

$$q = -a \frac{\mathrm{d}h}{\mathrm{d}t}$$

In the sample $q = Aki$

i at any time $t = \dfrac{h}{l}$

$$\therefore -\int_{h_0}^{h_1} \frac{\mathrm{d}h}{h} = \frac{Ak}{al} \int_{t_0}^{t_1} \mathrm{d}t$$

giving

$$\ln \frac{h_1}{h_0} = \frac{Ak}{al}(t_1 - t_0)$$

with $h = h_0$ at $t_0 = 0$ and $h = h_1$ at t_1.

This equation can be expressed as:

$$k = 2.3 \frac{al \log_{10} \dfrac{h_0}{h}}{At} \qquad (3.13)$$

If a graph of t versus $\log_{10}(h_0/h)$ is plotted a straight line should be obtained and k can be derived from the gradient.

SEE WORKED EXAMPLE 3.2

The main limitations of the test are the potential for leakages around the soil specimen, particularly if an undisturbed specimen is used. The effective stresses to which the soil is subjected *in situ* are not represented in the test so some clay soils could tend to swell.

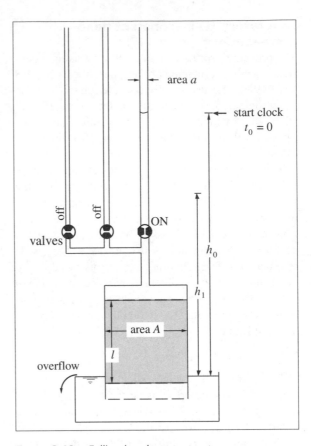

FIGURE 3.18   *Falling head permeameter*

Laboratory test – hydraulic cell – vertical permeability (Figure 3.19)

This constant head test is described in BS 1377:1990 as a means of determining k for an 'undisturbed', large volume (up to 250 mm diameter) and therefore representative sample of soil of low to intermediate permeability.

The sample is subjected to a vertical effective stress representing *in situ* stress levels and the stiff cell provides K_0 conditions. Water flows are induced and measured under the influence of a back pressure. This is an elevated pressure in the pore water to prevent air or gas bubbles coming out of solution and affecting the results. Organic clays for which this test would be appropriate are liable to allow gases dissolved in the pore water to form bubbles in the void system when the stresses on the sample are removed. A back pressure therefore models the *in situ* pore pressure.

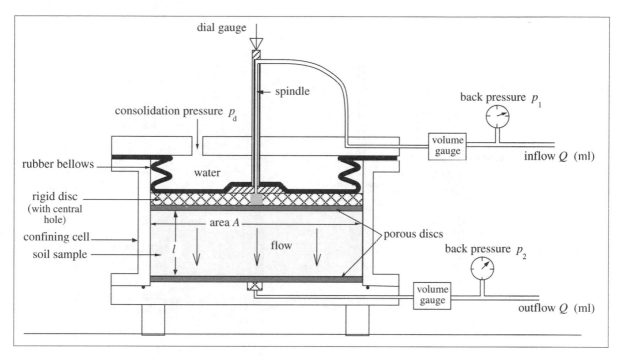

FIGURE 3.19 *Hydraulic cell – vertical permeability*

With the sample installed in the apparatus it is necessary to first saturate the sample, the cell, the pressure lines and the gauges. The sample is then consolidated under a pressure p_d to introduce an effective stress (p_d) into the sample. The inlet pressure p_1 is increased above the outlet pressure p_2 to induce a flow of water downwards through the sample and this difference of pressure is maintained until the flow becomes steady. p_1 must never exceed p_d.

For silty and sandy soils a head difference $(h_1 - h_2)$ of only a few centimetres may be necessary to produce measurable flow whereas for clay soils a head difference of up to 2 m or pressure difference of about 20 kN/m^2, may be required depending on the thickness of the sample and its permeability. When the difference between p_1 and p_2 is small it is best measured using a differential pressure gauge or manometer.

When the rate of flow is high (> 20 ml/minute) there may be a significant pressure or head loss in the system (p_c) due to friction losses in the pressure lines and valves etc. This pressure loss can be determined during calibration of the apparatus and sub-

tracted from the measured pressure loss.

The expression for the determination of the vertical coefficient of permeability k_v is obtained from:

$$q = k_v A i$$

$$q = Q/t$$

For full saturation and no leakages:

inflow Q = outflow Q

h = head difference across the sample = $h_1 - h_2$ where:

$$h_1 = \frac{p_1}{\gamma_w} \text{ and } h_2 = \frac{p_2}{\gamma_w}$$

$$k_v = \frac{q}{Ai} = \frac{q l \gamma_w}{A(p_1 - p_2)} R_T \qquad (3.14)$$

Laboratory test – hydraulic cell – horizontal permeability (Figure 3.20)

The test is set up in a similar fashion to the previous test but water is prevented from flowing vertically by plugging the central hole in the rigid disc and placing this disc on an impervious layer of latex rubber similar to the rubber membrane used in the

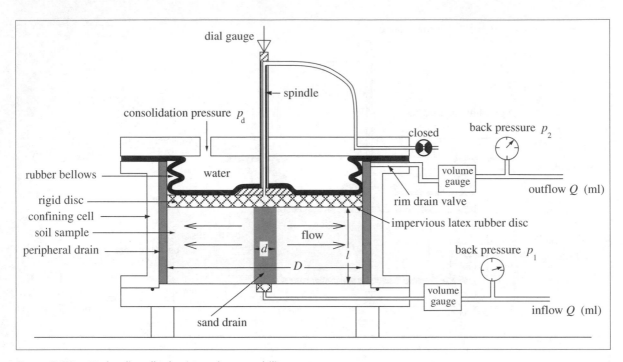

FIGURE 3.20 *Hydraulic cell – horizontal permeability*

triaxial test. The peripheral porous plastic drain is connected via the rim drain valve to a back-pressure system. A central sand drain formed by drilling a hole through the specimen and filling it with sand is connected to the other back-pressure system.

Care must be taken with the construction of the sand drain to minimise smear effects and to avoid the drain acting as a hard spot. Smearing can affect the results significantly where there is any anisotropic structure in the soil with a preference for water to flow horizontally such as with laminated or varved clays. For these reasons it is better to place the sand in a loose condition and keep the diameter of the sand drain small in relation to the sample diameter (less than 1:20).

After saturating the specimen and consolidation under the pressure p_d the inlet pressure p_1 is increased above the outlet pressure p_2 to induce water flow in a radial direction. This flow can either be inwards towards the central sand drain or outwards to the peripheral drain and should be measured when a steady flow is achieved.

The expression for the determination of the horizontal coefficient of permeability k_H is given by:

$$k_H = \frac{q\gamma_w}{2\pi l (p_1 - p_2)} \ln\left(\frac{D}{d}\right) R_T \qquad (3.15)$$

Borehole tests – open borehole
(Figures 3.21–3.23)

Permeability tests can be carried out in cased or lined boreholes providing the casing extends below the water table. Tests in cased boreholes above the water table can be carried out but interpretation of the results is doubtful. This is because the water flowing out of the borehole must first saturate the voids and suctions between the pore water and pore air in the voids will affect the hydraulic head.

The test is carried out by providing a head of water inside the borehole above the water table level to induce flow into the soil. Either a constant head can be maintained at H_c above the water table by introducing water at a steady flow rate q or a falling head test can be carried out by measuring the head of water, H above the water table at time intervals t as it falls.

Sometimes rising head tests have been used by pumping out water below the water table level but these can be prone to more uncertain results due to

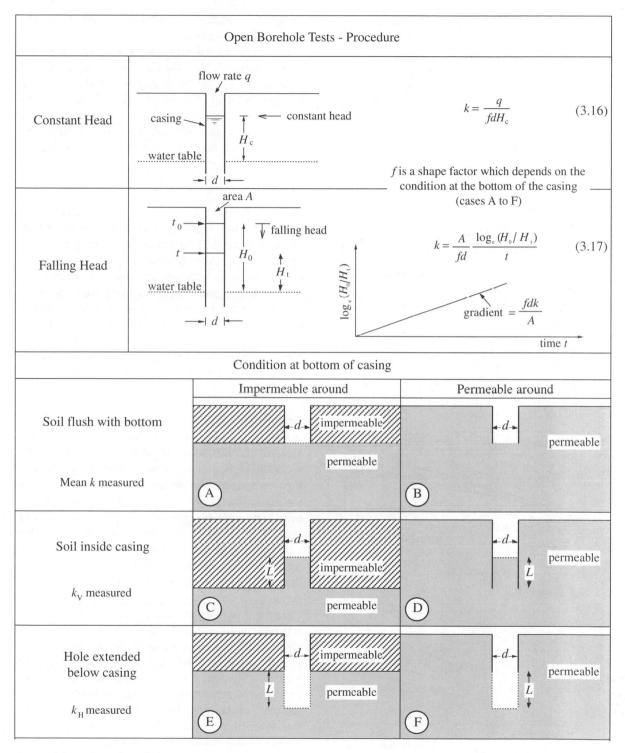

FIGURE 3.21 *Open borehole tests*

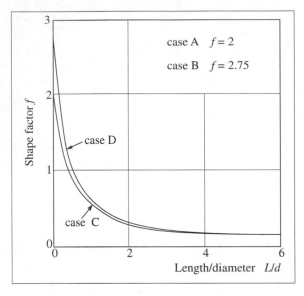

FIGURE 3.22 *Shape factor for cases A, B, C, D*

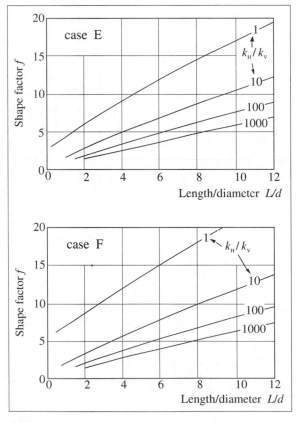

FIGURE 3.23 *Shape factors – cases E and F*

potentially unstable conditions (piping) at the bottom of the borehole.

Expressions for determining k are given on Figure 3.21; equation 3.16 is for the constant head test and equation 3.17 is for the falling head test. f is termed a shape factor or intake factor and is related to the condition at the bottom of the borehole. Shape factors for the six cases A–F have been derived from the work of Hvorslev (1951) and are given in Figures 3.22 and 3.23.

SEE WORKED EXAMPLE 3.5

These figures show that the condition at the bottom of the borehole has a significant effect, something that drilling and boring techniques cannot easily control. A small amount of soil inside the casing due to insufficient cleaning out or fines settling reduces f and it is more likely that only k_v is measured whereas extending the borehole slightly below the casing increases f and it is more likely that k_H is measured. Thus very different results can be obtained depending on the condition at the bottom of the casing and whether the soil is anisotropic ($k_H > k_v$).

Borehole tests – packer tests (Figure 3.24)
A packer is a rubber bag which is inflated against the sides of the borehole and around a tube to form a seal. The section of borehole beneath the packer can then be tested by applying a head of water H and measuring the flow rate q to maintain this constant head. The applied head can simply be the gravity head H_g of water in the tube above the water table although it is more common to apply a pressure p to increase the water head. The total head must be limited to ensure that hydraulic fracturing does not occur.

The test is more frequently carried out as drilling proceeds with a single packer sealed against the inside of the casing or it can be carried out in a completed borehole at selected levels between double packers. The packers are inflatable rubber bags expanded to achieve the necessary watertightness on the sides of the hole. Their length should be at least five times the hole diameter. The water table level should be determined before starting each test since the applied heads are related to this level. The length of the test section should be at least five times the

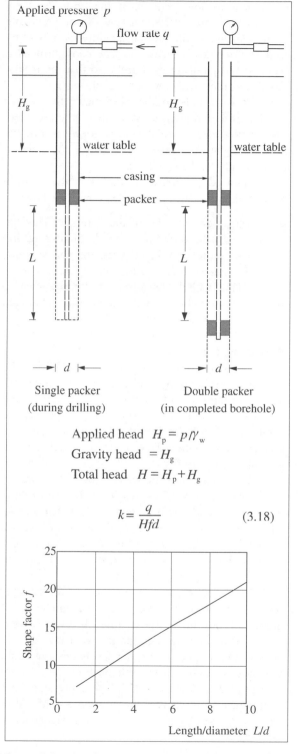

Applied head $H_p = p/\gamma_w$

Gravity head $= H_g$

Total head $H = H_p + H_g$

$$k = \frac{q}{Hfd} \qquad (3.18)$$

FIGURE 3.24 *Packer tests*

hole diameter and can be several metres long depending on the stratification and jointing pattern in the rock.

With the test carried out below the water table this requires the soil in the uncased section to be self-supporting so the test is usually restricted to determining the permeability of jointed bedrocks. The sides of the hole must be clean and free from smeared material and the water in the borehole must be clean so it is common practice to fill the hole with water, surge it, bail it out and fill with clean water.

At each test level water is pumped through the perforated pipe in the test section and a number of values of flow rate and applied head should be determined so that a graph of q versus H can be plotted to check that a straight line is obtained.

The test can be carried out above the water table where the gravity head H_g will be the head of water in the tube above the mid-point of the tested section but these results must be treated with caution as they will be less accurate. The formula given in Figure 3.24 is also less valid when the thickness of the deposit tested is less than 5 L.

Borehole tests – piezometers (Figure 3.25)

In this test a constant head is applied to the sand filter surrounding the piezometer so the dimensions of this zone must be known with reasonable accuracy. The length:diameter ratio of the sand filter should not exceed 5. It is presumed that the sand is much more permeable than the surrounding soil and it has been found (Gibson, 1966) that the permeability of the piezometer tip material itself must be at least 10 times more permeable than the surrounding soil otherwise the test will merely measure the permeability of the ceramic tip. For this reason the test is only suitable for soils of low permeability.

The constant head is maintained by an elevated water container with the tap to the piezometer turned on. This induces flow into the soil surrounding the sand filter and the flow is allowed to continue until a steady rate of flow, q_∞, is reached. However, in low permeability soils many hours are required to achieve this condition so q_∞ can be found by determining the flow rate q_t at various times t after the start of the test and plotting q_t versus $1/\sqrt{t}$. With sufficient readings a straight line can be extrapolated back to the origin to give q_∞ when $t = \infty$.

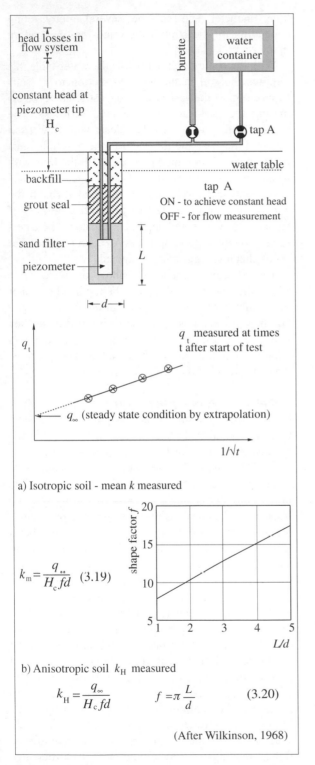

a) Isotropic soil - mean k measured

$$k_m = \frac{q_\infty}{H_c fd} \quad (3.19)$$

b) Anisotropic soil k_H measured

$$k_H = \frac{q_\infty}{H_c fd} \qquad f = \pi \frac{L}{d} \qquad (3.20)$$

(After Wilkinson, 1968)

FIGURE 3.25 *Piezometer test*

To measure flow at any stage the tap to the water container is turned off. Water flows from the burette to the piezometer tip and the quantity of flow Q ml in time t minutes is measured. The constant head is maintained by raising the burette so that the meniscus always remains at the same level.

Either two leads to the piezometer tip or two standpipes within the filter zone should be provided, one for flow measurement and one for measuring the head. At higher flow rates head losses in the tubes on the flow side can be significant.

The plot of q_t versus $1\sqrt{t}$ may not always be a good straight line. Curvature can be produced by:

● Smearing on the sides of the borehole. This can vary with the different methods of drilling and will be more significant when the soil contains well-defined macrostructure. Smearing is inevitable with the push-in type of piezometer.

● Air or gas bubbles in the sand filter. The sand should be de-aired and poured under water.

Pumping tests (Figure 3.26)
These tests are described fully in BS 5930:1981. They are carried out by constructing a pumping well to the full thickness of the aquifer and pumping out water from the bottom using suction pumps for depths less than about 5 m or submersible pumps for greater depths with a flow meter or notch tank to measure the rate of flow q.

Pumping will lower the water table and eventually a symmetrical cone of depression will form when a steady-state condition has been reached. This occurs when the rate of water pumped out of the well is comparable to the rate of water flow in the aquifer towards the well. This may take several days to achieve for low permeability soils.

The determination of k for the aquifer requires some information on the shape of the cone of depression. For this purpose, four observation wells are installed into the aquifer in two rows at right angles to each other and water levels observed until a steady state has been achieved or the rate of change of drawdown is small. The expression for k then depends on whether the aquifer is confined or unconfined, see Figure 3.2, and these are given in Figure 3.26.

SEE WORKED EXAMPLES 3.6 AND 3.7

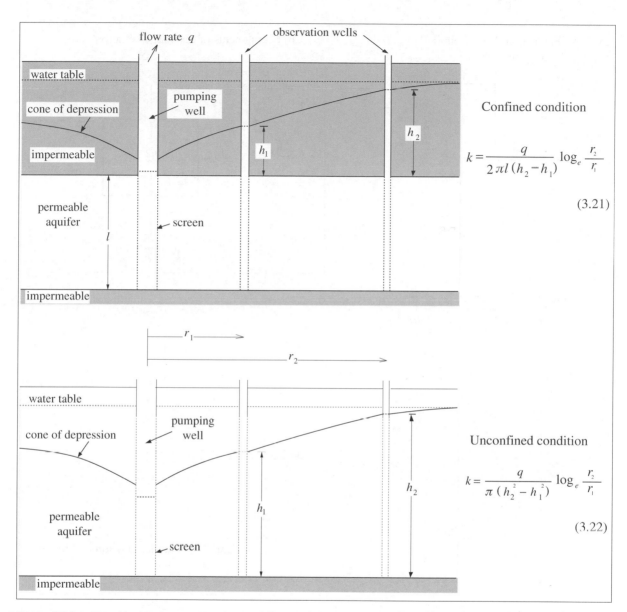

FIGURE 3.26 *Pumping tests*

Confined condition

$$k = \frac{q}{2\pi l (h_2 - h_1)} \log_e \frac{r_2}{r_1}$$

(3.21)

Unconfined condition

$$k = \frac{q}{\pi (h_2^2 - h_1^2)} \log_e \frac{r_2}{r_1}$$

(3.22)

Seepage

Seepage theory (Figure 3.27)

Seepage can occur in three directions or dimensions as in the case of flow to a pumping well. The theory considered here is for two-dimensional flow which is applicable for any cross-section through a long structure such as a sheet-pile cofferdam, earth dam or concrete weir.

Flow along the structure (perpendicular to Figure 3.27) cannot occur but flow can have vertical and horizontal components of discharge velocity v_x and v_z in the $x - z$ plane (Figure 3.27a). The permeability of the soil is considered initially to be homogeneous, or the same at all locations, and isotropic, the same in all directions, so that $k_H = k_v = k$. This theory only applies to fully saturated soils and considers water flow only below the water table.

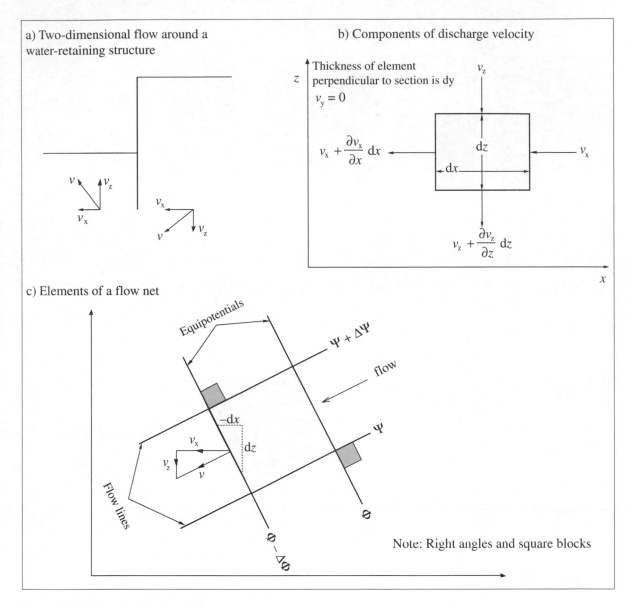

FIGURE 3.27 *Seepage theory*

Consider an element of soil with dimensions dx and dz in the plane of seepage (Figure 3.27b) and dy perpendicular to the cross-section with flow occurring in the $x - z$ plane. The discharge velocities (or components) are v_x and v_z and the velocity gradients are $\partial v_x/\partial x$ and $\partial v_z/\partial z$, respectively. Assuming the water to be incompressible and the mineral grain structure to be unaffected by seepage the quantity of water leaving the element must equal the quantity

entering it.

Quantity of flow:

$$q = Av$$

$$v_x \, dydz + v_z dydx =$$

$$\left(v_x + \frac{\partial v_x}{\partial x} \, dx\right) dydz + \left(v_z + \frac{\partial v_z}{\partial z} \, dz\right) dydx$$

Therefore:

$$\frac{\partial v_x}{\partial x} + \frac{\partial v_z}{\partial z} = 0 \tag{3.23}$$

This is the *continuity equation* in two dimensions.

The hydraulic gradient components for the element $i_x = \partial h/\partial x$ and $i_z = \partial h/\partial z$ can be inserted into Darcy's law to give:

$$v_x = ki_x = -k\,\frac{\partial h}{\partial x}$$

and

$$v_z = ki_z = -k\,\frac{\partial h}{\partial x}$$

where the total head h is decreasing in the direction of v_x and v_z.

Two functions:

$\Phi(x,z)$ – potential function
$\Psi(x,z)$ – flow or stream function

are introduced to give:

$$v_x = -k\,\frac{\partial h}{\partial x} = \frac{\partial \Phi}{\partial x} = \frac{\partial \Psi}{\partial z} \tag{3.24a}$$

$$v_z = -k\,\frac{\partial h}{\partial z} = \frac{\partial \Phi}{\partial z} = -\frac{\partial \Psi}{\partial x} \tag{3.24b}$$

From equation 3.23 it can be seen that:

$$\frac{\partial^2 \Phi}{\partial x^2} + \frac{\partial^2 \Phi}{\partial z^2} = 0$$

and

$$\frac{\partial^2 \Psi}{\partial x^2} + \frac{\partial^2 \Psi}{\partial z^2} = 0$$

which are the Laplace equations in two dimensions.

Integrating the potential function in equation 3.24 gives:

$$\Phi(x,z) = -kh(x,z) + c$$

such that:

$$\Phi = -kh$$

If this potential function is given a constant value, say Φ_1, then it will form a curve called an equipotential which can be plotted in the x–z plane and along which the total head, h will be constant.

Differentiating the stream function in equation

3.24 gives:

$$d\Psi = \frac{\partial \Psi}{\partial x}\,dx + \frac{\partial \Psi}{\partial z}\,dz$$

$$= -v_z dx + v_x dz$$

If the stream function is given a constant value, say Ψ_1, then $d\Psi = 0$ and:

$$\frac{dz}{dx} = \frac{v_z}{v_x} \tag{3.25}$$

so that the tangent to the stream function curve Ψ_1 plotted in the $x - z$ plane defines the direction of the resultant discharge velocity and this curve, of constant Ψ is called a flow line or stream line.

Differentiating the potential function in equation 3.24 gives:

$$d\Phi = \frac{\partial \Phi}{\partial x}\,dx + \frac{\partial \Phi}{\partial z}\,dz$$

$$= v_x dx + v_z dz$$

and for a particular equipotential where $d\Phi = 0$

$$\frac{dz}{dx} = -\frac{v_x}{v_z} \tag{3.26}$$

Comparing this equation with equation 3.25 shows that the flow lines intersect the equipotentials at right angles. The equipotentials act as 'contours' of equal potential or total head. Thus the direction of the maximum hydraulic gradient will be at right angles to the equipotentials and in the direction of the flow lines with flow occurring in the direction of decreasing values of potential Φ.

It can be shown that if the intervals between equipotentials $\Delta\Phi$ (potential drops) are assumed equal to the intervals between stream lines, $\Delta\Psi$ then they should be plotted on the $x - z$ plane at equal intervals thus forming 'square' shapes and producing a flow net.

Flow nets

When there is a difference in hydraulic head either side of a water-retaining structure such as a dam or sheet-pile wall water will flow beneath and around the structure. This flow can be represented mathematically by the Laplace equations of continuity given above but solutions require complex mathematical procedures.

In seepage problems it has to be accepted that only an estimate of the flows or resulting water pressures can be obtained, since this is the best our determinations of the coefficient of permeability will allow.

A simple approach to the seepage problem can be obtained by representing the Laplace equations in the form of a simple flow net sketched on a cross-section of the problem as carefully as possible following a number of rules.

A flow net consists of two sets of lines:

● *Flow lines* – these are paths along which water can flow through a cross-section. There are an infinite number of flow lines available but only a few (four to five) need to be selected for an adequate flow net. Drawing too many will complicate the result.

The intervals between adjacent flow lines (flow channels) represent a constant flow quantity, Δq so the total seepage flow is given by Δq multiplied by the number of flow channels. The number of flow channels need not be a whole number. This can be useful when plotting a flow net up to an impermeable boundary when only a part flow channel is left. The seepage quantity passing through this remaining part channel is proportional to its width relative to a full channel.

● *Equipotential lines* – these are lines of equal energy level or equal total head. As the water flows through the pore spaces its energy is dissipated by friction and the equipotential lines act like contours to show how the energy is lost. The intervals between adjacent equipotentials represent a constant difference in total head loss, Δh and the total head h lost around the structure is shared equally between equipotential drops.

It must be stressed that the head along an equipotential represents total head, not pressure head, see Figure 3.13.

A flow net is usually used to represent the steady-state condition. For example, on impounding a reservoir behind an earth dam the soil voids must first become saturated before a steady-state flow through the dam can develop.

Flow net construction (Figures 3.28–3.31)

A certain amount of skill is required in drawing a flow net but adequate results can quickly be obtained providing the following rules are observed:

● *Right angles* – flow lines and equipotential lines must cross at right angles.
● *Square blocks* – the areas formed by intersecting flow lines and equipotential lines must be as near square as possible, i.e. the central dimensions should be equal. A useful test is to visualise whether a circle can be placed inside the block and touch all four sides.
● *Impermeable boundaries* – these are flow lines. Examples are the surface of a clay layer, the ver-

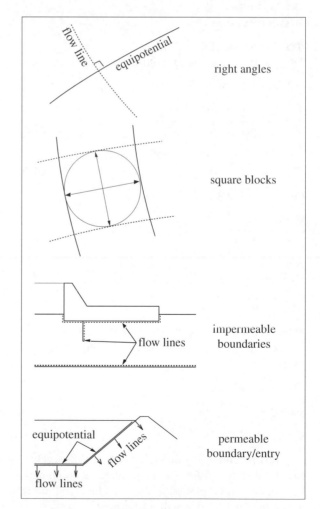

FIGURE 3.28 *Flow net rules*

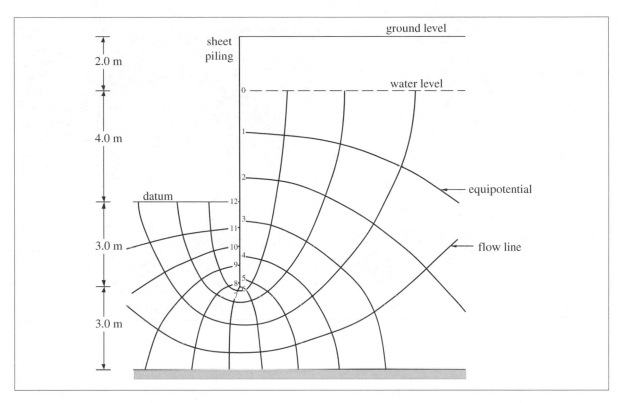

FIGURE 3.29 *Flow net – sheet piling*

tical surface of sheet piling or the underside of a concrete dam.

- *Permeable boundaries* – where a permeable soil boundary is in contact with open water as on the upstream face of an earth dam, this boundary will be an equipotential, i.e. the total head is constant on this boundary. This also applies to a horizontal water table within a permeable deposit when flow is occurring vertically downwards, as occurs alongside a sheet piled excavation.

An example of a flow net sketched (drawn, erased, edited and re-drawn) on a cross-section through a sheet-piled excavation is given in Figure 3.29.

SEE WORKED EXAMPLE 3.9

- *Entry requirements* – these apply for the construction of a flow net through an earth dam. The upstream face of the dam is an equipotential so flow lines must intersect at right angles.

SEE WORKED EXAMPLE 3.10

- *Deflection rule* (Figure 3.30) – when water flows across a boundary between soils of different permeabilities the flow lines bend, the flow channel width (distance between two flow lines) alters and the distance between equipotentials changes so that the blocks become rectangular. The quantity of flow in both deposits must be the same, i.e. $q = Aki$. When water flows from a soil of high permeability to one of low permeability A and i must increase so the flow channel width increases and the distance l between the equipotentials decreases.

This will apply within a zoned earth dam where water is flowing from the upstream shoulder fill into a central clay core. To the rear of the clay core a downstream shoulder fill is placed which is relatively permeable or sometimes a chimney drain is placed which is very permeable. Thus when water flows from a soil of low permeability such as a clay core to one of higher permeability A and i must decrease so the

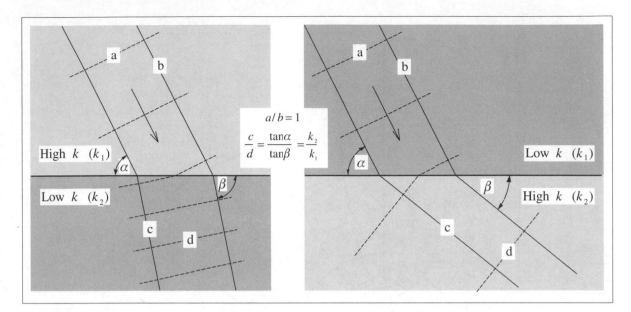

FIGURE 3.30 *Deflection rule*

rectangular blocks elongate to provide narrower flow channels and greater distance between equipotential drops.

- *Phreatic surface* – when water flows through an earth dam or flood bank the upper surface of the flowing water is the upper flow line and the flow is described as unconfined. The location of the phreatic surface is not known so a construction is adopted as shown below for a homogeneous earth structure.

- *Transformed sections* (anisotropic soils) (Figure 3.31) – flow nets are constructed as above on the assumption that permeabilities are equal in both the vertical and horizontal direction, i.e. they are isotropic. However, most natural soils and compacted fills display anisotropic permeabilities. To allow for this the cross-section is first drawn to a transformed scale and then the flow net is constructed following the above rules for isotropic conditions.

The transformed section is obtained by keeping the same vertical scale but multiplying the horizontal scale by:

$$\sqrt{(k_v / k_H)} \qquad (3.27)$$

Since k_H is usually greater than k_v this means that the horizontal dimensions must be reduced. For example, if $k_H = 9\ k_v$ all horizontal dimensions are divided by 3. This is illustrated in the Worked Example 3.9 for seepage beneath a concrete weir, Figure 3.31. If pore pressure or uplift pressures are required then the flow nets produced must be re-drawn to a natural scale and the flow net will consist of diamond shapes not squares.

SEE WORKED EXAMPLE 3.10

Seepage quantities

These can be determined from flow nets such as in Figures 3.29 and 3.31 using Darcy's law. Flow is assumed to be two-dimensional so a unit width of the cross-section is considered. The total flow around a structure will then depend on its overall length. Since the blocks on a flow net are square the width of a flow channel will be equal to its length Δl. Therefore:

Area $A = 1\ \Delta l$

The total flow $q = \Delta q$ in each channel × no. of channels

$= \Delta q \times n_f$

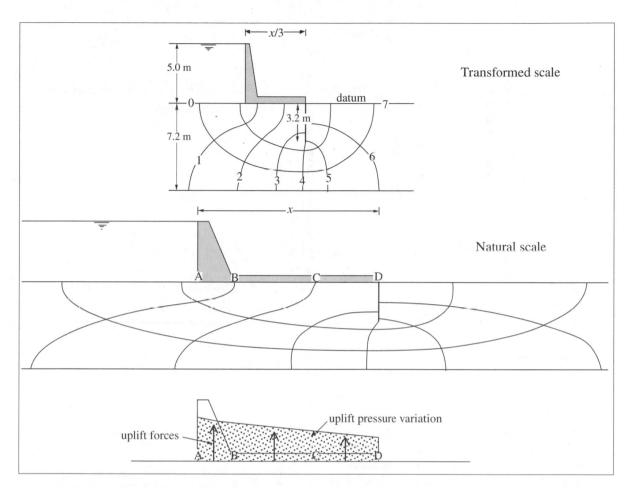

FIGURE 3.31   *Transformed section – Worked example 3.9 – seepage beneath a concrete weir*

head loss between equipotentials = Δh and

$$\Delta h = \frac{\text{total head lost}}{\text{number of intervals}} = \frac{h}{n_d}$$

The hydraulic gradient for any block

$$i = \frac{\Delta h}{\Delta l}$$

The flow in each channel = $\Delta q = Aki =$

$$1 \times \Delta lk \frac{h}{n_d} \frac{1}{\Delta l} = k \frac{h}{n_d}$$

The total flow is then given by:

$$q = kh \frac{n_f}{n_d} \qquad\qquad (3.28)$$

where:

n_f / n_d = shape factor

SEE WORKED EXAMPLES 3.8, 3.9 AND 3.10

From Figure 3.31 it can be seen that the shape factor is the same for isotropic and anisotropic permeabilities. For the isotropic case, k is obviously the isotropic k value but when k_H is different to k_v the seepage quantity must be determined using:

$$k = \sqrt{(k_H \times k_v)} \qquad\qquad (3.29)$$

Total head, elevation head and pressure head (Figure 3.32)

Total head includes position head (or elevation head) and pressure head and represents an energy level. Water will only flow if there is a difference in energy

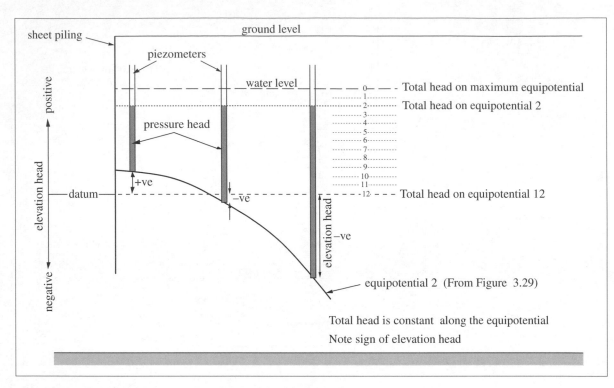

FIGURE 3.32 *Total head, elevation head and pressure head*

level but it need not flow if there is a difference in pressure level. For example, consider water at two different levels in a lake. There is a difference in pressure but there is no flow between these levels because they have the same energy or total head. Along an equipotential the total head is constant and can be determined from:

$$
\begin{array}{ll}
\text{total head on} & = \text{total head} \quad - \Delta h \times \quad (3.30) \\
\text{equipotential} & \quad \text{on maximum} \quad \text{number} \\
& \quad \text{equipotential} \quad \text{of drops}
\end{array}
$$

In Figure 3.29 the total head loss occurs over 12 equipotential drops (between 13 equipotentials) so the total head on the third equipotential (equipotential 2) is:

$4.0 - 4.0/12 \times 2 = 3.33$ or

$4.0 \times 10/12 = 3.33$ m

If a standpipe or piezometer is installed at any point on an equipotential then water will rise up the standpipe to the same level since the pressure head h_P is given by:

pressure head h_P = total head – elevation head

$$(3.31)$$

and the total head is constant along the equipotential. This is illustrated in Figure 3.32.

SEE WORKED EXAMPLE 3.8

Pore pressure and uplift pressure

Pore water pressure u is then:

$$u = \gamma_w h_P \qquad (3.32)$$

where γ_w is the unit weight of water, 9.81 kN/m³.

Since pore water pressure acts equally in all directions (it is hydrostatic) uplift pressure is the pore water pressure at the underside of an impermeable structure.

SEE WORKED EXAMPLES 3.8 AND 3.10

Seepage force (Figure 3.33)

As water flows through the voids of a soil it transfers some of its energy to the soil particles and a force is applied by the flowing water which in certain cir-

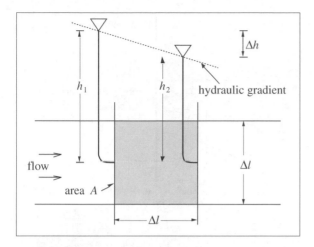

FIGURE 3.33 *Seepage force*

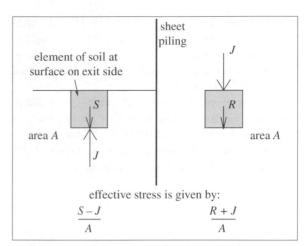

FIGURE 3.34 *The 'quick' condition and boiling*

cumstances can be detrimental to stability of the soil and any structure on the soil. This seepage force (and seepage pressure) can be derived by considering a block (Figure 3.33) in a flow net bounded by two flow lines and two equipotential lines:

water force on LHS = $\gamma_w h_1 A$
water force on RHS = $\gamma_w h_2 A$
area per unit width of section = $\Delta l \times 1$
volume affected by seepage force = $V = \Delta l^2 \, 1$
hydraulic gradient $i = \Delta h / \Delta l$

force applied to sand particles
= force on LHS – force on RHS
$= \gamma_w h_1 \Delta l \, 1 - \gamma_w h_2 \Delta l \, 1 = \gamma_w \Delta h \, \Delta l \, 1$
$= \gamma_w \Delta h / \Delta l \, \Delta l^2 \, 1 = \gamma_w \, i \, V$

Seepage force J (units of kN) is then:

$$J = \gamma_w \, i \, V \qquad (3.33)$$

Seepage pressure is given by:

$$\gamma_w \, i \qquad (3.34)$$

which is the seepage force per unit volume (units of kN/m^3). The term seepage pressure is misleading since it is not a pressure.

SEE WORKED EXAMPLE 3.9

Quick condition and boiling (Figure 3.34)

Consider a block in a flow net at the soil surface on the exit side, such as inside a sheet-pile cofferdam or downstream of a concrete or earth dam.

To the right of the sheet piling the seepage force J acts in the same direction as the gravity force R in an element of soil, so effective stresses and hence shear strength are increased. However, to the left of the sheet piling where the seepage exits from the soil the upward seepage force J is acting against the downward gravity force S. This downward force acts within and between the soil grains and represents the effective stresses in the element. Hence:

$$S = \gamma_{sub} \, V$$

However the upward seepage force J also acts on the soil grains thereby reducing the effective stresses and shear strength.

The 'quick condition' occurs when $S = J$ and the effective stresses are zero. Then the soil has no strength at all, interlock between the grains is removed and the soil is in a 'quick' or active, fluid state. Small upward seepages will be seen as localised 'boiling' carrying sand particles upwards which flow outwards to form small volcano-like mounds.

There is a critical hydraulic gradient, i_c at which this condition occurs, when:

$$S = J$$

$$\gamma_{sub} \, V = \gamma_w \, i_c \, V$$

$$i_c = \frac{\gamma_{sub}}{\gamma_w} = \frac{G_s - 1}{1 + e} \qquad (3.35)$$

This critical hydraulic gradient will depend, therefore, on the particle density and particle packing. For lightweight particles in a loose condition i_c could be as low as 0.6 but for densely packed quartz grains a value over 1.0 is likely.

A factor of safety against this 'quick' or 'boiling' condition can be obtained by considering the last block of the flow net on the exit side adjacent to the structure. The head loss is Δh and the length of this block is Δl so the exit hydraulic gradient i_e is:

$$i_e = \frac{\Delta h}{\Delta l}$$

and

$$F_{boiling} = \frac{i_c}{i_e} \qquad (3.36)$$

It can be seen that the soil immediately adjacent to a structure is the soil most prone to instability and 'piping' failures typically commence in this region. Care must also be taken in estimating the exit hydraulic gradient i_e since this will be affected by the accuracy with which the flow net is sketched.

SEE WORKED EXAMPLE 3.9

Piping adjacent to sheet piling
(Figure 3.35)

Piping failures are progressive and relentless, starting from:

● localised 'boiling' at the surface
● soil grains moving apart
● increased permeability
● increased flow
● quick condition
● loss of strength
● catastrophic collapse

Consider the portion of the flow net on the downstream side of the sheet pile excavation. With the depth of penetration of the sheet piling d it has been shown that the prism of soil of height d and width $d/2$ will be prone to failure due to upward seepage. For this prism its effective weight of prism is:

$$W = \gamma_{sub} \times d \times \frac{d}{2} = \frac{1}{2}\gamma_{sub}d^2$$

seepage force on prism $J = \gamma_w\, i\, V$

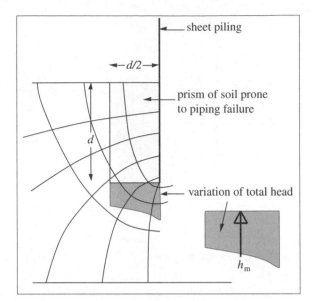

FIGURE 3.35 *Piping adjacent to sheet piling*

mean total head on base of prism $= h_m$
total head at top of prism $= 0$
mean head lost through prism $= h_m$
mean hydraulic gradient $= h_m/d = i_m$

$$\text{seepage force } J = \gamma_w iV = \gamma_w \frac{h_m}{d}\frac{d^2}{2} = \frac{1}{2}\gamma_w h_m d$$

$$\frac{W}{J} = \frac{\frac{1}{2}\gamma_{sub}d^2}{\frac{1}{2}\gamma_w h_m d} = \frac{\gamma_{sub}}{\gamma_w}\frac{d}{h_m}$$

$$F_{piping} = \frac{i_c}{i_m} \qquad (3.37)$$

From the flow net the total head is determined at a number of points at the base of the prism and plotted as shown in Figure 3.35. The mean head h_m can be obtained by estimating the area of this total head variation and dividing by $d/2$. It can be seen that the mean hydraulic gradient i_m for the prism of soil is greater than the exit hydraulic gradient i_e for the last block of soil so the factor of safety against 'piping' or 'heaving' will be lower than the factor of safety against 'boiling' and therefore, more critical.

SEE WORKED EXAMPLE 3.9

Seepage through earth dams

An earth dam is a mound of soil constructed to retain a fairly permanent reservoir level. The depth of water retained and hence the height of the dam can be considerable (20–50 m) and considering the catastrophic consequences, the risk of failure must be very low. Seepages through the dam must be small and the pore pressures produced by seepage inside the dams must not be allowed to produce instability.

For these reasons, earth dams are usually constructed with a composite cross-section with a core of impermeable clay incorporated to minimise seepage losses and filter zones to ensure that seepages are controlled and piping instability prevented. Flow nets through composite cross-sections require skill to construct and can be quite intricate. For more detailed study of this topic the reader is referred to Cedergren (1989).

Seepage through flood banks, levees
(Figure 3.36)

These are mounds of soil placed alongside a river in its flood plain or estuary to prevent flooding of large areas of flat land when river levels are high. A similar form of construction could be an irrigation dam or an earth sea wall. They may only be subject to water pressures for short periods of time and the risk of collapse may not be as unacceptable as for earth dams especially if the land protected is low-grade agricultural land.

For the above reasons flood banks, irrigation dams and sea walls are constructed in a simple fashion using local materials, preferably as impermeable as possible. These earth structures are of considerable length so zoning of materials is kept to a minimum to reduce the costs of transporting to site and placing more expensive clay cores, drainage layers and filter materials. The result is usually a simple homogeneous cross-section for which a flow net can be readily drawn.

Water flow through the cross-section can be analysed using seepage theory and a flow net can be constructed using the techniques described above and the following:

1. It is likely that the soil will have different permeabilities in the horizontal and vertical directions owing to placing and compacting the soil in layers so some degree of transformation of the section should be considered. If care is not taken with the method of embankment construction especially if smooth interfaces between compacted layers are produced then a high horizontal permeability may result leading to the destruction of the embankment by a piping failure.
2. The upstream soil surface in contact with the water is an equipotential.
3. The flow lines commence from this equipotential at right angles.
4. The top flow line or stream line is called the phreatic surface and its location is not known but can be constructed with sufficient accuracy by

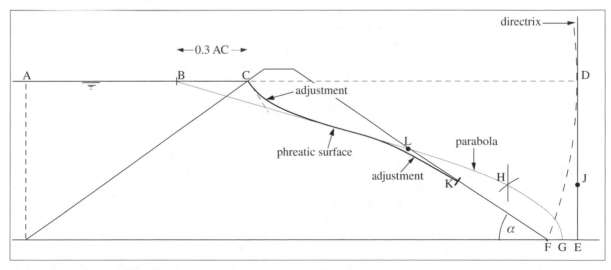

FIGURE 3.36 *Construction for phreatic surface*

first drawing a parabola and then applying some adjustments at the entry and exit points.

5. The parabola is located by assuming that it passes through point B where BC = 0.3 AC and has its focus at point F. This point F lies either at the downstream toe of a homogeneous embankment or the upper end of the filter drain, if present.

6. The property of a parabola is such that any point on it lies at the same distance from the focus F and from the directrix. The directrix DE is a vertical tangent through the point D which lies at the same level as B such that BD = BF. Points on the parabola are then obtained by drawing arcs of varying radius with the compass point at the focus F and intersecting lines parallel to the directrix at distances equal to the radius values. Then EG = FG and FH = JH.

7. The entry part of the parabola is corrected in accordance with (3) above commencing at the point C.

8. In a homogeneous earth dam where the parabola cuts the downstream slope above the toe the phreatic surface is adjusted downwards so that it is tangential to the slope cutting it at point K such that:

FK = 0.64 FL for $\alpha = 30°$

FK = 0.68 FL for $\alpha = 60°$

9. The pore pressure or pressure head along the phreatic surface is zero so total head equals elevation head along this surface. The equipotentials will then cut this surface at equal vertical distances which can be marked off and used in the flow net construction. For example, if the height of the phreatic surface is split equally into 10 vertical intervals then these will mark the starting points of 9 equipotentials.

Soil filters (Figures 3.37 and 3.38)

When water seeps out of a soil there will be a tendency for the hydraulic gradients at the soil surface to destabilise the soil and produce erosion or the boiling or piping condition, see above. To control this flow drains are provided to discharge the water rapidly so, of necessity these are highly permeable

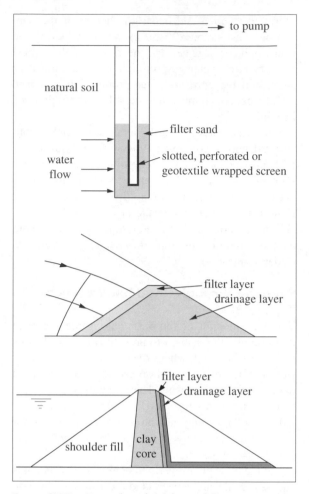

FIGURE 3.37 *Examples of the uses of filters*

and could be open pipes. However, to prevent the movement of the soil particles into the drains an intermediate filter acting as a transition zone is required. Some examples are given in Figure 3.37.

The main design criteria that a soil filter must meet are:

● *A filtering or piping criterion*
The pore sizes of the filter must be small enough to prevent the smaller natural soil particles from passing through the filter so that the soil is not eroded or allowed to develop piping. It has been shown that for uniformly graded granular soils the pore sizes are of the order of 0.1 to 0.2 times the grain sizes. Soil particles smaller than this

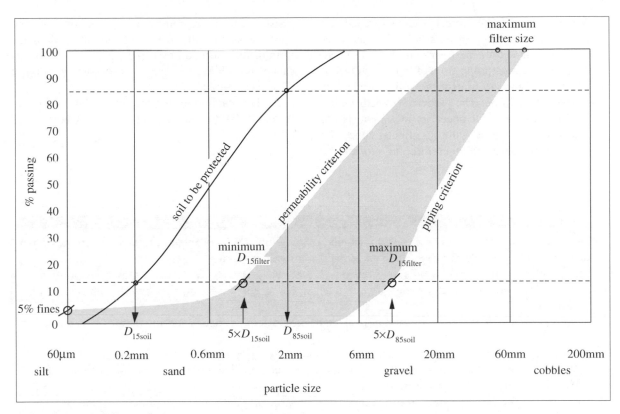

FIGURE 3.38 *Soil filter criteria*

could pass through a filter. In order to characterise the smaller sizes of a soil the particle size at the 15% percentile is commonly adopted, D_{15} and for the larger sizes the 85% percentile is used, D_{85}. These are shown in Figure 3.38. The 'tails' are not considered to be significant, although see below.

The filter must be coarser than the soil protected and a design rule commonly adopted to prevent migration of the soil into the filter is:

$$\text{(maximum) } D_{15filter} < 5 \times D_{85soil} \qquad (3.38)$$

Somerville (1986) suggested that the factor should be 4.

- *A permeability criterion*

The filter must be significantly more permeable than the soil to ensure that water can pass freely out of the soil and drain readily through the filter. A design rule to achieve this is:

$$\text{(minimum) } D_{15filter} > 5 \times D_{15soil} \qquad (3.39)$$

Somerville (1986) suggested that this factor should be 4.

To ensure a high permeability the filter material should contain no more than 5% fines (<60 μm) and should be cohesionless. An example of the application of the above rules is given in Figure 3.38.

- *Segregation must be avoided*

If the filter has a wide range of particle sizes there is a risk of the smaller sizes separating from the larger particles so that non-uniform filtering and permeation properties are obtained. To avoid this it is recommended that the filter should not be gap-graded and that the uniformity coefficient should be low.

For medium and high plasticity clays that are not prone to erosion the filter criteria can be more relaxed but for dispersive clays and silts which can erode more stringent design requirements and construction precautions should be exercised.

Where the soil to be protected is gap-graded and contains both fine and coarse particles then the filter should be designed for the finer particles and where it contains a large proportion of coarse gravel it should be designed to protect the soil particles less than about 20–25 mm. The particle size distribution for natural soils and fill materials in general is not unique; it usually varies within a grading zone. The filter must be designed, therefore, to protect the finest soil from erosion by using equation 3.38, but must be sufficiently permeable to allow drainage from the coarsest soil by using equation 3.39. If this cannot be achieved it may be necessary to adopt a two-layer or graded filter.

Further reading on this subject can be obtained in Anon (1983), Preene *et al.* (1997) and Sherard *et al.* (1984a and b).

SUMMARY

Groundwater flow through soils is the most common cause of instability problems on construction sites when excavating below the water table and in earth structures retaining water.

The fundamental law of groundwater flow in saturated soils is Darcy's law which relates the quantity of water flowing through a cross-sectional area to the hydraulic gradient causing flow by the coefficient of permeability, k.

k is related to various soil properties, particularly the void sizes and shapes and the mass or macrostructure within a soil deposit.

Laboratory tests can be carried out to determine values of k but they may not represent the *in situ* macrofabric effects. Tests carried out in boreholes are described and larger scale pumping tests are introduced.

From the theory of seepage the construction of a flow net can provide values of seepage quantity and seepage force and an assessment of the stability of earth structures with respect to piping, boiling and heaving can be made.

Given the accuracy with which *in situ* values of k can be determined the sketching method for a flow net may be sufficient.

The factors governing the performance of soil filters is described.

Piping erosion at Teton Dam, Idaho, USA

Case Objectives

This case illustrates:

- the phenomena of arching, silo action and hydraulic fracture
- the importance of preventing seepage beneath a dam
- the serious problem of erosion in some soils
- the catastrophic nature of geotechnical failures
- the high level of responsibility placed on the geotechnical engineer

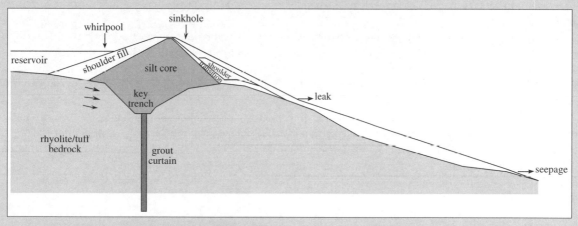

Teton dam was commenced in 1972 across the Teton River valley in Idaho, USA. The 93 m high dam was breached completely across the right abutment on 5 June 1976 when the reservoir level was about 9 m below the crest. On Thursday, 3 June small springs were seen in the right abutment about 450 m downstream of the toe of the dam. On 4 June another seepage was observed close to the toe of the dam. At about 7.30 a.m. on the morning of Saturday, 6 June a leak was observed issuing muddy water at a third of the height of the dam with a flow estimated at about 700 litres per second. This developed into a large leak and at about 10.30 a.m. an internal collapse showed that a 2 m diameter tunnel had formed in the fill. By 11.00 a.m. backsapping and piping were producing a large erosion channel carrying considerable volumes of soil-laden water. At the same time a whirlpool about 1 m diameter developed in the reservoir just upstream of the channel and at 11.45 a.m. a sinkhole was observed just

Hydraulic fracture

It has been known for some time that the core of an earth dam can be 'held up' by the adjoining shoulder fill as it attempts to settle more than the shoulders: a phenomenon known as arching. The rock surface in the narrow key trench was left rough and uncoated and a similar process called silo action is also likely to have contributed to a reduction in the total stresses. The result was that the total vertical stresses within the core, particularly in the key trench were less than the calculated overburden pressures. When the reservoir water pressure became greater than the minor principal stress, the vertical total stress in this case, the formation of horizontal cracks through the core would occur, a phenomenon known as hydraulic fracture. These cracks would obviously increase the permeability dramatically despite the lowering of the reservoir levels which would help to close them and reduce leakage rates (Penman, 1986).

downstream of the crest of the dam. Despite the efforts of dozers, 12 minutes later the dam was breached and six hours later the 27 km long reservoir was virtually empty.

It was estimated that 2.5 million m^3 of fill had been removed. Eleven people lost their lives and about 25,000 were made homeless. Much livestock was lost and considerable damage was caused to property. It was estimated that the claims for third-party damage would approach $400 million (Penman, 1977).

Beneath thick deposits of granular alluvium the bedrock comprised rhyolite and tuff with a high permeability due to a network of open joints and cavities. Grout trials showed high intakes so it was decided to replace the upper part of the grout curtain with a key trench excavated by blasting into the rock and filled with core material. However, the rock surfaces and the open joints in the trench were not sealed before placing the core. The core material was from a local aeolian silt to provide a low permeability even though it was known that the silt was relatively lightweight, brittle, prone to cracking and eroded easily. It was also found difficult to control the placement moisture content at the specified value of 0.6% dry of the optimum moisture content. A shoulder fill of sand and gravel was placed on each side of the core. Slip circle analysis gave adequate factors of safety of greater than 1.3.

It was found that a layer of slightly drier silt had been placed across the right trench close to the location of failure. Silt placed in a fairly dry condition would be quite strong and could support itself by arching over hydraulic fractures and subsequent pipes formed by erosion. When wetted the silt would be prone to a volume reduction akin to collapse settlement (Leonards et al., 1984). This would also reduce the total stresses and aggravate the hydraulic fracture condition above.

With a combination of hydraulic fracture of the core, copious supplies of water at high pressure through the rock upstream, poor erosion resistance of the core material and open joints in the rock downstream for the silt to enter, piping erosion through the core was inevitable.

Other relevant chapters are 12, Slope Stability and 13, Earthworks and Soil Compaction.

Worked Example 3.1   Constant head test

A constant head permeameter was set up containing fine sand with an effective size d_{10} of 0.12 mm. The times required to collect 250 ml of water were recorded at the following manometer readings. Determine the average value of k and compare this with Hazen's empirical relationship.

Diameter of sample = 100 mm

Distance between manometers, h = 100 mm

Area of sample, $A = \dfrac{\pi \times 100^2}{4} = 7854$ mm^2

$Q = 250$ ml $= 250 \times 10^3$ mm^3

Times to collect Q	2 min 25 s	3 min 15 s	4 min 55 s
Manometer A reading (mm)	157	191	262
Manometer B reading (mm)	24	89	195

$$k = \frac{Ql}{Aht} = \frac{250 \times 10^3 \times 100}{7854 \times h \times t} \left[\frac{\text{mm}^3 \times \text{mm}}{\text{mm}^2 \times \text{mm} \times \text{s}}\right] = \frac{3183}{ht} \text{ mm/s} = 3.183 \text{ m/s}$$

h (mm)	t (s)	k (m/s)
133	145	1.65×10^{-4}
102	195	1.60×10^{-4}
67	295	1.61×10^{-4}

∴ Average $k = 1.62 \times 10^{-4}$ m/s (report as $k = 1.6 \times 10^{-4}$ m/s)

From Hazen's relationship, equation 3.7, $k = (0.010 - 0.015) \times 0.12^2 = 1.4 - 2.2 \times 10^{-4}$ m/s. Since k can vary considerably these results are quite comparable. The test would give different values of k at different void ratios or densities which would not be reflected in Hazen's relationship.

Worked Example 3.2   Falling head test

The following data were obtained during a falling head test on a sample of clayey silt. Determine the average value of the coefficient of permeability k.

diameter of sample = 100 mm

length of sample, l = 100 mm

diameter of standpipe tube = 3 mm

Time after start, t (s)	0	15	30	49	70	96
Water level in tube (mm)	1000	900	800	700	600	500

Area of sample, $A = \dfrac{\pi \times 100^2}{4} = 7854$ mm^2

Internal area of tube, $a = \dfrac{\pi \times 3^2}{4} = 7.07$ mm^2

Time, t (s)	15	30	49	70	96
$\text{Log}_{10} h_0/h_t$	0.046	0.097	0.155	0.222	0.301

The gradient of a graph of $\log_{10} h_0/h_t$ versus t gives 3.14×10^{-3} [1/s]

$$k = \frac{3.14 \times 10^{-3} \times 2.3 \times 7.07 \times 100}{7854} = 6.5 \times 10^{-4} \text{ mm/s} = 6.5 \times 10^{-7} \text{ m/s}$$

Worked Example 3.3 Clay layer in sand

A sand deposit contains thin (10 mm) horizontal layers of clay 1.0 m apart. This is a fairly infrequent spacing. Determine the overall coefficients of permeability in the horizontal and vertical directions given the following:

For the sand L_s = 1000 mm $k_s = 1 \times 10^{-3}$ m/s
For the clay L_c = 10 mm $k_c = 1 \times 10^{-7}$ m/s

For horizontal flow, the overall coefficient of permeability is given by equation 3.9. The sequence of layers is consecutive so equation 3.8 can be used.

$$k_H = \frac{1.0 \times 10^{-3} + 0.01 \times 10^{-7}}{1.0 + 0.01} = 0.99 \times 10^{-3} \text{ m/s}$$

$\therefore k_H$ is hardly altered compared to k_s.

Also

$$\frac{q_c}{q_s} = \frac{L_c k_c}{L_s k_s} = \frac{10^{-2} \times 10^{-7}}{1.0 \times 10^{-3}} = 1 \times 10^{-6}$$

$\therefore$ Horizontal flow through the clay layers is negligible.

For vertical flow, use equation 3.11.

$$k_v = \frac{1.0 + 0.01}{\dfrac{1.0}{10^{-3}} + \dfrac{0.01}{10^{-7}}} = 1 \times 10^{-5} \text{ m/s}$$

$\therefore$ Clay thickness is only 1/100 that of sand but k_V overall is reduced by 100.

Worked Example 3.4 Sand layer in clay

A clay deposit contains thin (10 mm) horizontal layers of sand 1.0 m apart. This is a fairly infrequent spacing. Determine the overall coefficients of permeability in the horizontal and vertical directions.

For the sand L_s = 0.01 m $k_s = 1 \times 10^{-3}$ m/s
For the clay L_c = 1.0 m $k_c = 1 \times 10^{-7}$ m/s

For horizontal flow, use equation 3.8.

$$k_H = \frac{10^{-2} \times 10^{-3} + 1.0 \times 10^{-7}}{0.01 + 1.0} = 1.0 \times 10^{-5} \text{ m/s}$$

The sand thickness is 1/100 that of the clay but k_H is increased by 100.

Also, from equation 3.10:

$$\frac{q_s}{q_c} = \frac{10^{-2} \times 10^{-3}}{1.0 \times 10^{-7}} = 100$$

Horizontal flow through the sand is considerable, 100 times that through the clay.

For vertical flow, use equation 3.11.

$$k_v = \frac{0.01 + 1.0}{\dfrac{0.01}{10^{-3}} + \dfrac{1.0}{10^{-7}}} = 1.01 \times 10^{-7} \text{ m/s}$$

k_v is hardly altered compared to k_c.

Worked Example 3.5   Open borehole test

A falling head test has been carried out in a borehole sunk below the water table in a uniform deposit of silty fine sand. Details of the test measured as depths below ground level are as follows:

Bottom of casing = 5.7 m   Bottom of borehole = 6.3 m
Water table level = 4.5 m   Initial water level = 1.6 m
Internal diameter = 200 mm

The test observations and calculations are tabulated together for convenience.

| | *Test observations* | | | *Calculations* | |
Time t	Depth to water level	H	H_o/H	$x = \log_e H_o/H$	x/t ($\times 10^3$)
0	1.60	2.90	1.00	0	0
30 s	1.90	2.60	1.12	0.11	3.64
1 min	2.17	2.33	1.24	0.22	3.65
2 min	2.62	1.88	1.54	0.43	3.61
3 min	2.99	1.51	1.92	0.65	3.63
4 min	3.29	1.21	2.40	0.87	3.64
5 min	3.52	0.98	2.96	1.08	3.62
7 min	3.87	0.63	4.60	1.53	3.64
10 min	4.17	0.33	8.79	2.17	3.62

Ave = 3.63

$$\frac{L}{D} = \frac{0.6}{0.2} = 3 \text{ shape factor } f = 10.5 \text{ (Case F)} \quad \text{From Figure 3.23}$$

Assume $k_H = k_v$, from equation 3.17,

$$k = \frac{\pi \times 0.2^2 \times 3.63 \times 10^{-3}}{4 \times 10.5 \times 0.20} = 5.4 \times 10^{-5} \text{ m/s}$$

Worked Example 3.6   Pumping test in confined aquifer

Determine the value of k for a confined sand aquifer given the following:

Thickness of overlying clay = 5.5 m
Thickness of sand aquifer = 3.5 m (clay beneath)
Quantity of flow from pumping well = 0.30 m^3/min

Observation well	Distance from pumping well (m)	Depth to water level (m)
1	14	3.4
2	48	2.8

$h_2 - h_1 = 6.2 - 5.6$ or $3.4 - 2.8 = 0.6$ m

From equation 3.21, Figure 3.26:

$$k = \frac{0.3 \log_e \left(\frac{48}{14}\right)}{60 \times 2 \times \pi \times 3.5 \times 0.6} = 4.7 \times 10^{-4} \text{ m/s}$$

Worked Example 3.7   Pumping test in unconfined aquifer

An unconfined sand aquifer is 9.0 m thick with a water table 1.5 m below ground level and clay underneath. It is required to lower the water table at a point to 3.5 m by pumping from a well, 300 mm diameter, at a distance of 5 m away from the point. The coefficient of permeability of the sand k = 8 × 10^{-4} m/s. Determine the pumping rate required and the highest level the pump can be placed in the well.

Where drawdowns are a significant proportion of the saturated thickness they must be corrected using the equation:

$$S_c = S_0 - \frac{S_0^2}{2h_0}$$

S_c = corrected drawdown
S_0 = observed or required drawdown
h_0 = initial saturated thickness
$S_0 = 3.5 - 1.5 = 2.0$ m $h_0 = 9.0 - 1.5 = 7.5$ m
$$S_c = 2.0 - \frac{2.0^2}{2 \times 7.5} = 1.73 \text{ m}$$

$h_1 = 7.5 - 1.73 = 5.77$ m at a distance from the well $r_1 = 5.0$ m
r_0 is the radius when pumping has no influence (at $h = h_0$)
Assume $r_0 = 750$ m (BS 5930:1981)
From Figure 3.26, rearranging the expression for k, equation 3.22 gives

$$q = \frac{k\pi(h_0^2 - h_1^2)}{\log_e \left(\frac{r_0}{r_1}\right)}$$

$$q = \frac{8 \times \pi(7.5^2 - 5.77^2) \times 60}{10^4 \times \log_e \left(\frac{750}{5}\right)} = 0.69 \text{ m}^3/\text{min or } 690 \text{ litres/min}$$

Water level $= h_w$ at the side of the well where $r = r_w$
r_w = effective radius of well $= 0.15 \times 1.20 = 0.18$ m (BS 5930:1981)
$h_0 = 7.5$ m at $r = r_0 = 750$ m

$$0.69 = \frac{8 \times \pi(7.5^2 - h_w^2) \times 60}{10^4 \times \log_e \left(\frac{750}{0.18}\right)} \qquad \text{giving } h_w = 4.3 \text{ m}$$

Drawdown $S_c = 7.5 - 4.3 = 3.2$ m

$\therefore 3.2 = S_0 - S_0^2/15 \qquad$ giving $S_0 = 4.63$ m

Allowing for well losses $S_w = 4.63 \times 1.33 = 6.2$ m
The pumps must be placed at least 6.2 m below the water table or 7.7 m below ground level. This procedure is useful in assessing dewatering requirements.

Worked Example 3.8   Pressure head, hydraulic gradient and heaving

A 4 m thick layer of sand (sand A) overlies a 9 m thick layer of clay which in turn overlies a sand deposit (sand B). The water table in sand A is 1 m below ground level and the water table in sand B is 2 m above ground level. Determine:

(i)   *the pore water pressure at the top, middle and bottom of the clay layer*
(ii)   *the hydraulic gradient and the seepage through the clay layer*
(iii)   *the maximum depth to which an excavation could be taken in the clay layer without the risk of heave*

The permeability of the clay is 5×10^{-8} m/s and its unit weight is 21.5 kN/m^3.

i)   Take the base of the clay layer as the datum
  At the top of the clay
    Total head $= 9 + 3 = 12$ m
    Elevation head $= 9$ m
    Pressure head $= 3$ m
    Pore water pressure $= 3 \times 9.8 = 29.4$ kN/m^2.
  At the bottom of the clay
    Total head $= 0 + 15 = 15$ m
    Elevation head $= 0$
    Pressure head $= 15$ m
    Pore water pressure $= 15 \times 9.8 = 147.0$ kN/m^2
  At the middle of the clay
    The head difference between the top and the bottom is 3 m, assuming that this is lost uniformly through the clay
    Total head $= 1/2 \, (15 + 12) = 13.5$ m
    Elevation head $= 4.5$ m

Pressure head = 13.5 – 4.5 = 9 m
Pore water pressure = 9 × 9.8 = 88.2 kN/m^2

(ii) The hydraulic gradient, $i = 3/9 = 0.33$
From Darcy's law the flow rate in litres per day per square metre of the clay surface is
$q = Aki = 1 \times 5 \times 10^{-8} \times 0.33 \times 60 \times 60 \times 24 \times 1000 = 1.43$ litres/day/m^2

(iii) If the thickness of clay remaining beneath the excavation is D the balance of pressures at the base of the clay gives:
downward pressure from clay = upward pressure from water in sand B
$21.5 \times D = 147$
$\therefore D = 6.84$ m
giving a maximum excavation depth of $13 – 6.84 = 6.16$ m
This is only just over 2 m into the clay. If the site investigation had not penetrated the clay and discovered the artesian pressures in sand B then serious instability would have occurred if the excavation had been taken any deeper. Note that the groundwater in sand A would also require control either by a cut-off such as sheet piling or dewatering.

Worked Example 3.9   Seepage around a sheet pile cofferdam

A flow net has been constructed on Figure 3.29 for a cross-section through a sheet pile wall excavation in a uniform sand deposit. Determine the following:

(i) the quantity of flow in litres/minute.
(ii) the pore pressure distribution on both sides of the sheet piling.
(iii) the factors of safety against boiling and heaving.

For the sand $k = 3 \times 10^{-5}$ m/s, $G_s = 2.65$, $e = 0.60$
Total head loss $h = 4.00$ m
Number of equipotential drops $n_d = 12$
Number of flow channels $n_f = 4.5$

$q = kh \dfrac{n_f}{n_d} = \dfrac{3}{10^5} \times 4.0 \times \dfrac{4.5}{12} \times 60 \times 1000 = 2.7$ litres/min per m length of wall

Pressure head h_w = total head – elevation head   Pore pressure $u = \gamma_w h_w$

Location	Total head (m)	Elevation head (m)	Pressure head (m)	Pore pressure (kN/m^2)
0	4.0	+4.0	0	0
1	11/12 × 4.0 = 3.67	+2.5	1.17	11.5
2	10/12 × 4.0 = 3.33	+0.9	2.43	23.8
3	9/12 × 4.0 = 3.00	–0.6	3.60	35.3
4	8/12 × 4.0 = 2.67	–1.9	4.57	44.8
5	7/12 × 4.0 = 2.33	–2.6	4.93	48.4
6	6/12 × 4.0 = 2.00	–3.0	5.00	49.0
7	5/12 × 4.0 = 1.67	–3.0	4.67	45.8
8	4/12 × 4.0 = 1.33	–2.8	4.13	40.5
9	3/12 × 4.0 = 1.00	–2.2	3.20	31.4
10	2/12 × 4.0 = 0.67	–1.5	2.17	21.3
11	1/12 × 4.0 = 0.33	–0.9	1.23	12.1
12	0	0	0	0

Factor of safety against boiling $F_b = i_c/i_e$

Critical hydraulic gradient, $i_c = \dfrac{2.65 - 1}{1 + 0.60} = 1.03$

Head loss between equipotentials, $\Delta h = \dfrac{4.0}{12} = 0.33$

Δl for last block on exit $(11 \rightarrow 12) = 0.90$ m

$i_e = 0.33/0.90 = 0.37 \qquad \therefore F_b = 1.03/0.37 = 2.8$

Factor of safety against piping $\qquad F_h = i_c/i_m \qquad i_m = h_m/d$

$h_m = \dfrac{3.5 \times 4}{12} = 1.17 \qquad i_m = \dfrac{1.17}{3.00} = 0.39$

$F_h = \dfrac{1.03}{0.39} = 2.6$

Worked Example 3.10   Seepage beneath a concrete weir

A concrete weir is to be placed on a sand deposit as shown on Figure 3.31. The coefficients of permeability are $k_H = 9 \times 10^{-5}$ m/s and $k_V = 1 \times 10^{-5}$ m/s. Determine the quantity of flow in litres/minute emanating downstream and the uplift force (upthrust) acting on the underside of the structure.

The soil is anisotropic with respect to permeability so the cross-section must be drawn to a transformed scale by reducing the horizontal dimensions by:

$x_T = \sqrt{(k_V / k_H)}\, x = 1/3\, x$

and maintaining the vertical dimensions. A flow net is then drawn on this transformed cross-section (see Figure 3.31), following the rules for construction.

Total head loss $h = 5.0$ m
Overall $k = \sqrt{(k_V \times k_H)} = \sqrt{(1 \times 10^{-5} \times 9 \times 10^{-5})} = 3 \times 10^{-5}$ m/s
Number of equipotential drops $n_d = 7$
Number of flow channels $n_f = 3$

$q = \dfrac{khn_f}{n_d} = \dfrac{3 \times 5.0 \times 3 \times 60 \times 1000}{10^5 \times 7} = 3.9$ litres/min per metre length of wall

Elevation head on underside of weir $= 0$
$\therefore$ total head $=$ pressure head

Location	Total head (m)	Uplift pressure (kN/m^2)
A	5.0	49.0
B	$6/7 \times 5.0 = 4.3$	42.2
C	$5/7 \times 5.0 = 3.6$	35.3
D	$4.5/7 \times 5.0 = 3.2$	31.4

Width of weir = 15 m
Upthrust A $\rightarrow$ B = 1/2 (49.0 + 42.2) × 3.0 = 136.8 kN/m
Upthrust B $\rightarrow$ C = 1/2 (42.2 + 35.3) × 6.7 = 259.6 kN/m
Upthrust C $\rightarrow$ D = 1/2 (35.3 + 31.4) × 5.3 = 176.8 kN/m

All are forces per metre length of weir.

EXERCISES

3.1   A constant head permeameter test has been carried out on a sample of sand. With a head difference of 234 mm, 200 ml of water was collected in 3 minutes 45 seconds. The diameter of the sample is 75 mm and the distance between the manometer points is 100 mm. Determine the coefficient of permeability of the sand.

3.2   A falling head test has been carried out on a sample of soil 120 mm long in a permeameter 100 mm diameter, with a 4 mm diameter standpipe tube attached. The initial head of water in the standpipe was 950 mm and fell to 740 mm after 30 minutes. Determine the coefficient of permeability of the soil.

3.3   A constant head permeameter test has been carried out on the same sand as in Exercise 3.1 but with a layer of silt, 5 mm thick, placed within the sand between the manometer points. With a head difference of 672 mm, 100 ml of water was collected in 12 minutes 25 seconds. Assuming the value of the coefficient of permeability of the sand as obtained from Exercise 3.1 determine the coefficient of permeability of the silt.

3.4   A layer of clay 5.0 m thick overlies a thick deposit of sand. The water table in the sand is sub-artesian at 2.0 m below the top of the clay. The coefficient of permeability of the clay is 5×10^{-8} m/s. A reservoir, 100 m square and 5.0 m deep is to be impounded above the clay layer. Determine the initial seepage quantity through the clay.

3.5   A constant head test has been carried out in an open cased borehole, 150 mm diameter, in a uniform sand. A flow rate of 2.5 litres per minute was required to maintain the water level inside the casing at

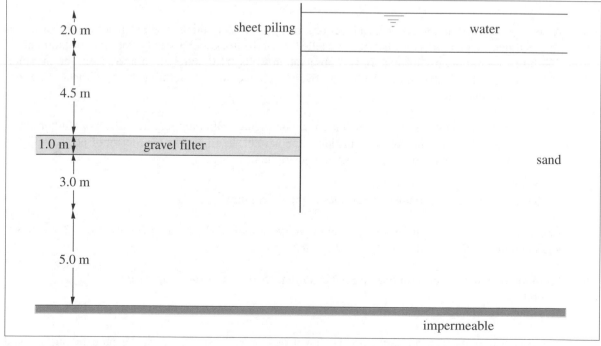

FIGURE 3.39   *Exercise 3.7*

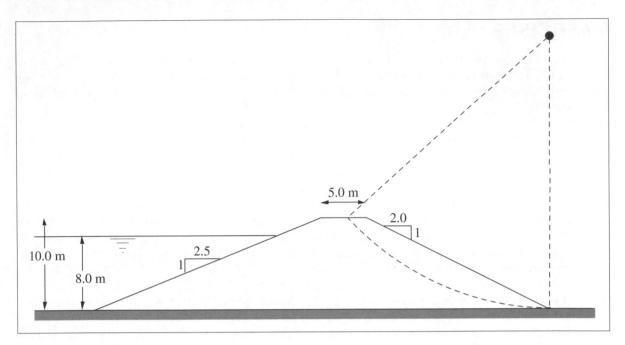

FIGURE 3.40 *Exercise 3.10*

0.5 m below ground level. The water table is at 4.0 m below ground level and the depth of casing and depth of borehole are at 7.5 m below ground level. Determine the coefficient of permeability of the sand.

3.6 A layer of clay, 6.0 m thick overlies a layer of sandy gravel 4.5 m thick which is underlaid by imperme-able bedrock. A pumping well has been installed to the bedrock and a steady state rate of pumping of 540 litres per minute established. In two observation wells, 11.0 and 37.0 m away from the pumping well, water levels of 2.65 and 2.20 m, respectively, were recorded. Determine the coefficient of perme-ability of the sandy gravel.

3.7 A cross-section through a sheet pile wall is shown on Figure 3.39. Pumping from the gravel filter main-tains the water level on the downstream side at the base of the filter. The coefficient of permeability of the sand is 3×10^{-5} m/s. Sketch a flow net and determine:

1. the rate of flow per metre of wall
2. the water pressure variation on both sides of the sheet piling.

3.8 Repeat Exercise 3.7 but with the coefficient of permeability in the vertical direction of $k_V = 3 \times 10^{-5}$ m/s and in the horizontal direction of $k_H = 6 \times 10^{-4}$ m/s.

3.9 From the flow net sketched in Exercise 3.7 determine the factor of safety against:
1. boiling
2. piping or heaving.

3.10 A cross-section through an earth dam is shown on Figure 3.40. Construct the phreatic surface and sketch a flow net. The coefficient of permeability of the soil forming the dam is 5×10^{-6} m/s. Determine the rate of flow through the dam per metre length.

3.11 From the flow net sketched in Exercise 3.10 determine the variation of pore pressure along the slip surface shown on Figure 3.40. The variation of pore pressure is required for the analysis of slope stability described in Chapter 12.

3.12 For the soil tested in Example 2.2 construct the grading envelope for a suitable filter to protect this soil.

Effective stress and pore pressure

- To understand the fundamental significance of effective stress in the study of soil mechanics
- To determine values of total stress, pore water pressure and effective stress.
- To appreciate the effects of geological and stress history on the behaviour of a soil and the difference between normally consolidated and overconsolidated soils
- To understand the state of stresses that can exist in the ground and how they are changed due to engineering processes.
- To introduce the theory of pore pressure parameters.
- To appreciate the pore pressure and effective stress state above a water table.
- To appreciate the effects of clay desiccation in some soils.
- To be aware of the problem of frost heave.

Total stress (Figure 4.1)

Total stress is the stress acting on a plane assuming the soil to be a solid material. For the small soil element shown in Figure 4.1 at a depth z below ground level the vertical total stress, σ_V, would be the stress acting on the horizontal surface of the element.

This soil element will remain in equilibrium in the ground (no consolidation or shear failure) because of a horizontal total stress, σ_H, acting on the vertical surface of the element. The stresses in soils are not isotropic, i.e. usually σ_V does not equal σ_H. In this chapter the stresses considered mostly are the vertical stresses but it must always be remembered that horizontal stresses also act. Above the water table total stress is obtained from the bulk unit weight γ_b for a partially saturated soil, i.e. possibly with some air present in the void spaces. Below the water table the saturated unit weight is used since all of the void space is fully saturated with water.

SEE WORKED EXAMPLE 4.1

Pore pressure below the water table (Figure 4.2)

This is the pressure in the water in the void spaces or pores which exist between and around the mineral grains. In the ground the level at which the water pressure is the same as atmospheric pressure is called the water table. If a tube was inserted below the water table (and no seepage was occurring) then water would rise up the tube to the water table level. See Chapter 3 for further discussion on water tables and pore pressure conditions.

The pores in the soil below the water table are fully saturated with water so this ground is often referred to as the zone of saturation. Water below the water table is called phreatic water so the phreatic surface is another term for the water table.

If seepage is not occurring only gravitational forces act on the pore water so the hydrostatic pressure within it is given by:

$$u_w = \rho\, g\, h_w \text{ or } \gamma_w\, h_w \tag{4.1}$$

where h_w is the depth below the water table. The

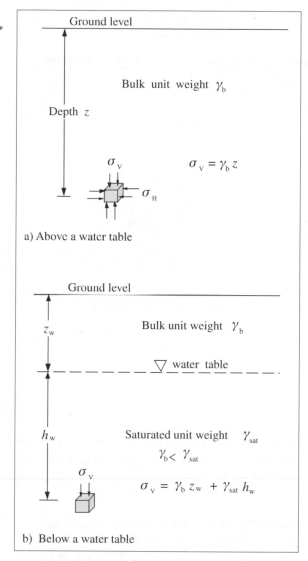

a) Above a water table

Ground level

Bulk unit weight γ_b

Depth z

σ_v

σ_H

$\sigma_v = \gamma_b z$

Ground level

z_w

Bulk unit weight γ_b

▽ water table

h_w

Saturated unit weight γ_{sat}

$\gamma_b < \gamma_{sat}$

σ_v

$\sigma_v = \gamma_b z_w + \gamma_{sat} h_w$

b) Below a water table

FIGURE 4.1 *Total stress*

pore water pressure, u_w is hydrostatic meaning that it has the same value in all directions and that it increases uniformly or hydrostatically with depth below the water table. For engineering purposes the pore pressure above the water table is often treated as being zero since this is a conservative and simplifying approach. However, pore pressures above the water table can be significantly negative and are considered later in this chapter.

SEE WORKED EXAMPLE 4.1

Effective stress (Figure 4.3)

The principle of effective stress strictly applies only to fully saturated soils. It equates the total stress (which can be imagined to exist external to an element of soil) to the stresses within the two components of the soil, i.e. the mineral grain structure (effective stress) and the water in the pores (pore water pressure).

The effective stress, therefore, is a measure of the stress existing within the mineral grain structure. This is provided by interparticle forces either between mineral grains in direct contact as with sands or as electrochemical forces between finer grained clay particles which are not in direct mineral–mineral contact.

The effective stress is strictly not the direct contact stress between the soil particles as shown in Figure 4.3 but represents the contribution provided by the soil structure or soil skeleton to support the total stress.

Since the major soil mechanics phenomena (shear strength, consolidation) are dependent on the soil structure this principle is of fundamental importance.

Effective stress in the ground (Figure 4.4)

Above the water table pore water pressures are usually assumed to be zero (rather than negative) and total stresses then equal effective stresses. Below the water table the effective stress is obtained from the total stress minus the pore water pressure or directly by using the submerged unit weight, as shown in Figure 4.4.

Total stress = effective stress + pore water pressure

$$\sigma = \sigma' + u_w \qquad (4.2)$$

SEE WORKED EXAMPLE 4.1

Stress history

The stresses to which a soil has been subjected during its formation up to the present time are referred to as its stress history. During deposition effective stresses in the soil will increase as more soil particles are placed and during erosion effective stresses will decrease as soil is removed.

Soils existing in and around river estuaries at the present time may be only centuries or even decades old and have fairly straightforward stress histories of

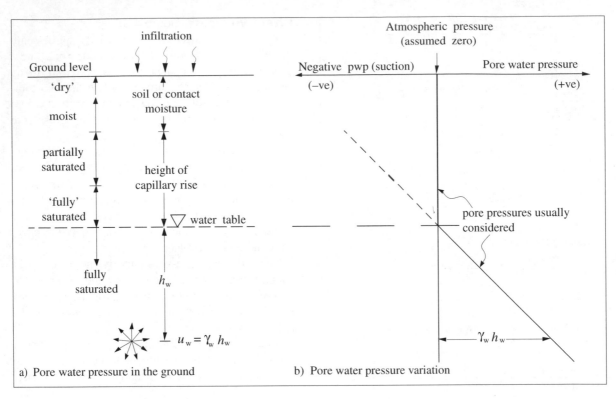

FIGURE 4.2 *Pore water pressure*

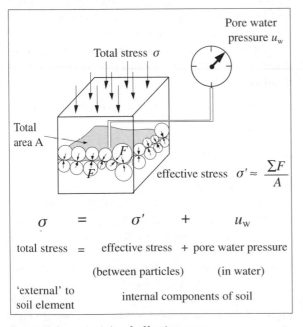

FIGURE 4.3 *Principle of effective stress*

deposition only. Older clays such as London Clay are millions of years old and have undergone more complex stress histories, particularly with significant thicknesses removed due to erosion.

Normally consolidated clay (Figure 4.5)

This is a clay that has undergone deposition only.

In Figure 4.5 two stages are shown during deposition above a soil element. It is assumed that the water level in the sea or lake remains above the soil level. As deposition occurs the vertical (and horizontal) effective stress in the soil element increases and due to the process of consolidation the void ratio of the soil is reduced. The plot of void ratio versus stress on a logarithmic scale is usually a straight line. It is denoted as the deposition line or, since there has been no other process involved, the virgin compression curve. See Chapter 6 on Consolidation to understand the processes producing volume change and the relationship between void ratio and effective stress.

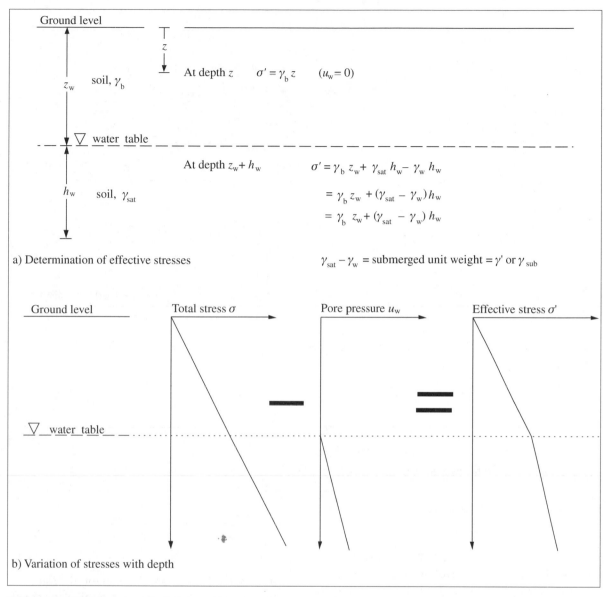

FIGURE 4.4   *Effective stress in the ground*

During deposition the mineral grain structure of the soil element will be adjusting to the changes in void ratio mostly by structural rearrangement of the particles. This process, where the particles move closer together, is an irreversible one so that the original particle arrangement could not be recovered even if the stresses were removed.

It is also postulated that some of the volume change of the soil skeleton occurs due to factors such as elastic strains in the particles themselves and compression of the layers of water molecules attracted around each particle. These volume changes are reversible or recoverable.

From stage 1 to stage 2 in Figure 4.5a, from the early part of the deposition history to a later stage, the void ratio is decreasing throughout the deposit as effective stresses increase. Note that effective stresses only increase as soil particles are deposited

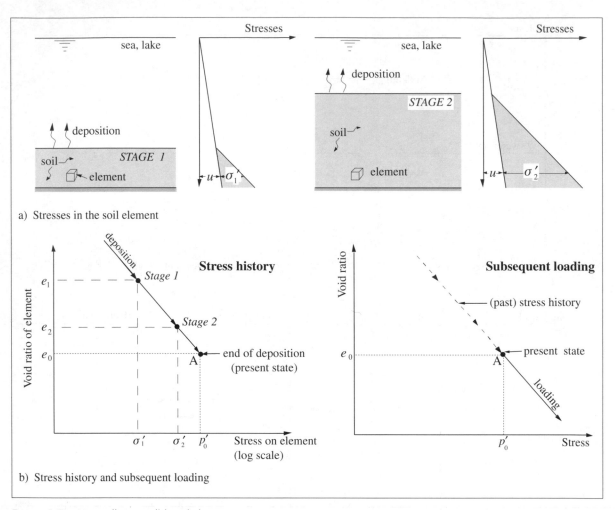

FIGURE 4.5 *Normally consolidated clay*

with the pore pressure remaining hydrostatic and associated with the water level. It should be appreciated that even if the water level fluctuates this in itself will not produce a change in effective stress providing the water level remains above the soil level.

At a certain stage, depending on the geological environment, deposition and hence consolidation will cease and the soil will remain at this state of void ratio and effective stress for ever, providing no other geological processes occur. The effective stress at this stage will be the maximum stress ever placed on the soil, p_c' and it will also be the present overburden stress or pressure p_0'. Thus:

$$p_0' = p_c' \qquad (4.3)$$

If at some time after the end of deposition a sample of the soil at the level of the soil element was taken from say a borehole with minimal disturbance, placed in a laboratory consolidation apparatus and subjected to a vertical stress p_0' it would exist at point A shown in Figure 4.5b. If the pressure on the sample is then increased the loading line would continue from point A on the original deposition line and would be expected to follow the deposition line. This would also be the case for the soil deposit *in situ* when subjected to a stress increase from a structure or an embankment.

A truly normally consolidated clay would only

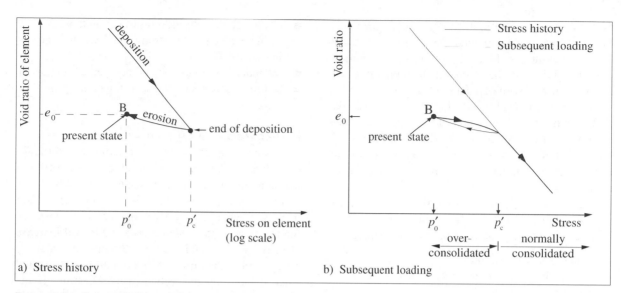

FIGURE 4.6 *Overconsolidated clay*

exist in a current deposition environment such as a river, estuary, lake or coastal region where the water level has always remained above the soil level and where it would not have been affected by any other processes which might produce overconsolidation.

Overconsolidated clay (Figure 4.6)

This is a clay that has been subjected to an effective stress in its past stress history, p_c' larger than the effective stress existing at the present time, p_0'. The most common cause of overconsolidation is erosion but it can be produced by other processes as described below. On removal of stress the soil skeleton swells but only the reversible components of volume change are recovered so the void ratio increases are smaller and the 'erosion' line follows a flatter path.

The present void ratio and pressure state for an overconsolidated clay are shown as point B in Figure 4.6. If at some time after the end of erosion a sample of this soil was taken from a borehole and reloaded from p_0' it would follow a fairly flat reloading line by overcoming the 'elastic' components of volume change until it reached the previous maximum pressure (the preconsolidation pressure, p_c'). Then it would follow the steeper deposition or virgin compression line when irreversible structural rearrangements would recommence.

At stress levels below the preconsolidation pres-

sure the soil is in an overconsolidated state but when the stress levels are higher than p_c' the soil returns to a normally consolidated state.

SEE WORKED EXAMPLE 4.2

Overconsolidation ratio
This is a measure of the degree of overconsolidation the soil has been subjected to in the past, as shown in Figure 4.6.

The overconsolidation ratio is given by:

$$OCR = \frac{p_c'}{p_0'} \tag{4.4}$$

Note that this is a ratio of vertical stresses. The effects of erosion and removal of stress on the horizontal stresses are discussed below. Typical values of the overconsolidation ratio (OCR) are:

Soil type	OCR
Normally consolidated	1
Lightly overconsolidated	1.5–3
Heavily overconsolidated	>4

Some very heavily overconsolidated clays can have *OCR* values of greater than 50 indicating that several tens of metres of soil have been removed by erosion up to the present time.

SEE WORKED EXAMPLE 4.3

Overconsolidation ratio in the ground (Figure 4.7)
For a normally consolidated clay the present over-burden pressure is the same as the past maximum pressure therefore the overconsolidation ratio will remain constant with depth and $OCR = 1$.

For an overconsolidated clay the overconsolidation ratio does not have a constant value. It decreases with depth if due to erosion, as shown in Figure 4.7, so that a soil may be heavily overconsolidated in its upper regions but tend to a normally consolidated state at depth.

The overconsolidation ratio will also vary according to other processes which the soil may have undergone since deposition. These could include:

● water table fluctuations
● desiccation due to emergence, evaporation, vegetation, roots

● physicochemical processes of cementation, cation exchange, thixotropy, leaching, weathering
● secondary compression and delayed compression
● tectonic forces, ice sheets, sustained seepage forces.

Desiccated crust (Figure 4.8)
This is produced in the upper horizons of a normally consolidated clay due to water table lowering from a to b and then back to c as shown in Figure 4.8.

The effective stresses throughout the deposit are increased when the water table drops to level b. Suctions above the lowered water table will increase effective stresses and produce further consolidation. At this stage the soil is still normally consolidated since it has only undergone further loading. If the water table rises to level c the effective stresses

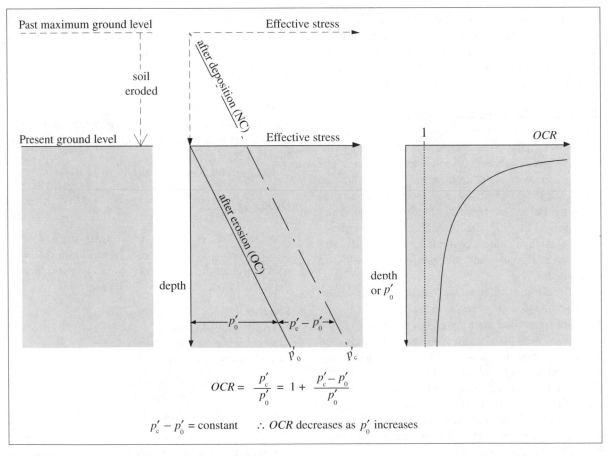

FIGURE 4.7 *Overconsolidation ratio versus depth*

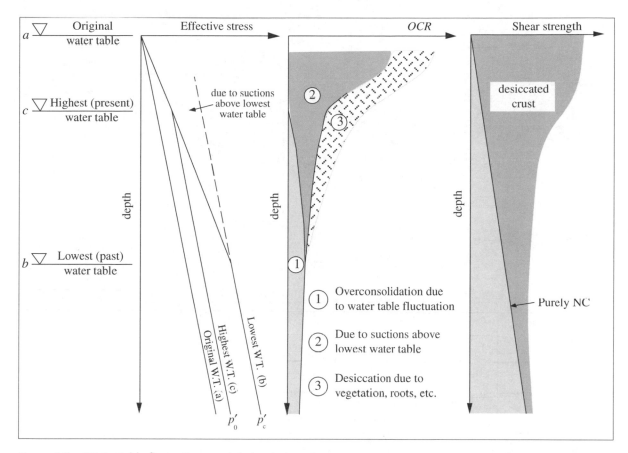

FIGURE 4.8   *Water table fluctuations and desiccated crust*

decrease below this level and above level *b*, and the soil between these levels then becomes overconsolidated. When this is followed by desiccation from evaporation, plant transpiration and other physicochemical processes a stiffer crust at the top of an otherwise very soft clay is produced. This is beneficial for agricultural and some building purposes.

Successful development of low lying marshland, particularly for agricultural purposes can be achieved by land drainage and the growth of vegetation with deep roots.

Present state of stress in the ground
(Figure 4.9)

For an element of soil in the ground to remain at equilibrium, total stresses on three orthogonal axes are required, σ_1, σ_2 and σ_3. For simplicity of analysis these stresses are usually chosen to act on planes

(principal planes) on which the shear stresses are zero and these stresses are referred to as principal stresses. There will, therefore, be three effective stresses (major, intermediate and minor principal stresses) maintaining equilibrium in the soil structure given as σ_1', σ_2' and σ_3'.

For most studies of soil mechanics it is assumed that $\sigma_2' = \sigma_3'$ (axi-symmetric conditions) and most test procedures are carried out to model this condition. There are situations where $\sigma_2' \neq \sigma_3'$ and plane strain conditions then apply, such as beneath or behind a retaining wall or beneath a slope. These conditions are discussed in more detail in Chapter 7.

Mohr's circle of stress (Figure 4.10)

The Mohr circle is a useful way of determining the shear stress and normal stress acting on a plane at any angle within a soil element given the values of

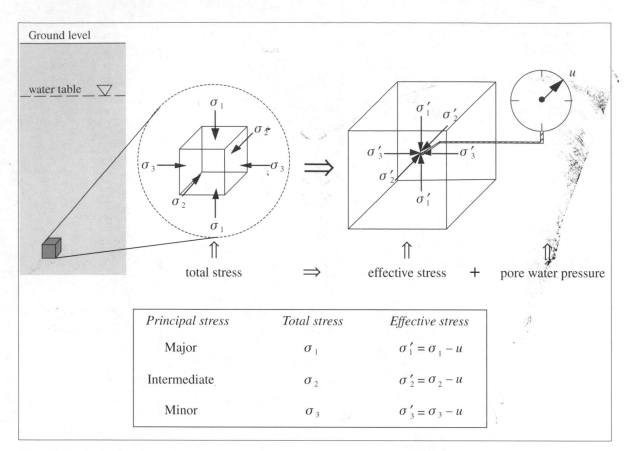

FIGURE 4.9 *In situ stresses*

the principal stresses. In Figure 4.10 the shear stress τ and normal stress σ_N acting on a plane at an angle θ to one of the principal planes can be derived or plotted from a Mohr circle given the two values of the principal stresses.

This method can depict the state of stresses for total as well as effective stresses with the circles having the same diameter but separated horizontally by the value of pore water pressure, u_w.

More detail on the use of this method is given in Chapter 7 on Shear Strength.

In situ horizontal and vertical stresses
(Figures 4.11 and 4.12)

The state of a soil *in situ* at equilibrium with no vertical or horizontal strains occurring is referred to as the 'at-rest' condition. The ratio of horizontal (σ_H') to vertical (σ_V') effective stresses in this state is denoted by K_o, the coefficient of earth pressure at rest:

$$K_0 = \frac{\sigma_H'}{\sigma_V'}$$

(4.5)

For a normally consolidated clay with isotropic stresses where $\sigma_H' = \sigma_V'$ then $K_o = 1$ and this is shown as the line of equality in Figure 4.11a. For a normally consolidated clay with anisotropic stresses i.e. $\sigma_H' < \sigma_V'$ the ratio of σ_V' and σ_H' is found to remain constant therefore K_o is constant and represented by the line AC in Figure 4.11a. σ_H' will increase uniformly with depth for this type of clay as shown in Figure 4.12a. The overconsolidation ratio *OCR* is constant and unity throughout the depth of the deposit since p_c' is equal to p_0'.

For an overconsolidated clay as the vertical stress is reduced due to erosion from point B the horizontal stresses do not reduce by the same amount but remain 'locked in' so the stress state does not return down the line AC in Figure 4.11a. An overconsoli-

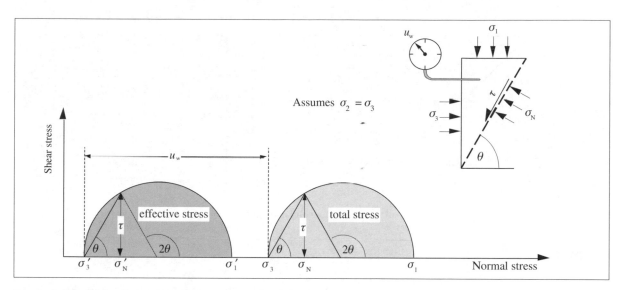

FIGURE 4.10   *Mohr circle of stress*

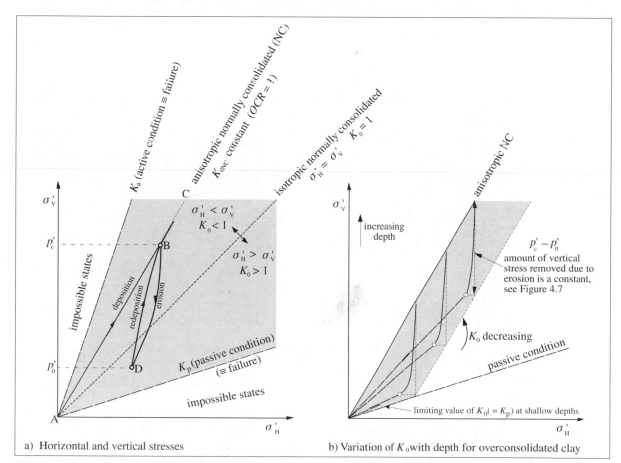

a) Horizontal and vertical stresses

b) Variation of K_0 with depth for overconsolidated clay

FIGURE 4.11   *Horizontal and vertical stresses*

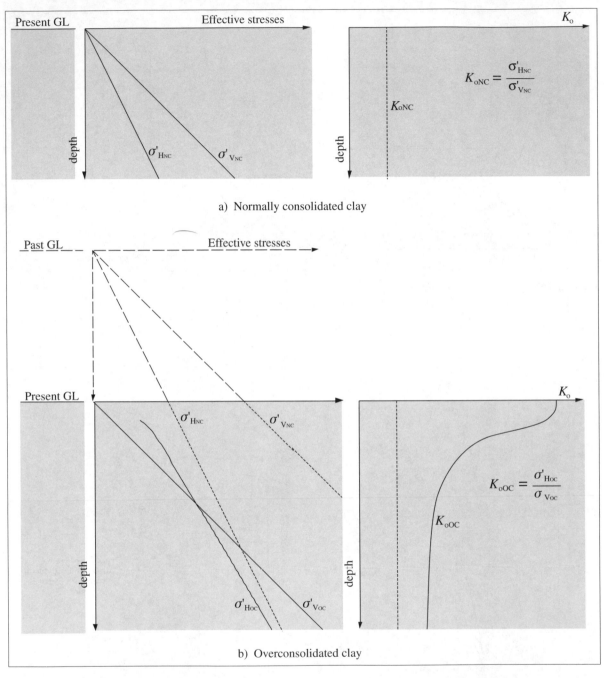

a) Normally consolidated clay

b) Overconsolidated clay

FIGURE 4.12 *Earth pressures at rest*

dated soil will have been a normally consolidated soil during its period of deposition represented by the line AB. Point B is at the end of deposition when the vertical effective stress is the preconsolidation pressure. Erosion then reduces the stresses to the point D where the vertical stress is the present overburden pressure p_0'. It can be seen that the coefficient of earth pressure at rest K_0 is increasing and the

TABLE 4.1 *Coefficients of earth pressure at rest K_0*

Soil type	Expression	Author
Elastic material	$K_o = \nu/(1-\nu)$ (ν – Poisson's ratio)	–
Normally consolidated clay	$K_{oNC} = 0.95 - \sin\phi'$ $K_{oNC} = 0.19 + 0.233 \log_{10} I_p$	Brooker and Ireland (1965) Alpan (1967)
Overconsolidated clay	$K_{oOC} = K_{oNC} \times OCR^{\sin\phi'}$	Mayne and Kulhawy (1982)
Normally consolidated sand	$K_{oNC} = 1 - \sin\phi'$	Jaky (1944)
Overconsolidated sand	$K_{oOC} = K_{oNC} \times OCR^{\sin\phi'}$	Mayne and Kulhawy (1982)

overconsolidation ratio *OCR* is also increasing as the stress state moves from B to D.

The variation of vertical and horizontal stresses with depth is shown in Figure 4.12b where it can be seen that the K_0 value for an overconsolidated soil is greater than unity and can have a large value in the upper horizons of the soil deposit and then decreases gradually with depth to less than unity.

If further deposition or reloading took place then the vertical stress would increase but the horizontal stress would not increase as much. The stress state would move from point D to B in Figure 4.11a with K_0 and *OCR* decreasing until the stress state reaches point B when the soil returns to a normally consolidated condition.

As shown later in Chapter 7 on shear strength there are values of the vertical and horizontal stresses that can produce a state of failure in the ground. These are represented by the active and passive conditions beyond which the stress state cannot exist and they, therefore, represent bounds for the values of K_0.

The variation with depth of σ_H' and σ_V' for an overconsolidated soil is shown in Figure 4.12b and the effect on K_0 and *OCR* is illustrated in Figure 4.11b. As shown in Figure 4.7 erosion reduces the vertical effective stress throughout the deposit by $p_c' - p_0'$. Since the horizontal stresses are not reduced by the same amount the value of K_0 as given by the gradient of the line from the origin in Figure

4.11b to each final stress point is decreasing with depth. However, the upper limit to this value at which passive failure occurs cannot be exceeded. This is denoted by the K_p value.

Some expressions for the determination of K_0 are given in Table 4.1. The expressions given by Mayne and Kulhawy (1982) are for the unloading cycle along the line B→D in Figure 4.11a. Lower values of K_0 will be obtained for subsequent reloading along the line D→B.

SEE WORKED EXAMPLE 4.4

Changes in stress due to engineering structures (Figure 4.13)

A soil element in the ground will remain at equilibrium supporting the total overburden stresses and with a pore water pressure u_w related to the water table level, condition A on Figure 4.13a, until a change of stress occurs. The soil will behave differently depending on whether it subsequently undergoes loading or unloading.

Loading

Figure 4.13a represents the consolidation analogy. The 'tap' on the soil element represents the facility for drainage from the soil and the degree of opening of the tap represents the permeability of the soil or the rate at which water can flow out of the soil. The spring represents the soil skeleton.

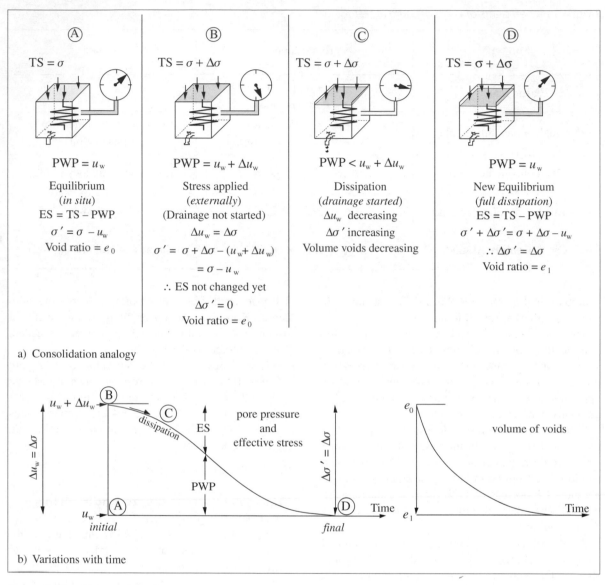

FIGURE 4.13 *Changes in stress*

When the engineering structure produces an increase in stress (total stress) on the soil element it is assumed that this stress increase is applied quickly. In relation to the permeability of the soil which is assumed to be low (such as for a clay) it is as though the tap has not yet been turned on therefore the soil element has to respond in an undrained manner.

When lateral strains in the soil are prevented (one-dimensional strain condition or K_0 consolidation) and water cannot be squeezed out of the soil (undrained conditions) the soil particles cannot rearrange themselves to develop stronger interparticle forces and cannot support the increase in total stress. In this situation the pore water has to support the total stress entirely and the increase in pore water pressure will be equal to the increase in total stress, i.e. $\Delta u_w = \Delta \sigma$, as in condition B in Figure 4.13a. At this stage the spring cannot compress since volume changes cannot occur so the soil skeleton is largely unaffected.

If lateral strains were permitted (three-dimensional strain conditions) then some particle rearrangement could take place, the interparticle forces could provide some support to the total stress (by an increase in effective stress) and the pore pressure would be lower than the applied total stress, i.e. $\Delta u_w < \Delta\sigma$, see the section on pore water pressure parameters, below. This would be represented at condition B by the sides of the element moving outwards allowing the spring to compress but without any volume change.

Drainage will occur from the clay towards the nearest permeable boundary (such as a sand layer) so with the 'tap' now considered open, flow of water will occur from the soil element under the induced hydraulic gradient and the pore pressure will decrease. This is referred to as dissipation, condition C in Figure 4.13a. As the pore pressure dissipates and the void volume decreases the soil structure responds with the particles moving closer together, increasing the interparticle forces and hence the effective stress. This would be represented by the spring shortening under increased loading.

Dissipation takes time because of the low permeability of clay soils and the distances the water has to travel to move out of the soil deposit. Eventually the pore pressure throughout the deposit returns to the original steady-state pore water pressure, u_w, condition D in Figure 4.13a. At this stage the new particle arrangement is in a stronger state to support the change in (external) total stress such that the change in effective stress becomes the change in total stress, $\Delta\sigma' = \Delta\sigma$.

This process is called consolidation and is described in more detail in Chapter 6. The relationships between pore pressure, effective stress and void ratio with time are illustrated in Figure 4.13b.

Unloading

When the engineering structure produces a decrease in total stress on the soil element such as beneath a basement or below an excavation the soil structure will have a tendency to swell due to the recoverable components of volume change but if this stress decrease is applied quickly and the soil is of low permeability the soil will again respond in an undrained manner, i.e. the 'tap' remains closed. This will produce a negative pressure in the pore water but the effective stress will remain unchanged.

As water is then drawn towards the soil element the negative pore water pressure change dissipates and increases until it finally reaches the original steady-state pore water pressure u_w or a new pore pressure value produced after steady-state or equilibrium conditions have been achieved such as in a cutting slope. The change in effective stress will now be a reduction. As the effective stress decreases the particles move further apart and the volume of voids increases, a process referred to as swelling.

At the end of the swelling process the new effective stress will be less and hence the shear strength will be reduced. Thus when a stress decrease occurs such as around basements, excavations and cuttings the long-term condition produces the most critical case.

Pore pressure parameters – theory
(Figure 4.14)

It would be useful to determine the pore water pressure change (internal) in a soil element when the total stresses (external) are changed. It is assumed that for this purpose the soil behaves in an undrained manner with no water movement to or from the voids.

An explanation for the pore pressure parameters is given in Figure 4.14. From this various cases can be considered as follows.

1. For fully saturated soils, C_w is the compressibility of the water alone (with no air present) and is very small so that B tends to 1 and equation 4.6 in Figure 4.14 becomes:

$$\Delta u = \Delta\sigma_m \qquad (4.7)$$

2. For axisymmetrical conditions (as in the triaxial test) during isotropic consolidation where only the cell pressure ($\Delta\sigma_3$) is increased:

$$\Delta\sigma_1 = \Delta\sigma_2 = \Delta\sigma_3 \text{ so } \Delta\sigma_m = \Delta\sigma_3 \text{ and}$$
$$\Delta u = B\,\Delta\sigma_3 \qquad (4.8)$$

B can then be found by applying increments of cell pressure $\Delta\sigma_3$ and recording the increase in Δu. For partially saturated soils where the degree of saturation S_r is less than 1, B will be less than 1, see pore pressure parameters, below.

SEE WORKED EXAMPLE 4.5

3. For axisymmetrical conditions during anisotropic consolidation where lateral or horizontal strains are prevented (see K_o conditions above) as in the ground, an oedometer test or a triaxial test where lateral strain is not permitted it can be shown that for a fully saturated soil:

$$\Delta u = \Delta\sigma_1 \qquad (4.9)$$

$\Delta\sigma_1$ is usually the vertical applied stress.

SEE WORKED EXAMPLE 4.6

4. In the triaxial compression test where $\Delta\sigma_1 > \Delta\sigma_3$,

when the deviator stress $(\Delta\sigma_1 - \Delta\sigma_3)$ is applied the general expression can be written:

$$\Delta u = B\left[\frac{1}{3}(\Delta\sigma_1 + 2\Delta\sigma_3)\right] \quad \text{or}$$

$$\Delta u = B\left[\Delta\sigma_3 + \frac{1}{3}(\Delta\sigma_1 - \Delta\sigma_3)\right] \qquad (4.10)$$

5. In a triaxial extension test the cell pressure represents the major $(\Delta\sigma_1)$ and intermediate $(\Delta\sigma_2)$ principal stresses with $\Delta\sigma_1 = \Delta\sigma_2$ and these are greater than the axial stress $(\Delta\sigma_3)$ giving:

---

1) The mean total stress $\Delta\sigma_m = 1/3(\Delta\sigma_1 + \Delta\sigma_2 + \Delta\sigma_3)$

$$\Delta\sigma_1' = \Delta\sigma_1 - \Delta u$$
$$\Delta\sigma_2' = \Delta\sigma_2 - \Delta u$$
$$\Delta\sigma_3' = \Delta\sigma_3 - \Delta u$$

2) The mean effective stress

$$\Delta\sigma_m' = 1/3(\Delta\sigma_1' + \Delta\sigma_2' + \Delta\sigma_3')$$
$$= 1/3(\Delta\sigma_1 + \Delta\sigma_2 + \Delta\sigma_3 - 3\Delta u)$$

3) The pore water pressure change (internal) Δu in a soil element due to total stress change (external) is obtained as follows. Undrained conditions (no water entry or exit) are assumed.

4) The soil structure behaves as an elastic isotropic material with compressibility C_s:

$$\frac{\Delta V_s}{V_T} = C_s\Delta\sigma_m' = \frac{1 - 2v'}{E'}\Delta\sigma_m'$$

E' and v' are the modulus and Poisson's ratio of the soil structure.
The decrease in volume of the soil structure is:

$$-\Delta V_s = V_T C_s \Delta\sigma_m'$$

5) The relationship between volume change (ΔV_w) of the pore water itself and its pressure (Δu) is linear, i.e.

$$\frac{\Delta V_w}{V_w} = C_w\Delta u$$

where C_w is the compressibility of the pore water. The decrease in volume of the water is:

$$-\Delta V_w = V_w C_w \Delta u = -n V_T C_w \Delta u$$

6) The soil particles themselves are incompressible.

7) The decrease in volume of water = the decrease in volume of the soil structure:

$$-\Delta V_w = -\Delta V_s$$
$$n V_T C_w \Delta u = V_T C_s \Delta\sigma_m'$$

$$\Delta u = \frac{C_s}{nC_w}\Delta\sigma_m' \quad \text{in terms of effective stress}$$

$$\Delta u = \frac{C_s}{nC_w}\left[\frac{1}{3}(\Delta\sigma_1 + \Delta\sigma_2 + \Delta\sigma_3) - \Delta u\right]$$

in terms of total stress

8) Rearranging gives:

$$\Delta u = \frac{1}{3}\frac{1}{\left(1 + \dfrac{nC_w}{C_s}\right)}(\Delta\sigma_1 + \Delta\sigma_2 + \Delta\sigma_3)$$

9) Putting pore pressure parameter B as:

$$B = \frac{1}{\left(1 + \dfrac{nC_w}{C_s}\right)}$$

$$\boxed{\Delta u = B \Delta\sigma_m} \qquad (4.6)$$

FIGURE 4.14 *Pore pressure parameters – theory*

$$\Delta u = B\left[\frac{1}{3}(2\Delta\sigma_1 + \Delta\sigma_3)\right] \quad \text{or}$$

$$\Delta u = B\left[\Delta\sigma_3 + \frac{2}{3}(\Delta\sigma_1 - \Delta\sigma_3)\right] \tag{4.11}$$

6. In a plane strain test where $\Delta\sigma_2' = v'(\Delta\sigma_1' + \Delta\sigma_3')$ it can be shown (Bishop and Henkel, 1957) that:

$$\Delta u = B\left[\Delta\sigma_3 + \frac{1}{2}(\Delta\sigma_1 - \Delta\sigma_3)\right] \tag{4.12}$$

Pore pressure parameters A and B
(Figures 4.15 and 4.16)

In reality, soil is not elastic or isotropic so the above equations have been generalised to:

$$\Delta u = B[\Delta\sigma_3 + A(\Delta\sigma_1 - \Delta\sigma_3)] \tag{4.13}$$

after Skempton (1954) where A and B are the pore pressure parameters. This expression is useful for triaxial tests since it separates the effect of changes of cell pressure ($\Delta\sigma_3$) and changes of deviator stress ($\Delta\sigma_1 - \Delta\sigma_3$) on the changes in pore water pressure.

The pore pressure parameter B decreases as the degree of saturation S_r decreases since an increase in air content increases the overall compressibility of the pore fluids. A typical relationship is shown in Figure 4.15.

SEE WORKED EXAMPLE 4.7

From above, it can be seen that the parameter A has particular values for certain conditions of test and stress systems. It will also vary during a shear strength test as pore pressures develop with shearing so the value at failure, A_f, is usually reported. For the triaxial compression test on clays, A_f has been shown to depend on the overconsolidation ratio, see Figure 4.16.

SEE WORKED EXAMPLE 4.8

For sands the parameter A will depend on the initial density. For a loose sand A will be high due to contraction of the soil structure during shear. For a dense sand A will be low and probably negative due to dilatancy of the soil structure during shear. Typical values are:

	A_f
Loose sand	2.0–3.0
Medium dense sand	0–1.0
Dense sand	–0.3–0

Values of A_f should not be used to predict pore water pressure changes at stress levels before failure and since this is the condition required for most design purposes, in practice the A_f parameter is of limited value.

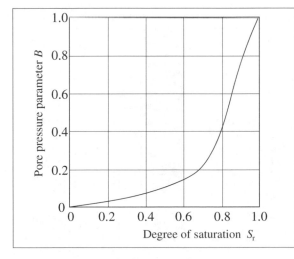

FIGURE 4.15 *Typical relationship for B*

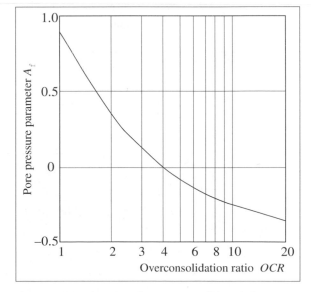

FIGURE 4.16 *A_f values for overconsolidated clays*

Capillary rise above the water table
(Figure 4.17)

The region above the water table is called the aeration zone. If water did not display the property of surface tension then the soil above the water table would be dry apart from water percolating downwards from the soil surface. In the soil belt much of this percolating water may be lost by evaporation and plant transpiration and will not, therefore reach the water table.

In the region immediately above the water table the pores may still be completely saturated because water is held within them due to the surface tension of the water. The voids in the soil form an intricate network of continuous channels which decrease in size as the soil particle size decreases and these channels can be imagined as fine capillary tubes. The height to which water will rise in a capillary tube due to surface tension effects increases as the diameter of the capillary tube decreases so this capillary zone will be greater for finer-grained soils.

Although soils contain voids (or capillary tubes) of varying sizes there will be a zone immediately above the water table where surface tension can hold water in all of the voids, even the largest. This will be a fully saturated zone of thickness h_s, as shown in Figure 4.17 and discussed below. Thus in this region the soil voids can sustain a negative pore water pressure without drawing in air but above this level the water will start to recede into the pore spaces and air will enter some of the voids. The limiting negative pore water pressure at which this occurs is called the air entry value.

Above the fully saturated capillary zone the surface tension cannot hold water in the larger sized voids and the soil becomes partially saturated, with air filling some of these voids. At higher levels the surface tension in only the finer capillaries can sustain water in them.

The height h_c to which water will rise in a capillary tube is theoretically proportional to the surface tension force T and the diameter of the tube d:

$$h_c \propto \frac{T}{d} \tag{4.14}$$

This is because the surface tension force on the air/water menisci produces a pressure in the pore water u_w greater than in the air u_a. The maximum height h_c of the capillary zone or capillary fringe as shown in Figure 4.17 will also be affected by the cleanliness of the water; it could be much less for polluted water. Therefore, h_c will represent the highest level to which a continuous channel of water will reach and this will be related to the smallest pore sizes and hence the smallest particle sizes.

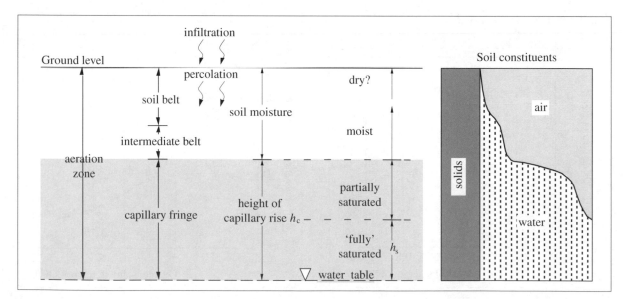

FIGURE 4.17 *Capillary rise above the water table*

According to Terzaghi and Peck (1967):

$$h_c = \frac{C}{ed_{10}}$$ (4.15)

where:

h_c = maximum height of capillary rise (mm) related to the minimum pore size
e = void ratio
d_{10} = effective size (mm)
C = constant = 10–50mm^2 (for clean water)

For open gravels the capillary rise will be negligible whereas for clays it will be considerable. Approximate values from the above expression are:

Soil type	$h_c m$
Gravel	0.01–0.05
Sand	0.1–1.0
Silt	2–10
Clay	10–30

These values apply for the smallest voids in the soil and are therefore the maximum values.

The zone of full saturation will be given by the maximum height of capillary rise in the largest voids, h_s. An approximate value for h_s could be obtained by using the d_{60} value of the soil instead of d_{10} in the above equation for h_c:

$$h_s = \frac{C}{ed_{60}}$$ (4.16)

In fine-grained soils in wet temperate climates where water tables are usually at shallow depths the capillary fringe will be thick enough to maintain moisture in the upper regions and the soil belt will not be very thick. In coarser soils in more arid climates where the water tables are much deeper and the capillary fringe is thinner the high surface temperatures and lack of rain will develop a thicker soil belt. With deeper water tables a region between the soil belt and capillary fringe can develop called the intermediate belt. The soil here is deep enough to be unaffected by surface temperatures and plants but contains a certain amount of held water suspended above the capillary fringe and prevented from moving downwards by suctions and other chemical forces.

Effective stresses above the water table
(Figure 4.18)

Within the zone of complete saturation pore water pressures will be negative and can be obtained from:

$$u_w = -\gamma_w z_a$$ (4.17)

where z_a is the elevation above the water table.
Effective stresses will then be given by:

$$\sigma' = \sigma + \gamma_w z_a$$ (4.18)

where σ is the total stress obtained in the normal way.

SEE WORKED EXAMPLE 4.9

In the zone of partial saturation there will be a pressure in the water u_w and in the air voids u_a and the difference between these is defined as suction, $u_a - u_w$. Bishop *et al.* (1960) proposed the following relationship for effective stress in a partially saturated soil:

$$\sigma' = \sigma - u_a + \chi (u_a - u_w)$$ (4.19)

where σ is the total stress and χ (chi) represents the proportion of a unit cross sectional area occupied by water. For dry soils $\chi = 0$ and for saturated soils $\chi = 1$. $u_a - u_w$ is called the matrix suction or soil water suction.

A reasonable approximation for effective stresses in the partially saturated zone can be obtained assuming $u_a = 0$ and $\chi = S_r$ (%) which gives:

$$\sigma' = \sigma + \gamma_w z_a \frac{S_r}{100}$$ (4.20)

This shows that effective stresses above the water table are enhanced and these will reduce the instability problems posed by partially saturated soils. On the other hand, if the suctions are cancelled by percolation following heavy rainfall then the large reduction in effective stress can cause instability especially in slopes.

Although the expressions given above demonstrate the link between effective stress and suction in a partially saturated soil they may not be strictly applicable for the assessment of strength and compressibility unless the degree of saturation is fairly high.

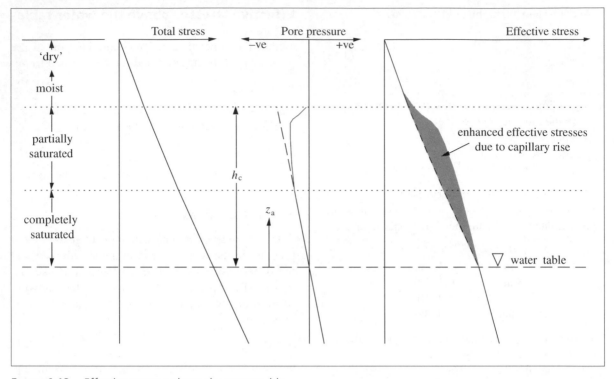

Figure 4.18 *Effective stresses above the water table*

Desiccation of clay soils
(Figures 4.19 and 4.20)

Soils above the water table can be prone to changes of moisture content and suction due to climatic effects of wet and dry seasons and due to the effects of the roots of vegetation. A typical example in Figure 4.19 for the soil in an open grass field illustrates the low generally unchanged suction levels for both seasons below the zone of seasonal variations, but significant seasonal changes occur within this zone. In the area adjacent to the trees the soil was permanently desiccated at the lower depths and mainly unaffected by seasonal variations. However, in the zone of seasonal variations large differences were observed.

Suction variations cause changes in the moisture content of a soil which in turn produces volume changes. This results in swelling of the soils during wet seasons and shrinkage of the soils during dry seasons.

If these soils are then covered by an engineering structure such as a building, a highway or a slab the

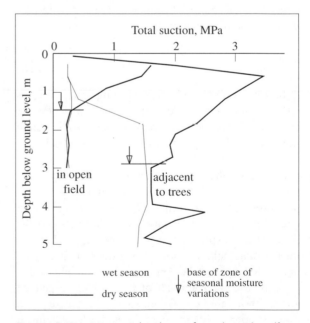

Figure 4.19 *Measured values of total suction (from Richards et al. 1983)*

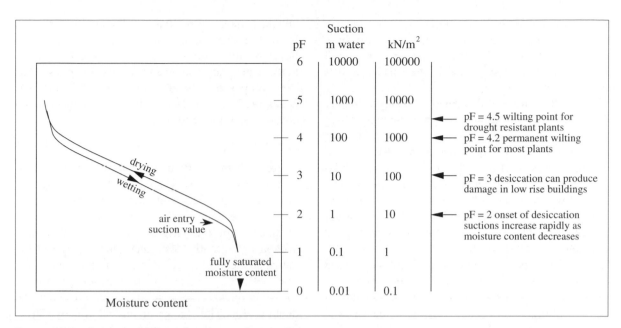

FIGURE 4.20 *Suction and the soil moisture characteristic*

natural processes are interrupted and either swelling or shrinkage may occur depending on the state of desiccation when the soil was covered. These effects can be further altered by additional water entering the soil from poor surface drainage, leaking pipes or by additional removal of moisture from heated buildings.

It has been estimated that in the United States the cost of damage to structures placed on expansive soils exceeds the financial losses caused by earthquakes, tornadoes, hurricanes and floods combined (Holtz, 1983). The severe drought in the summer of 1976 in the UK caused a dramatic increase in insurance claims for subsidence damage with costly repair bills as a result of expensive underpinning of the damaged foundations (Driscoll, 1983).

The major factor producing shrinkage is the desiccation effect caused by transpiration from vegetation roots which can extend to considerable depths. Evaporation provides a smaller contribution and extends to much shallower depths except for less vegetated areas in arid regions.

Suctions are characterised by the tensions that the soil can exert on the pore water in the pore spaces when it is not fully saturated. It comprises two components: matrix suction and osmotic suction. Matrix suction arises from the pressures generated by the capillary menisci and adsorption forces on particle surfaces and therefore depends on the moisture content and is the most important component. Osmotic suction arises from differences in concentrations of cations in the pore water and is not dependent on water content. Suctions vary considerably in value so a logarithmic scale is often adopted using the pF value given as:

$$pF = \log_{10} \text{(suction head measured in cm)} \qquad (4.21)$$

Values of pF and the equivalent suctions are given in Figure 4.20.

The soil moisture characteristic is the relationship between the moisture content and the matrix suction for a given soil, a typical example is given in Figure 4.20. When the soil is fully saturated there will be no suction in the soil but the soil will be able to sustain negative pore pressures (albeit small) without drawing in air, until the air entry value is reached. From then on the suctions increase dramatically as the moisture content decreases. The hysteresis is caused mainly by the variation in pore sizes. Consider a large void connected to a smaller void such as in the shape of an ink bottle. As the soil is drying the surface tension in the water in the small void can maintain the water in the large void and can sustain a higher suction. At the same moisture content as the soil is wetting up the larger void will be filled first and the suction in the smaller void is not as effective.

The oedometer test, described in detail in Chapter 6 can be used to determine the swelling pressure, p_s of a specimen of soil. As water is added to the container the load on the specimen is increased until an equilibrium is reached with no heave or consolidation. The initial total stress is zero and the sample suction will be equal to the mean effective stress. As the suctions are dissipated with the addition of water a total stress, the swelling pressure, must be applied to balance the effective stress.

The filter paper method is being increasingly used (Crilly and Chandler, 1993) to provide a simple test procedure for the determination of the soil suction. It entails cutting a soil sample into four equal discs at least 10 mm thick and placing a piece of Whatman's No 42 filter paper between the soil discs, sealing the discs together and allowing the filter paper to reach equilibrium with the suction in the soil by absorbing some moisture. After several days the filter papers are removed and their moisture content determined. From a correlation between the filter paper moisture content and soil suction the latter is obtained. The volume change potential of a clay soil is related to its mineralogy. There are several classifications for this potential such as given in Table 8.1 in Chapter 8. O'Neill *et al.* (1980) consider the USAEWES System to be reliable and this is reproduced in Table 4.2.

If the profile of suction with depth can be obtained such as by the filter paper test then a method of determining heave of the soil surface based on the oedometer approach (see Chapter 9, Settlements) and the results of swelling tests has been proposed (Chandler *et al.*, 1992 and BRE Digest 412, 1996). Heave can be calculated from equation 9.10 in the form:

$$\rho_h = \Sigma \, m_v \, \Delta\sigma_v{}' \, \Delta h \tag{4.22}$$

where m_v is the coefficient of compressibility (equation 6.39) for a stress change on the unloading part of the void ratio versus effective stress relationship. $\Delta\sigma_v{}'$ is the change in vertical effective stress obtained from the difference between the excess suction or currently desiccated suction, p_k and the anticipated equilibrium value based on a water table re-establishing itself at a shallow depth, given time. Δh is the thickness of a sub-layer of soil within the depth profile. The change in vertical effective stress is related to the change in suction Δp_k by:

$$\Delta p_k = \Delta\sigma_v{}' \left[\frac{1 + 2K_0}{3} \right] \tag{4.23}$$

The methods typically over-predict the heave by a factor of about 4 so clearly more research is required.

Frost action in soils (Figures 4.21 and 4.22)
Heat flow through soils occurs mostly by conduction through the solid phase of the soil since the thermal conductivity of the solids is greater than that of water and much greater than that of air. Heat will be conducted more easily through:

TABLE 4.2 *USAEWES Classification System (After O'Neill et al., 1980)*

Liquid limit %	Plasticity index %	Initial *in situ* suction kN/m²	Potential swell %	Potential swell classification	Likely volume change	Foundation design
< 50	< 25	< 145	< 0.5	low	minimal	normal construction procedures
50–60	25–35	145–385	0.5–1.5	marginal	significant	rely on previous experience with the soil or treat as high potential
> 60	> 35	> 385	> 1.5	high	substantial	quantify the amount of swell and cater for its effects

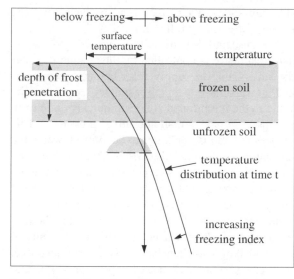

FIGURE 4.21 *Variation of temperature with depth*

- a denser soil
- a soil with a higher degree of saturation
- quartz grained sandy soils than silt and clay soils
- frozen ground than unfrozen ground since the thermal conductivity of ice is greater than water
- well-graded soils than poorly graded soils due to greater particle contact.

The temperature required to cause freezing in the ground is well below the normal freezing point of water, 0°C because of the pore water chemistry, its dissolved ions and the negative pore pressures existing within the pore water above the water table.

A typical variation of temperature with depth is

TABLE 4.3 *Frost penetration (in Mitchell, 1993)*

Freezing index (degree-days)	Frost penetration (mm)
30	220
300	630
3000	1850

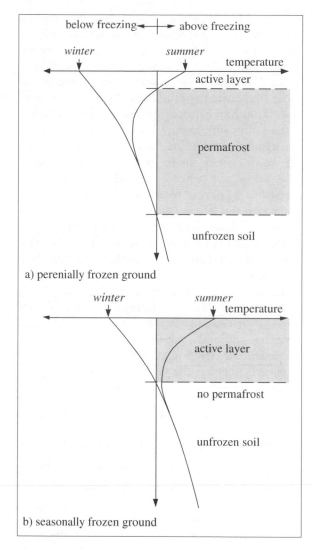

a) perenially frozen ground

b) seasonally frozen ground

FIGURE 4.22 *Permafrost conditions*

shown in Figure 4.21. The intensity of freezing is denoted by the freezing index which is the duration of the freezing period multiplied by the average surface temperature below freezing and has units of degrees × time such as degree-days. As the freezing index increases the temperature distribution will penetrate further into the ground and the depth of frozen ground or the depth of frost penetration will increase. Typical depths of frost penetration are given in Table 4.3.

As water freezes it increases in volume by nearly 10% so there will be some heaving of the ground surface and if the cooling rate is high then the soil water freezes before additional water can migrate through the soil to form an ice lens and the amount of heave is limited. However, if the temperature is lowered gradually, as in nature, small pockets of water freeze and crystallise and they give off latent heat. If this heat balances the heat lost by conduction to the earth's surface the freezing front becomes stationary and ice lenses start to form. Water molecules are drawn towards the ice nuclei from the capillary zone below and lenses of virtually pure ice interspersed with layers of frozen soil form perpendicular to the direction of the cold front. The growth of the ice lens continues until the pore water tension produced causes cavitation or the water supply from below is depleted and ice lenses then start to form at lower levels.

Ice lenses form discrete layers and completely disrupt the soil structure causing the ground surface to rise or heave considerably up to several decimetres, and with heave pressures measured in MN/m^2 the destructive effects are significant and any surface or buried structures can be lifted from their original positions. Frost heave is usually non-uniform so any

structure such as a surface road pavement will be severely affected by differential heave.

The susceptibility of soils to ice lens formation and frost heave is controlled by the amount of water the soil is able to supply from below. Clean sands and gravels have a small capillary zone and the water in the unsaturated zone will quickly drain from the pores under small negative pressures. This reduces the ability for water to be drawn towards the ice lenses in these soils. Although clay soils can sustain high negative pore pressures and still remain saturated they cannot supply water to the ice lenses quickly enough because of their low permeability. When they contain fissures or are laminated the permeability may be sufficient to permit water migration and they can then be highly frost susceptible. Silts can sustain high negative pore pressures and maintain a water supply so they have the severest frost susceptibility.

According to Mitchell (1993) soils that contain more than 3% of fine and medium silt (less than 0.02 mm) are potentially frost susceptible. The frost susceptibility classification proposed by the US Army Corps of Engineers is based on this criterion, exten-

TABLE 4.4   *Frost susceptibility classification (US Army Corps of Engineers, 1965)*

Frost susceptibility	Frost group	Soil type	% finer than 0.02mm
Non-frost susceptible	none	gravels sands	0–1.5 0–3
Possibly frost susceptible*	–	gravels sands	1.5–3 3–10
very low to high	F1	gravels	3–10
medium to high negligible to high	F2	gravels sands	10–20 10–15
medium to high low to high very low to very high	F3	gravels sands, except very fine sands clays, PI > 12	> 20 > 15 –
low to very high very low to high low to very high very low to very high	F4	all silts very fine silty sands clays, PI < 12 varved clays and other fine-grained banded sediments	– > 15 – –

sive laboratory tests and field observations and is summarised in Table 4.4.

A laboratory test to determine the susceptibility of the soil to frost action is described in BS 812:Part 124:1989 based on a procedure given by Roe *et al.*, 1984. To account for the variability of the results obtained the materials are tested in sets of three specimens each 102 mm diameter and about 150 mm high compacted to their optimum moisture content and maximum dry density using vibratory compaction or at a moisture content that provides a stable specimen when extruded from the mould. The sides of the specimen are wrapped with waxed paper, a porous disc is placed underneath and the base of the specimen is immersed in a constant level water bath maintained at a temperature of +4°C. With an air temperature at the top of the specimen maintained at −17°C the temperature gradient and access to water from the base may lead to the formation of ice lenses. Any increase in height is measured on brass rods impinging on the tops of the specimens, every 24 hours for a minimum period of 96 hours. The maximum heave observed within the 96 hours is determined for each specimen and the average of these maxima is calculated as the frost-heave value. The method of classification given in TRRL SR 829 is summarised as:

Frost-heave value	Initial classification
≤ 9.0 mm	non frost-susceptible
≥ 15.0 mm	frost-susceptible
9.1–14.9 mm	not proven

For the 'not proven' results further testing is carried out and the mean values then classified as follows:

≤ 12.0 mm	non frost-susceptible
≥ 12.1 mm	frost-susceptible

Permafrost is perennially frozen ground, even during summer months and forms when mean annual temperatures remain at least 3°C below freezing for several years (Andersland, 1987) such as in large parts of Canada, Russia and Alaska. In the northern-most latitudes the permafrost layer can extend to a depth of several hundreds of metres. Seasonally frozen ground forms when significant freezing with a high freezing index occurs during winter months but mean annual temperatures are above 3°C below freezing and development of per-

mafrost does not take place as shown in Figure 4.22. In most parts of Europe where milder climates are provided by the Gulf Stream the depth of frost penetration is small. In the UK it is usually recommended that no frost-susceptible soils should exist within 450 mm of the surface of a road pavement and that foundations are placed 600 mm below the ground surface.

When frozen ground thaws, as in the active layer in Figure 4.22 the ice lenses return to liquid water, which in the presence of the thawed soil layers and with drainage beneath prevented by the already frozen soil, a layer of very wet soil with low strength and poor bearing capacity results. If the soil is on a sloping surface then excess pore pressures set up during the thaw can produce instability in the form of a plane translational slide.

As the water in the pore spaces thaws then the volume increase that occurred on freezing is recovered. As the ice lenses melt the water produced is slowly squeezed out of the soil under the overburden pressures. These two processes combined with the additional compression of a weakened soil structure will cause large thaw settlements. It is unlikely that these phenomena will occur uniformly throughout a soil deposit so differential settlements are likely.

Ground freezing is a technique used in a positive way to provide a stabilised mass of soil for temporary support to the sides of an excavation, around a shaft or around a tunnel. A wall of frozen ground or a freeze wall will have a high strength to support the surrounding ground and will be impermeable so that groundwater is excluded. Freeze pipes are installed in the ground around the structure and a cooling medium or refrigerant is circulated down a feed pipe and back up a return pipe with the diffusion of heat from the surrounding ground causing freezing of the pore water and with time the formation of a column of frozen ground.

With freeze pipe spacings around 1 m when these columns coalesce a continuous ice wall is formed. Several days to weeks are required to achieve this. Longer freezing times are required with clay soils, higher moisture contents and higher coolant temperatures. The coolant is usually brine (calcium chloride) or for much lower temperatures liquid nitrogen is used to reduce freezing times although it is more expensive.

SUMMARY

The principle of effective stress is fundamental to the understanding of soil mechanics. It equates the internal stresses within the mineral grain structure (the effective stress) and within the water (the pore water pressure) to the external stress (the total stress).

All soils during their geological formation have been subjected to a stress history, comprising deposition or loading, erosion or unloading and other environmental processes.

A normally consolidated soil is one that has undergone deposition only. An overconsolidated soil is one that has undergone unloading usually due to erosion but other processes can cause overconsolidation.

The horizontal effective stress in the ground is not the same as the vertical effective stress, they are anisotropic. They are related by the coefficient of earth pressure at rest, K_o.

When a soil element is subjected to a change of stress it will undergo consolidation if loaded or swelling if unloaded.

The change of pore pressure caused by a change of total stress can be determined using the pore pressure parameters.

Above a water table there is a zone of full saturation where the surface tension in the pore water can sustain water in all of the voids. Above this level the soil becomes partially saturated where the finest capillaries can sustain water up to the capillary fringe.

The effective stresses are enhanced above the water table due to the negative pore pressures.

The effects of swelling and shrinkage during wet and dry seasons cause considerable damage to properties especially where they are close to tree roots.

Frost action on soils and the formation of ice lenses causes considerable heave problems during the cold period followed by weakening and settlements during the thaw period. Nevertheless ground freezing is used in a positive way to provide temporary support to excavations.

CASE STUDY

Construction pore pressures – Usk dam, South Wales

Case Objectives:

This case illustrates:

- the change in pore pressure caused by a change in total stress
- the significance of drainage boundaries (sand drains, drainage blankets) in the dissipation of pore water pressures
- the importance of the observational method in monitoring the stability and performance of an earth structure

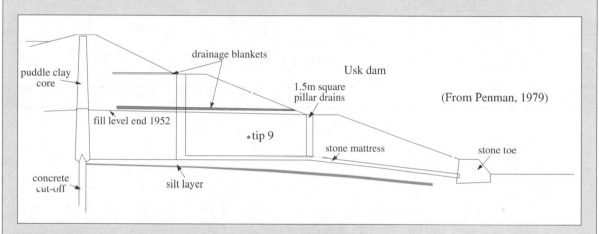

The Usk dam was constructed up to 33 m high across the River Usk in South Wales with an impermeable concrete cut-off and a puddle clay core. The shoulder fill comprised a boulder clay with typical properties comprising a maximum dry density of 1.96 Mg/m³, a placement moisture content of 12.2% and an optimum moisture content (light compaction) of 9.8%. The fill was, therefore, being placed in a condition wet of optimum. It would be fully saturated and prone to the development of excess pore water pressures when the total stresses were increased. Fill was placed in the summer of 1952 up to the level shown in the diagram below. The rate of placement was restricted to 0.9 m per month but because the contractor elected not to place fill during the six-month winter period 1952–53 he was allowed to double this rate during the summer of 1953.

A waterbearing silt layer had been discovered about 3 m below ground level within the clay soil beneath the middle part of the dam. Because of its low permeability the detrimental effects of high pore pressures generated in the foundation soil would require special measures. Vertical sand drains connecting with the silt bed and the stone mattress were installed to dissipate pore water pressures. The Building Research Station (now the Building Research

Artesian pore pressures

At the same time BRE took the opportunity to place three tips in the 1952 fill at about its mid-height to monitor construction pore pressures in the fill and the results from tip no 9 are presented below (Penman, 1979). Initially the tip recorded a pore pressure equal to 150% of the total overburden pressure, a result met with some concern. Following advice from Professor Skempton and Dr Bishop of Imperial College, this artesian condition was checked by the installation of some steel stand-pipe piezometers driven into the fill on an adjacent cross-section which recorded water levels up to 2.4m above the fill surface.

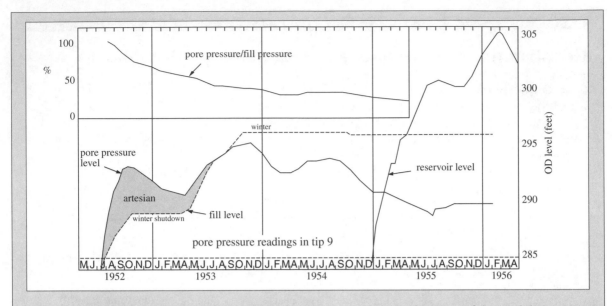

Establishment) were involved in placing piezometer tips to check on the performance of the sand drains. No significant pore pressures were recorded in the tips in the silt layer showing that the sand drains were effective (Sheppard *et al.*, 1957).

When clay soil is placed wet of its optimum value it would be virtually fully saturated giving it a high pore pressure parameter B, see equation 4.6, and with a fairly soft consistency it would tend to have a positive A value, see equation 4.13, so that increases in the total stresses would produce high positive pore water pressures.

During the winter shut-down the pore pressures dissipated but due to a fairly high fines content the boulder clay had a low permeability and dissipation was not sufficient at the end of the shut-down to allow further fill placement without the risk of slope instability. It was decided to place horizontal sand and gravel drainage blankets at the locations shown in the diagram above, to reduce the drainage path length, to speed up the rate of dissipation and permit the desired rate of filling. The blankets comprised 300 mm of river gravel, then about 200 mm of crushed stone followed by 450 mm of river gravel.

The lower drainage layer was successful in reducing the rate of increase of pore pressures during the 1953 earthmoving season (May–October) and increasing the rate of dissipation during winter, 1953–54 (November–February). It is interesting that the pore water pressures increased during the 1954 earthmoving season (March–August) with no increase of fill and no change in vertical overburden pressure above tip 9. The increase in pore water pressure could have been due to:

1. increased horizontal stresses at tip 9 from the placement of the fill near the crest of the dam and
2. the horizontal dissipation towards tip 9, of the pore water pressures induced upstream of the tip.

The puddle clay core was effective in providing an impermeable barrier since when the reservoir level was increased the pore pressures on the upstream side responded immediately but on the downstream side they were not significantly affected. The efficiency of the horizontal drains was confirmed by the line of seepage observed on the downstream face of the dam during dry weather.

Other relevant chapters are

- 6   Consolidation – dissipation;
- 12   Slope Stability – the effect of pore water pressures on the stability of a slope and
- 13   Earthworks and Soil Compaction – fill placement moisture content.

Worked Example 4.1　Effective stress in the ground

For the ground conditions given below determine the variation of stresses with depth.

The simplest way to proceed is to determine values of total stress and pore water pressure at particular depths assuming a linear variation between these points. Effective stress is obtained by subtracting pore pressure from total stress.

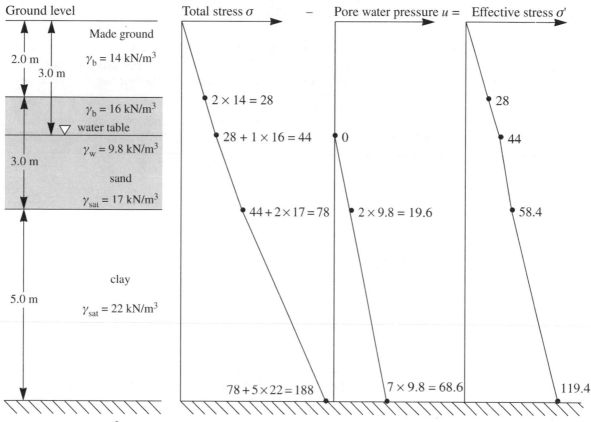

Note: $\gamma_w = 10$ kN/m^3 is often used and gives a sufficiently accurate result

Worked Example 4.2　Overconsolidated clay

A sample of clay has been taken from 6 m below the bed of a river. The saturated unit weight of the clay $\gamma_{sat} = 21.5$ kN/m^3 *and the overconsolidation ratio of the clay has been found to be OCR = 2.5. Determine how much soil has been removed by erosion.*

Assume $\gamma_w = 9.8$ kN/m^3

$\gamma_{sub} = 21.5 - 9.8 = 11.7$ kN/m^3

The present effective stress or overburden pressure $p_o' = 6 \times 11.7 = 70.2$ kN/m^2
From equation 4.3, the past maximum pressure $p_c = 70.2 \times 2.5 = 175.5$ kN/m^2
Assuming the saturated unit weight of the soil removed to be 21.5 kN/m^3 and the original maximum thickness of soil $= z$ then

$z \times 11.7 = 175.5$ $\qquad \therefore z = 15.0$ m

The thickness of soil removed $= 15.0 - 6.0 = 9.0$ m

Worked Example 4.3 Overconsolidation ratio

An embankment 8 m high and with unit weight 21.5 kN/m^3 was placed over a normally consolidated clay many years ago. As part of a redevelopment it is to be removed. Determine the overconsolidation ratio of the clay at 1 m and 10 m below ground level after removal of the embankment. The unit weight of the clay is 18.5 kN/m^3.

Assuming that the negative pore pressures generated as the embankment was removed are fully dissipated and the water table remains at 1 m below ground level.
The previous maximum effective stresses were

at 1 m $8 \times 21.5 + 18.5 \times 1 = 190.5$ kN/m^2
at 10 m $8 \times 21.5 + 10 \times 18.5 - 9.8 \times 9 = 268.8$ kN/m^2

The present effective stresses are

at 1 m $18.5 \times 1 = 18.5$ kN/m^2
at 10 m $18.5 \times 10 - 9.8 \times 9 = 96.8$ kN/m^2

The overconsolidation ratios are

at 1 m $\quad OCR = \dfrac{190.5}{18.5} = 10.3$

at 10 m $\quad OCR = \dfrac{268.8}{96.8} = 2.8$

The clay at shallow depth has become very heavily overconsolidated but the clay at depth is only lightly overconsolidated.

Worked Example 4.4 $\quad K_o$ condition

A sample of clay has been taken from 5 m below ground level with the water table at 1.5 m below ground level. The unit weight of the clay above and below the water table is $\gamma_{sat} = 20.7$ kN/m^3 and the K_o value has been determined as $K_o = 0.85$. Determine the total and effective vertical and horizontal stresses at this depth.

Assume $\gamma_w = 9.8$ kN/m^3
Pore water pressure $u_w = 3.5 \times 9.8 = 34.3$ kN/m^2
Total vertical stress $\sigma_v = 5 \times 20.7 = 103.5$ kN/m^2
Effective vertical stress $\sigma_v' = 103.5 - 34.3 = 69.2$ kN/m^2

Effective horizontal stress $\sigma_H' = 69.2 \times 0.85 = 58.8$ kN/m^2 (From equation 4.4)
Total horizontal stress $\sigma_H = 58.8 + 34.3 = 93.1$ kN/m^2

Worked Example 4.5   Changes in stress

In a triaxial apparatus a specimen of fully saturated clay has been consolidated, by allowing drainage of pore water from the specimen, under an all-round pressure of 600 kN/m^2 and a back pressure of 200 kN/m^2. The drainage tap is then closed and the cell pressure increased to 750 kN/m^2. Determine the effective stress and pore water pressures before and after increasing the cell pressure.

(i) *Before increasing cell pressure*
     back pressure = pore water pressure = 200 kN/m^2
     The consolidation is isotropic (same stresses all-round) so the total stresses are
     $\sigma_1 = \sigma_3 = 600$ kN/m^2
     The effective stresses are
     $\sigma_1 = \sigma_3' = 600 - 200 = 400$ kN/m^2

(ii) *After increasing the cell pressure*
     Since the clay is fully saturated, $B = 1$ so the excess pore pressure will be
     $\Delta u = \Delta\sigma_3 = 750 - 600 = 150$ kN/m^2 (From equation 4.7)
     The pore pressure will now be $200 + 150 = 350$ kN/m^2
     The effective stress = $750 - 350 = 400$ kN/m^2. It has remained unchanged in the undrained and fully saturated condition. No consolidation or increase in shear strength due to this increase in cell pressure will take place until the excess pore pressure u is allowed to dissipate by drainage from the specimen. An explanation of the use of back pressure is given in Chapter 7 on shear strength.

Worked Example 4.6   Changes in stress in an oedometer

In an oedometer apparatus a specimen of fully saturated clay has been consolidated under a vertical pressure of 75 kN/m^2 and is at equilibrium. Determine the effective stress and pore water pressure immediately on increasing the vertical stress to 125 kN/m^2.

Under the applied pressure of 75 kN/m^2, the pore pressure will be zero since consolidation is complete.
Total vertical stress, σ_v or $\sigma_1 = 75$ kN/m^2
Pore water pressure $u = 0$
Effective vertical stress σ_v' or $\sigma_1' = 75$ kN/m^2
The horizontal stress is not known but for this test condition
$\Delta u = \Delta\sigma_1 = 125 - 75 = 50$ kN/m^2 (From equation 4.8)
so the pore pressure will rise immediately to 50 kN/m^2 on increasing the vertical stress to 125 kN/m^2.
The initial effective stress will be unchanged, i.e.
$\sigma_v' = 125 - 50 = 75$ kN/m^2
However, the specimen will immediately commence consolidating with the pore pressure decreasing and the effective stress increasing. The oedometer test for consolidation properties of soils is described in Chapter 6.

Worked Example 4.7   Pore pressure parameter *B*

In a triaxial test a soil specimen has been consolidated under a cell pressure of 400 kN/m^2 and a back pressure of 200 kN/m^2. The drainage tap is then closed, the cell pressure increased to 500 kN/m^2 and the pore pressure measured as 297 kN/m^2. Determine the pore pressure parameter B.

The change in pore pressure $\Delta u = 297 - 200 = 97$ kN/m^2
The change in total stress $= \Delta\sigma_3 = 500 - 400 = 100$ kN/m^2
$B = \Delta u/\Delta\sigma_3 = 97/100 = 0.97$ (From equation 4.7)

Worked Example 4.8   Pore pressure parameter *A*

From the Worked Example 4.7, with the specimen remaining under undrained conditions and after the increase in cell pressure to 500 kN/m^2, an axial load is applied to give a principal stress difference of 645 kN/m^2 when the pore pressure is measured as 435 kN/m^2. Determine the pore pressure parameter A at this stage.

From above, $B = 0.97$ $\Delta\sigma_3 = 100$ kN/m^2
$\Delta\sigma_1 - \Delta\sigma_3 = 645$ kN/m^2
$\Delta u = 435 - 200 = 235$ kN/m^2
From equation 4.12,
$235 = 0.97[100 + A(645)]$ $\therefore A = 0.22$
Note that the value of *A* changes throughout the test.

Worked Example 4.9   Pore pressures and effective stresses above the water table

For the soil conditions in Worked Example 4.1 determine the pore water pressure and effective stress within the sand assuming that the sand is sufficiently fine to maintain a zone of full saturation.

From equation 4.17 the pore water pressure at the top of the sand would be
$-9.8 \times 1.0 = -9.8$ kN/m^2
and from equation 4.18 the effective stress would be
$28 + 9.8 = 37.8$ kN/m^2
Assuming the made ground to be granular and loose with large void spaces the capillary rise will be terminated at the base of the made ground and this deposit will act as a capillary break.

EXERCISES

4.1 A river, 5 m deep, flows over a sand deposit. The saturated unit weight of the sand is 18 kN/m³. At a depth of 5 m below the river bed determine:

1. total vertical stress
2. pore water pressure
3. effective vertical stress.

Assume the unit weight of water is 9.8 kN/m³.

4.2 In Exercise 4.1 the river level falls to river bed level. Determine the stresses at 5 m below river bed level.

4.3 A layer of clay, 5 m thick, overlies a deposit of sand, 5 m thick, which is underlaid by rock. The water table in the sand is sub-artesian with a level at 2 m below ground level. The saturated unit weights of the clay and sand are 21 and 18 kN/m³, respectively. Determine the total stress, pore water pressure and effective stress at the top and bottom of the sand.

4.4 In Exercise 4.1 the river dries up and the water table lies at 3 m below ground level. Capillary attraction maintains the soil 1.0 m above the water table in a saturated state. The saturated unit weight of the sand is 18 kN/m³ and the bulk unit weight (above the saturated zone) is 16.5 kN/m³. Determine the effective stress at 2 m and 5 m below ground level.

4.5 A layer of sand, 4 m thick, overlies a layer of low permeability clay, 5 m thick. The bulk unit weight of the sand is 16.5 kN/m³ and its saturated unit weight is 18 kN/m³. The saturated unit weight of the clay is 21 kN/m³. The water table exists initially at 1.0 m below ground level but pumping will permanently and rapidly lower the water table to 3.0 m below ground level. Determine the effective stress at the top, middle and bottom of the clay:

1. before pumping
2. immediately after lowering the water table
3. in the long-term.

Determine the change in effective stress in the clay caused by the pumping.

4.6 In Exercise 4.5, if the unit weight of the sand above and below the water table is assumed to be the same, what is the overall change in effective stress in the clay caused by pumping?

4.7 A clay soil deposited in an estuary and originally normally consolidated has been subjected to water table fluctuations. The lowest water table level was 6 m below ground level and the present water table is at 2 m below ground level. Assume the clay to be fully saturated with a unit weight of 19.5 kN/m³. Determine the overconsolidation ratio at 2 m, 6 m and 20 m below ground level.

4.8 A pressuremeter test carried out at a depth of 6.0 m below ground level in clay soil has determined the horizontal total stress to be 120 kN/m². The water table lies at 1.5 m below ground level. Assume the clay to be fully saturated with a unit weight of 20.5 kN/m³. Determine the coefficient of earth pressure at rest, K_0 at the depth of the test. The angle of shearing resistance of the clay is 25°. Is the clay normally consolidated or overconsolidated? The pressuremeter test is described in Chapter 14.

4.9 The results of the saturation stage of a triaxial test are given below. Determine the pore pressure parameter B at stages a) to f). The triaxial test is described in Chapter 7.
Hint: tabulate the data for stages a) to f) as Δu $\Delta \sigma_3$ B.
All pressures are in kN/m^2

	Cell pressure	Back pressure valve	Back pressure	Pore pressure
	0	closed	0	–5
a)	50	closed	–	12
	50	open	40	38
b)	100	closed	–	67
	100	open	90	89
c)	150	closed	–	126
	150	open	140	140
d)	200	closed	–	184
	200	open	190	190
e)	300	closed	–	285
	300	open	290	290
f)	400	closed	–	388

4.10 The results of the shearing stage of a consolidated undrained triaxial test are given below. Determine the pore pressure parameter A for each value. This test is described in Chapter 7.
Cell pressure = 450 kN/m^2 Back pressure = 300 kN/m^2

Deviator stress (kN/m^2)	Pore pressure (kN/m^2)
0	300
68	315
117	319
146	312
171	301
190	287
198	275

CHAPTER 5

Contact pressure and stress distribution

OBJECTIVES

- To understand the effects of stiffness of the soil and stiffness of the foundation on the variation of contact pressures beneath a uniformly loaded area and an area loaded by a point load
- To appreciate the assumptions made when using stress distribution methods
- To be able to determine the stresses beneath uniformly loaded areas of flexible and rigid characteristics and of a variety of shapes

Contact pressure

A foundation is the interface between a structural load and the ground. The stress q applied by a structure to a foundation is often assumed to be uniform. The actual pressure then applied by the foundation to the soil is a reaction, called the contact pressure p and its distribution beneath the foundation may be far from uniform.

This distribution depends mainly on:

- stiffness of the foundation, i.e. flexible → stiff → rigid
- compressibility or stiffness of the soil
- loading conditions – uniform or point loading.

Contact pressure – uniform loading
(Figure 5.1)
The effects of the stiffness of the foundation (flexible or rigid) and the compressibility of the soil (clay or sand) are illustrated in Figure 5.1.

Stiffness of foundation
A flexible foundation has no resistance to deflection and will deform or bend into a dish-shaped profile when stresses are applied. An earth embankment would comprise a flexible structure and foundation.

A stiff foundation provides some resistance to bending and will deform into a flatter dish-shape so that differential settlements are smaller. This forms the basis of design for a raft foundation placed beneath the whole of a structure.

A rigid foundation has infinite stiffness and will not deform or bend, so it moves downwards uniformly. This would apply to a thick, relatively small reinforced concrete pad foundation.

Stiffness of soil
The stiffness of a clay will be the same under all parts of the foundation so for a flexible foundation a fairly uniform contact pressure distribution is obtained with a dish-shaped (sagging) settlement profile.

For a rigid foundation the dish-shaped settlement profile must be flattened out so the contact pressure beneath the centre of the foundation is reduced and beneath the edges of the foundation it is increased. Theoretically, the contact pressure increases to a very high value at the edges although yielding of the soil would occur in practice, leading to some redistribution of stress.

The stiffness of a sand increases as the confining pressures around it increase so beneath the centre of the foundation the stiffness will be at its greatest whereas near the edge of the foundation the stiffness of the sand will be smaller. A flexible foundation will, therefore, produce greater strains at the edges than in the centre so the settlement profile will be

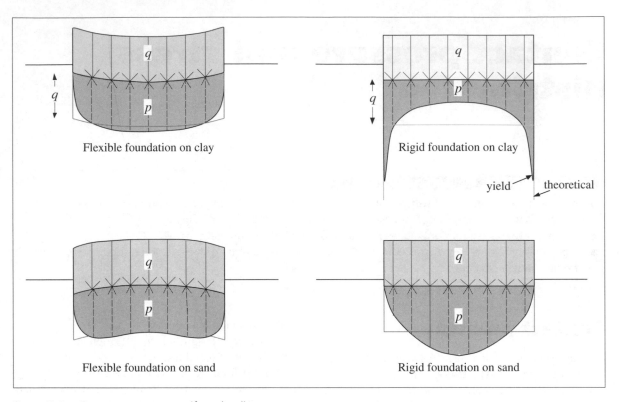

FIGURE 5.1   *Contact pressure – uniform loading*

dish-shaped but upside-down (hogging) with a fairly uniform contact pressure.

For a rigid foundation this settlement profile must be flattened out so the contact pressure beneath the centre would be increased and beneath the edges it would be decreased.

Contact pressure – point loading
(Figure 5.2)

An analysis for contact pressure beneath a circular raft with a point load W at its centre resting on the surface of an incompressible soil (such as clay) has been provided by Borowicka, 1939 (in Poulos and Davis, 1974).

This shows that the contact pressure distribution is non-uniform irrespective of the stiffness of the raft or foundation. For a flexible foundation the contact pressure is concentrated beneath the point load which is to be expected and for a stiff foundation it is more uniform.

For a rigid foundation the stresses beneath the

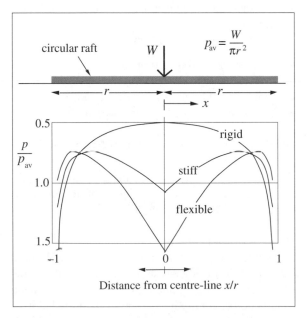

FIGURE 5.2   *Contact pressure – point loading (After Borowicka, 1939, in Poulos and Davis, 1974)*

edges are very considerably increased and a pressure distribution similar to the distribution produced by a uniform pressure on a clay (cf. Figure 5.1) is obtained. This suggests that a point load at the centre of a rigid foundation is comparable to a uniform pressure.

Stress distribution

The stresses that already exist within the ground due to self-weight of the soil are discussed in Chapter 4. Any element of soil in the ground will be at equilibrium under three normal stresses σ_x, σ_y and σ_z (or σ_1, σ_2 and σ_3) acting on three orthogonal axes x, y and z. This element will also be subjected to a system of shear stresses acting on the surfaces of the element.

When a load or pressure from a foundation or structure is applied at the surface of the soil this pressure is distributed throughout the soil and the original normal stresses and shear stresses are altered. For most civil engineering applications the changes in vertical stress are required so the methods given below are for increases in vertical stress only. A comprehensive review of solutions for stress distribution is given by Poulos and Davis (1974).

Stresses beneath point load and line load (Figure 5.3)

Boussinesq published in 1885 a solution for the stresses beneath a point load on the surface of a material which had the following properties:

- *semi-infinite* – this means infinite below the surface therefore providing no boundaries to the material apart from the surface
- *homogeneous* – the same properties at all locations
- *isotropic* – the same properties in all directions
- *elastic* – a linear stress–strain relationship.

Expressions for the stresses beneath a point load and line load are given in Figure 5.3.

SEE WORKED EXAMPLE 5.1

Assumptions

It must be pointed out that the stresses obtained by the methods given below may differ from the stresses obtained in real soils by a significant amount due to the various assumptions made:

- *Infinite layer thickness*
  Soil deposits should not be considered as infinitely thick. With a rigid stratum beneath such as a bedrock it is found that higher stresses are obtained, as shown in Figure 5.3. A rigid stratum is one which does not strain so it does not contribute to settlements or distortions.

  Where several different layers exist on a site stresses are often determined for each layer assuming infinite thickness and similar stiffnesses. This is erroneous. At least using methods which adopt a finite thickness will provide for more conservative (larger) stress estimates.

- *Homogeneous*
  For most soils the modulus or stiffness is not constant or homogeneous. It usually increases with depth and is then described as heterogeneous. This has been found to concentrate settlements and stresses beneath a loaded area with minimal stress dispersed beyond the loaded area. On most sites there are several soil layers with different stiffnesses and these will influence the distribution of stresses. The presence of a stiff upper layer has a particularly marked effect. For a soil whose stiffness increases uniformly with depth the vertical stresses tend to be greater than for the Boussinesq case.

- *Isotropic*
  Overconsolidated clays and rocks can be much stiffer in the horizontal direction than in the vertical direction, i.e. they are anisotropic with $E_H > E_v$.

- *Elastic* (Figure 5.4)
  The linear elastic assumption allows stresses to be calculated which far exceed the yield stress of the soil when redistribution of stresses occurs. For overconsolidated clays when a reasonable factor of safety is applied, say 3, the linear elastic assumption is acceptable but for soft normally consolidated clays and loose sands it may be considerably in error.

- *Foundation depth*
  For foundations at shallow depths assuming the load is applied at the surface of the soil provides stresses which are on the conservative side.

Descriptive figure	Expression

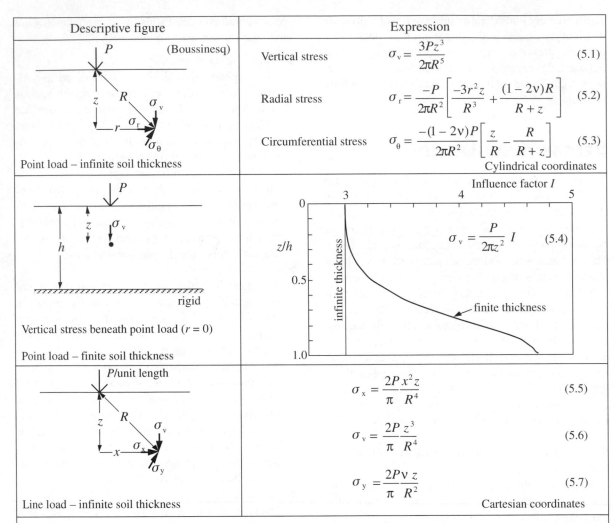

(Boussinesq) Vertical stress	$\sigma_v = \dfrac{3Pz^3}{2\pi R^5}$	(5.1)
Radial stress	$\sigma_r = \dfrac{-P}{2\pi R^2}\left[\dfrac{-3r^2 z}{R^3} + \dfrac{(1-2v)R}{R+z}\right]$	(5.2)
Circumferential stress	$\sigma_\theta = \dfrac{-(1-2v)P}{2\pi R^2}\left[\dfrac{z}{R} - \dfrac{R}{R+z}\right]$	(5.3)

Point load – infinite soil thickness

Cylindrical coordinates

Vertical stress beneath point load ($r = 0$)

Point load – finite soil thickness

$$\sigma_v = \frac{P}{2\pi z^2} I \qquad (5.4)$$

Line load – infinite soil thickness

$$\sigma_x = \frac{2P}{\pi}\frac{x^2 z}{R^4} \qquad (5.5)$$

$$\sigma_v = \frac{2P}{\pi}\frac{z^3}{R^4} \qquad (5.6)$$

$$\sigma_y = \frac{2Pv}{\pi}\frac{z}{R^2} \qquad (5.7)$$

Cartesian coordinates

Assumptions
- The soil is homogeneous, isotropic with linear stress-strain (elastic) properties
- The line load is flexible and infinitely long. It is the integration of the point load case.

Notes
- For the finite layer thickness with a rigid stratum beneath the stresses are larger than the infinite thickness case.
- Westergaard material – this is the same as the Boussinesq case except that lateral strain is prevented. This is depicted as a homogeneous mass reinforced horizontally with thin but rigid reinforcement and is an extreme case of anisotropy producing stresses which are smaller than for the Boussinesq case. For real soils stresses probably lie between these two cases but Boussinesq produces more conservative values.

Figure 5.3 *Stresses beneath a point load and line load*

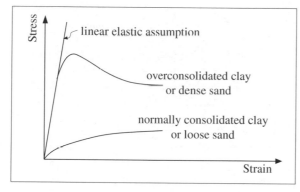

FIGURE 5.4 *Linear elastic assumption*

results for vertical stresses but horizontal stresses can be significantly in error as a result of the simplifying assumptions.

Stresses beneath uniformly loaded areas (Figure 5.5)

A uniform applied pressure can be represented as a large number of point loads. Each of these loads will produce stresses at a point within the soil mass so integration of the Boussinesq equations will give the stress under a uniform pressure.

Figure 5.5 gives expressions for the stresses at points beneath a flexible strip and a circular loaded area.

If the uniform pressure is applied at some depth below the soil surface then the methods described for a surface foundation can be used assuming the

Burland *et al.* (1978) have stated that for most ground conditions stress distribution methods based on the Boussinesq analysis give reasonably accurate

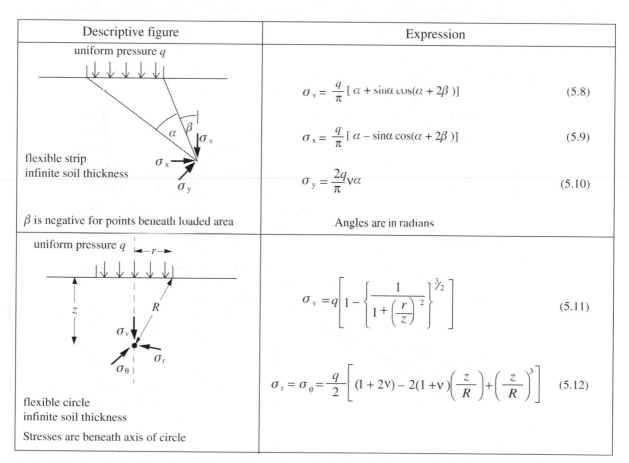

Descriptive figure	Expression
uniform pressure *q* flexible strip infinite soil thickness *β* is negative for points beneath loaded area	$\sigma_v = \dfrac{q}{\pi}[\,\alpha + \sin\alpha\cos(\alpha + 2\beta)\,]$ (5.8) $\sigma_x = \dfrac{q}{\pi}[\,\alpha - \sin\alpha\cos(\alpha + 2\beta)\,]$ (5.9) $\sigma_y = \dfrac{2q}{\pi}\nu\alpha$ (5.10) Angles are in radians
uniform pressure *q* flexible circle infinite soil thickness Stresses are beneath axis of circle	$\sigma_v = q\left[1 - \left\{\dfrac{1}{1 + \left(\dfrac{r}{z}\right)^2}\right\}^{3/2}\right]$ (5.11) $\sigma_r = \sigma_\theta = \dfrac{q}{2}\left[(1 + 2\nu) - 2(1 + \nu)\left(\dfrac{z}{R}\right) + \left(\dfrac{z}{R}\right)^3\right]$ (5.12)

FIGURE 5.5 *Stresses beneath uniformly loaded areas*

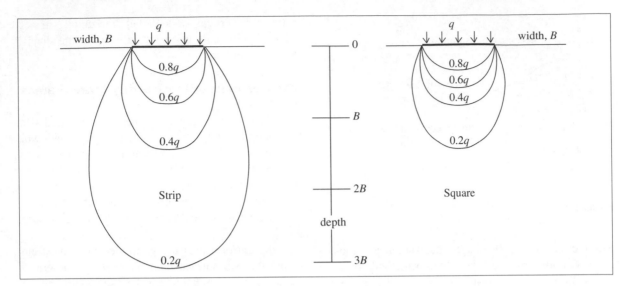

FIGURE 5.6 *Bulbs of pressure*

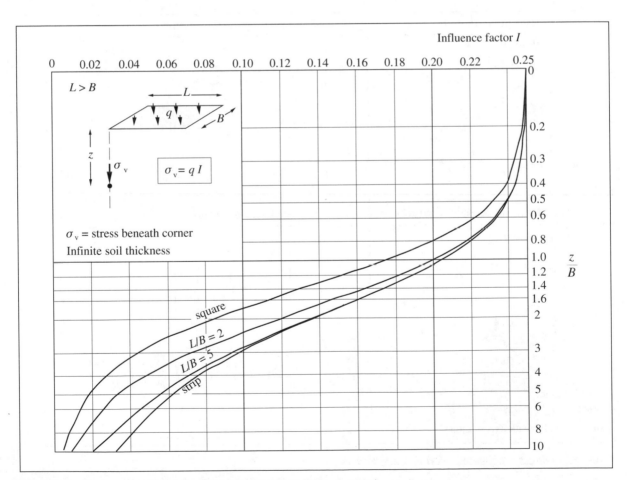

FIGURE 5.7 *Stresses beneath a flexible rectangle (From Giroud, 1970)*

The pressure applied is uniform but the stress distributed in the ground varies beneath the loaded area. The stress distribution methods give the stress at the *corner* of a loaded area so for points other than the corner the principle of superposition should be used. For the stress at the point × split the area into rectangles with their corners at the point ×.

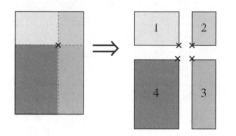

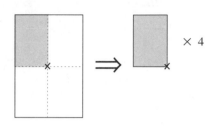

Stress at × = Sum of stresses at the corners of rectangles 1 + 2 + 3 + 4

Maximum stress (at centre) = stress at the corner of a quarter foundation, multiplied by 4

FIGURE 5.8 *Principle of superposition*

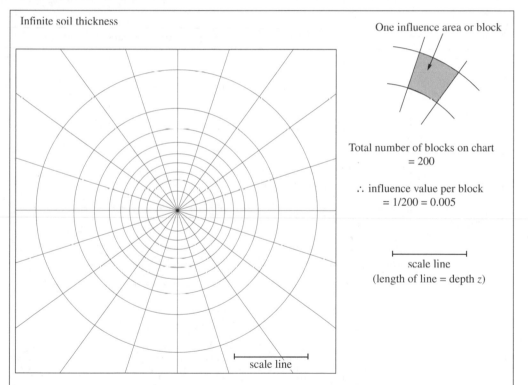

For the vertical stress σ_v at a depth z beneath any point × on or outside a loaded area:

1 Draw a plan sketch of the building outline on tracing paper such that the length of the scale line equals the depth z where the stress is required.

2 Place the scale drawing on the chart with the point × at the centre of the chart.

3 Count the number of blocks N covered by the scale drawing. Group together part blocks.

4 The vertical stress at the depth z and beneath the point × is given by $\sigma_v = 0.005\,N\,q$

5 The tracing can then be moved to other locations to obtain the stress beneath other points.

FIGURE 5.9 *Stress beneath flexible area of any shape (Newmark's chart)*

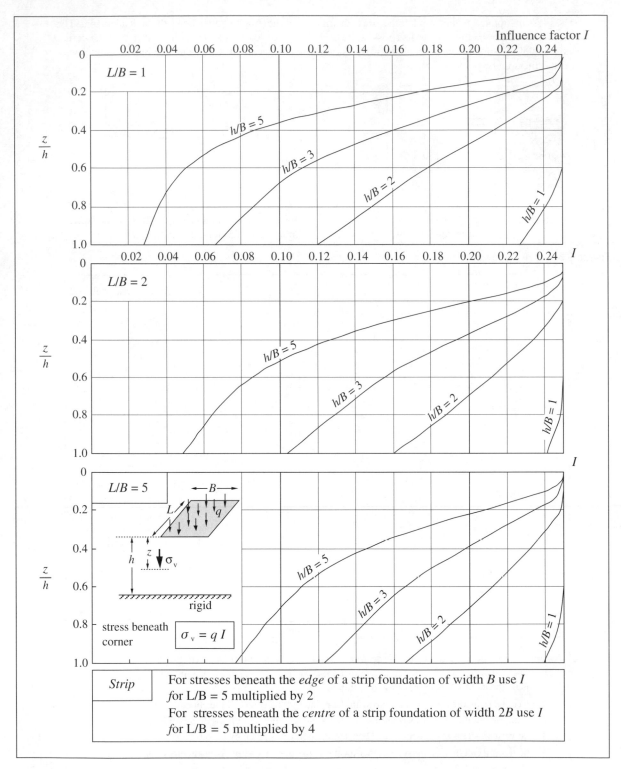

Figure 5.10 *Stress beneath flexible rectangle – finite soil thickness (From Milovic et al. 1971)*

soil surface to be at foundation level. This may give higher stresses than the theoretical values but considering the assumptions made it is prudent to adopt a conservative approach. The applied pressure q should, however, be the net applied pressure q_{net} obtained from:

q_{net} = gross applied pressure – pressure of soil removed

SEE WORKED EXAMPLE 5.2

Bulbs of pressure (Figure 5.6)

Lines or contours of equal stress increase can be plotted from the equations available, given in Figure 5.5. Because of their shape they are referred to as bulbs of pressure.

They form the basis of the rule of thumb for depth of site investigation since if a borehole is sunk to within the $0.2q$ contour, say, stresses can be expected to be small below this depth.

Stresses beneath a flexible rectangle
(Figure 5.7)

The vertical stress σ_v at a depth z beneath the corner of a flexible rectangle supporting a uniform pressure q has been determined using:

$$\sigma_v = q\,I \qquad (5.13)$$

and influence factors I, given by Giroud (1970) are presented in Figure 5.7. They are for an infinite soil thickness.

These curves are equivalent to the commonly used charts of Fadum (1948) but are easier to use.

SEE WORKED EXAMPLE 5.3

Principle of superposition (Figure 5.8)

For stresses beneath points other than the corner of the loaded area the principle of superposition should be used, as described in Figure 5.8.

SEE WORKED EXAMPLE 5.3

Stresses beneath flexible area of any shape (Figure 5.9)

Newmark (1942) devised charts to obtain the vertical stress at any depth, beneath any point (inside or outside) of an irregular shape. Use of the charts is explained in Figure 5.9.

SEE WORKED EXAMPLE 5.4

Stresses beneath a flexible rectangle – finite soil thickness (Figure 5.10)

Figure 5.7 gives the stresses beneath a flexible rectangle for a deep soil layer (of infinite thickness). Where a rigid underlying deposit exists providing a finite soil thickness the stresses given by Milovic *et al.* (1971) can be used, see Figure 5.10. These give higher stresses than in Figure 5.7, particularly in the lower levels of the compressible stratum.

The solutions of Fox (1948a) for a two-layer problem suggest that this 'rigid' stratum need only be about 10 times stiffer than the compressible stratum for the finite soil layer case to apply.

SEE WORKED EXAMPLE 5.5

Stresses beneath a rigid rectangle
(Figures 5.11 and 5.12)

Although the stresses within the soil immediately beneath a rigid foundation (contact pressures) vary considerably, see Figure 5.1, the stresses at depth become more uniform as shown in Figure 5.11. Vertical stresses beneath the centre of a rigid rectangle given by Butterfield and Banerjee (1971) are presented in Figure 5.12 with a modification applied for the stresses at shallow depths as suggested in Figure 5.11. If these stresses are taken as 'mean' stresses

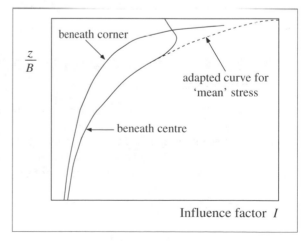

FIGURE 5.11 *Stresses beneath a rigid rectangle*

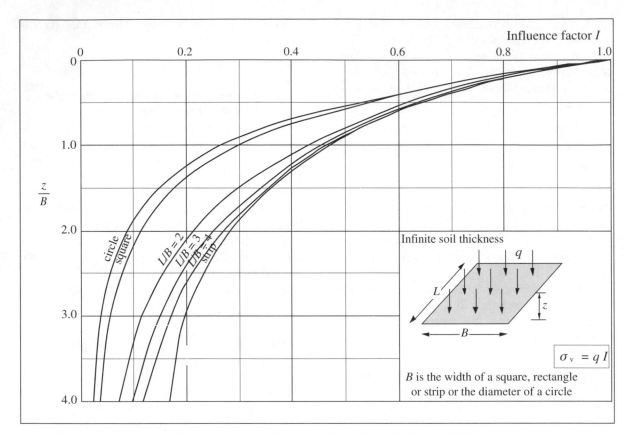

FIGURE 5.12 *Stress beneath the centre of a rigid rectangle (From Butterfield et al. 1971)*

beneath a rigid foundation they will probably be on the safe side.

SEE WORKED EXAMPLE 5.6

SUMMARY

In this chapter it has been shown that:

- the contact pressure beneath a loaded foundation area depends on the compressibility of the soil and the stiffness of the foundation
- the stresses obtained by the stress distribution methods may vary considerably from those experienced in situ due to the simplifying assumptions made
- most methods give the stress at the corner of the loaded area so the principle of superposition must be used to determine stresses at other locations
- the stresses within a stratum of finite thickness are greater than those within an infinitely thick deposit

Worked Example 5.1 Stress beneath a point load

Determine the change in vertical stress at 2.5 m below ground level directly beneath the 870 kN point load, as shown on Figure 5.13, using the Boussinesq analysis.

Assuming the soil layer to be of infinite thickness the vertical stress is given by:

$$\sigma_v = \frac{3Pz^3}{2\pi R^5} \quad \text{(see Figure 5.3)}$$

$$\sigma_v \text{ from 870 kN load} = \frac{3 \times 870 \times 2.5^3}{2 \times \pi \times 2.5^5} = 66.5 \text{ kN/m}^2$$

$$\sigma_v \text{ from 640 kN load} = \frac{3 \times 640 \times 2.5^3}{2 \times \pi \times 4.30^5} = 3.2 \text{ kN/m}^2$$

$$\sigma_v \text{ from 560 kN load} = \frac{3 \times 560 \times 2.5^3}{2 \times \pi \times 5.15^5} = 1.2 \text{ kN/m}^2$$

$$\text{Total} = 70.9 \text{ kN/m}^2$$

Note the influence from each load.

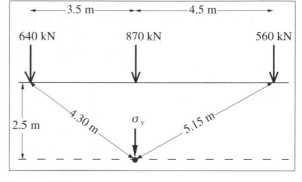

FIGURE 5.13

Worked Example 5.2 Stresses beneath a strip load

A strip foundation, 6 m wide, is uniformly loaded with a pressure of 100 kN/m². Determine the vertical stress distribution beneath and outside the strip at a depth of 3 m below ground level.

Values of α and β have been obtained by first determining ω, see Figure 5.14, for particular values of x, the distance from the centre-line.

x	$\omega°$	$\beta°$	$\alpha°$	$\alpha°$ radian	σ_v kN/m²
0	45.0	−45.0	90.0	1.571	81.8
1.0	53.1	−33.7	86.8	1.515	78.2
2.0	59.0	−18.4	77.4	1.351	66.6
3.0	63.4	0	63.4	1.107	48.0
4.0	66.8	18.4	48.4	0.845	28.9
5.0	69.4	33.7	35.7	0.623	15.6
6.0	71.6	45.0	26.6	0.464	8.4
7.0	73.3	53.1	20.2	0.353	4.7
8.0	74.7	59.0	15.7	0.274	2.8

The variation of stress is plotted on Figure 5.14.

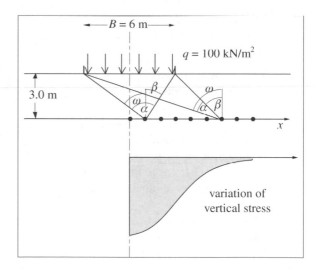

FIGURE 5.14

Worked Example 5.3   Stresses beneath a flexible rectangle – infinite thickness

Determine the vertical stress beneath the point O on a rectangular area, 6 m × 3 m, see Figure 5.15, at a depth of 2 m below ground level when uniformly loaded with a pressure of 50 kN/m².

Using the principle of superposition (Figure 5.8) the stress is required beneath the corner of four smaller areas 1 to 4. The stress values are determined from Figure 5.7.

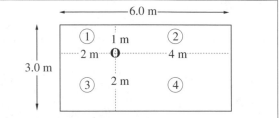

Area	L m	B m	L/B	z/B	I	σ_v kN/m²
1	2	1	2	2	0.120	6.0
2	4	1	4	2	0.134	6.7
3	4	2	2	1	0.200	10.0
4	2	2	1	1	0.175	8.8

Total 31.5 kN/m²   **Figure 5.15**

Worked Example 5.4   Stresses beneath a flexible area of any shape

The Charity Hospital, New Orleans is described as a case study in Chapter 10, Pile Foundations. Assuming it imposes a uniform pressure of 250kN/m² over its full area determine the applied stress at the middle of the soft compressible clay, beneath the point A.

The depth from the underside of the foundation to the middle of the clay is 35 m. A plan sketch of the building is drawn with the scale line on Newmark's chart, see Figure 5.9, equal to 35 m. Place the point A on the plan sketch at the centre of the chart, as in Figure 5.16 and count the number of blocks on the chart, combining part blocks. Note that the building is symmetrical so it is only necessary to count one side. The total number of blocks covered by the building is 39.

The vertical stress at the middle of the clay layer is:

$\sigma_v = 0.005 \times 39 \times 250 = 48.8$ kN/m².

Newmark's method assumes that the area loaded is flexible which may not seem appropriate for a multi-storey building.

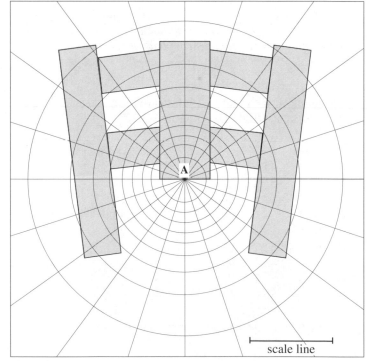

Figure 5.16

However, settlement measurements of the building showed that it had settled into a classic dish-shaped profile with a differential settlement between the centre and edges of the building of over 60 mm at the end of construction.

Worked Example 5.5   Stresses beneath a flexible rectangle – finite thickness

For the same foundation used in Example 5.3, use Figure 5.10 to determine the stress at the depth of 2 m below ground level assuming the layer thickness to be 4 m.

$z/h = 2/4 = 0.5$

Area	L m	B m	L/B	h/B	I	σ_v kN/m^2
1	2	1	2	4	0.138	6.9
2	4	1	4	4	0.148	7.4
3	4	2	2	2	0.224	11.2
4	2	2	1	2	0.196	9.8

Total = 35.3 kN/m^2

Linear interpolation between the curves for h/B and the charts of L/B is acceptable. This has been adopted for areas 1 and 2.

Worked Example 5.6   Stresses beneath a rigid rectangle

Determine the vertical stress beneath a rigid rectangular foundation, 6 m × 3 m, at a depth of 2 m below ground level when uniformly loaded with a pressure of 50 kN/m^2.

$L/B = 6/3 = 2$
$z/B = 2/3 = 0.67$
From Figure 5.12, $I = 0.55$
vertical stress = $0.55 \times 50 = 27.5$ kN/m^2
(beneath centre)

EXERCISES

5.1 Three point loads each of 500 kN are applied at ground level in a straight line 4.0 m apart. Determine the increase in vertical stress at 2.0 m below ground level beneath the middle load.

5.2 Three parallel line loads each of 500 kN/m are applied at ground level 4.0 m apart. Determine the increase in vertical stress at 2.0 m below ground level beneath the middle load.

5.3 A flexible strip foundation 8.0 m wide lies at ground level and applies a uniform pressure of 80 kN/m². Determine the increase in vertical stress at 5.0 m below ground level:

a) beneath the centre of the foundation b) beneath its edge and c) 4.0 m away from the edge.

5.4 A flexible rectangle is 12 m long and 6 m wide lies at ground level and applies a uniform pressure of 105 kN/m². Determine the increase in vertical stress at 6.0 m below ground level:

a) beneath the centre of the foundation and b) beneath its corner.

Assume the soil to be infinitely thick.

5.5 An L-shaped raft as shown in Figure 5.16 lies at ground level and applies a uniform pressure of 80 kN/m². Determine the increase in vertical stress at 8.0 m below ground level:

a) beneath the point A and b) beneath the point B.

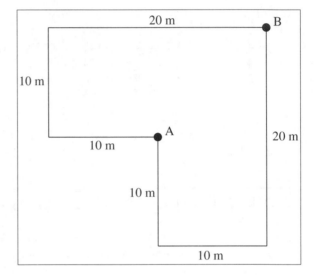

Figure 5.17

5.6 Repeat Exercise 5.4, assuming the soil deposit to be 12 m thick.

5.7 In Exercise 5.4, assume the foundation to be rigid and determine the increase in vertical stress beneath the centre of the foundation.

CHAPTER 6

Compressibility and consolidation

Introduction

Compressibility and consolidation can be distinguished as:

- *compressibility* – volume changes in a soil when subjected to pressure – giving *AMOUNTS* of settlement
- *consolidation* – rate of volume change with time – giving *TIME* to produce an amount of settlement required

These are distinct from:

1. compaction which is the expulsion of air from a soil by applying compaction energy and

2. immediate or undrained settlement which is the resultant deformation of a soil under applied stresses without any volume change taking place.

Due to the insurmountable problems of obtaining good quality samples of sands from the ground this chapter concentrates on the behaviour of clays. Clay soils usually produce large amounts of settlement over a long period of time after the end of construction so their effects on a structure are more significant. Methods for the determination of the amount of settlement from this volume change process are given in Chapter 9.

Sands generally produce smaller amounts of settlement in a much quicker time, often during the construction period, so they may be of less concern. Nevertheless some empirical methods of estimating settlements of structures on sands are given in Chapter 9.

Compressibility

Void ratio/effective stress plot
(Figure 6.1)

Volume changes in a soil occur because the volume of voids changes. They are defined by the void ratio:

$$\text{void ratio} = \frac{\text{volume of voids}}{\text{volume of solids}} \qquad (6.1)$$

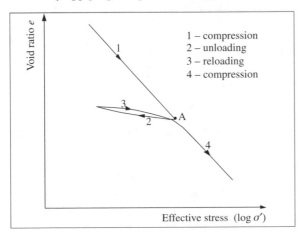

1 – compression
2 – unloading
3 – reloading
4 – compression

FIGURE 6.1 *Void ratio – effective stress*

and it is assumed that the volume of solids does not alter. The consolidation analogy in Chapter 4 describes the changes that occur in an element of soil when it is subjected to a change in effective stress.

Effective stress is the stress seated in the mineral grain structure or soil skeleton so if it changes the soil skeleton will respond by decreasing or increasing in volume as water is squeezed out of or drawn into the void spaces.

There are recoverable and irrecoverable components to the volume changes of the soil skeleton and it is postulated that these are due to:

- rearrangement of the soil particles (*a*) – this is permanent or irrecoverable
- elastic strains in the particles (*b*) – these are recoverable
- compression of bound water layers (*c*) – this is recoverable

All three components will produce volume decrease but only *b* and *c* will allow volume increase.

A typical plot of void ratio versus effective stress (plotted as log σ') for a clay soil is shown on Figure 6.1. As the effective stress increases during compression components *a*, *b* and *c* are occurring and these produce volume decreases but on unloading only *b* and *c* are recovered.

When the soil is reloaded only components *b* and *c* take place until the previous maximum pressure at A is reached. Further loading causes particle rearrangement, component *a*. Therefore, particle rearrangement only occurs when the soil lies on the steeper portions and this can be seen as a yielding (and hardening) process so that the soil becomes a little stronger and stiffer with each increment of effective stress. Elastic and plastic strains occur when on the steeper portion but only elastic recoverable strains occur when on the flatter portion.

The soil is described as normally consolidated when its state exists on the steeper line (1 and 4) and the effective stress can only be increased with subsequent reduction in volume because decreasing the stress takes the soil state away from the normally consolidated line. This line is often referred to as the virgin compression curve or line as any change of effective stress along it will be for the first and only time whereas any number of unload/reload paths

could be followed. The soil is described as overconsolidated when it occurs on the flatter portions (2 and 3) and volume changes can increase or decrease with changes in effective stress.

These consolidation processes occur during the formation of a soil and the current state of the soil is related to its past deposition and erosion history, as described in Chapter 4. Overconsolidated clays will generally have lower moisture contents and higher shear strengths. Settlements of structures placed on or in the soil are related to the volume changes and these will be much smaller for an overconsolidated clay when stresses lie on the flatter part of the plot, therefore it is important to determine the present state of overconsolidation. Structures placed on normally consolidated clays will undergo much larger settlements.

SEE WORKED EXAMPLE 6.1

Reloading curves (Figure 6.2)
The past geological and stress history has brought the soil to its present condition. If a sample is taken from the ground with no moisture content change and loaded in a laboratory consolidation apparatus a reloading curve can be obtained by plotting the void ratio produced after the soil has consolidated to a new equilibrium for each change in effective stress.

The shape of the reloading curve (as well as geological information about the soil at the site) will

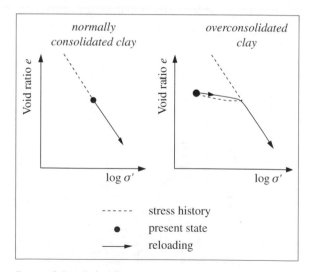

FIGURE 6.2   *Reloading curves*

help to determine whether the soil is normally consolidated or overconsolidated. If the soil is normally consolidated the reloading curve will continue on the virgin compression line from its present condition and will follow the straight line on a log σ' plot.

The reloading curve for an overconsolidated clay will have two portions, one commencing from its present condition and following a flatter path until it reaches the virgin compression line at the preconsolidation pressure, followed by a steeper line corresponding to the virgin compression curve.

Preconsolidation pressure p_c' and overconsolidation ratio *OCR*

The preconsolidation pressure is the previous maximum effective stress to which the soil has been subjected in the past and is described in Chapter 4. It enables an assessment of the degree of overconsolidation using the overconsolidation ratio (*OCR*) defined as:

$$OCR = \frac{p_c'}{p_0'} \qquad (6.2)$$

$$= \frac{\text{previous maximum effective stress}}{\text{present effective stress}}$$

For a normally consolidated clay the present effective stress is also the previous maximum so the *OCR* = 1. For a heavily overconsolidated clay the *OCR* may be 4 or more therefore this type of soil has been subjected to a much greater stress in the past compared to its present condition.

The significance of p_c' for an overconsolidated clay is that if stresses are kept below this value then settlements can be expected to be small but if the applied stresses due to loading exceed this value then large settlements will occur as consolidation will take place along the virgin compression line.

SEE WORKED EXAMPLE 6.2

Casagrande method for p_c' (Figure 6.3)

From Figures 6.1 and 6.2 it is clear that p_c' for an overconsolidated clay occurs just to the right of the change in slope of the reloading curve of void ratio versus effective stress (plotted as log σ').

Casagrande (1936) proposed a simple graphical method to determine p_c' from the laboratory reloading curve as follows:

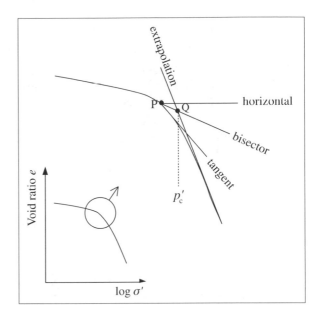

FIGURE 6.3   *Casagrande construction for p_c'*

1. locate the point of maximum curvature P
2. from point P draw a horizontal line and a tangent line and draw a bisector line between them
3. extrapolate upwards from the lower straight line to cut the bisector
4. the point of intersection Q of the extrapolated line and the bisector gives the value of p_c'

SEE WORKED EXAMPLE 6.2

Effect of sampling disturbance
(Figure 6.4)

Soil disturbance during sampling from the ground will alter and partially destroy the stable arrangement of soil particles and the forces or bonds acting between them. If the soil skeleton is disturbed it is found that a flatter void ratio–effective stress plot is obtained and the more disturbed the soil the flatter the curve. Since disturbance during sampling from the ground is inevitable, especially for normally consolidated soils, no laboratory test will truly reflect the *in situ* condition.

In situ curve for normally consolidated clay (Figure 6.5)

From Figure 6.4 it can be inferred that the *e* versus log σ' relationship for the *in situ* condition will lie to

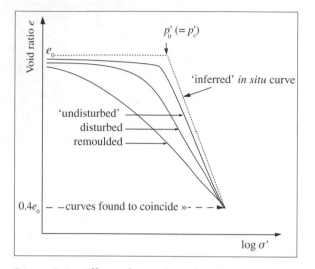

Figure 6.4 *Effect of sampling disturbance on normally consolidated clay*

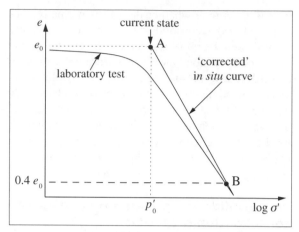

Figure 6.5 *In situ curve for normally consolidated clay*

the right of the curve for the least disturbed soil. Schmertmann (1955) found that irrespective of the degree of disturbance all curves coincided at the value of $0.4e_o$ so it is reasonable to assume that the *in situ* curve also passes through this point.

To plot the assumed '*in situ*' *e* versus log σ' curve:

1. Carry out a consolidation test to produce a reloading curve with pressures applied which reduce the void ratio to $0.4e_o$ or at least provide sufficient values to enable extrapolation to $0.4e_o$.
2. Determine the present void ratio e_o from the moisture content and particle density values and the effective stress p_o' at the depth of the sample and plot this as point A on the graph.
3. Plot the point B at $0.4e_o$ on the reloading curve.
4. The '*in situ*' curve is then obtained as the line A–B.

In situ curve for overconsolidated clay
(Figure 6.6)

The *in situ* reloading curve for an overconsolidated clay will commence at the point e_o, p_o' and will have a flatter portion up to p_c' followed by a steeper portion when the soil returns to being normally consolidated. The construction of the *in situ* curve is based on the observations that:

1. All curves irrespective of degree of disturbance

pass through the point $0.4e_o$ (Schmertmann, 1955), see Figure 6.4.
2. All unload/reload loops are parallel, irrespective of where they occur.

To plot the *in situ* void ratio versus log σ' curve it

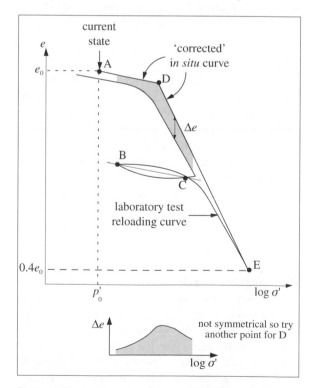

Figure 6.6 *In situ curve for overconsolidated clay*

is necessary to carry out a consolidation test to produce a reloading curve with pressures applied beyond the preconsolidation pressure p_c', followed by an unload/reload loop (BC in Figure 6.6) and then applying pressures which reduce the void ratio to $0.4e_0$ or at least sufficient values to enable extrapolation to $0.4e_0$. The *in situ* curve is then constructed from this reloading curve as follows:

1. Plot the present state of the soil at point A at e_0, p_0' and from A draw a line parallel to the average unload/reload loop BC up to the p_c' value at D. p_c' may be obtained from the Casagrande construction, illustrated in Figure 6.3.
2. Plot the point E at $0.4e_0$ on the reloading curve and draw the line DE to complete the two-limb *in situ* curve.

Alternatively, if p_c' cannot be determined very accurately choose a reasonable point for D and draw the trial lines AD and DE. Δe is the vertical difference between the trial lines and the laboratory test result and is plotted beneath versus $\log \sigma'$. When the Δe–$\log \sigma'$ plot is symmetrical then it is considered that the point D represents the most likely value of p_c'.

Effect of load increments (Figure 6.7)

It has been found that the shape of the void ratio–log σ' plot can be made more distinct by reducing the increments between applied loads or pressures, especially for normally consolidated clays. This is very important when determining the preconsolidation pressure of lightly overconsolidated clays and plotting the *in situ* curve.

The best quality results will be obtained with the least sample disturbance and small load increments. For some schemes it may be justified to carry out constant rate of strain (CRS) or constant rate of loading (CRL) tests although these would require modified versions of the consolidation test, usually using the Rowe hydraulic cell.

Effect of duration of load (Figure 6.8)

The void ratio–log σ' plot represents the equilibrium condition for the soil at the end of the primary consolidation process when the excess pore water pressure set up by the applied total stress has fully dissipated. For soils that do not display secondary compression properties there will be no further volume change and the void ratio will remain at its equilibrium value under the new applied stress.

For soils that display secondary compression, volume changes occur after the end of primary consolidation so that the void ratio will decrease if each load increment is left in place for a longer period. This phenomenon is most common with normally consolidated and lightly overconsolidated soils since heavily overconsolidated soils are less prone to secondary compressions. From Figure 6.8 it can be seen that the shape of the curves becomes less distinct with longer load duration and it is more difficult to determine p_c'.

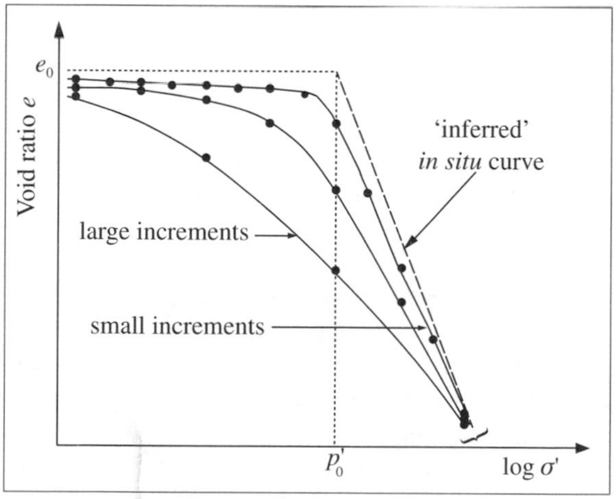

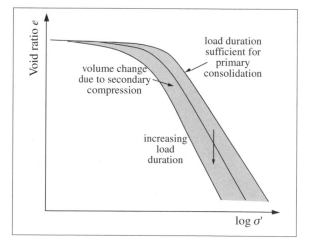

FIGURE 6.7 *Effects of load increments*

FIGURE 6.8 *Effects of load duration*

Isotropic compression (Figure 6.9)

This occurs when the three principal stresses are equal:

$$\sigma_1 = \sigma_2 = \sigma_3 \tag{6.3}$$

and hence

$$\sigma_1' = \sigma_2' = \sigma_3' \tag{6.4}$$

Since there are no shear stresses applied ($q' = \sigma_1' - \sigma_3' = 0$) it is sufficient to display isotropic compression as a plot of volume change versus mean stress. Here, the parameters used in critical state soil mechanics terminology will be used, namely for volume changes:

$$\text{specific volume, } v = 1 + e \tag{6.5}$$

and for mean stress or pressure p':

$$p' = 1/3\ (\sigma_1' + \sigma_2' + \sigma_3') \tag{6.6}$$

These plot for both compression and swelling on the flat $p' - v$ plane in Figure 6.9. This type of compression occurs in a triaxial compression apparatus when the cell pressure alone is applied.

Anisotropic compression (Figure 6.10)

This occurs when the principal stresses are not equal:

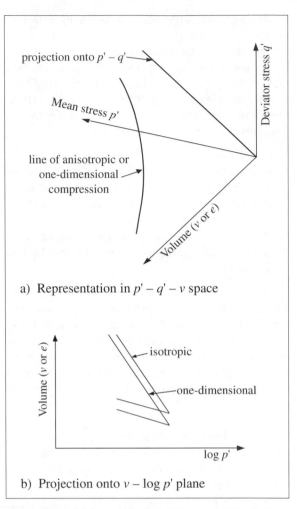

a) Representation in $p' - q' - v$ space

b) Projection onto $v - \log p'$ plane

FIGURE 6.10 *Anisotropic or one-dimensional compression*

$$\sigma_1' \neq \sigma_2' \neq \sigma_3' \tag{6.7}$$

producing shear stresses represented for axisymmetric conditions by the deviator stress q':

$$q' = \sigma_1' - \sigma_3' \tag{6.8}$$

Soil existing *in situ* will have undergone anisotropic or one-dimensional compression during deposition or loading since any element of the soil will not have been able to strain horizontally because of the lateral confinement within the soil deposit. Strains, therefore, only occur in the vertical direction (one-dimensional). The relationship between the vertical and horizontal effective stresses σ_V' and σ_H' is

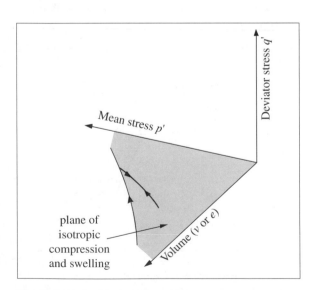

FIGURE 6.9 *Isotropic compression*

given by the coefficient of earth pressure at rest K_o:

$$\sigma_H' = K_o \sigma_V' \tag{6.9}$$

so the ratio of q' and p' is given by:

$$\frac{q'}{p'} = \frac{3(\sigma'_V - \sigma'_H)}{\sigma'_V + 2\sigma'_H} = \frac{3(1 - K_o)}{1 + 2K_o} \tag{6.10}$$

which will be a constant value for a normally consolidated clay, see Figure 4.11, since K_o is a constant for this type of soil. The plot of anisotropic normal compression is then represented by the inclined line in Figure 6.10a. This type of compression occurs in an oedometer or consolidation cell where the confinement from the apparatus prevents horizontal strain so this test is appropriate for modelling one-dimensional *in situ* strain conditions. An explanation of the relationship between the *in situ* stresses σ_H' and σ_V' is given in Chapter 4.

It can be seen that shear stresses ($q' > 0$) and compression stresses p' are both applied to the soil during anisotropic or one-dimensional loading whereas shear stresses are zero for isotropic compression. Under the same mean stress p' a soil which has undergone one-dimensional compression will have a smaller volume, as shown in Figure 6.10b.

Consolidation

Terzaghi theory of one-dimensional consolidation (Figure 6.11)

The process of consolidation comprises the gradual reduction of the volume of a fully saturated soil with time as water is squeezed out of the pore spaces under an induced or excess pore water pressure. This is coupled with the gradual increase with time of the effective stress within the soil skeleton. A simple model illustrating the process is described in Chapter 4. The main distinction from compressibility is that consolidation is concerned with the time aspects and is, therefore, fundamentally related to the permeability of the soil.

The theory considers the rate at which water is squeezed out of an element of soil and can be used to determine the rates of:

1. volume change of the soil with time
2. settlements at the surface of the soil with time

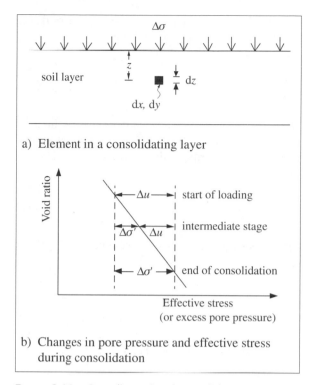

a) Element in a consolidating layer

b) Changes in pore pressure and effective stress during consolidation

FIGURE 6.11 *One-dimensional consolidation*

3. pore pressure dissipation with time.

Several assumptions are made such as the soil is homogeneous and fully saturated and the solid particles and the pore water are incompressible. The tenuous assumptions are that:

1. Compression and flow are one-dimensional, i.e. they are vertical only, whereas significant horizontal flow can occur in layered deposits.
2. Darcy's law is valid at all hydraulic gradients but deviation may occur at low hydraulic gradients.
3. k and m_v remain constant. However, they both usually decrease during consolidation.
4. No secondary compression or creep occurs. If this occurs the void ratio–effective stress relationship is not solely dependent on the consolidation process, see Figure 6.8
5. The load is applied instantaneously and over the whole of the soil layer. However, loads are applied over a construction period and usually do not extend over a wide area in relation to the thickness of the consolidating deposit.

Consider an element of soil in a consolidating layer (Figure 6.11a). The hydraulic gradient across the element is:

$$-\frac{\partial h}{\partial z} = -\frac{1}{\gamma_w} \frac{\partial u_e}{\partial z} \tag{6.11}$$

where u_e is the excess pore water pressure induced by the applied total stresses. The average velocity of water passing through the element, from Darcy's law:

$$v = -\frac{k}{\gamma_w} \frac{\partial u_e}{\partial z} \tag{6.12}$$

The velocity gradient across the element:

$$\frac{\partial v}{\partial z} = -\frac{k}{\gamma_w} \frac{\partial^2 u_e}{\partial z^2} \tag{6.13}$$

From the equation of continuity (equation 3.23) if volume changes in the soil element are occurring the volume change per unit time can be expressed as:

$$\frac{dV}{dt} = \frac{\partial v}{\partial z} \, dx \, dy \, dz \tag{6.14}$$

$$\frac{dV}{dt} = -\frac{k}{\gamma_w} \frac{\partial^2 u_e}{\partial z^2} \, dx \, dy \, dz \tag{6.15}$$

This can now be equated to the volume change of the void space in the element. The total volume of the element $= dx \, dy \, dz$ so the proportion of voids in the element is:

$$\frac{e}{1+e_0} \, dx \, dy \, dz \tag{6.16}$$

The rate of change of void space with respect to time is then:

$$\frac{1}{1+e_0} \frac{\partial e}{\partial t} \, dx \, dy \, dz \tag{6.17}$$

$$= \frac{1}{1+e_0} \frac{\partial e}{\partial \sigma'} \frac{\partial \sigma'}{\partial t} \, dx \, dy \, dz \tag{6.18}$$

$$= -m_v \frac{\partial u_e}{\partial t} \, dx \, dy \, dz \tag{6.19}$$

where

$$m_v = \frac{\partial e}{1+e_0} \frac{1}{\partial \sigma'} \tag{6.20}$$

and

$$\frac{\partial \sigma'}{\partial t} = -\frac{\partial u_e}{\partial t} \tag{6.21}$$

which considers the rate of increase of effective stress being equal to the rate of dissipation of the excess pore pressure u_e, as illustrated on Figure 6.11b.

Equating 6.15 and 6.19, and dividing by $dx \, dy \, dz$ gives:

$$\frac{\partial u_e}{\partial t} = \frac{k}{m_v \gamma_w} \frac{\partial^2 u_e}{\partial z^2} = c_v \frac{\partial^2 u_e}{\partial z^2} \tag{6.22}$$

where:

c_v is the coefficient of consolidation
m_v is the coefficient of compressibility
k is the coefficient of permeability

Solution of the consolidation equation
(Figures 6.12 and 6.13)

The basic differential equation of consolidation (equation 6.22) gives the relationship between three values, the excess pore pressure u_e, the depth below the surface of the soil layer z and the time t that has elapsed since the instantaneous application of the load. This equation can be expressed in dimensionless terms as:

$$\frac{dU_v}{dT_v} = \frac{d^2U_v}{dZ^2} \tag{6.23}$$

Z is a dimensionless depth given by:
$$Z = z/d \tag{6.24}$$

where z defines the depth below the layer surface of the soil element under consideration. d is the length of drainage path and represents the *maximum* distance a molecule of water would have to travel to escape from the soil layer and depends on the permeability at the boundaries. This is illustrated in Figure 6.12 where for a half-closed layer d is the full thickness of the soil but for an open layer d is half of the layer thickness.

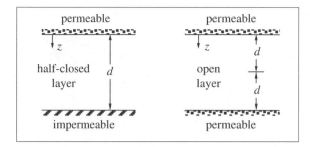

FIGURE 6.12 *Drainage path length d*

T_v is a dimensionless time factor given by:

$$T_v = \frac{c_v t}{d^2} \qquad (6.25)$$

where c_v is the coefficient of consolidation, given by equation 6.22, t is the time elapsed after instantaneous loading and d is the drainage path length.

U_v is the degree of consolidation and represents the proportion of the consolidation process that has taken place at a particular time at a particular location within the layer. It can be defined in three ways. One is the amount of void ratio change that has occurred at the time t compared to the final void ratio change.

$$U_v = \frac{e_0 - e_t}{e_0 - e_f} \qquad (6.26)$$

where:

e_0 = initial void ratio before the instantaneous loading increment was applied

e_t = void ratio at time t from the point of instantaneous loading

e_f = final void ratio at equilibrium under the applied load increment.

It can also be defined as the amount of the excess pore water pressure that has dissipated at the time t compared to the initial excess pore pressure:

$$U_v = \frac{\Delta u_i - \Delta u_t}{\Delta u_i} \qquad (6.27)$$

SEE WORKED EXAMPLE 6.5

where:

Δu_i = initial excess pore pressure

Δu_t = excess pore pressure at time t

and as the effective stress increase at the time t compared to the final increase in effective stress:

$$U_v = \frac{\Delta \sigma_t'}{\Delta \sigma_f'} \qquad (6.28)$$

where:

$\Delta \sigma_t'$ = increase in effective stress at time t

$\Delta \sigma_f'$ = final increase in effective stress.

The solution to the dimensionless one-dimensional consolidation equation is given as:

$$U_v = 1 - \sum_{m=0}^{m=\infty} \frac{2}{M} \sin(MZ) e^{-M^2 T_v} \qquad (6.29)$$

where $M = \frac{\pi}{2} (2m + 1)$

Although it appears a daunting expression it merely relates the three parameters U_v, Z and T_v and these are conveniently represented as a graph of U_v versus Z for different T_v values, Figure 6.13. The curved lines refer to constant values of time (or T_v) and are called isochrones.

From $Z = 0$ to 2 the diagram represents the state or degree of consolidation at any point in an open soil layer whereas half of the diagram from $Z = 0$ to $Z = 1$ would represent a half-closed layer with an impermeable lower boundary and from $Z = 1$ to $Z = 2$ would represent a half-closed layer with an impermeable upper boundary. The diagram also applies to a uniform distribution of initial excess pore pressure Δu set up throughout the soil (Case A in Figure 6.14). For triangular distributions (Cases B and C) the isochrones would not be symmetrical.

The line $T_v = 0$ in Figure 6.13 represents the instantaneous loading condition (time $t = 0$, see equation 6.25) where $U_v = 0$ since consolidation has not yet commenced. Very soon after commencement of the consolidation process, say $T_v = 0.05$, an element of soil adjacent to the permeable boundaries will have been able to fully consolidate (when $U_v = 1.0$) but for a soil element at the middle of the soil layer consolidation will hardly have started.

SEE WORKED EXAMPLE 6.7

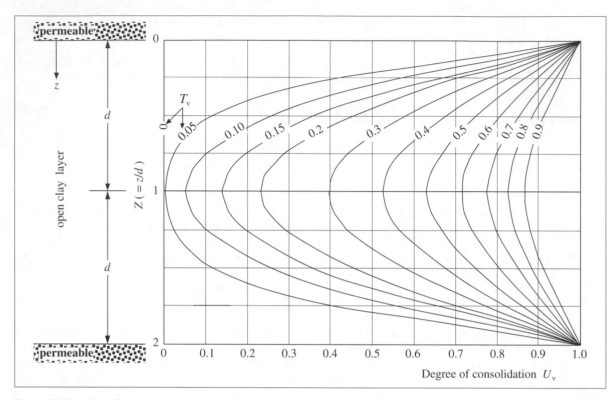

FIGURE 6.13 *One-dimensional consolidation – Isochromes*

Average degree of consolidation

(Figure 6.14)

When dealing with settlements of loaded areas placed on the surface of a soil layer the average degree of consolidation at a particular time is required, independent of depth z. This is obtained by integrating equation 6.29 with respect to z for a particular T_v value and gives:

$$\overline{U}_v = 1 - \sum_{m=0}^{m=\infty} \frac{2}{M^2} e^{-M^2 T_v} \tag{6.30}$$

which is an expression relating the average degree of consolidation $\overline{U}_v$ to T_v. The average degree of consolidation defined in terms of settlement is:

$$\overline{U}_v = \frac{\rho_t}{\rho_c} \tag{6.31}$$

where:

ρ_t = settlement at time t
ρ_c = final consolidation settlement

There are three relationships between $\overline{U}_v$ and T_v (Curves 1, 2 and 3 in Figure 6.14) depending on the variation of initial excess pore water pressure set up within the soil layer and the permeability of the soil boundaries. The three cases could be produced by:

- Case A – a wide or extensive applied pressure compared to the thickness of the soil layer, such as an embankment, or general lowering of the water table to give a uniform pressure distribution throughout the soil layer.
- Case B – an applied pressure over a small area on the surface of the layer such as a foundation.
- Case C – this could be due to the self-weight of the soil forming an embankment or a hydraulic fill deposit.

Values of $\overline{U}_v$ and T_v are given in Table 6.1 for the three curves and good approximations for the most commonly adopted cases (Curve 1) are:

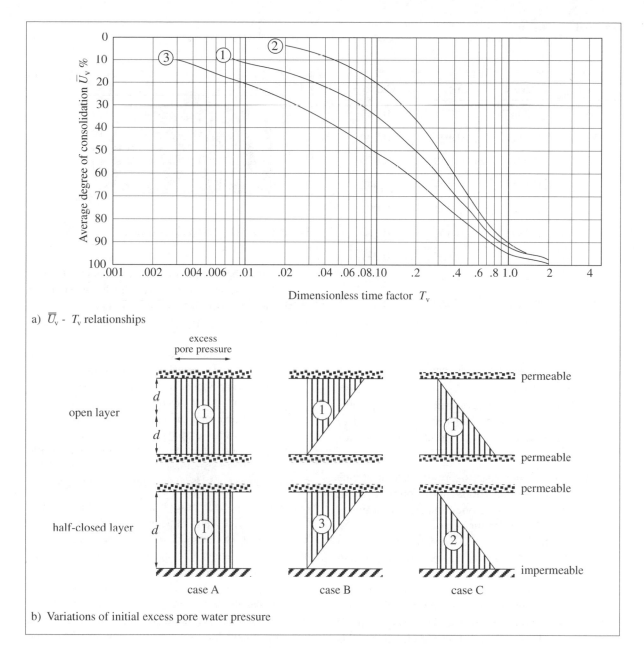

a) $\bar{U}_v$ - T_v relationships

b) Variations of initial excess pore water pressure

FIGURE 6.14 *Average degree of consolidation $\bar{U}_v$ versus T_v*

$$T_v = \frac{\pi}{4}\left\{\frac{\bar{U}_v\%}{100}\right\}^2 \qquad (6.32)$$

for $\bar{U}_v < 60\%$

and

$$T_v = 1.781 - 0.933\log_{10}\left\{100 - \bar{U}_v\%\right\} \qquad (6.33)$$

for $\bar{U}_v > 60\%$

SEE WORKED EXAMPLE 6.6

TABLE 6.1 *Values of* T_v

U_v %	T_v		
	Curve 1	Curve 2	Curve 3
10	0.008	0.047	0.003
20	0.031	0.100	0.009
30	0.071	0.158	0.024
40	0.126	0.221	0.048
50	0.196	0.294	0.092
60	0.287	0.383	0.160
70	0.403	0.500	0.271
80	0.567	0.665	0.440
90	0.848	0.940	0.720

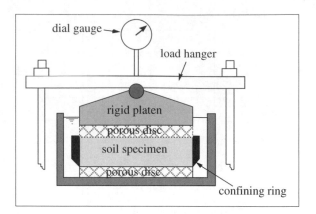

FIGURE 6.15 *Oedometer test apparatus*

Oedometer test (Figure 6.15)

The recommended test procedure is described in BS 1377:1990 and a cross-section through the apparatus is shown in Figure 6.15. The test was designed to reproduce one-dimensional consolidation by providing:

1. A rigid confining ring to prevent lateral strains and produce K_o conditions and to prevent lateral drainage thus ensuring one-dimensional drainage conditions. The inner surface must be smooth and coated with low friction material or grease to minimise wall friction as the specimen reduces in thickness.

2. Porous discs top and bottom to act as permeable boundaries. Coupled with a relatively thin specimen this will provide an open layer (Figure 6.12) and hence the smallest drainage path length. This should enable consolidation under each increment of applied stress to be completed in a manageable time of about 24 hours. A typical specimen height is 19 or 20 mm with diameters of 75 or 100 mm. With thicker specimens the confining ring surface could produce excessive friction during loading and excessive disturbance during specimen preparation and the time required for consolidation would be increased.

On the other hand, the specimen should be at least five times thicker than the largest particle diameter to avoid interference effects.

3. A rigid loading platen to provide equal settlement.

4. A water container to immerse the specimen and porous stones to ensure full saturation. All of the metallic components must therefore be non-corrodible. The water is added after installation of the specimen and assembling the load frame since some soils may have a tendency to swell in the presence of water and others may settle rapidly due to structural collapse on wetting. This behaviour should be investigated separately.

Loading is applied through a loading yoke or load hanger and a counter-balanced lever system with a load ratio of around 10:1 so that a relatively small mass on the hanger can produce the large stresses required on the specimen.

An initial pressure is applied to return the specimen to near its *in situ* effective stress $p_0{}'$ to act as a starting point for reloading or swelling. This stress will depend on the stress history the soil has been subjected to.

Each pressure increment above $p_0{}'$ requires about 24 hours of application so to make the test manageable and economical 4 to 6 increments of loading and 1 or 2 larger unloading increments are normally chosen using the following suggested sequence:

6, 12, 25, 50, 100, 200, 400, 800, 1600, 3200 kN/m^2

Smaller increments may be adopted for samples

taken from shallow depths and for small applied stresses from the structure. If the soil is overconsolidated loading should continue into the normal compression region so that a measure of the preconsolidation pressure can be obtained.

The change in thickness of the specimen (either compression or swelling) is measured using a dial gauge or a displacement transducer impinging on the loading yoke. Readings must be taken at frequent intervals initially when the specimen is rapidly changing in thickness followed by less frequent intervals. This should give sufficient data when plotting thickness against square root or log time.

Nowadays, it is much more convenient to use displacement transducers attached to an automatic recording device so that continual readings are obtained irrespective of the working hours of the laboratory. If the test is only required to measure primary consolidation then the readings can be plotted to assess whether this process has been completed when it will be permissible to apply the next increment of pressure. This may occur within the same day for some soils.

From the oedometer test results a number of soil properties can be determined:

1. The initial and final moisture contents, bulk density, dry density and degree of saturation.
2. Void ratio at the end of each pressure increment.
3. Compression index, C_c for normally consolidated soil.
4. Swelling index, C_s for overconsolidated soil.
5. Coefficient of volume compressibility, m_v, for each loading increment (not unloading) above p_0'.
6. Coefficient of consolidation, c_v, for each loading increment above p_0'.

The initial moisture content $w_o\%$ can be obtained from the trimmings around the specimen and the final moisture content $w_f\%$ can be obtained from the specimen itself at the end of the test.

The initial bulk density is given by:

$$\rho = \frac{m_0 \times 1000}{A\,H_0} \tag{6.34}$$

where:

m_0 = initial wet mass of specimen (g)
A = area of specimen (mm^2)

H_o = initial height of specimen (mm)

The initial dry density is obtained from:

$$\rho_d = \frac{\rho \times 100}{100 + w_0\,(\%)} \tag{6.35}$$

The initial void ratio is given by:

$$e_0 = \frac{\rho_s}{\rho_d} - 1 \tag{6.36}$$

where:

ρ_s = particle density (Mg/m^3)

The initial degree of saturation is obtained from:

$$S_{r0} = \frac{w_0\rho_s}{e_0} \tag{6.37}$$

The void ratio at the end of each load increment is determined from:

$$e_f = e_i - \frac{\Delta H}{H_i}\left(1 + e_i\right) \tag{6.38}$$

where, for *each pressure increment*:

e_i = initial void ratio
e_f = equilibrium void ratio at the end of consolidation
H_i = initial thickness for the pressure increment (mm)
H = change in thickness over the pressure increment (mm)

The coefficient of volume compressibility m_v, with units of m^2/kN, is obtained *for each pressure increment* from:

$$m_v = \frac{e_i - e_f}{1 + e_i}\frac{1}{\Delta\sigma} = \frac{\Delta e}{1 + e_i}\frac{1}{\Delta\sigma} \tag{6.39}$$

or

$$m_v = \frac{\Delta H}{H_i}\frac{1}{\Delta\sigma} \tag{6.40}$$

where $\Delta\sigma$ is the pressure increment (kN/m^2).

SEE WORKED EXAMPLE 6.1

The compression index C_c is given by:

$$C_c = (e_1 - e_2) \log_{10} \frac{\sigma_1'}{\sigma_2'} \qquad (6.41)$$

The swelling index C_s is given by:

$$C_s = (e_1 - e_2) \log_{10} \frac{\sigma_1'}{\sigma_2'} \qquad (6.42)$$

where e_1 is the void ratio at the effective stress σ_1' and e_2 is the void ratio at σ_2'.

C_c represents the gradient of the void ratio–effective stress plot for the *normally consolidated* line only and C_s represents the gradient of the *overconsolidated* portion of the reloading line, see Figure 6.2. C_s is referred to as the swelling index because it is used to determine drained heave movements when stresses are removed from a soil. It is not necessary to determine values for each pressure increment, it is

preferable to plot void ratio versus effective stress and obtain C_c or C_s from the gradient of each line. Typically, the ratio of the two values is approximately

$$C_c \approx 0.1 \text{ to } 0.2 \, C_s \qquad (6.43)$$

SEE WORKED EXAMPLE 6.1

Coefficient of consolidation, c_v – root time method (Figure 6.16)

Taylor (1948) showed that the Terzaghi theory of one-dimensional consolidation gave a straight line when $\bar{U}_v$ was plotted against $\sqrt{T_v}$, at least up to $\bar{U}_v = 60\%$, and that at $\bar{U}_v = 90\%$ the theoretical curve occurred at 1.15 times the extrapolated straight-line portion. In Figure 6.16a the length AC is equal to 1.15 × length AB. The theory of consolidation relates only to pore water pressure dissipation and is

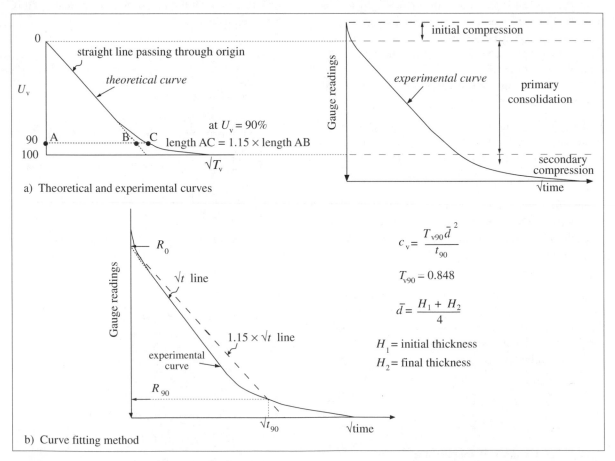

a) Theoretical and experimental curves

$$c_v = \frac{T_{v90} \bar{d}^2}{t_{90}}$$

$$T_{v90} = 0.848$$

$$\bar{d} = \frac{H_1 + H_2}{4}$$

H_1 = initial thickness
H_2 = final thickness

b) Curve fitting method

FIGURE 6.16   *Root time curve fitting method*

referred to as *primary consolidation*. The experimental curve often produces a straight-line portion when dial gauge readings are plotted against the square root of time, $\sqrt{t}$ and this can be adapted to apply to the theory, as shown on Figure 6.16a. However, the experimental curve deviates from the theoretical curve at the beginning and end due to:

1. At the beginning an initial compression is produced by factors such as bedding of the porous discs and compression of air or gas bubbles that came out of solution following sampling.
2. Towards the end any secondary compression will be recorded as continuing volume decrease even after all measurable pore pressures have fully dissipated.

In order to apply the consolidation theory to the laboratory test result the 'root time' curve-fitting method can be used as follows:

1. On the plot of dial gauge readings versus square root of time (Figure 6.16b) draw a best fit line through the straightest portion and extrapolate back to a corrected thickness reading origin R_0 at $t = 0$ to eliminate the initial compression.
2. Draw a straight line from R_0 at a gradient 1.15 times the gradient of the straight portion of the experimental plot. Where this line cuts the experimental curve is considered to be the point at 90% degree of consolidation.
3. Determine the value of $\sqrt{t_{90}}$ and hence t_{90}, the time required to produce 90% degree of consolidation
4. The coefficient of consolidation, c_v, is then obtained from equation 6.25 as:

$$c_v = \frac{T_{v90}\,\bar{d}^{\,2}}{t_{90}} \qquad (6.44)$$

where:

T_{v90} is the value of T_v at $\bar{U}_v = 90\%$, i.e. from Table 6.1, $T_{v90} = 0.848$
$\bar{d}$ is the average length of drainage path for the load increment ($\bar{H}/2$), as shown in Figure 6.16b.

The conventional units for c_v are m²/year so the above equation should be modified from mm²/min. Since c_v is dependent on permeability a correction for temperature should be applied such as that given

in Chapter 3, Figure 3.14. However, the British Standard states that this correction is not justified in view of the inaccurate and unrepresentative value obtained from the test, see *in situ* c_v values, below.

In practice, the root time method is attempted first because it requires readings over a much shorter time period therefore allowing the next load increment to be placed sooner. Also a reasonable straight line is usually obtained so less judgement is required. The log time method is used when a straight-line portion cannot easily be deduced from the root time method.

SEE WORKED EXAMPLE 6.3

Coefficient of consolidation, c_v – log time method (Figure 6.17)

This method is due to Casagrande who observed that the plot of $\bar{U}_v$ versus log T_v had three portions:

1. an initial portion with a shape very similar to a parabola
2. a linear middle portion
3. a final portion asymptotic to the horizontal.

This relationship applies only to primary consolidation, as shown in Figure 6.17a, therefore Casagrande devised two constructions to eliminate the initial compression and the secondary compression from the experimental curve of dial gauge readings versus log time.

Using the properties of a parabola the initial part is corrected by choosing two points A and B (any two points) at values of time t in the ratio 1:4 such as 10 seconds and 40 seconds, as shown in Figure 6.17b. The vertical distance between these points D is marked above point A to give the initial reading R_0 for primary consolidation. This is the most uncertain part of the method since the experimental curve is not always parabolic and often the number of data points which can be obtained in the short initial time available are insufficient to accurately define the curve. The Taylor method gives a more definite initial reading.

For soils which display very little secondary compression such as heavily overconsolidated clays the final part of the experimental curve will be a horizontal line which will then define the end reading for primary consolidation R_{100}.

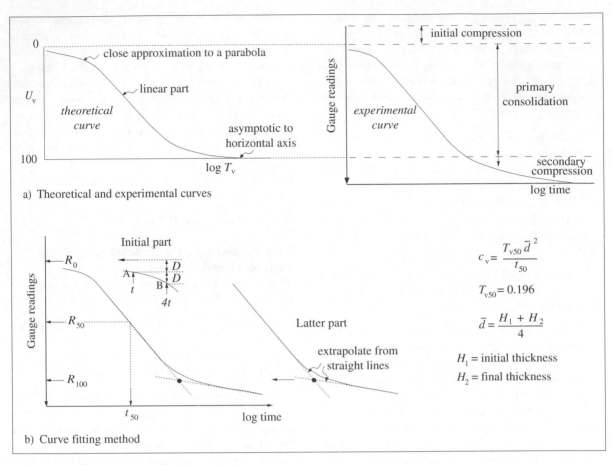

a) Theoretical and experimental curves

b) Curve fitting method

FIGURE 6.17 *Log time curve fitting method*

For soils displaying secondary compression it is commonly found that compressions continue beyond primary consolidation in a straight line on the log plot so this line is extrapolated back to meet the extrapolated straight middle portion, Figure 6.17b. Where these lines intersect is the final reading for primary consolidation R_{100} at a degree of consolidation of 100%. This enables the point R_{50} corresponding to 50% degree of consolidation to be located mid-way between R_o and R_{100}. The value of t_{50} the time required to produce 50% degree of consolidation is obtained from the graph and the coefficient of consolidation, c_v is determined from equation 6.25 as:

$$c_v = \frac{T_{v50}\,\overline{d}^{\,2}}{t_{50}} \qquad (6.45)$$

where:

T_{v50} is the value of T_v at $\overline{U}_v = 50\%$, i.e. from Table 6.1, $T_{v50} = 0.196$
$\overline{d}$ is the average length of drainage path.

Allowing each load increment to produce secondary compressions will require longer load duration and will distort the void ratio–effective stress plot as shown in Figure 6.8.

SEE WORKED EXAMPLE 6.4

In situ c_v values
Calculations for rates of settlement of structures on clay soils based on laboratory c_v values obtained from the oedometer test have been found to be grossly in error when compared with the actual observed rates of settlement (Rowe, 1968). These

differences are due to the influence of the macro-fabric found in most clay soils but which is not represented in the small-scale oedometer test specimen. Macro-fabric or structure in a clay can be:

- horizontal – laminations, layers of peat, silt and fine sand, varves
- vertical – root holes
- inclined – fissures, silt-filled fissures.

This structure will provide many permeable boundaries, decrease the drainage path lengths for the consolidating clay and will overall increase the mass permeability of the soil. The time required for consolidation will then be reduced considerably.

To overcome this problem of unrepresentative sampling, the Rowe cell was developed for testing larger samples, see below. Alternatively, *in situ* permeability tests such as the constant head piezometer test (Chapter 3) to give an *in situ k* value can be combined with a laboratory oedometer m_v value to give an *in situ* c_v value from equation 6.22 as:

$$c_{v\,in\,situ} = \frac{k_{in\,situ}}{m_{v\,lab}\,\gamma_w} \tag{6.46}$$

since it has been found that these values are more reliable (Bishop and Al-Dhahir, 1969).

Rowe consolidation cell (Figure 6.18)

This test was developed at the University of Manchester (Rowe and Barden, 1966) to improve the quality of data obtained from a consolidation test and to eliminate several of the disadvantages of the oedometer test. The test procedure is detailed in BS 1377:1990 where the apparatus is described as a hydraulic consolidation cell. The details of the apparatus are illustrated in Figure 3.19.

The specimen is enclosed by a cell base, the cell body around and a convoluted rubber membrane above, which moves up or down as the specimen changes in thickness. A total stress increment is applied outside the specimen by the constant pressure supply above the membrane. This supply must be able to compensate for leakages and changes in specimen thickness. Internally, within the specimen a back pressure may be applied and the pore water pressure can be measured. Drainage from the specimen can be either vertical or radial (Figure 6.18) and

as the specimen consolidates the change in thickness is recorded by a dial gauge or a displacement transducer which follows the movement of the central hollow drainage spindle. The main advantages compared to the oedometer are:

1. Large specimens can be tested to ensure that *in situ* macro-fabric is represented in the test. Specimens 250 mm diameter and 100 mm thick are considered to be sufficiently representative to give a good measure of the *in situ* c_v.
2. Pore water pressure can be measured. This is useful to ensure that pore pressures have uniformly increased within the specimen following application of the external total stress increment with the drainage valve closed, i.e. before consolidation commences. During consolidation pore pressure will dissipate and the plot of degree of consolidation U (= pore pressure dissipated/initial pore pressure) versus log time can be used directly to obtain c_v. In the study of secondary compression, the complete dissipation of pore pressures can be taken as the commencement of this process.
3. A back pressure can be applied to the pore water in the specimen to ensure full saturation and to prevent gases coming out of solution, particularly for organic soils.
4. The drainage paths can be controlled (Figure 6.18). The drainage can be vertical only (one-way or two-way) or radial only (inwards to a central sand drain or outwards to a peripheral porous lining).
5. A 'free-strain' or 'equal strain' loading condition can be applied. The former is provided by applying the pressure to the specimen directly via the flexible diaphragm. This produces a more uniform pressure distribution of known value (equal to the applied stress) and reduces the effects of side friction.

Flexible loading means that the settlement at the top of the specimen will be non-uniform and since displacements are measured at the centre of the specimen the maximum values will be obtained. More appropriate determinations of m_v should be obtained from volume change measurements rather than thickness change measurements although they may not be as accurate.

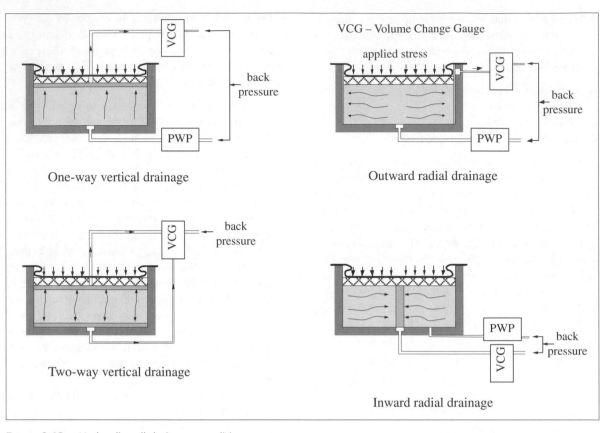

FIGURE 6.18 *Hydraulic cell drainage conditions*

Equal strain loading is produced by inserting a rigid disc above the specimen to produce conditions similar to the oedometer. m_v can then be determined from changes in thickness which will be more accurate.

Two- and three-dimensional consolidation (Figure 6.19)

One-dimensional consolidation will occur beneath a wide load or where the soil layer thickness is small compared to the width of a load such as beneath a large fill area or when the water table is lowered generally. Two-dimensional consolidation will occur beneath a strip load and three-dimensional consolidation will occur beneath a circle, square or rectangle.

As shown in Figure 6.19a one-dimensional consolidation means no horizontal drainage or horizontal strains in any direction (i.e. vertical only) and the one-dimensional consolidation theory described above can be used to obtain the settlement-time relationship. With two-dimensional consolidation vertical drainage and strains will be accompanied by horizontal drainage and strains in one direction only. Three-dimensional consolidation results in horizontal drainage and strains in all directions so it can be seen that the geometrical shape and size of the load will have a major influence on the rate of settlement.

Conventionally rates of settlement are determined from the one-dimensional theory but Davis and Poulos (1972) have shown that the practical effect of two and three-dimensional conditions is that the time required for consolidation decreases:

1. as the footing area decreases
2. as the soil layer thickness decreases
3. more for square or circular loaded areas (three-dimensional) than for strip loading (two-dimensional).
4. as the horizontal permeability increases ($c_H > c_V$)

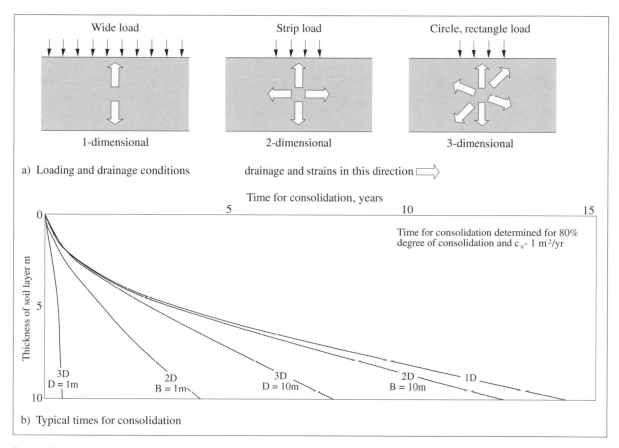

Figure 6.19 *One-, two- and three-dimensional conditions*

5. if the underside of the foundation can be considered permeable, e.g. a thin sand layer exists beneath.

Some typical results shown in Figure 6.19b illustrate the above.

Correction for construction period
(Figure 6.20)
The one-dimensional consolidation theory assumes that loads are placed instantaneously at time $t = 0$ whereas in reality they are applied over a construction period. Assuming that the net load is applied uniformly over the construction period t_e Terzaghi proposed a simple empirical method for correcting the 'instantaneous' curve for this effect.

After the end of construction
The assumption is that the load is placed instantaneously at $t_e/2$ so that the theory can be used to obtain the instantaneous value of time and this time is then assumed to commence at time $t_e/2$.

During the construction period
It is assumed that the settlement ρ_1 which occurs at time t_1 when the load P_1 has been applied is obtained from the instantaneous curve at $t_1/2$ but it must also be reduced in the ratio P_1/P_e since the instantaneous curve is determined from the full load applied P_e. Thus by similar triangles:

$$\frac{P_1}{P_e} = \frac{t_1}{t_e} = \frac{\rho_1}{\rho_e} \tag{6.47}$$

The procedure would be to find ρ_e at the time $t_1/2$ on the instantaneous curve. Then from the above ratio determine ρ_1 and plot it at t_1. By repeating this

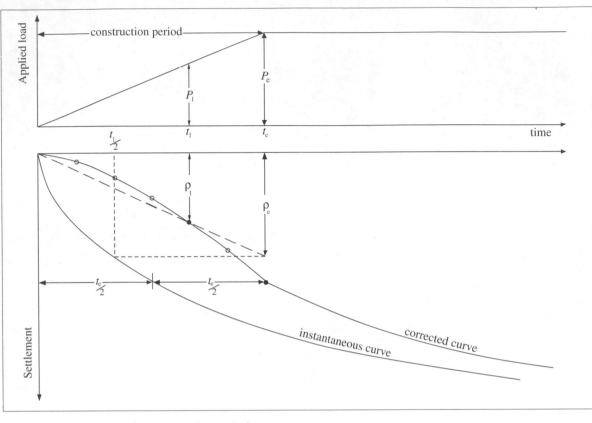

FIGURE 6.20 *Correction for construction period*

procedure for a number of points the corrected curve for the construction period is obtained.

A further adjustment can be made for the immediate settlement produced assuming that it is proportional to the load applied and is complete at the end of construction.

SEE WORKED EXAMPLE 6.8

Precompression by surcharging
(Figure 6.21)

When large post-construction settlements are expected or when it is desirable to achieve the settlements within a practical time-scale it is possible to accelerate the rate of consolidation by applying an additional load called surcharge or preload. As surcharge is usually applied as an extra amount of soil fill this approach is most convenient for earth structures. However, soil fill has been placed as a surcharge to produce the desired settlements prior to

constructing other structures, particularly those that are not so sensitive to settlements.

Other means of applying additional load include water loading in tanks and groundwater lowering using a range of pore pressure reduction techniques. The former produce an initial excess pore pressure that when subsequently dissipating causes an increase in effective stress accompanied by strengthening and decrease in volume accompanied by settlements and decrease in compressibility. The latter would be most successful where the water table in a sand layer above or below the compressible layer can be lowered to produce an initial reduced pore pressure within the layer. Surcharging can also be successful in removing some of the secondary compressions.

On the settlement–time relationship in Figure 6.21 the desired settlement ρ_f under the final construction load P_f will occur at a time t_f. If this time is unacceptable and it is required to achieve the desired set-

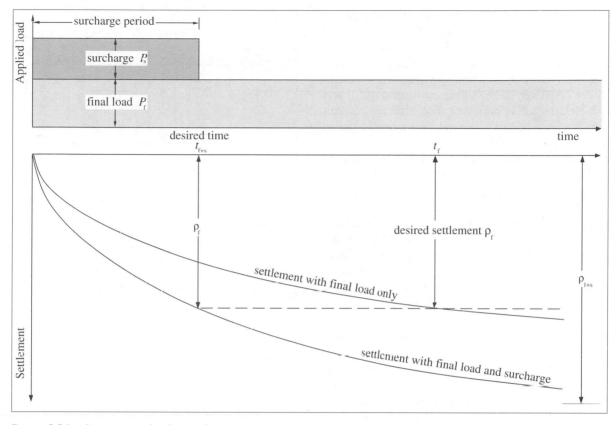

FIGURE 6.21 *Precompression by surcharging*

tlement in a time t_{f+s} then an additional loading, surcharge P_s can be placed to increase the rate of consolidation.

The calculation procedure usually entails determining the amount of surcharge required. This can be done by deciding on the amount of settlement to achieve, ρ_f and the time available to produce this settlement, t_{f+s}. From equation 6.25 this time will give a dimensionless time factor T_v and from Figure 6.13 the degree of consolidation that must be achieved by the total load $P_f + P_s$ is obtained. The final amount of settlement ρ_{f+s} produced by the fill and the surcharge is:

$$\rho_{f+s} = \frac{\rho_f}{U_{vf+s}} \qquad (6.48)$$

By using the equations in Figure 9.9 for normally consolidated soil and equation 9.10 for lightly over-consolidated soil the amount of pressure required to achieve this settlement can be determined and from

this, knowing the bulk unit weight of the soil fill the height of the surcharge can be calculated.

When the surcharge load is removed the compressible stratum is only partially consolidated and for a uniform deposit the middle of the layer will be less consolidated than the upper and lower horizons as shown in Figure 6.13. For wide surcharge loading the soil beneath the central portion will also be less consolidated than near the edges. Use of the average degree of consolidation as the criterion for surcharge removal would then be inappropriate and it is suggested (Johnson, 1970) that the more conservative value of the degree of consolidation at the centre of the deposit is used.

SEE WORKED EXAMPLE 6.9

Factors to consider when choosing precompression as a viable process are associated with the benefits of improving poor quality compressible ground compared to the most economical means of providing the additional load, leaving the site with a period of down-time and then removing the surcharge load on completion. Placing additional load on these weaker soils will also increase the risk of instability with the possibility of foundation failure at the edges of the surcharge due to increased heights of fill and greater excess pore pressures. The latter can be dealt with by incorporating vertical or horizontal drains within the compressible soil.

A more detailed ground investigation will be required with more testing and design input since the geological stratification, mass permeability and compressibility characteristics of the ground are more important. *In situ* values of the coefficient of consolidation c_V and c_H should be determined. This should be followed up by an instrumentation program to monitor settlements, pore pressures and horizontal displacements of the structure. There is, therefore, less certainty of the outcome with this technique.

Radial consolidation for vertical drains
(Figure 6.22)
The time for one-dimensional consolidation to take place is inversely proportional to d^2, therefore for thick layers of clay settlements of earth structures placed above this type of consolidation take a long time to be completed. By inserting vertical drains at fairly close spacings much shorter horizontal drainage paths are created allowing faster dissipation of pore pressures, removal of pore water and accelerated consolidation settlements.

The original 'sand drains' were formed by sinking boreholes 200–400 mm diameter using cable percussion or flight auger methods through the soil and backfilling with a suitable filter sand. However, these could be slow to construct, produce large amounts of spoil and surface damage and be prone to 'waisting'. This is caused by soft clays squeezing into the borehole on removal of the casing leaving a reduced diameter which could prevent adequate drainage. Other techniques include driving a closed end steel pipe to form a hole although this is prone to producing a disturbed or smear zone around the hole which can reduce the permeability of the soil around the drain.

Prefabricated drains comprising a continuous filter stocking filled pneumatically with sand called sandwicks and installed in a pre-drilled hole can

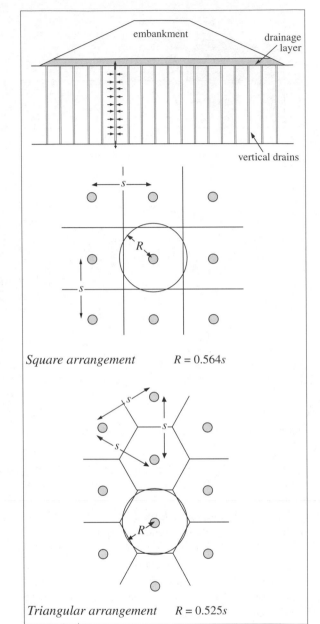

Figure 6.22 *Radius of influence of drains*

provide a more reliable and cost-effective solution. Since the drain diameter is not a particularly important design parameter, smaller diameters of about 70 mm could be used.

Nowadays, the plastic 'band' drain is commonly used. This consists of a continuous flat plastic core about 60–100 mm wide and 2–5 mm thick which is corrugated to provide vertical drainage channels and wrapped around by a geotextile fabric to act as a filter. The band drain is installed by attaching one end of the band at the bottom of a rectangular steel mandrel with a clip and driving or pushing the mandrel vertically into the soil to the depth required. As the mandrel is withdrawn the clip retains the band drain in the soil which will then squeeze around the drain holding it in place. It is essential that the band drains are maintained vertical within the drainage blanket and not pushed over or constricted otherwise drainage would be impaired. These drains are cheaper, lighter, more robust and quicker to install.

A solution for radial consolidation has been obtained (Barron, 1948) by considering Terzaghi's equations for three-dimensional consolidation in polar coordinates as:

$$\frac{\partial u}{\partial t} = c_v \frac{\partial^2 u}{\partial z^2} + \left[c_H \frac{\partial^2 u}{\partial r^2} + \frac{1}{r} \frac{\partial u}{\partial r} \right] \quad (6.49)$$

overall drainage = vertical drainage + radial drainage

where:

$$c_v = \frac{k_v}{m_v \gamma_w} \quad \text{and} \quad c_H = \frac{k_H}{m_v \gamma_w} \quad (6.50)$$

The circular drains of radius r_d will have a circular area of influence of radius R (Figure 6.22) with water moving horizontally towards each drain. In practice, drains placed in a square grid pattern will have a square area of influence of size s^2 for a spacing s. This is equated to a circular area of influence such that $s^2 = \pi R^2$ and:

$$R = 0.564s \quad (6.51)$$

For a triangular pattern:

$$R = 0.525s \quad (6.52)$$

The overall degree of consolidation can be obtained from the combination of vertical and radial drainage as:

$$1 - U_c = (1 - U_v)(1 - U_R) \quad (6.53)$$
where:

U_c = overall or combined degree of consolidation

U_v = average U_v for one-dimensional vertical flow only

U_R = average U_R for radial flow only

Settlements are now related to the overall degree of consolidation as:

$$U_c = \frac{\rho_t}{\rho_c} \quad (6.54)$$

Values of U_v are obtained from Curve 1 in Figure 6.13 or Table 6.1, related to the vertical time factor T_v. Values of U_R are related to a (radial) time factor T_R:

$$T_R = \frac{c_H t}{4 R^2} \quad (6.55)$$

but also to the ratio:

$$n = \frac{R}{r_d} \quad (6.56)$$

and are given in Table 6.2.

For design the size or radius of the drain cannot be varied much but this does not matter as it is not as critical as the spacing between the drains. The equivalent diameter of a band drain is determined assuming that it has the same perimeter as a circle.

SEE WORKED EXAMPLE 6.10

It has been found that the time t for consolidation varies approximately as:

$$t \propto s^{2.5} \quad (6.57)$$

so the spacing is critical.

Smearing of the soil during installation will affect the horizontal permeability and hence c_H. To allow for this a reduced value, as low as $c_H = c_V$ could be used or a smaller drain radius of $r_d/2$ could be used. It has been found that the cost of a drain installation is inversely proportional to the value of c_V or c_H so care during the site investigation in obtaining a representative value is essential.

Vertical drains can be ineffective:

1. With thin soil layers where sufficient consolidation may be achieved in one dimension (vertical) alone.

2. If c_H is much larger than c_V. Where macro-fabric

TABLE 6.2 *Radial consolidation – Values of T_R*

Degree of consolidation U_R %	Time factor T_R										
	$n = 5$	10	15	20	25	30	40	50	60	80	100
10	0.012	0.021	0.026	0.030	0.032	0.035	0.039	0.042	0.044	0.048	0.051
20	0.026	0.044	0.055	0.063	0.069	0.074	0.082	0.088	0.092	0.101	0.107
30	0.042	0.070	0.088	0.101	0.110	0.118	0.131	0.141	0.149	0.162	0.172
40	0.060	0.101	0.125	0.144	0.158	0.170	0.188	0.202	0.214	0.232	0.246
50	0.081	0.137	0.170	0.195	0.214	0.230	0.255	0.274	0.290	0.315	0.334
55	0.094	0.157	0.197	0.225	0.247	0.265	0.294	0.316	0.334	0.363	0.385
60	0.107	0.180	0.226	0.258	0.283	0.304	0.337	0.362	0.383	0.416	0.441
65	0.123	0.207	0.259	0.296	0.325	0.348	0.386	0.415	0.439	0.477	0.506
70	0.137	0.231	0.289	0.330	0.362	0.389	0.431	0.463	0.490	0.532	0.564
75	0.162	0.273	0.342	0.391	0.429	0.460	0.510	0.548	0.579	0.629	0.668
80	0.188	0.317	0.397	0.453	0.498	0.534	0.592	0.636	0.673	0.730	0.775
85	0.222	0.373	0.467	0.534	0.587	0.629	0.697	0.750	0.793	0.861	0.914
90	0.270	0.455	0.567	0.649	0.712	0.764	0.847	0.911	0.963	1.046	1.110
95	0.351	0.590	0.738	0.844	0.926	0.994	1.102	1.185	1.253	1.360	1.444
99	0.539	0.907	1.135	1.298	1.423	1.528	1.693	1.821	1.925	2.091	2.219

will permit sufficient drainage horizontally, vertical drains may not be required.

3. If large secondary compressions are likely. Drains only accelerate pore pressure dissipation or primary consolidation so if:

$$\frac{\rho_c}{\rho_{total}} = \frac{\text{consolidation settlement}}{\text{total settlement}} < 0.25 \qquad (6.58)$$

then drains may not be viable.

4. Where they are severed due to large horizontal shear deformations around an incipient slip surface.

SUMMARY

The compressibility of a soil is usually represented by a void ratio–effective stress relationship. From the pre-consolidation pressure the soil can be distinguished as either normally consolidated and likely to produce large settlements or overconsolidated with much smaller settlements.

The process of consolidation comprises the volume reduction with time of a fully saturated soil as water is squeezed out of the pore spaces under the application of a load. From the solution of the one-dimensional theory of consolidation the pore pressure within a soil layer and the settlement of the layer can be determined at any time after the application of the load.

The rate of settlement is determined by the *in situ* permeability which depends on the macro-fabric of the soil. Laboratory values of the coefficient of consolidation, c_v should be used with caution.

Two- and three-dimensional consolidation will increase the rate of settlement.

Construction techniques such as precompression by surcharging and the installation of vertical drains have been successfully adopted to increase the rate of settlements.

CASE STUDY

Sinking cities - Flooding of Venice, Italy

Case objectives
This case illustrates:
- that reduced pore water pressures and increased effective stress produces consolidation and pumping groundwater from aquifers results in regional settlements
- settlements are irreversible even when groundwater levels recover, in the normally and lightly overconsolidated soils
- the economical and social consequences of disregarding well-known soil mechanics principles
- the delicate balance that exists for sea-level cities and their surrounding ecosystems

The historic city of Venice lies on an island within a lagoon, offshore from the north coast of Italy with the Adriatic Sea to the south. The unique aspect of the city is that it 'lies in the water' and even modest changes in water levels during the present century have caused increasingly severe flooding damage. An exceptional storm surge on 4 November 1966 breached the coastal defences and caused the highest ever recorded level of 1.94 m above mean sea level, compared to the normal spring tide level of +0.5 to 0.6 m. The average tidal range at Venice is only about 1 m so even small changes of water level superimposed on the spring high tides can have serious effects.

> **Drinking in wellies!**
> During the acque alte, the popular Venetian bar, Alla Rampa, just to the east of the Piazza San Marco, quickly floods. Undaunted, the patron and locals simply don high wellington boots and carry on drinking as the water rises (*The Observer*, 1999).

Sedimentary deposits of considerable thickness, up to 1000 m exist beneath the city. The top 350 m comprise six sand aquifers interbedded with deposits of silt/clay which act as aquitards. The soils mainly comprise fine to medium sands and fine to coarse silts and with some very silty and silty clays and occasional peaty intercalations of fluvial, lacustrine and marsh origin. The clay minerals are mainly illite and chlorite with small amounts of kaolinite and montmorillonite. The clayey soils are of low to intermediate plasticity with moisture contents typically below the plastic limit. The few thin layers of peat present are highly compressed (Ricceri *et al.*, 1974)

The subsidence is considered to be due to three causes: natural subsidence, groundwater lowering induced by pumping and eustatic changes (Carbognin *et al.*, 1986). The natural subsidence is caused by regional geotectonic processes and in the present century has been estimated to be about 0.4 mm per year (Bortolami *et al.*, 1986) contributing up to 30 mm from 1908 to 1980. Eustacy refers to the world-wide rising of sea levels and has been calculated as 1.27 mm per year so this has contributed a further 90 mm to the difference in the relative land/sea level.

The main cause of the subsidence has been the exploitation of the aquifer system. Prior to 1900 artesian heads were recorded up to 6 m above sea level in Venetian wells sunk to the shallow aquifers. More intensive pumping mainly for industrial purposes began around 1930 and intensified after the Second World War. The lowering of the average piezometric level in the industrial zone, north of the city was up to 20 m while in the city area it amounted to about 10 m. In 1970 severe pumping restrictions and the provision of alternative water supplies allowed the water levels to recover to ground level. However, during the period of lowering consolidation of the more compressible strata has caused irreversible volume changes in the soil resulting in a permanent consolidation settlement of about 100 mm. About 20 mm of rebound was measured during the recovery period.

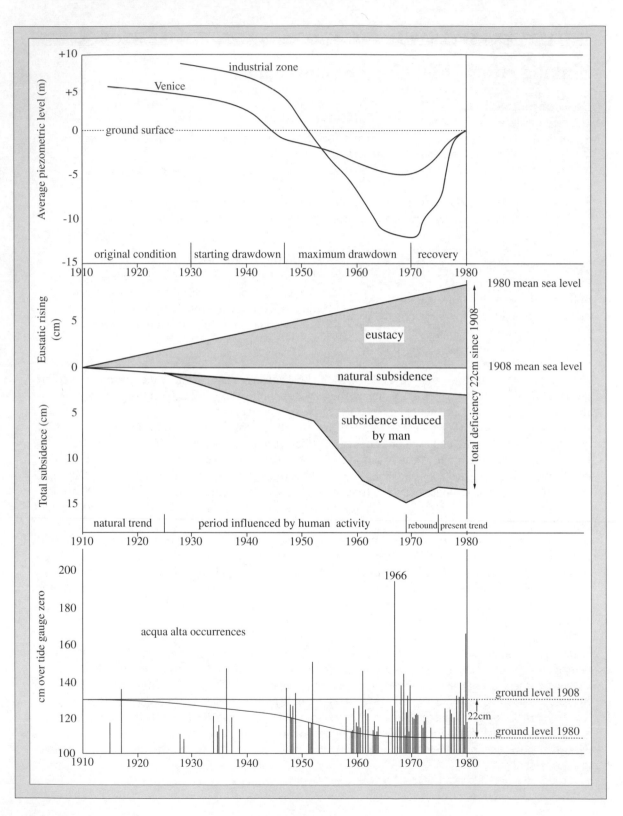

The compressibility of the soil column calculated from the back analysis of surface settlements and piezometric level reductions is less than that indicated by oedometer tests (Ricceri *et al.*, 1974). This is considered to be due to sampling disturbance and stress relief affecting the void ratio–effective stress plots, a lower clay content than anticipated and the presence of some light overconsolidation, particularly in the deposits at the higher levels.

In most parts of the world this magnitude of subsidence would have had little consequence. However, for this sea-level city whereas the spring high tides (acqua alte) of 90 years ago would not have caused flooding, the frequency of flooding has now increased to many times a year, St Mark's Square was flooded 79 times in 1997.

In 1984 the Italian government initiated a £2 billion scheme to safeguard the city, involving major civil engineering works with the installation of mobile barriers across the lagoon inlets, reinforcement of coastal defences, reconstruction of the marshlands and works to improve the ecosystem of the lagoon (Bandarin, 1994). The three inlets into the lagoon will be protected by fixed barriers and mobile flap gates leaving sufficient openings for shipping traffic but allowing for withholding up to 2 m in level difference between the sea and the lagoon.

The other relevant chapter is 9, Settlements.

CASE STUDY

Sinking cities - Mexico City, Mexico

Case objective

- This case has strong similarities to Venice and many other cities of the world such as Shanghai, China, Bangkok, Thailand and many parts of Japan and the United States. The economic, social and environmental consequences are considerable.

Mexico City lies within a large basin at an elevation of 2250 m above sea level surrounded by volcanic mountains. Parts of the city are underlaid by one of the most difficult and unusual soils in the world. Mexico City clay is a quaternary lacustrine clay deposited in the former Lake Texcoco. It is up to 50 m thick and contains some intercalations of silts and sands with high yield sand and gravel aquifers below and around the margins of the deposit (Figueroa Vega, 1984). The constituents of the clay have been found to be quite variable (Mesri *et al.*, 1975) but the origins appear to be fine volcanic ash and micro-fossils such as shells, diatoms, ooliths and ostracods. These provide the sand and silt particles while the clay sizes have been described variously as montmorillonite, allophane and amorphous gel type materials. An organic content up to 10% appears to be prevalent.

Pumping of the groundwater from the aquifers commenced around 1850, rapidly increased since the 1940's and in 1974 was estimated to be 12 m³/s (Figueroa Vega, 1984). With over 3000 wells into the shallow aquifers and 200 wells into the deeper aquifers (>100 m) abstraction has far exceeded recharge and piezometric levels have fallen, up to 30 m in parts of the city. However, it was only in the late 1940s that the resulting subsidence* was attributed to groundwater lowering.

Settlement records from 1891 to 1973 have shown that settlements up to 8.7 m had occurred in the Old City with a maximum rate of 460 mm/year (nearly 1.3 mm per day) in 1950. It has been estimated that settlements could potentially reach 20 m. The majority of the settlement has been due to consolidation and secondary compression of the clays down to 50 m.

The consequences of the settlements have been the disruption of the surface infrastructure such as roads, bridges, pavements, services and drainage. In particular, the loss of water supply from dislocated leaking pipes has been considerable. Buildings which were supported on piles taken through the compressible soils have not settled as much but with settlements of the surrounding areas they appear to be raised out of the ground.

In 1951 at a time when the city was sinking at its fastest a severe flood caused the city authorities to obtain water from outside sources, to implement a major sewerage system and to control the groundwater abstraction especially by industrial users. From 1960 to 1973 the settlement rate had reduced to between 50 and 70 mm/year.

The other relevant chapters are 5, Stress Distribution and 9, Settlements.

* Subsidence is a term used for regional settlements.

Unusual soil properties

Mexico City clay has the unusual soil properties of a natural moisture content up to 650%, liquid limit up to 500%, plasticity index up to 350%, void ratio up to 15 and variable but generally low specific gravity. These properties impart to the soil its most distinct characteristic of extremely high compressibility both during primary consolidation with very high values of the compression index, C_c and high values of the coefficient of secondary compression, ε_α. It is also of low strength and sensitive. Following remoulding it has been found to be thixotropic in that with time its stiffness and strength increase.

The soil is considered to be a normally consolidated aged deposit. Since its deposition thixotropic processes have imparted a quasi-preconsolidation to the soil such that it has a yield stress (or quasi-preconsolidation pressure) somewhat higher (about 50%) than its present overburden pressure. Thus the soil is not truly normally consolidated until this stress has been exceeded.

Regardless of its high void ratio and compressibility it can display a high angle of internal friction of 35–45° most likely as a result of the high angular diatom content.

Worked Example 6.1   Compressibility

The following results were obtained from an oedometer test on a specimen of fully saturated clay. Determine the void ratio at the end of each pressure increment and plot void ratio versus log σ'. Calculate m_v for each loading increment and C_c for the normally consolidated portion.

Initial thickness $H_0 = 20$ mm
Initial moisture content = 24%
Particle density = 2.70 Mg/m^3

Applied pressure (kN/m^2)	Thickness of specimen (mm)
0	20
25	19.806
50	19.733
100	19.600
200	19.357
400	18.835
800	18.167

Initial void ratio $e_0 = m_0 \, \rho_s = 0.24 \times 2.70 = 0.648$

From $\dfrac{\Delta H}{H_0} = \dfrac{\Delta e}{1 + e_0}$

The void ratio at the end of each pressure increment is given by:

$$e_f = e_0 - \Delta e = e_0 - \frac{\Delta H_f}{H_0}\left(1 + e_0\right)$$

where ΔH_f is the change in thickness of the specimen from the initial thickness. Compare this with equation 6.38.

$$e_f = 0.648 - \frac{1.648}{20}\,\Delta H_f$$

Pressure	Thickness (mm)	ΔH_f	Δe	e_f
0	20	0	0	0.648
25	19.806	0.194	0.016	0.632
50	19.733	0.267	0.022	0.626
100	19.600	0.400	0.033	0.615
200	19.357	0.643	0.053	0.595
400	18.835	1.165	0.096	0.552
800	18.167	1.833	0.151	0.497

From equation 6.40:

Pressure increment (kN/m^2)	ΔH_i	H_i	$\Delta \sigma_i$	m_v (m^2/MN)
0–25	0.194	20	25	0.388
25–50	0.073	19.806	25	0.147
50–100	0.133	19.733	50	0.135
100–200	0.243	19.600	100	0.124
200–400	0.522	19.357	200	0.135

The graph of void ratio versus log pressure or stress (Figure 6.23) shows the soil to be overconsolidated since there are two distinct lines. The m_v value for the first pressure increment is probably unrepresentative due to initial bedding errors and m_v should only be measured for pressure increments above the present overburden pressure. m_v for the pressure increment 200–400 kN/m² is larger because it includes volume changes on the normally consolidated limb.

The pressure increment 400–800 kN/m² lies on the normally consolidated line so the compression index can be obtained as:

$$C_c = \frac{0.552 - 0.497}{\log_{10}(800 \div 400)} = 0.18$$

Worked Example 6.2   Preconsolidation pressure and overconsolidation ratio

From the plot of void ratio versus log σ' in Worked Example 6.1 (Figure 6.23) use the Casagrande construction to estimate the preconsolidation pressure of the clay. If the specimen of soil was taken from 3 m below ground level in a soil with unit weight of 19 kN/m³ and a water table at ground level determine the overconsolidation ratio.

The accuracy of this method is largely affected by the position chosen for the point of maximum curvature. It is estimated that the preconsolidation pressure is about 240 kN/m².
The present overburden pressure is $3 \times (19 - 9.8) = 27.6$ kN/m²

$\therefore$ The overconsolidation ratio $OCR = \dfrac{240}{27.6} = 8.7$

This result shows that the soil is heavily overconsolidated.

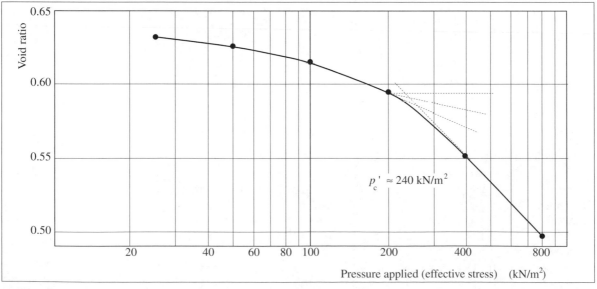

FIGURE 6.23 *Compressibility and the Casagrande construction*

Worked Example 6.3 c_v – root time method

An oedometer test on a specimen of fully saturated stiff clay gave the following results for the pressure increment from 100 to 200 kN/m². The initial thickness of the specimen under no pressure was 19 mm. Determine m_v, c_v and k.

Time from start of loading (minutes)	$\sqrt{t}$	Specimen compression (mm)
0	0	0.61
0.25	0.50	0.96
0.50	0.71	1.06
0.75	0.87	1.16
1.00	1.00	1.24
1.50	1.22	1.35
2.25	1.50	1.45
4.00	2.00	1.60
5.0	2.24	1.66
7.0	2.65	1.73
11.0	3.32	1.79
16.0	4.00	1.82
30.0	5.48	1.86
90.0	9.49	1.92

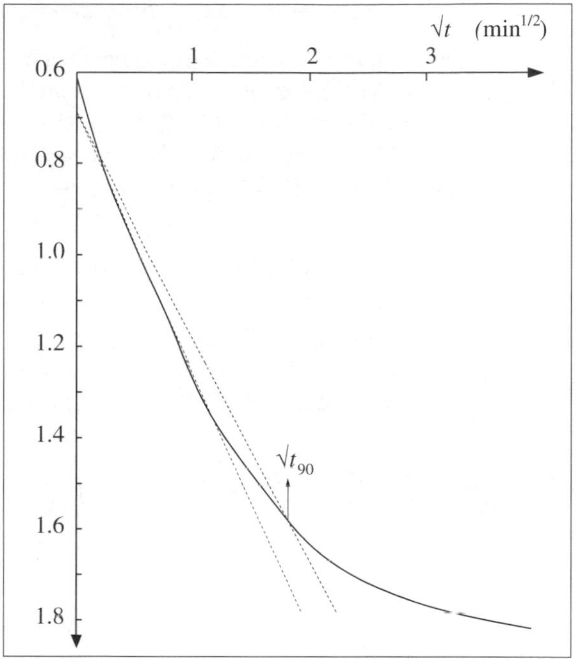

FIGURE **6.24**

From the graph of $\sqrt{t}$ versus specimen compression (Figure 6.24)

$R_o = 0.69$ mm $\therefore$ $H_o = 19.00 - 0.69 = 18.31$ mm
$R_{100} = 1.92$ mm $\therefore$ $H_f = 19.00 - 1.92 = 17.08$ mm

$$\therefore \; H_{ave} = \frac{1}{2}\,(18.31 + 17.08) = 17.70 \text{ mm}$$

$$d = \frac{17.70}{2} = 8.85 \text{ mm}$$

From the graph, $\sqrt{t_{90}} = 1.85$ $\therefore$ $t_{90} = 1.85^2 = 3.42$ minutes

From equation 6.44

$$c_v = \frac{0.848 \times 8.85^2}{3.42} = 19.4 \text{ mm}^2/\text{minute}$$

$$c_v = \frac{19.4 \times 60 \times 24 \times 365}{1000 \times 1000} = 10.2 \text{ m}^2/\text{year}$$

$$m_v = \frac{1.92 - 0.61}{19 - 0.61} \times \frac{1}{200 - 100} \times 1000 = 0.71 \text{ m}^2/\text{MN}$$

$$k = c_v m_v \gamma_w = \frac{10.2}{60 \times 60 \times 24 \times 365} \times \frac{0.71}{1000} \times 9.81 = 2.2 \times 10^{-9} \text{m/s}$$

Worked Example 6.4 c_v – log time method

With the same data in Worked Example 6.3 determine c_v and k using the log time curve fitting method.

From the initial part of the graph of log time versus specimen compression (Figure 6.25):
For times of 0.1 and 0.4 minutes the readings are 0.92 and 1.01 mm.
difference $D = 0.09$ ∴ $R_o = 0.92 - 0.09 = 0.81$ mm
For times of 0.16 and 0.64 minutes the readings are 0.93 and 1.12 mm
difference $D = 0.19$ ∴ $R_o = 0.93 - 0.19 = 0.74$ mm

Take the average: $R_0 = \dfrac{1}{2} (0.81 + 0.74) = 0.78$ mm

From Figure 6.25 $R_{100} = 1.79$ mm

$R_{50} = \dfrac{1}{2} (0.78 + 1.79) = 1.28$ mm ∴ $t_{50} = 1.17$ minutes

$H_o = 19 - 0.78 = 18.22$ mm
$H_f = 19 - 1.79 = 17.21$ mm

$H_{ave} = \dfrac{1}{2} (18.22 + 17.21) = 17.72$ mm ∴ $d = \dfrac{17.72}{2} = 8.86$ mm

From equation 6.45

$c_v = \dfrac{0.196 \times 8.86^2}{1.17} = 13.2$ mm²/minute

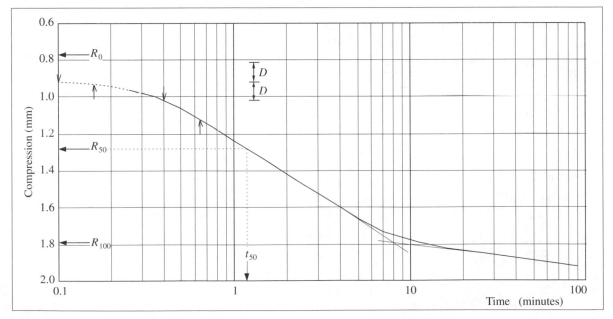

FIGURE 6.25   *Log time method*

$$= \frac{13.2 \times 60 \times 24 \times 365}{1000 \times 1000} = 6.9 \text{ m}^2/\text{year}$$

From before, $m_v = 0.71 \text{ m}^2/\text{MN}$

$$k = c_v m_v \gamma_w = \frac{6.9}{60 \times 60 \times 24 \times 365} \times \frac{0.71}{1000} \times 9.81 = 1.5 \times 10^{-9} \text{m/s}$$

Worked Example 6.5   Excess pore pressure

A wide sand fill area 3 m thick and with unit weight of 19 kN/m³ is spread over a layer of clay 8 m thick with a water table at its surface. The coefficient of consolidation of the clay is 8 m²/yr and its unit weight is 22 kN/m³. Determine the pore water pressure and effective stress at the middle of the clay layer 3 months after placing the fill. It is assumed that the sand fill is placed instantaneously and a permeable stratum underlies the clay.

Initial pore pressure before placing the fill $= 4 \times 9.8 = 39.2 \text{ kN/m}^2$
Initial effective stress $= 22 \times 4 - 9.8 \times 4 = 48.8 \text{ kN/m}^2$
Excess pore pressure at the middle of the clay $= 3 \times 19 = 57 \text{ kN/m}^2$

After 3 months $T_v = \dfrac{8 \times 3}{4 \times 12} = 0.125$

From Figure 6.13, U_v at the middle of the clay is 0.10
The excess pore pressure dissipated $= 57 \times 0.1 = 5.7 \text{ kN/m}^2$
The excess pore pressure remaining $- 57 \times 0.9 = 51.3 \text{ kN/m}^2$
The pore pressure in the middle of the clay at this time $= 39.2 + 51.3 = 90.5 \text{ kN/m}^2$
The total stress after placing the sand fill $= 4 \times 22 + 3 \times 19 = 145 \text{ kN/m}^2$
The effective stress after 3 months $= 145 - 90.5 = 54.5 \text{ kN/m}^2$
Note that this is the initial effective stress plus the amount of excess pore pressure dissipated.

Worked Example 6.6   Time required for consolidation settlement

An embankment for a highway, 5 m high above existing ground level is to be placed on a deposit of soft clay, 8 m thick with $m_v = 0.50 \text{ m}^2/\text{MN}$ and $c_v = 10 \text{ m}^2/\text{year}$. If the final road pavement on top of the layer can tolerate 50 mm settlement after it is constructed calculate how soon the pavement can be placed.

The bulk density of the embankment fill is 2200 kg/m³. Assume permeable boundaries top and bottom of the clay, the load is placed instantaneously and $\rho_{oed} = \rho_c$.
The amount of consolidation settlement is given by:
$$\rho_c = m_v \Delta\sigma H$$

For 5 m height of fill $\Delta\sigma = 5 \times 2200 \times \dfrac{9.81}{1000} = 107.9 \text{ kN/m}^2$

$$\therefore \ \rho_c = \frac{0.50}{1000} \times 107.9 \times 8 = 0.432 \text{ m}$$

If only 5 m of fill was placed, the top of the embankment would be $5.00 - 0.432 = 4.57$ m above existing ground level. Since it is necessary for the top of the embankment to be 5.0 m above existing ground level the amount of fill to be placed must be $5.0 + \rho_c$ m. Then:

$$\rho_c = \frac{0.50}{1000} \times (5 + \rho_c) \times 2200 \times \frac{9.81}{1000} \times 8$$

giving $\rho_c = 0.472$ mm
Check: for 5.472 m of fill:

$$\rho_c = \frac{0.50}{1000} \times 5.472 \times 2200 \times \frac{9.81}{1000} \times 8 = 0.472 \text{ m}$$

At the time required to place the pavement the degree of consolidation must be

$$U_v = \frac{472 - 50}{472} = 0.894$$

From equation 6.33, $T_v = 0.824$

$$t = \frac{T_v d^2}{c_v} = \frac{0.824 \times 4^2}{10} = 1.32 \text{ years or } 15.8 \text{ months}$$

Worked Example 6.7   Settlement achieved after a particular time

For the embankment described in Worked Example 6.6 determine the amount of settlement after 5 months.

From equation 6.25 the dimensionless time factor for this time is:

$$T_v = \frac{10 \times 5}{4^2 \times 12} = 0.260$$

From equation 6.32

$$U_v\% = 100 \times \sqrt{\frac{4}{\pi} T_v} = 100 \times \sqrt{\frac{4}{\pi} \times 0.260} = 57.5\% \quad (\bar{U}_v < 60\%)$$

The settlement after 5 months = $472 \times 0.575 = 272$ mm

Worked Example 6.8   Construction period

In the previous examples it is assumed that the load is placed instantaneously. This example assumes that it takes 3 months for construction of the embankment. From Terzaghi's empirical rule it is assumed that the load is placed instantaneously halfway through the construction period.

In Worked Example 6.6 the time when the pavement can be placed would be

$$t = \frac{3}{2} + 15.8 = 17.3 \text{ months from the start of construction}$$

In Worked Example 6.7, 272 mm of settlement would have occurred at

$$t = \frac{3}{2} + 5 = 6.5 \text{ months from the start of construction}$$

Worked Example 6.9   Precompression by surcharging

In Worked Examples 6.6 and 6.8 it is required to place the pavement 15 months from the start of construction. Determine the amount of surcharge required to achieve this.

The surcharge period, $t_{f+s} = 15 - \frac{3}{2} = 13.5$ months

$$T_v = \frac{10 \times 13.5}{4^2 \times 12} = 0.703$$

From Figure 6.13, in the middle of the clay layer the degree of consolidation $U_v = 0.775$
This must be achieved by the total load P_f and surcharge load P_s so from equation 6.48:

$$\rho_{f+s} = \frac{0.472}{0.775} = 0.609 \text{ m}$$

The applied pressure required to achieve this settlement is

$$\Delta \sigma = \frac{0.609 \times 1000}{0.5 \times 8} = 152.3 \text{ kN/m}^2$$

From Worked Example 6.6 the final stress applied by 5.472 m of fill is 118.1 kN/m² so the surcharge pressure is 152.3 − 118.1 = 34.2 kN/m²
Assuming the same fill is used the height of the surcharge would be

$$= 34.2 \times \frac{1000}{2200 \times 9.81} = 1.585 \text{ m}$$

Alternatively, the height of embankment and surcharge would be

$$= 152.3 \times \frac{1000}{2200 \times 9.81} = 7.057 \text{ m}$$

giving a height of surcharge of 7.057 − 5.472 = 1.585 m
Note that, practically, heights of fill should not be calculated to the nearest millimetre.

Worked Example 6.10   Vertical drains

An embankment is to be constructed up to 7 m above existing ground level on a clay deposit, 15 m thick, over a period of 1.5 months. The bulk unit weight of the embankment fill is 21.5 kN/m³ and the properties of the clay are:

$m_v = 0.25 \text{ m}^2/\text{MN}$
$c_v = 2.5 \text{ m}^2/\text{year}$
$c_H = 5.5 \text{ m}^2/\text{year}$

Determine the spacing required for band drains with a cross-section 112 mm × 6 mm installed in a square grid pattern to meet the requirement that only a further 50 mm settlement will occur 2.5 months after the end of construction.

Actual height of embankment fill = $7.0 + \rho_c$

$$\therefore \rho_c = \frac{0.25}{1000} = (7.0 + \rho_c) \times 21.5 \times 15$$

giving $\rho_c = 0.614$ m

The overall degree of consolidation required

$$U_c = \frac{614 - 50}{614} = 0.919$$

at an 'instantaneous' time of $\frac{1.5}{2} + 2.5 = 3.25$ months

assuming the load to be placed instantaneously halfway through the construction period.
For vertical (one-dimensional) consolidation, from equation 6.25

$$T_v = \frac{2.5 \times 3.25}{7.5^2 \times 12} = 0.012 \quad \text{giving } U_v = 0.124$$

For radial consolidation, from equation 6.53
$1 - 0.919 = (1 - 0.124)(1 - U_R)$ giving $U_R = 0.908$
The 'radius' of the band drain is obtained assuming equivalent perimeters:
$2\pi r_d = 2(6 + 112)$
$\therefore r_d = 37.6$ mm
The radius of influence of each drain $R = 0.564\ s$
From equation 6.56

$$n = \frac{R}{r_d} = \frac{0.564s}{0.0376} = 15.0s \quad \therefore \text{ spacing } s = \frac{n}{15.0}$$

The radial time factor T_R, from equation 6.55

$$T_R = \frac{5.5 \times 3.25}{4 \times (0.564s)^2 \times 12} = \frac{1.17}{s^2}$$

$$\therefore \text{ spacing } s = \sqrt{\frac{1.17}{T_R}}$$

Since U_R is a function of both T_R and n it is necessary to find for $U_R = 0.908$ when:

$$s = \frac{n}{15.0} = \sqrt{\frac{1.17}{T_R}} \quad \text{i.e. spacing in terms of } n = \text{spacing in terms of } T_R$$

Interpolating from Table 6.2, values of n and T_R are obtained when $U_R = 0.908$

n	T_R	$s = \dfrac{n}{15.0}$	$s = \sqrt{\dfrac{1.17}{T_R}}$
10	0.477	0.67	1.57
15	0.594	1.00	1.40
20	0.680	1.33	1.31 ←
25	0.746	1.67	1.25

These values show that a spacing of about 1.3 m would be required.

EXERCISES

6.1 The results of an oedometer test on a sample of fully saturated clay are given below. At the end of the test the moisture content was determined to be 27.3% and the particle density was 2.70 Mg/m^3. Determine values of the void ratio at the end of each pressure increment and plot them against the logarithm of pressure.

Pressure (kN/m^2)	Thickness (mm)
0	19.000
25	18.959
50	18.918
100	18.836
200	18.457
400	17.946
800	17.444
200	17.526
25	17.669
0	17.782

6.2 From the graph of void ratio versus log pressure in Exercise 6.1 estimate the preconsolidation pressure. If the sample of clay was taken from 5 m below ground level at a site where the water table exists at ground level determine the overconsolidation ratio of the clay at this depth. The saturated unit weight of the clay is 21.2 kN/m^3.

6.3 In Exercise 6.1 determine the coefficient of compressibility, m_v in m^2/MN for each pressure increment (loading only). From the graph of void ratio versus log pressure determine the coefficient of compressibility for the pressure increments p_o' + 50 and p_o' + 100 kN/m^2.

6.4 For the pressure increments on the normally consolidated limb of the pressure-void ratio curve in Exercise 6.1 determine the compression index, C_c.

6.5 The results of an oedometer test on a sample of fully saturated clay are given below for the pressure increment from 200 to 400 kN/m^2. Using the root time curve fitting method determine the coefficient of consolidation, c_v. Determine the coefficient of compressibility, m_v and derive the coefficient of permeability, k.

Time (minutes)	Thickness (mm)	Time (minutes)	Thickness (mm)
0	18.500	6	17.930
0.25	18.341	9	17.861
0.50	18.282	12	17.822
1	18.223	16	17.794
2	18.135	25	17.771
3	18.073	36	17.759
4	18.017		

6.6 A clay layer, 10 m thick, with sand beneath is to be loaded with a wide layer of fill, 2.5 m thick and with unit weight 20 kN/m^3. The coefficient of compressibility of the clay decreases with depth z in the form $m_v = 0.24 - 0.02z$. Determine the settlement due to consolidation of the clay.

6.7   A clay layer, 6 m thick with a water table at 2m below ground level is to be loaded over a wide area with a pressure of 50 kN/m^2. The coefficient of consolidation of the clay layer c_v is 12 m^2/year. Assuming the load to be placed instantaneously determine the excess pore water pressure after 6 months

  a)   at the middle of the clay layer, assuming permeable strata above and below.

  b)   at the bottom of the clay layer, assuming a permeable stratum at the top and an impermeable stratum beneath the clay.

  Assume the bulk unit weight of the clay is 21 kN/m^3.

6.8   A 6 m layer of sand overlies a 5 m thick layer of clay with a sand deposit beneath. The water table is to be lowered permanently from 1 m below ground level to 4 m below ground level by pumping from the sand over a period of 6 weeks. Determine the settlement due to consolidation of the clay layer 6 months from the start of pumping. For the clay,

  $m_v = 0.45$ m^2/MN and $c_v = 5.5$ m^2/year.

6.9   In Exercise 6.8 determine the time from the start of pumping to achieve 90% of the final consolidation settlement.

6.10 A motorway embankment is to be constructed over a layer of soft compressible clay, 10 m thick. At the time required to place the road pavement on top of the embankment it has been calculated that the average degree of consolidation is only 30% assuming 1- dimensional vertical consolidation only.

  To reduce the settlements that would occur after pavement construction it is proposed to install sand drains through the soft clay. Determine the overall degree of consolidation which will be achieved when the pavement is constructed if sand drains, 200 mm diameter, are installed at 3.5 m centres on a square grid pattern.

  Assume that sand exists above and below the clay layer and the coefficients of horizontal and vertical consolidation are in the ratio $c_H = 3.5\ c_v$.

6.11 An embankment is to be constructed up to 6 m above a layer of clay, 13 m thick, over a period of 1 month. The properties of the clay are :

  $m_v = 0.30$ m^2/MN

  $c_v = 2.5$ m^2/year

  $c_H = 7.0$ m^2/year

  The bulk unit weight of the embankment fill is 20.5 kN/m^3. It is required to construct the road pavement on top of the embankment 2 months after the end of construction but the pavement can only tolerate a settlement of 50 mm. Using vertical band drains with a rectangular cross-section of 8 mm × 70 mm installed in a triangular grid pattern determine the required spacing of the drains to meet the settlement criterion.

  Assume the settlement within the embankment itself is minimal.

Shear strength

Introduction (Figure 7.1)

In soils failure occurs as a result of mobilising the maximum shear stress the soil can sustain, therefore an understanding of shear strength is fundamental to the behaviour of a soil mass. The shear strength of the soil, allied with a particular method of analysis, will determine the maximum or ultimate (failure) load that can be applied on a foundation resting on soil or the ultimate force required to cause failure of a soil mass forming a slope.

These values permit the determination of the overall factor of safety, which is the ratio of the ultimate loads or forces available to the load or force applied. The overall factor of safety is a measure of the intensity of loading or the degree of mobilisation of the ultimate load. The designer must be satisfied

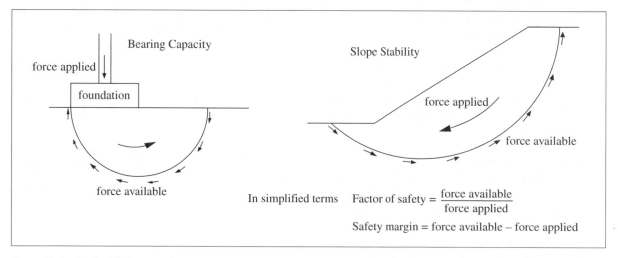

FIGURE 7.1 *Typical failure modes*

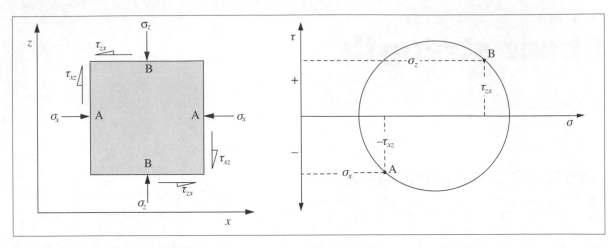

Figure 7.2 *Stresses in two dimensions*

that there is an adequate value of this ratio to ensure that the earth structure will remain stable. Assessment of the appropriate soil shear strength value is, therefore, of prime importance for an ultimate limit state design approach.

Stresses and strains in soils

Representation of stresses (Figure 7.2)

Although stresses in soils occur in three dimensions (x, y, z) for an initial understanding the state of stress in a two-dimensional plane (x, z) is considered. The normal and shear stresses acting on the sides A and B of a plane element as shown in Figure 7.2 can be represented as points A and B on a $\tau - \sigma$ plot assuming compressive stresses and anti-clockwise shear stresses are positive. The stress states can be analysed by using the Mohr circle of stresses which passes through points A and B with the distance AB being the diameter of the circle.

Pole (Figures 7.3 and 7.4)

In the x–z plane the line from the point A or B plotted to the pole P is parallel to the plane on which the stresses act. Thus the pole lies vertically above the point A, and similarly the pole lies horizontally in line with B as shown in Figure 7.3. AP defines the angle of the plane on which the stresses σ_x and τ_{xz}

act which is parallel with the z axis (plane A) in Figure 7.2 and similarly BP is parallel with the x axis.

Reversing the definition given above to obtain the stresses acting on any other plane at an angle α to the x axis (Figure 7.4) draw a line from the pole to a point on the circle at the angle α from the horizontal (or x axis) to give the stresses acting on that plane, σ_α, τ_α. Similarly, for the plane at the angle β to the vertical or z axis (numerically the same as α in this case) the stresses acting on the plane are σ_β and τ_β.

Principal stresses (Figure 7.5)

A principal stress acts on a principal plane defined as a plane on which the shear stress is zero. By plot-

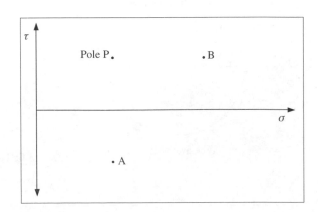

Figure 7.3 *Location of the pole*

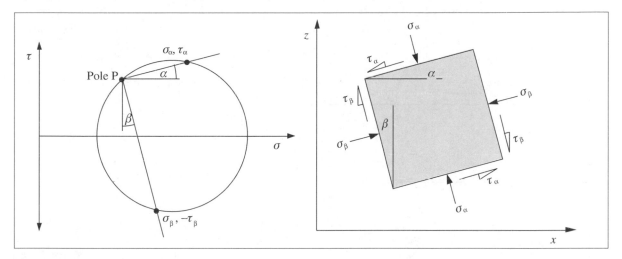

FIGURE 7.4 *Stresses on any plane*

ting a Mohr circle passing through A and B the points C and D can be found where the shear stresses are zero. The principal stresses are then σ_1 (major principal stress) and σ_3 (minor principal stress). Lines drawn from the pole P to the points C and D give the angles of the planes on which the principal stresses act. Thus the line PD is parallel to the plane on which σ_1 (and $\tau = 0$) acts and similarly the line PC is parallel to the plane on which σ_3 (and $\tau = 0$) acts.

The above representations are for total stresses, for a consideration of effective stresses see Figure 4.10.

Axial symmetry (Figure 7.6)

In the ground in the three-dimensional state there are three principal stresses σ_1, σ_2 and σ_3 on three principal planes where σ_2 is called the intermediate principal stress with a value between the minor and major stresses. The condition of axial symmetry eliminates the need to consider this intermediate value since the stresses σ_2 and σ_3 are radial and equal. The condition of axial symmetry is employed in soil testing such as in the triaxial test so that the test can be analysed in two dimensions only using σ_1 and σ_3.

The condition is also relevant to field situations such as beneath the centre of a circular foundation and in circular excavations. In this situation the strains will occur in three dimensions but in one plane, usually the horizontal plane, the strains will

be equal so again a two-dimensional representation of strain can be adopted.

Plane strain (Figure 7.6)

This condition occurs when in one plane or on one of the orthogonal axes the strain is zero, e.g. $\varepsilon_y = 0$. This is the condition found beneath a long slope or behind a long wall. The strains can be analysed in two dimensions. It is to be noted that the intermediate principal stress σ_2 is not zero.

K_o condition (Figure 7.6)

This occurs when the strains on two axes are constrained to zero, e.g. $\varepsilon_y = \varepsilon_x = 0$. This is a condition found in the laboratory oedometer test where the soil specimen is surrounded by a very stiff confining ring. The condition also exists in the ground in its natural undisturbed state and behind a wall that is not allowed to deflect inwards or outwards so it is often referred to as the 'at rest' condition and in this state the ground is obviously not at failure. To take the soil to a state of failure a number of stress paths can be followed as will be illustrated later.

Normal and shear strains (Figure 7.7)

Consider an element of soil undergoing normal and shear strains without displacement and rotation so that the diagonals AC and AC' are coincident. The normal strains ε_z and ε_x are obtained from the changes in length of the x and z dimensions of the

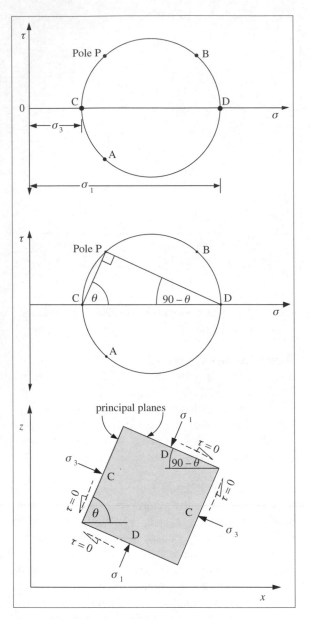

Figure 7.5 *Principal stresses*

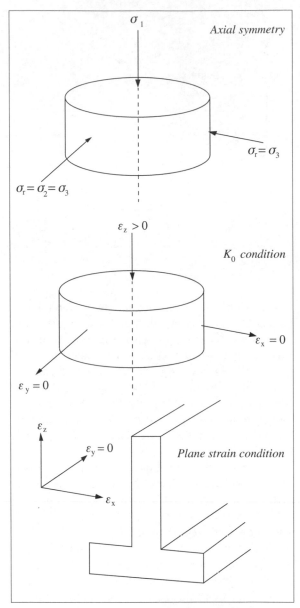

Figure 7.6 *Axial symmetry, plane strain and K_0 condition*

element and the pure shear strains ε_{zx} and ε_{xz} are produced by rotation of the sides of the element. Engineer's shear strain γ is given by the change in the angle between the two sides of the element so since $\varepsilon_{zx} = \varepsilon_{xz}$

$$\gamma_{zx} = 2\varepsilon_{zx}$$

or $\quad \varepsilon_{zx} = \dfrac{1}{2}\gamma_{zx}$ $\hfill (7.1)$

For a uniaxial compression or extension test where the radial stress is held constant the normal axial strain ε_z is related to the axial stress σ_z by

$$\varepsilon_z = E' \sigma_z \tag{7.2}$$

where E' is the Young's modulus.

For the case of axial symmetry it can be shown that the shear strain ε_s is related to the deviator stress q' by

$$\varepsilon_s = \frac{2(1+v')\,q'}{3E} \tag{7.3}$$

or $\varepsilon_s = 1/3\,G'\,q'$ (7.4)

where G' is the shear modulus.

Mohr circle of strain (Figures 7.8, 7.9 and 7.10)

The Mohr circle of strain is constructed using the axes of normal strain ε and pure shear strain $1/2\gamma$. Consider a plane element of soil as shown in Figure 7.8 with compressive normal strains positive and tensile strains and clockwise shear strains are negative. The point A for plane A of the element is plotted as ε_x, $1/2\,\gamma_{xz}$ and point B is given by ε_z, $-1/2\gamma_{zx}$. Note that the strains do not occur on the planes (as the stresses do) but that they occur between the planes.

The pole P is located in the same way as for the circle of stress, i.e. vertically below A and horizontally in line with B. The principal normal strains ε_1 and ε_3 will then occur between principal planes on which the shear strains are zero (Figure 7.9). The angles of the principal planes θ and $90 - \theta$ both to the horizontal are obtained by drawing a line from P to D and C, respectively, to give the angles of the planes between which the major (ε_1) and minor (ε_3) principal normal strains act.

Strains acting between other planes such as at angles α to the horizontal and β to the vertical as shown in Figure 7.10 can be obtained by drawing lines from the pole P at these angles to cut the Mohr circle at the points ε_β, $1/2\gamma_\beta$ and ε_α, $-1/2\gamma_\alpha$.

Volumetric strains

If it is assumed that the mineral grains and the pore water are incompressible then for a fully saturated soil volumetric strains (changes of volume) can only occur by squeezing water out of or drawing water into the void spaces. For the case of axial symmetry it can be shown that the volumetric strain ε_v is

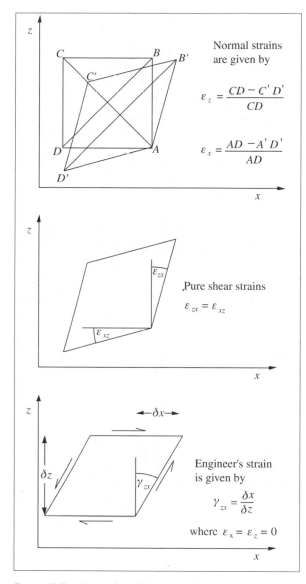

Normal strains are given by

$$\varepsilon_z = \frac{CD - C'D'}{CD}$$

$$\varepsilon_x = \frac{AD - A'D'}{AD}$$

Pure shear strains

$$\varepsilon_{zx} = \varepsilon_{xz}$$

Engineer's strain is given by

$$\gamma_{zx} = \frac{\delta x}{\delta z}$$

where $\varepsilon_x = \varepsilon_z = 0$

FIGURE 7.7 *Normal and shear strains*

related to the mean effective stress p' by

$$\varepsilon_v = \frac{3(1-2v')\,p'}{E} \tag{7.5}$$

or $\varepsilon_v = \frac{1}{K'}q'$ (7.6)

where E' is the Young's modulus, v' is the Poisson's ratio given by:

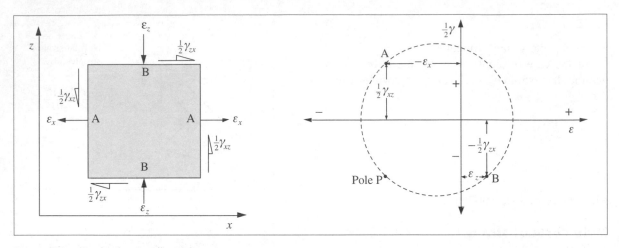

FIGURE 7.8 *Strains in two dimensions*

$$v' = \frac{\varepsilon_{\text{horiz}}}{\varepsilon_{\text{vert}}} \qquad (7.7)$$

and K' is the bulk modulus. When soil is strained in an undrained manner so that water cannot move into or out of the voids and volumetric strains are zero it can be seen from equation 7.5 that v_u (v for undrained conditions) must be 0.5.

Shear strength

Effect of strain (Figure 7.11)
Shear stresses in a soil mass are only produced when shear strains can occur so it is said that placing a foundation load on a soil *mobilises* the available shear strength (or a part of it). Similarly, for a slope, gravity forces mobilise the shear strength available within the slope, although it is usually assumed (for analytical simplicity) that shear failure occurs along a simple, single slip surface.

Strictly speaking, it is the shear strain γ which produces shear stress τ (or vice versa) but this is not easy to determine either *in situ* or in laboratory tests so direct strains, or in some cases, just displacement δL, have to be used.

A typical stress-strain curve is given in Figure 7.11. As shear stress is applied the soil structure dis-

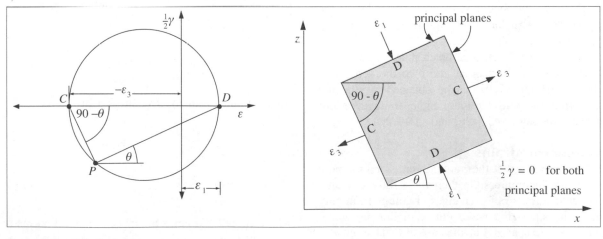

FIGURE 7.9 *Principal strains*

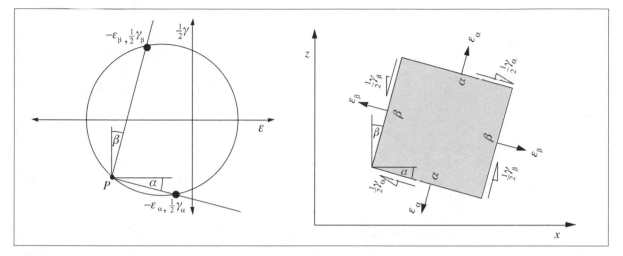

FIGURE 7.10 *Strains on any plane*

torts. Initially, this distortion is proportional to the stress applied and if the stress is removed the distortions are recovered. These distortions are probably associated with small rotations at the numerous particle contacts and some elastic compression of the particles themselves. The soil in this region is said to behave in an elastic manner and only elastic strains occur.

At a certain stress level, depending on the soil type, the soil structure will deform in a plastic manner by rearrangement of the particle locations and strains from this point (yield) will comprise both elastic and plastic components. However, the plastic strains will not be recovered on removal of the stress since the soil particles have moved into a new arrangement.

As the soil is strained or 'worked' further additional shear stress can be sustained due to a process described as work-hardening.

In dense sands and stiff overconsolidated clays this would be due to expansion of the mineral grain structure (dilatancy) as more stress is required to achieve further strain. After reaching a peak shear stress these soils typically display work-softening since strains beyond this value are being applied to a soil structure which has been weakened. This phenomenon would also apply to a soil that has developed cementation or chemical bonding since deposition. A reduction of strength beyond the peak

value is referred to as brittleness and some materials such as sensitive clays and 'collapsing' soils will suddenly crush or collapse due to breakdown of interparticle bonds.

For loose sands and soft normally consolidated clays work-hardening would commence at much lower stress and strain levels since their relatively open mineral grain structures will be contracting as shearing occurs making the structure progressively more able to support more stress. These materials do not tend to display a peak value followed by work-softening or brittleness.

Figure 7.11 shows that choosing the point at which the soil has 'failed' requires a definition. This could be:

1. *Yield*
    Although not the maximum shear stress available, if the soil is stressed any further beyond the point Y in Figure 7.11 the strains and movements of the earth structure (foundation, slope, etc.) could be so large and irrecoverable that they may be deemed to have failed as a serviceability limit state. τ_y represents a yield stress.

2. *Peak shear strength*
    This is the maximum shear stress which can be sustained. It may be dangerous to rely on this value for some brittle soils due to the rapid loss of strength that occurs when the soil is strained beyond this point.

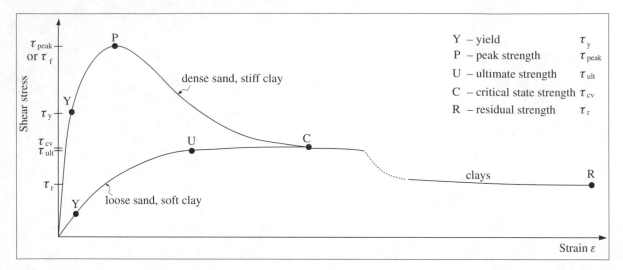

FIGURE 7.11 *Definition of failure*

3. *Ultimate strength*

   For loose sands and soft clays work-hardening may continue to increase the shear stress that can be sustained even at very large strains so a maximum stress is not achieved. A maximum strain limit must then be imposed, usually related to the performance of the earth structure, say 10 to 20% strain, point U in Figure 7.11.

4. *Critical state strength*

   This is sometimes referred to as the ultimate strength. After a considerable amount of shear strain a soil will achieve a constant volume state (by the soil structure expanding or contracting) and it will continue to shear at this constant volume without change in volume or void ratio. These shear strains must be uniform throughout the soil and not localised. It is sometimes referred to as the constant volume strength (ϕ_{cv}).

5. *Residual strength*

   This is also sometimes referred to as ultimate strength. After a considerable amount of strain on a single slip zone or surface (point R in Figure 7.11) the particles on each side of this surface will rearrange to produce a more parallel orientation and this will produce the lowest possible or residual strength. This strength is important in the re-activation of old landslides and is obviously more significant for platy minerals such as clays.

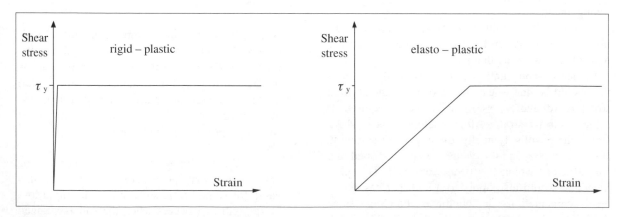

FIGURE 7.12 *Idealised stress–strain relationships*

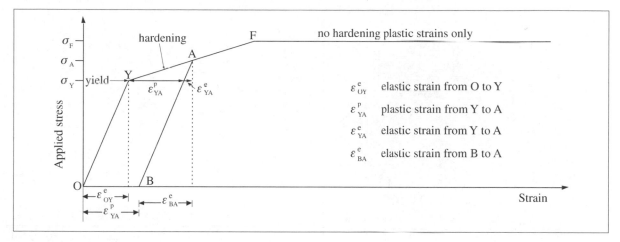

FIGURE 7.13 *Elasto-plastic soil behaviour*

Friction between the surfaces of particles where they are in contact with each other is related to the effective stress and is represented by the angle of interparticle friction, ϕ_μ. This friction forms the basis for the above observed shear strengths the differences being produced by rolling friction and dilatancy (Rowe, 1962).

Idealised stress–strain relationships
(Figure 7.12)

These are adopted to assist the methods of analysis used in soil mechanics. The most common of these is the rigid-plastic model although the elasto-plastic form is a more realistic relationship. Compare these with Figure 7.11.

Yield and plasticity (Figures 7.13, 7.14 and 7.15)

As stresses are applied to a soil only elastic strains develop until a yield stress at the point Y in Figure 7.13 is achieved. As the stress then increases both elastic and plastic strains will occur simultaneously as the soil undergoes strain or work-hardening. If at A the stress is removed the strains recovered are wholly elastic since the plastic strains are irrecoverable.

Increasing the stress from B will recover the elastic strains but because of the previous strain-hardening a higher yield stress is achievable at A. As long as the soil can undergo strain-hardening there will be an infinite number of yield points between Y

and F. The relationship between the stresses after the first yield and the plastic strains is known as a hardening law. At a final stage, F, referred to as failure, the soil can sustain no more stress and will continue to strain but only plastically.

In the above it is assumed that only one stress σ_z' is applied, say axially, with no confining stress. Where there is a combination of stresses σ_z' and σ_x' in two planes with associated strains ε_z^P and ε_x^P the first yield curve can be represented by the curve in Figure 7.14 with an infinite number of subsequent yield curves producing a yield surface until the

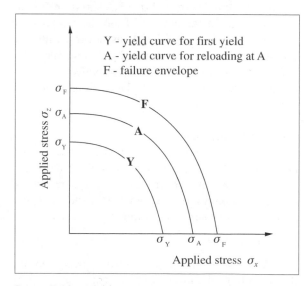

FIGURE 7.14 *Yield curves*

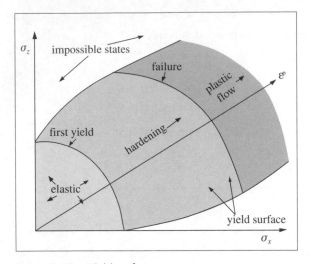

FIGURE 7.15   *Yield surfaces*

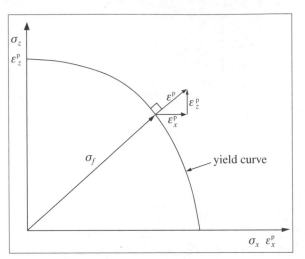

FIGURE 7.16   *Flow rule and normality*

failure envelope is achieved. Figure 7.15 shows a complete plot of this surface with respect to plastic strains. The yield surface represents a boundary for soil states since the state cannot be outside the surface. For a state inside the surface the behaviour will be elastic until it reaches the surface when simultaneous elastic and plastic strains will occur.

Flow rule and normality   (Figure 7.16)

At failure when the soil is behaving in a perfectly plastic manner and the failure envelope for the combination of stresses σ_z and σ_x is as shown in Figure 7.16 then the vector obtained from the components of plastic strain ε_z^p and ε_x^p, associated with the vector of stress σ_f' are related by a flow rule. The normality condition applies to a perfectly plastic material where the vector of plastic strain is normal to the failure envelope.

Failure criterion   (Figure 7.17)

The Mohr–Coulomb relationship is the most appropriate strength criterion adopted in soil mechanics. It simply relates the shear stress at failure (shear strength, τ_f) on a failure plane or slip surface to the normal effective stress σ_N' acting on that plane:

$$\tau_f = c' + \sigma_N' \tan\phi' \qquad (7.8)$$

A different relationship is obtained depending on the definition of failure adopted, e.g. peak, critical state or residual, as shown in Figure 7.17, and the

drainage conditions applicable, see Effects of drainage below.

Failure of soil in the ground   (Figure 7.18)

Failure will be produced by changing the natural *in situ* stress state which would lie below the failure criterion to a state coinciding with the criterion. This will occur when the *in situ* Mohr circle is enlarged or moved (see stress paths below) and becomes tangential with the failure criterion.

If σ_H and σ_V are the horizontal and vertical principal stresses then in the two-dimensional representation of the stress state, the pole occurs at the point P where a vertical plane (on which σ_H acts) through the point σ_H, $\tau = 0$ cuts a horizontal plane (on which σ_V acts) through the point σ_V, $\tau = 0$.

The most critical condition in the element of soil will occur when the combination of σ and τ coincides with the failure envelope as given by the Mohr–Coulomb relationship. This condition occurs at points A and B in Figure 7.18. By drawing lines to A and B from the pole P the angles θ and β of the planes on which the critical conditions occur are obtained and these are the planes along which failure will occur. The normal stress σ_a on the failure plane at the angle θ to the horizontal is:

$$\sigma_a = \frac{1}{2}\left(\sigma_V' + \sigma_H'\right) + \frac{1}{2}\left(\sigma_V' - \sigma_H'\right)\cos 2\theta \qquad (7.9)$$

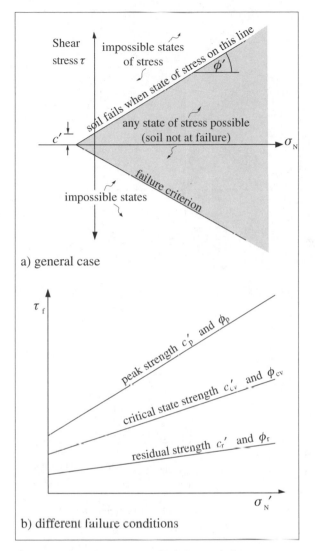

a) general case

b) different failure conditions

FIGURE 7.17 *Mohr–Coulomb failure condition*

and the shear stress on this failure plane is:

$$\tau_a = \frac{1}{2}\left(\sigma'_V - \sigma'_H\right)\sin 2\theta \qquad (7.10)$$

The normal stress σ_b and the shear stress $-\tau_b$ acting on a failure plane at the angle β to the horizontal are obtained in a similar manner. In Figure 7.18:

$$2\theta = 90 + \phi' \text{ so } \theta \text{ (and } \beta) = 45 + \phi/2 \qquad (7.11)$$

For a plastic soil that does not achieve a peak strength such as a soft clay or a loose sand and only

hardening occurs during straining after the first yield these planes will occur throughout the soil undergoing shear so a uniform shear mode will occur. This is because throughout the soil there will be non-uniform stresses and strains and those zones which are strained most will be hardening so that the less strained zones will be strained more.

For a more brittle soil that does achieve a peak strength and then undergoes softening on further straining a single shear plane or a narrow shear zone would tend to form along one surface. This is because the more stressed zones are becoming weaker after reaching their peak strength and further straining will be concentrated around these softer zones.

Stress paths (Figures 7.19 and 7.20)

The behaviour of a soil will depend on the initial (*in situ*) and final (failure) states as described above but will also depend on the route taken between these states. These routes are referred to as stress paths and may be plotted as either effective stresses or total stresses. For plane strain conditions they are plotted as t' and s' or t and s and for axial symmetry conditions they are plotted as q' and p' or q and p (Figure 7.19). Plane strain conditions and axial symmetry are described above.

t' and s' refer to the apex of a Mohr circle of stress while q' and p' are the deviator stress and mean stress, respectively. In both cases, the total and effective stress paths are separated horizontally by an amount equal to u_w, the pore water pressure since:

$$t' = t \text{ and } s' = s - u_w \qquad (7.12)$$

and

$$q' = q \text{ and } p' = p - u_w \qquad (7.13)$$

SEE WORKED EXAMPLES 7.5 AND 7.6

A soil element will take a different stress path to failure from the *in situ* equilibrium state depending on the type of loading. The *in situ* state is represented by the K_o condition described above. From this point the soil can reach a state of failure following a path to the failure envelope by a combination of compression or extension and loading or unloading. Some examples are given in Figure 7.20. Compare this figure with Figure 7.18.

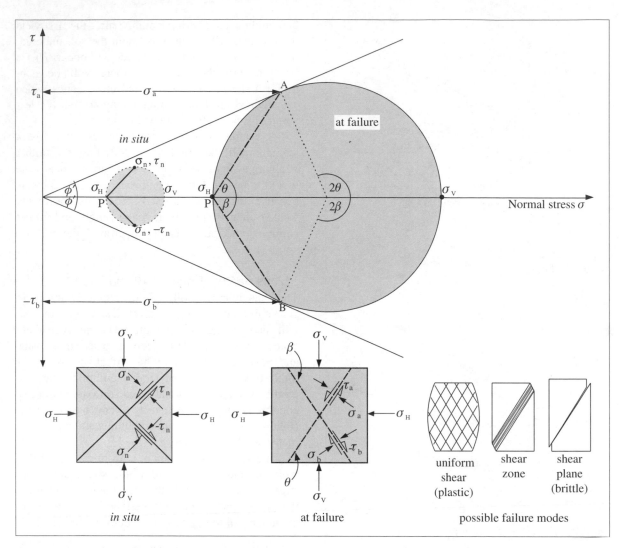

FIGURE 7.18 *Failure of soil in the ground*

Effects of drainage (Figures 7.21 and 7.22)

It has been shown in Chapter 4 that when stresses (loading or unloading) are applied to the ground the immediate response of the soil is for all of this stress to be supported by the pore water with a consequent change in pore water pressure Δu_w (an excess above or below the static pore water pressure already existing in the pores). The terms drained or undrained are used in soil mechanics to denote whether dissipation of this pore water pressure change can occur or not. If dissipation cannot occur then undrained conditions will apply.

Dissipation means the return of the altered pore water pressure to its original static value and can be a decrease if pore pressures were raised above the static value (e.g. beneath a foundation) or an increase if they were lowered below the static value, e.g. beneath a basement or cutting. The rate of dissipation will largely depend on the permeability of the soil, the proximity of permeable boundaries for water to be forced towards or obtained from and the time allowed for dissipation in relation to the rate of loading or unloading.

This aspect of 'drainage' refers to the degree of

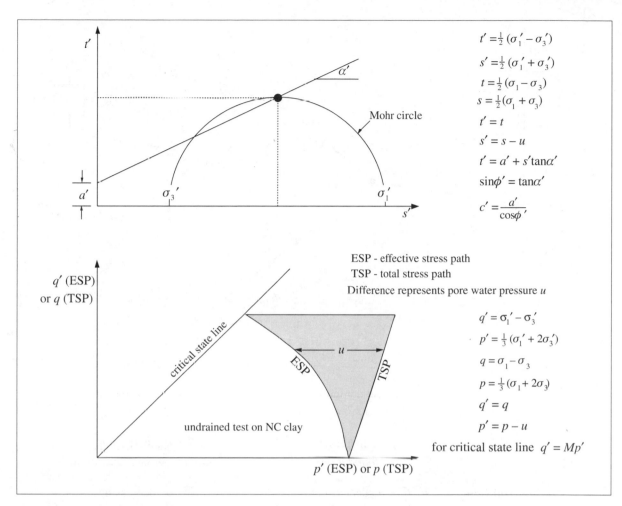

The equations shown in the figure:

$$t' = \tfrac{1}{2}(\sigma_1' - \sigma_3')$$
$$s' = \tfrac{1}{2}(\sigma_1' + \sigma_3')$$
$$t = \tfrac{1}{2}(\sigma_1 - \sigma_3)$$
$$s = \tfrac{1}{2}(\sigma_1 + \sigma_3)$$
$$t' = t$$
$$s' = s - u$$
$$t' = a' + s'\tan\alpha'$$
$$\sin\phi' = \tan\alpha'$$
$$c' = \frac{a'}{\cos\phi'}$$

ESP - effective stress path
TSP - total stress path
Difference represents pore water pressure u

$$q' = \sigma_1' - \sigma_3'$$
$$p' = \tfrac{1}{3}(\sigma_1' + 2\sigma_3')$$
$$q = \sigma_1 - \sigma_3$$
$$p = \tfrac{1}{3}(\sigma_1 + 2\sigma_3)$$
$$q' = q$$
$$p' = p - u$$

for critical state line $q' = Mp'$

FIGURE 7.19 *Stress paths*

dissipation of the excess pore water pressures and is different from the flow of free water under gravity towards a drain or sump. The terms used are:

- *Undrained*
  Dissipation of the excess pore water pressure u_w is prevented. This condition is produced when a soil of low permeability such as a clay is loaded quickly.
- *Drained* (fully)
  Dissipation of any excess pore water pressure is permitted fully at all times so that effectively there is no measurable excess pore water pressure, i.e. $\Delta u_w = 0$. This condition is produced when a soil of high permeability such as sand is loaded slowly.

- *Drained* (partially)
  An excess pore water pressure develops to a certain extent due to loading, but not fully since dissipation is proceeding at the same time and reducing this pore water pressure. This is probably the situation in many engineering applications but the assumptions of fully drained or fully undrained conditions are adopted for simplicity in applying test results and analytical procedures.

Engineering works change the total stresses, $\Delta\sigma_1$ and $\Delta\sigma_3$ or $\Delta\sigma_V$ and $\Delta\sigma_H$, in the ground in various ways, see Figure 7.20. The excess pore water pressure Δu_w, produced by these changes, can be considered as comprising two components, a consolidation or mean stress change occurring before or during

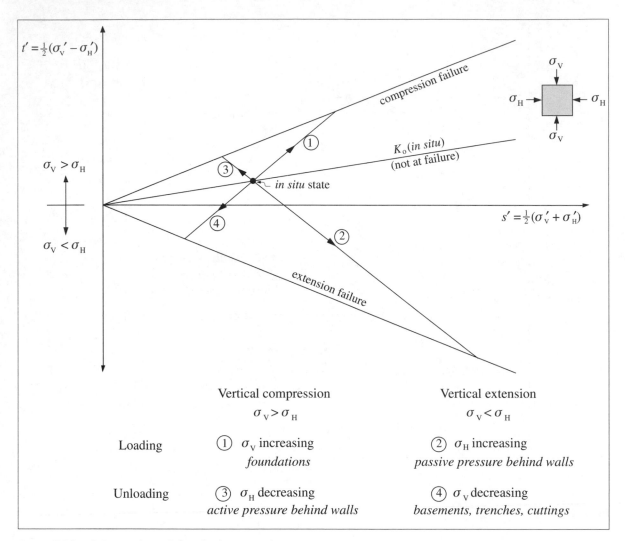

FIGURE 7.20 *Stress paths to failure in the ground*

shear and a deviator stress change during shear. For a fully saturated soil ($B = 1$) the pore pressure parameter expression (equation 4.12) can be rearranged to:

$$\Delta u_w = 1/3 \ (\Delta\sigma_1 + 2\Delta\sigma_3) + (A - 1/3) \ (\Delta\sigma_1 - \Delta\sigma_3)$$

due to mean stress   due to deviator stress
(consolidation stage)   (shear stage)

$$(7.14)$$

and these stages may occur separately or concurrently.

The various engineering works will stress the ground in different ways and three common test procedures have been devised to model or represent these applications:

Test procedure	Consolidation stage	Shear stage
UU	unconsolidated	undrained
CD	consolidated	drained
CU	consolidated	undrained

Some examples of the drainage behaviour and the appropriate types of test for sands and clays are given in Figures 7.21 and 7.22. It is obvious that unconsolidated and undrained mean no volume change or no moisture content change while consolidated and drained involve volume changes

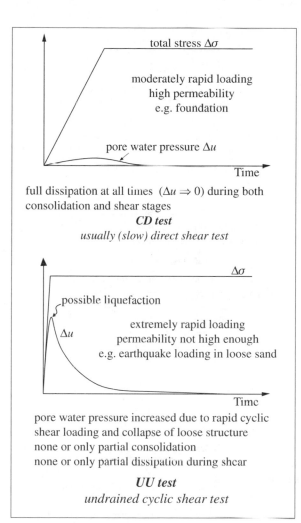

FIGURE 7.21 *Effect of drainage conditions – sand*

during both the consolidation and shear stages, respectively.

Where the permeability of the soil is low the consolidated drained condition (which is the most critical for unloading situations) will require a long time to achieve, several decades in the case of long-term instability of London Clay cutting slopes (Skempton, 1964). However, where clays contain macro-fabric such as fissures, silt partings, etc. making the mass permeability much higher the critical consolidated drained case can be obtained very soon after unloading, within hours in the case of trench or trial pit sides. This risk is often unappreciated and is one of

the largest causes of fatalities in excavations on construction sites. This can only be avoided by the immediate insertion of adequate temporary supports.

Test procedures (Figure 7.23)

A laboratory test on a soil sample is intended to represent or model the conditions that the engineering works will impose on the soil mass. The major determinants of shear strength controlled by the laboratory apparatus are:

1. *The mode of drainage*

   Is a consolidation stage provided and are the pore water pressures allowed to dissipate during shear or not? This aspect of the test procedure attempts to model the effects of drainage as described above.

2. *The need for strain control*

   There are three ways in which the strain is controlled by the engineering works: axial symmetry, plane strain and direct shear conditions. These are illustrated in Figure 7.23. The triaxial system applies to the condition of axial symmetry where $\sigma_2 = \sigma_3$ and $\varepsilon_2 = \varepsilon_3$ and would apply beneath the centre of square or circular foundations. The plane strain condition relates to two-dimensional shearing where strain in the intermediate direction is prevented, i.e. $\varepsilon_2 = 0$ and $\sigma_2 \neq \sigma_3$. This condition would apply beneath a long foundation or long slope. The direct shear condition relates to shearing on a slip plane or a narrow shear zone where the soil is strained in a fairly pure shear manner.

3. *The means of applying stress changes*

   There are also three ways in which the stresses are changed: by compression, extension and direct shear. With the first two the shear stress is applied indirectly by changes in the principal stresses (σ_V and σ_H or σ_1 and σ_3) while the latter occurs where the shear stress is applied directly. Where the engineering works increases the vertical stress above the horizontal stress the stress change is referred to as compression and extension occurs where the horizontal stress exceeds the vertical stress change.

These strain and stress conditions are important since it has been found that the strength of a soil differs according to the strain control and stress

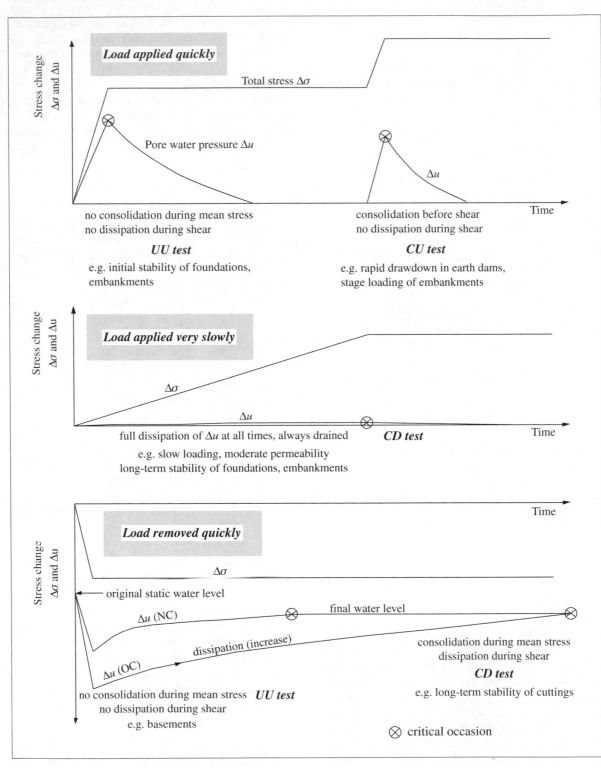

FIGURE 7.22   *Effect of drainage conditions – clay*

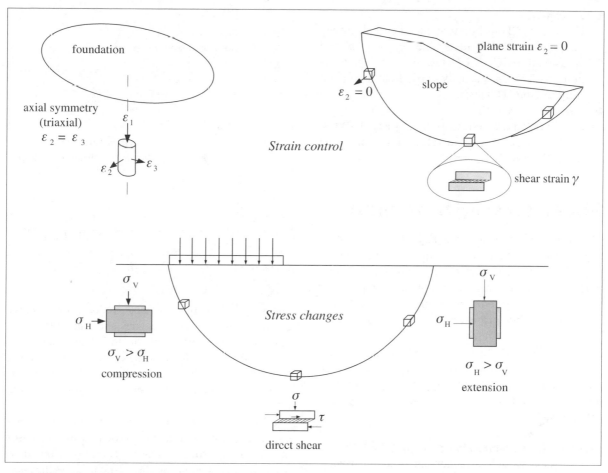

FIGURE 7.23   *Strain control and stress changes*

change and test procedures have been designated accordingly:

- DS – direct shear
- UC – uniaxial compression
- TC – triaxial compression
- TE – triaxial extension
- PSC – plane strain compression
- PSE – plane strain extension.

Plane strain tests are not easy to perform and remain in the laboratories of research institutes. Research has been carried out to determine the different strengths obtained by these procedures and fortunately, these differences are not particularly large with axial symmetry tests giving more conservative values than plane strain tests.

In most commercial soils laboratories the strength tests available are:

- *Shear box test*
  Consolidated drained (CD) direct shear (DS) on reconstituted samples of sand to determine the peak or ultimate angle of internal friction
- *Unconfined compression test*
  Unconsolidated undrained (UU) uniaxial compression (UC) test on reconstituted, remoulded or undisturbed samples of clay to determine c_u.
- *Vane test*
  Unconsolidated undrained (UU) direct shear (DS) test in boreholes in the ground or on remoulded or undisturbed samples of clay in the laboratory to determine c_u.

- *Triaxial test*
  UU, CU or CD test on remoulded or undisturbed samples of clay to determine c_u, c' or ϕ' and occasionally CU or CD tests on sand to determine ϕ'. They are usually carried out as triaxial compression (TC) tests.
- *Ring shear test*
  Consolidated drained (CD) direct shear (DS) test on remoulded samples of clay to determine the residual strength, c'_r and ϕ'_r.

Shear strength of sand

The following discusses the behaviour of sand as found from reconstituted specimens in the laboratory. The virtually insurmountable problems of stress relief and fabric changes following sampling of sands from the ground have made the laboratory study of the *in situ* properties of sand very difficult.

Emphasis has, therefore, been placed on the use of *in situ* testing techniques to assess the state of the sand. These can only give an indirect measure of the relevant parameters although correlations have been developed empirically, for example between ϕ' and SPT 'N' value.

Stress–strain behaviour (Figure 7.24)
When a sand particle arrangement is confined laterally strains can only occur in the vertical direction, such as in an oedometer test. As the vertical stress is increased small groups or arrays of particles in a loose state will collapse into the surrounding voids producing volumetric (or vertical) strains. This provides a more tightly packed arrangement with a larger number of particle contacts and each particle becomes more fixed in place because it is given less freedom of movement, a phenomenon known as 'locking'. Thus the vertical stress can increase with less increase of strain and the stress – strain curve is concave upwards (Figure 7.24).

The term stiffness represents the gradient of this curve so it can be seen that the soil is becoming stiffer. As the vertical stress is increased the contact stresses between particles increase and the particles will begin to fracture and crush producing yield and allowing vertical strains to increase. An increase in the numbers of particles due to crushing then pro-

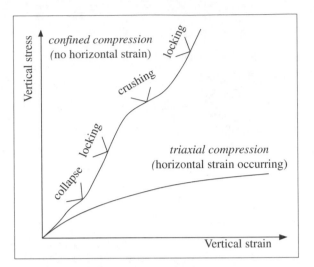

FIGURE 7.24   *Stress – strain behaviour of sand*

duces further particle contacts, reduces the average contact stress between particles and causes the stiffness to continue increasing due to further 'locking'.

The stress levels at which structural collapse, locking and yielding occur will depend on the initial overall density and the inherent strength or crushability of the particles.

Under confined compression where horizontal strains are restricted, as the vertical stress increases the horizontal stress increases. The horizontal stress need not be as large as the vertical stress because part of the latter will be supported by the shearing resistance of the sand. The ratio of horizontal stress to vertical stress is given by K_o, the 'at rest' or no lateral strain condition, see above (equation 4.4). An established relationship (Jaky, 1944):

$$K_o = 1 - \sin\phi \qquad (7.15)$$

(see Table 4.1) shows that as the angle of internal friction ϕ increases the horizontal stress decreases.

With triaxial compression as the vertical stress increases the horizontal stress does not increase. It is usually kept constant, allowing some of the sand particles to move horizontally which in turn allows the particles above to move downwards under the vertical stress so producing larger vertical strains. When the sand is initially loose large vertical strains can occur for a given horizontal strain since there are void spaces for the grains to move into and there will be a net volume reduction or contraction.

In a dense sand where the particles are in close contact with each other a vertical strain will be accompanied by a large horizontal strain as the particles are forced outwards under the low confinement. This results in a net volume increase, referred to as dilatancy. A simple test to demonstrate the phenomenon of dilatancy is described in Chapter 2.

If the pores are full of water and dissipation is prevented (undrained condition) this volume increase cannot take place and this causes a reduction in the pore water pressure which will increase effective stresses and, in turn, increase the shear strength.

Shear box test (Figure 7.25)

When a direct measure of the shear strength of a granular soil is desired a shear box test can be carried out. It must be remembered, however, that the results obtained will be for reconstituted specimens with densities and particle arrangements different to those found *in situ* and allowance should be made for this.

Nowadays, the shear box test tends to be used for the investigation of the shear strength properties of the more unusual granular materials where correlations between ϕ and *in situ* tests such as the SPT are not available or are unreliable. These include crushable sands (e.g. calcareous, vesicular sands), granular fills (fragmented rock particles, both soft and hard), waste materials (e.g. colliery spoil) and the shear strength of interfaces between two construction materials, e.g. steel and sand (steel piles) or plastic and clay (geomembrane and clay liner).

Other applications which have utilised this test are for the quick undrained strength of clay and cut-plane or reversal tests (returning the split specimen to its starting position) for the determination of the drained residual strength. These have generally been superseded by the triaxial and ring shear tests, respectively.

The basic principle of the test and typical results are illustrated in Figure 7.25. Three specimens are prepared at the same density and moisture content and each is tested at a different normal stress to provide values of the peak shear stress and if possible the ultimate (or constant volume) shear stress so that the relationship between τ_f and σ_N' can be obtained from:

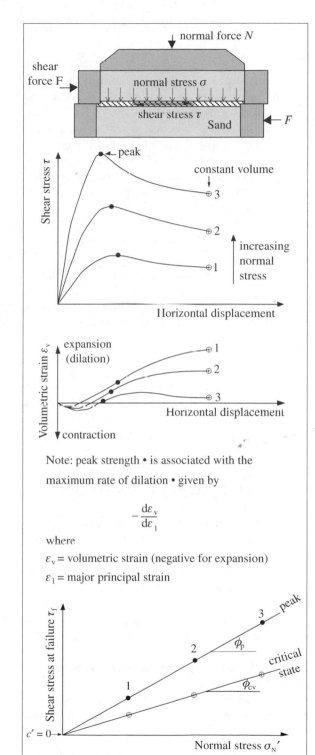

Note: peak strength • is associated with the maximum rate of dilation • given by

$$-\frac{d\varepsilon_v}{d\varepsilon_1}$$

where

ε_v = volumetric strain (negative for expansion)

ε_1 = major principal strain

FIGURE 7.25 *Shear box test on dense sand*

$$\tau_{\text{fpeak}} = \sigma_N' \tan \phi_p' \tag{7.16}$$

$$\tau_{\text{fcv}} = \sigma_N' \tan \phi_{cv}' \tag{7.17}$$

In the standard shear box a 60 mm square specimen is tested although apparatus for 100 mm and 300 mm specimens are available.

The test is strain controlled in that the shear force is applied at a constant rate of strain. Different rates of strain are available so that undrained (quick) or drained (slow) conditions can be assumed in a clay specimen. These rates may vary from 1 mm/min requiring about 10 minutes to conclude a quick undrained test in a clay to less than 0.001 mm/min requiring several days to perform a test to ensure that drained conditions apply in a fairly low permeability clay. For a clean free-draining sand the 'quick' test is appropriate to ensure drained conditions but where the permeability is reduced by the presence of fines a slower rate of strain should be adopted.

See Worked Example 7.1

For granular soils the test is fast and simple and, therefore, relatively inexpensive. The test is also appropriate in that shear planes or thin shear zones are often encountered in failed soil in nature. However, there are significant disadvantages which include:

1. The poor, uncertain control of drainage conditions and the inability to measure pore water pressures.
2. The distribution of stresses on the shear plane are non-uniform and complex with stress concentrations at the sample boundaries.
3. The specimen is forced to fail along a predetermined plane which may not be the weakest zone.
4. As the shear stress is applied the planes on which the principal stresses act will rotate, and will not accurately model the *in situ* loading conditions.

Effect of packing and particle nature
(Figures 7.26 and 7.27)
The shear strength (as given by the angle of shearing resistance ϕ) of a granular soil is obtained from two components:

● friction between the grains at particle contacts
● interlocking of the particles.

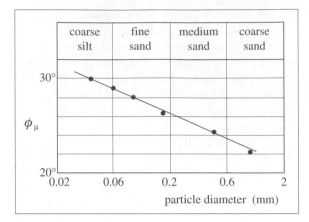

Figure 7.26 *Interparticle friction angle ϕ_μ of quartz sands (after Rowe, 1962)*

The former is denoted by the interparticle friction angle ϕ_μ, which depends on the mineral type. For hard quartz grains this value has been found (Rowe, 1962) to be a function of grain size (Figure 7.26) and can be considered as the absolute lowest possible shear strength of the soil, see the constant volume condition below. For other weaker or platy minerals a lower interparticle friction angle can be expected, e.g. ϕ_μ for mica will be less than 15°.

In order to shear a mass of dense sand it will be necessary to move the particles up and over each other. This will require shear stresses in addition to those required to overcome interparticle friction. The additional shear stress will be larger for a greater degree of interlock.

Shearing a dense sand will, therefore, entail volume increases and greater shear strength (Figure 7.27). However, as the peak shear strength approaches, the particles continue to move further apart and are unable to maintain the same degree of interlock so the shear strength after the peak decreases (strain-softening).

A loose sand will contain groups of particles which can collapse on shearing so a volume decrease is likely but as the particles move closer together there is a tendency for greater interlock and the shear stress is continually increasing (strain-hardening).

The angle of shearing resistance ϕ of a granular soil is generally found to increase with:

● *Increasing particle size*
However, larger rock particles can contain more natural fractures than well-worn sands and with larger contact stresses because of the smaller number of particles present, crushing is much more likely to occur. This could cause additional settlements in granular backfills and rockfill structures

● *Increasing angularity*
Angular particles may give a ϕ value up to 5° more than rounded particles but these too could be more prone to crushing at high stresses.

● *Increasing uniformity coefficient*
More well-graded soils will produce better interlock and have more interparticle contacts so that the shear strength is increased (by up to 5°) and the risk of crushing is reduced.

● *Stronger mineral particles*
Weak particles are prone to crushing which reduces the ϕ value, especially at higher confining stresses.

Constant volume condition (Figure 7.28)

Figure 7.28 is a three-dimensional representation of the two parts of Figure 7.27 and shows that if the tests are continued to large strains the void ratios will eventually coincide at the constant volume or constant void ratio (CVR) condition. This is also referred to as the critical state condition, see later.

The initially loose sand is increasing its strength as it densifies and the dense sand is losing its ability to support shear stress as it loosens until a common shear strength is obtained at the constant volume condition.

The constant volume strength can be used to define an angle of shearing resistance, ϕ_{cv} (Figure 7.17) which can be adopted as the lowest possible value for a dense granular soil if there is some concern that strains may exceed those at the peak value. ϕ_{cv} is always greater than ϕ_r' because there will still be some interlocking even when the constant volume condition is attained.

Effect of density (Figure 7.29)

The effect of density on the angle of shearing resistance ϕ' either side of the constant volume condition is illustrated in Figure 7.29. At initial densities lower than the CVR condition the stress–strain curve

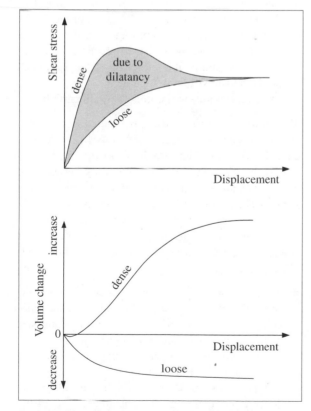

FIGURE 7.27 *Shear strength – effect of packing*

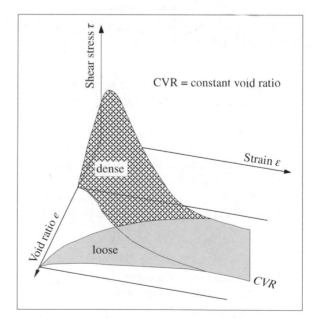

FIGURE 7.28 *Constant volume condition*

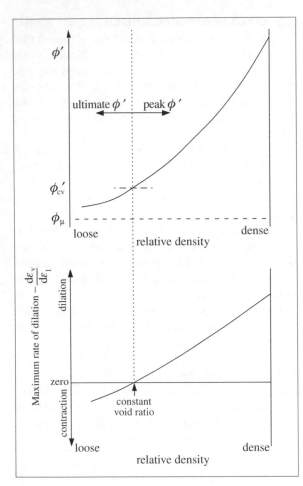

Figure 7.29   *Effect of density on ϕ'*

would display no peak effect and the strength would be determined as an ultimate value at some arbitrary strain level. At initial densities higher than the CVR condition the soils would display dilatancy giving a 'peaky' stress–strain curve. The peak above the constant volume level would be greater for higher initial densities giving higher ϕ_p' values.

Bolton (1986) has shown that for dense sands the peak angle of shearing resistance can be given in the form:

$$\phi_p' = \phi_{cv}' + 0.8\psi \qquad (7.18)$$

where ψ is the angle of dilatancy and the following empirical relationships were obtained.

For plane strain:

$$\phi_p' - \phi_{cv}' = 5I_R \qquad \text{degrees} \qquad (7.19)$$

For axial symmetry:

$$\phi_p' - \phi_{cv}' = 3I_R \qquad \text{degrees} \qquad (7.20)$$

where I_R is the relative dilatancy index given by:

$$I_R = I_D(Q - \ln p') - 1 \qquad (7.21)$$

It can be seen that the peakiness of a stress strain curve will depend on the initial relative density I_D and a measure of the crushing strength Q of the grains but it will be subdued by a higher mean effective stress p'.

Shear strength of clay

Effect of sampling

Sampling clay soils from the ground is much more successful than sampling of sands because of:

● The low permeability of the clay preventing free drainage of the pore water. Removal of the *in situ* stresses following sampling transfers to a decrease in the pore water pressure (suction) which maintains the effective stress state and holds the soil structure together.
● The bonds between the mineral grain particles.

Various laboratory test procedures have been developed successfully for clay soils in the knowledge that they would resemble the *in situ* condition.

The shear strength of a soil can only be provided by the resistance to shearing of the soil structure, the water in the pores having no shear strength at all. To obtain the shear strength of a clay soil as it exists *in situ* its structure must not be altered before it is sheared in the laboratory apparatus. To achieve this it is necessary to:

● Take samples in a manner which produces the least disturbance such as using thin walled sampling tubes, laboratory specimens the same size as the sample and careful hand trimming.
● Avoid moisture content changes after sampling due to drying out or wetting up.
● Adopt a test procedure which controls water leaving or entering the specimen, i.e. ensure undrained or drained conditions apply.

Undrained cohesion, c$_u$

This parameter is determined for a clay in its *in situ* state ensuring no moisture content change since c_u is uniquely related to the moisture content. The unconsolidated undrained (UU) test provides a good measure of *in situ* shear strength and is appropriate for methods of analysis where the rate of loading is fast enough to prevent pore pressure dissipation (drainage) such as with bearing capacity of foundations and piles and trench stability. The tests carried out are:

- unconfined compression test
- vane test
- quick undrained triaxial test.

Unconfined compression test

In this test the soil is taken to failure by increasing the axial load only, with no surrounding confining stresses. The test is carried out on cylindrical specimens, usually 38 mm diameter and 76 mm long but larger diameters (100 mm) can be accommodated in larger compression machines.

The autographic apparatus (BS 1377:1990) is most commonly used for this test using 38 mm diameter specimens. The apparatus is portable, self-contained and hand-operated so it lends itself to use for on-site determination of clay strengths. The apparatus can be easily adapted to provide greater accuracy and sensitivity of results when clays of different strengths are tested and automatic adjustment for area corrections as the sample changes shape by barrelling is included to give directly the unconfined compressive strength σ_1.

The undrained shear strength of the clay is:

$$c_u = 1/2 \times \text{compressive strength} \qquad (7.22)$$

The test has the advantage of being fast, simple, compact and, therefore, inexpensive. However, there are limitations which include:

1. The specimen must be fully saturated, otherwise compression of any air voids and expulsion of air will produce an increased strength and excessive movement is recorded on the autograph. This should be borne in mind when testing compacted clays particularly if they are at or below optimum moisture content.
2. If the specimen contains any macro-fabric such as

fissures, silt partings, varves, gravel particles or defects such as cracks or air voids, then premature failure may result because of these inherent weaknesses.
3. If the specimen has a low clay content then premature failure is likely since it will have poor cohesion.
4. The drainage conditions are not controlled. The test must not be carried out too slowly, otherwise undrained shear conditions may not exist.

Vane test (Figure 7.30)

This test consists of rotating a cruciform shaped vane in a soil and producing a direct shear test on a cylindrical plane surface formed by the vane during rotation. The torque is applied at a constant rate of rotation and is measured by a spring balance or calibrated spring. The test provides a direct measure of the shear strength on the cylindrical failure surface.

No consolidation is permitted before shearing and the soil is sheared quickly so the test is presumed to be unconsolidated undrained (UU) and, therefore, only suitable for clay soils. In sands or clays with sand layers, dilatancy during undrained or partially drained shear produces very high torque requirements and the test provides inappropriate shearing conditions for these soil types.

The test was devised for use in the field (BS 1377:1990) carried out at the bottom of a borehole. The vanes are typically 50–75 mm diameter and 100–150 mm long (height:diameter ratio of 2) with an area ratio less than 15% to minimise disturbance when pushed into the soil. The latter is the ratio of the cross-sectional area of the blades themselves to the area of the circle they produce. The determination of c_u is derived in Figure 7.30.

Smaller versions exist such as the laboratory vane (12.7 mm diameter × 12.7 mm high) used to determine the undrained shear strength of soft clays in tube samples and the hand vane (19 mm or 33 mm diameter). This is very useful as a laboratory vane, but also in the field such as in trial pits and for control of compacted clay fill strength in embankments and earth dams.

The test has the advantage of being fast and simple and is, therefore, inexpensive. It is also useful for obtaining an *in situ* undrained shear strength profile with depth for soft clays and clays which are

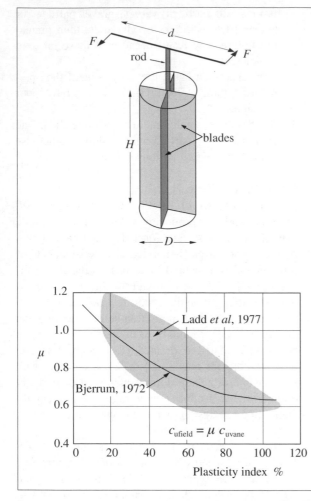

Torque applied $= T = F \times d$

Resisting torque

1) from surface of cylinder

$$\pi D H c_u \frac{D}{2}$$

2) from circular end areas

$$\frac{2 \pi D^2}{4} \frac{2}{3} \frac{D c_u}{2}$$

$$\therefore \; T = \frac{\pi D^2}{4} \left(2H + \frac{2D}{3}\right) c_u$$

for $\dfrac{H}{D} = 2 \quad T = \dfrac{7 \pi D^3 c_u}{6}$

Undisturbed strength c_u from T_u (or F_u)

Remoulded strength c_r from T_r (or F_r)
(after several turns of vane)

$$\text{Sensitivity} = \frac{c_u}{c_r} = \frac{F_u}{F_r}$$

$c_{ufield} = \mu \; c_{uvane}$

FIGURE 7.30 *Vane test*

difficult to sample such as sensitive clays. For these clays thinner blades should be used to minimise disturbance.

However, there are disadvantages which include:

1. Uncertain control of drainage – undrained conditions are assumed but may not occur if permeable macro-fabric (silt layers) exists.
2. A fairly fast rate of rotation is adopted to provide undrained conditions but the strength can be overestimated if the rate of strain is too high.
3. Uncertainty of shear stress distribution on the cylindrical shear plane.
4. The shear surface tested differs from the field loading condition especially if the clay strength is anisotropic, i.e. $c_{uH} \neq c_{uV}$.
5. The results are affected by macro-fabric effects such as fissures, stones, silt partings, fibrous inclusions, roots.
6. The major disadvantage is that the test results show little correlation with the strengths (c_{ufield}) back-analysed from embankments placed on soft clays and which have failed. The vane usually overestimates the strength obtainable *in situ* and although a correction factor has been proposed by Bjerrum (1972) the scatter of the available data, shown in Figure 7.30, demonstrates its uncertain value (Ladd *et al.*, 1977).

Triaxial test (Figure 7.31)

This apparatus was first developed in the 1930s and has largely replaced the direct shear test in commercial laboratories. The test consists of applying shear stresses within a cylindrical sample of soil by changing the principal stresses σ_1 and σ_3, the commonest procedure being to keep the triaxial cell pressure σ_3 constant and increasing the axial or vertical stress σ_1 until failure is achieved. The essential features of the apparatus are illustrated in Figure 7.31.

Specimen sizes in the UK are standardised at 38 mm and 100 mm diameter with a height:diameter ratio of 2:1 to ensure that the middle section of the specimen is free to shear. If this ratio is less than 2:1 then shear stresses at the ends of the sample in contact with the platens will affect the results by constraining the failure planes.

The specimen is surrounded by a rubber membrane to prevent the cell fluid (water) entering the specimen and altering its moisture content. For weaker soil specimens a correction to account for the restraint provided by the membrane should be applied. This correction will be small if the specimen deforms into a barrel-shape and can often be ignored particularly for higher strength soils but if a single plane develops the membrane restraint can be significant.

The axial stress is applied by a motorised drive which raises the specimen (and cell) against a piston reacting on a load frame. A proving ring between the piston and load frame or internal load transducer measures the axial force F from which the principal stress difference or deviator stress, $\sigma_1 - \sigma_3$, is calculated, as shown in Figure 7.31. The strength of the soil is then obtained from the Mohr circle plot.

As the stress is applied the specimen often becomes barrel-shaped so the vertical stress in the middle of the specimen must be determined from the force measured and this increased area by applying an area correction to each reading. The 'corrected' area A in the middle of the specimen is obtained for a drained test from:

$$A = A_0 \frac{(1 - \varepsilon_v)}{(1 - \varepsilon_a)} \qquad (7.23)$$

where A_0 is the initial cross-sectional area, ε_a is the vertical or axial strain and ε_v is the volumetric strain.

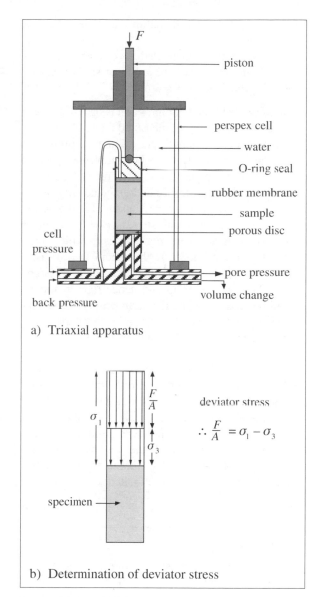

a) Triaxial apparatus

b) Determination of deviator stress

FIGURE 7.31 *Triaxial test*

For an undrained test where the volumetric strain is zero ($\varepsilon_v = 0$) the corrected area is given by:

$$A = \frac{A_0}{(1 - \varepsilon_a)} \qquad (7.24)$$

A more convenient approach for the undrained test is to plot the proving ring readings on a graph corrected for this effect. This method should not, however, be adopted if failure develops along a single plane.

The test is strain controlled as a constant rate of compression is applied and a rate of strain must be chosen for the following reasons:

1. In quick undrained compression tests (UU) where pore pressures are not measured a rate of strain of 2% of specimen length per minute is commonly adopted so that a test can be completed in about 10 minutes. Impermeable end platens and the rubber membrane ensure undrained conditions.
2. In a consolidated drained test (CD) the shear stresses (via the axial stress) must be applied slowly so that excess pore pressure which may develop in the middle of the specimen can dissipate to ensure fully drained conditions throughout. The rate of strain and hence time to failure will be determined by the permeability of the clay and the proximity of permeable boundaries. For many clays several days will be required to reach failure.
3. In a consolidated undrained test (CU) where during the undrained shearing stage the pore water pressure is measured at the base of the specimen sufficient time must be allowed to ensure that the pore pressure produced in the middle of the specimen is the same as at the base (equalisation) where it is measured. Otherwise, the strength will be overestimated.

For both the CD and CU tests the consolidation stage and the time to failure in shear can be reduced considerably by providing filter paper strip drains around the specimen and porous discs at the ends of the specimen permitting radial drainage or radial equalisation.

The major advantage of the triaxial test is that drainage conditions can be controlled so that conditions pertaining in the field can be modelled in the laboratory, see Effects of drainage, above.

Triaxial unconsolidated undrained (UU) test (Figure 7.32)

The unconfined compression test and the vane test are versions of this test approach but they have their limitations, described above. The UU test or quick-undrained test in the triaxial apparatus is one of the most common tests carried out in practice to determine the undrained shear strength of a clay at its *in situ* natural moisture content.

Its main application is in the design of shallow and pile foundations and in assessing initial stability of embankments on soft clays and suitability of clay fill for earthworks construction.

Standard procedure is to prepare three 38 mm diameter specimens from a 100 mm diameter 'undisturbed' sample (U100) and to test each specimen under a different confining cell pressure. The use of solid end platens and a rubber membrane ensure that no consolidation is permitted and that the specimen is undrained during shearing. If the specimens are fully saturated, of the same moisture content and

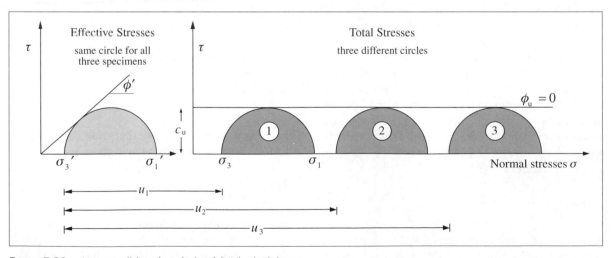

FIGURE 7.32 *Unconsolidated undrained (UU) triaxial test*

have similar soil structure then similar shear strength (c_u) values should be obtained.

With fully saturated specimens ($B = 1$) the application of the confining cell pressure will simply mean that the pore water pressure in each specimen is increased by the amount of confining pressure. No change in effective stresses occurs so the Mohr circle at failure in effective stress terms will be the same and the shear strength measured will be the same irrespective of confining pressure (Figure 7.32).

The Mohr–Coulomb envelope for this test is, therefore, represented in total stress terms with c_u being the radius of the Mohr circle at failure:

$$c_u = \frac{(\sigma_1 - \sigma_3)_f}{2} \qquad (7.25)$$

All of these circles will have the same radius so the Mohr–Coulomb envelope is a horizontal straight line with intercept c_u and gradient $\phi_u = 0$. Thus there is an apparent cohesion but no apparent friction.

For clays even of the same moisture content different strengths may be obtained due to the presence of macro-fabric such as inclined joints in fissured clays or gravel particles in glacial clays. Difficulties in preparing smaller diameter specimens from a

U100 may also be experienced, particularly with stones and fissures present.

It is usually recommended that the test then be carried out on 100 mm diameter specimens (i.e. straight out of the U100) to avoid further disturbance and to be more representative of the *in situ* mass fabric. However, there is insufficient material in a U100 (usually 450mm long) to prepare three separate specimens (200 mm long) so either testing of one specimen at one cell pressure (single stage) or testing the same specimen at three different cell pressures (multi-stage) can be adopted. The former will only give one Mohr circle which may be considered a disadvantage but if the material can be assumed to be fully saturated then this should not be so.

Multi-stage (UU) test (Figure 7.33)

This consists of applying the shear stresses under the first confining pressure at a somewhat slower rate to enable more readings to be taken until the load–strain plot starts to flatten. At this somewhat arbitrary point the cell pressure is increased and shearing is continued for the second stage. It is then increased again for the third stage. The load–strain plots are extrapolated to a similar strain value (such

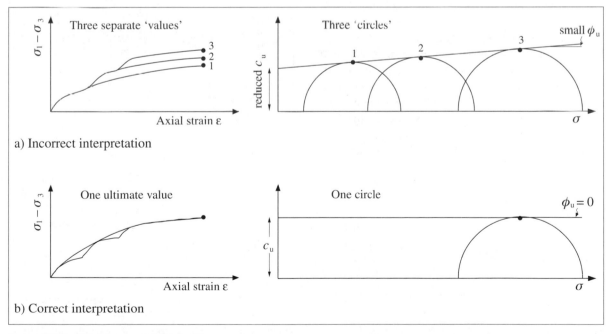

FIGURE 7.33 *Multi-stage (UU) triaxial test*

as 20%) and the deviator stresses calculated from these loads for each confining pressure. The test is only really suitable for plastic soils requiring large strains to failure and with no pronounced brittleness or peak shear strength, as is the case for many normally consolidated clays.

If the soil is tested at constant moisture content it should strictly only have one failure strength irrespective of cell pressure. The 'steps' in the stress–strain curve may be caused by an initial increase in stiffness when the cell pressure is increased. There is a tendency to imagine three different Mohr circles when really the stress–strain plot is continuous. Thus results are often reported with three different Mohr circles giving a reduced cohesion intercept c_u and with a ϕ_u value greater than zero, up to 10°, which is erroneous. Using these small ϕ_u values in a shallow foundation or immediate slope stability analysis can lead to a dangerous overestimate of stability.

Effect of clay content and mineralogy

Although the undrained shear strength of a clay is related to its moisture content, decreasing as the moisture content increases, it also depends on the mineralogy of the clay (see Figure 12.5 in Chapter 12) and the amount of clay present (see Figure 12.6).

Partially saturated clays (Figure 7.34)

If the triaxial test is carried out on a partly saturated specimen, equation 4.6 shows that the application of the cell pressure σ_3 produces a pore water pressure smaller than the value of σ_3 ($B<1$) so there is an increase in effective stress. This causes the soil structure to compress slightly and become stronger so when the sample is sheared the deviator stress at failure is increased.

When these Mohr circles at failure are plotted in terms of total stresses an inclined or curved Mohr–Coulomb envelope is obtained and the results then tend to be reported as an intercept c_u and an angle of $\phi_u > 0°$. Again care should be exercised in using these values in say a bearing capacity analysis since the state of partial saturation may be erroneous (due to the sample drying) or it may be transient (sampling carried out above a fluctuating water table or during a dry season).

A check on the degree of saturation (from moisture content and bulk density values, see Chapter 2) should be made to confirm the state of the specimen and hence its relevance to *in situ* conditions.

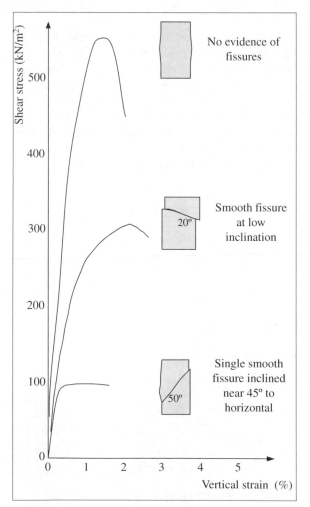

FIGURE 7.35 *Shear strength of fissured clays (from Marsland, 1971)*

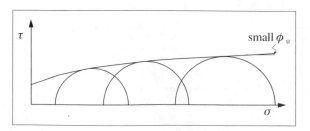

FIGURE 7.34 *Unconsolidated undrained triaxial test on partially saturated clay*

Fissured clays (Figure 7.35)

Fissures exist in most overconsolidated clays producing planes of weakness. The strength on these fissures is described as the fissure strength and is much lower (as low as 10%) than the intact strength of the clay between the fissures. The fissures, therefore, dictate the strength en masse of the clay so that the mass strength or design strength should be nearer the fissure strength but not necessarily equal to it.

The spacing between fissures usually increases with depth from typically 10–50 mm in a highly fissured upper zone to more than 100 mm at depth. It has been found that to represent the mass strength of the clay, specimens with a diameter at least equal to the fissure spacing are required for testing.

Some tests reported by Marsland (1971) are reproduced in Figure 7.35 and show the effect of the presence and inclination of fissures in 38 mm diameter specimens. If several of these results are then plotted at the depth of sampling a wide scatter is obtained. Testing larger samples (100 mm diameter) should produce less scatter since the effect of the fissures will be included and the strength obtained will be nearer to the mass strength although there may still be discrepancies at greater depths where fissure spacing increases.

If the samples are not tested soon after sampling, opening and extension of the fissures and small micro-cracks will cause a reduction of strength which can be significant even within hours after sampling. The samples should also be tested under confining pressures at least comparable to the *in situ* stress, otherwise reduced strengths may be recorded because the fissures have not been entirely reclosed.

Variation with depth
(Figures 7.36 and 7.37)

From the discussion in Chapter 4 on the compressibility of a clay soil it can be seen that for a normally consolidated clay as depth increases the effective stress increases and the void ratio decreases.

For a truly normally consolidated clay the soil at the surface (present ground level) will be of a mud consistency with a moisture content close to its liquid limit. Assuming a constant compression index C_c with depth the variation of moisture content for low and high plasticity clays has been plotted in Figures 7.36 and 7.37. The undrained shear strength c_u of a normally consolidated clay is related to moisture content and has been found by Skempton (1953) to increase linearly with effective stress p_o':

$$c_u / p_o' = 0.11 + 0.0037 \, I_p \qquad (7.26)$$

where I_p is the plasticity index. This is illustrated in the upper part of Figure 7.36. Note that high plasticity clays have higher moisture contents but also higher strengths.

For a lightly overconsolidated clay, say with 5 m of soil removed due to erosion, the clay at all depths will swell and its moisture content will increase, but to a lesser extent, determined by the swelling index C_s. This is the gradient of the void ratio–log effective stress curve on the unloading/reloading line, described in Chapter 9. Assuming the swelling index to be $0.25 \times C_c$ and the water table remains at ground level the moisture content profile has been plotted on the lower part of Figure 7.36.

The undrained shear strength of a lightly overconsolidated clay is the same as a normally consolidated clay providing they are at the same moisture content. With these assumptions the variation of c_u with depth for both clays has been plotted in Figure 7.36. It can be seen that a site comprising lightly overconsolidated clay will have shear strengths greater than a site comprising normally consolidated clay.

For a heavily overconsolidated clay, say with 50 m of soil removed due to erosion, Figure 7.37, the shear strengths will be significant largely because the soil has been substantially normally consolidated in the past. Note the higher moisture contents and lower shear strengths that are possible close to present ground level.

The above does not include for the effects of desiccation, cementation, secondary compressions and other processes which may have occurred during and following the erosion period.

Frictional characteristics
(Figure 7.38)

It cannot be stressed enough that the shear strength of clay is determined by the effective stresses and the frictional characteristics, both of which reside in the mineral grain structure. As effective stresses increase the shear strength increases and clay minerals with lower plasticity tend to give greater fric-

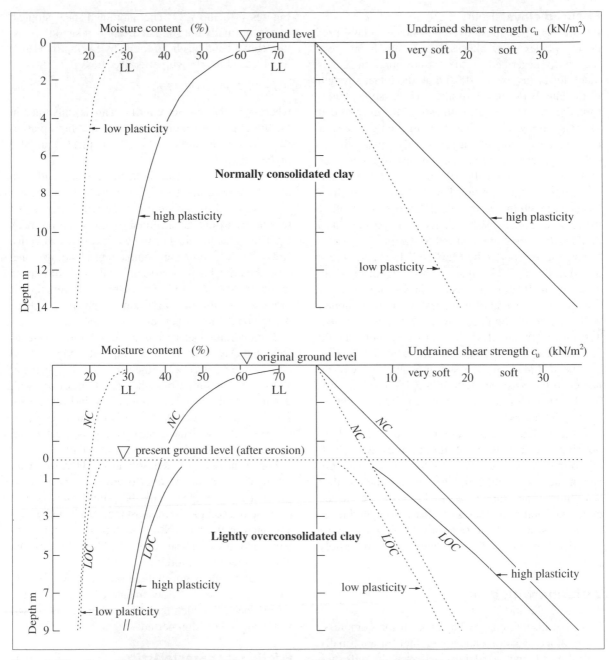

FIGURE 7.36 *Strength variation with depth – normally and lightly overconsolidated clays*

tional characteristics, represented by a higher ϕ value (Figure 7.38).

Test procedures

The Mohr–Coulomb envelope is represented by the angle of shearing resistance ϕ' and a small cohesion intercept c' (if present) when plotted in effective stress terms. To increase the effective stress in a clay specimen it is necessary to increase the total stress around it and allow the pore water pressure increase

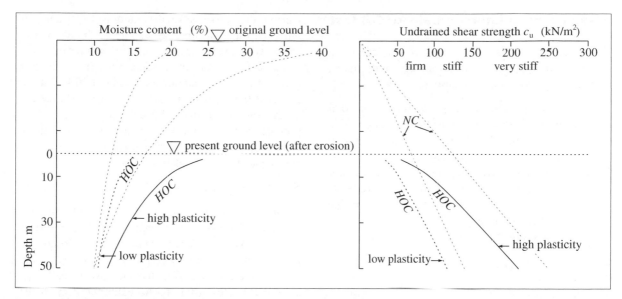

FIGURE 7.37 *Strength variation with depth – heavily overconsolidated clays*

to dissipate so each specimen must be consolidated to a different effective stress state before shearing.

The shearing stage is then carried out by changing the external total stresses on the specimen with either:

● *Undrained conditions*
No volume changes are permitted. The pore water pressure must then be measured to obtain effective stresses. This test procedure is described as consolidated undrained (CU); or

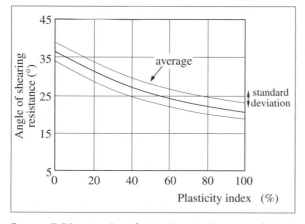

FIGURE 7.38 *Angle of shearing resistance of clays (from NAVFAC, 1971)*

● *Drained conditions*
Volume changes are permitted and shearing is carried out at a rate slow enough to ensure no pore water pressure changes occur so that effective stresses then equal the total stresses. The test procedure is described as consolidated drained (CD).

The stress history dictates the behaviour during shear in that the mineral grain structure of:

1. *Normally consolidated clays* contracts in a drained test resulting in reducing volumes. It attempts to contract in an undrained test but cannot since volume changes are prevented and this results in increased pore water pressures.
2. *Overconsolidated clays* expands or dilates in a drained test resulting in increased volumes. It attempts to expand in an undrained test resulting in decreased pore water pressures.

These effects are illustrated in Figures 7.39 and 7.40 which show that normally consolidated and overconsolidated clays of the same moisture content give the same strength when tested undrained but when tested drained the normally consolidated clay gives a much greater strength since it is becoming stronger during contraction whereas the overconsolidated clay is becoming weaker during expansion.

Triaxial consolidated undrained (CU) test
(Figure 7.39)

The procedure carried out for this test relates to a number of field applications, see Effects of drainage, above. It has two main applications:

1. For the undrained shear strength (c_u) in total stress terms following a consolidation stage. This gives the relationship between c_u and confining stress σ_3 or overburden pressure p_o'.
2. For the effective stress parameters c' and ϕ'. For this purpose pore water pressure measurements are taken during the shear stage to determine effective stresses at failure.

The test consists of three stages:

- saturation
- consolidation
- compression (shear).

For the accurate measurement of pore water pressures it is essential that air is prevented from entering the pressure system so all air voids must be eliminated to produce a saturated specimen. The B value is then virtually unity and changes of confining pressure $\Delta\sigma_3$ are reflected by equal changes in pore water pressure Δu_w.

Saturation is produced by increasing the pore water pressure in the specimen to remove the air. This is achieved by increasing the back pressure to the pore fluid in increments while at the same time increasing the cell pressure in increments so that changes in the effective stress are kept to a minimum.

During this process small amounts of water will move into the specimen and some of the air will be dissolved in the water to achieve full saturation.

The cell pressure increments must be larger than the back pressure increments, by about 10 kN/m^2, so that effective stresses are always positive. At the end of the saturation stage the cell pressure and back pressure will be quite large values but their difference will be small. The specimen is now ready to commence the consolidation stage.

SEE WORKED EXAMPLE 7.2

The consolidation stage is commenced by increasing the cell pressure and keeping the back pressure constant. With drainage prevented (valve closed) this induces an increase in pore pressure in the specimen which should be equal to the increase in cell pressure, if B is close to unity. An example of the determination of the pore pressure parameter B is given in Worked Example 4.7.

The difference between this pore pressure and the back pressure is the excess pore pressure to be dissipated during consolidation once the drainage valve is opened. During this stage the pore pressure then dissipates (reduces) to the back pressure value and water is squeezed out of the specimen. At the same time the effective stress in the specimen is increasing until at the end of the consolidation stage the effective stress is given by the difference between the cell pressure and the back pressure and is the effective stress present at the start of the shear stage.

The sample volume change and pore pressure dis-

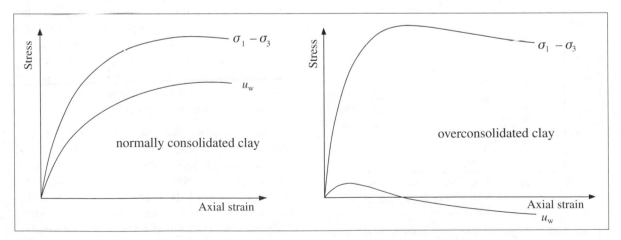

FIGURE 7.39 *Consolidated undrained test on clay*

sipation are plotted against time (square root or log) and when 95% dissipation is reached the consolidation stage can be terminated.

The volume change versus root time plot will give a measure of the permeability of the soil and using methods derived by Bishop and Henkel (1962) the time to failure for a particular test (with full drainage in the drained shear test or full equalisation in the undrained shear test) can be determined and assuming an amount of strain required to cause failure the rate of strain can be calculated. The specimen is now ready to commence the shear stage.

With the drainage valve closed to prevent movement of water into or out of the specimen (undrained condition) the axial stress σ_1 is increased while keeping the cell pressure σ_3 constant. This produces shear stresses in the specimen which give rise to changes in the pore water pressure and these are measured, see Figure 7.39. Thus the effective stresses σ_1' and σ_3' at failure (for the Mohr circle) and at any time during the shear stage (for stress paths) are obtained.

SEE WORKED EXAMPLE 7.3

Three specimens should be tested at different cell pressures (three different effective stresses at the start of the shearing stage). One specimen should be tested at the *in situ* effective stress level with the others at two, three or four times this value.

SEE WORKED EXAMPLE 7.4

To obtain three specimens from one U100 sample it is necessary to use 38 mm diameter specimens each one placed in a separate triaxial apparatus. Smaller specimens also provide quicker drainage in the drained test and quicker equalisation of pore pressures in the undrained case so the test can be completed in an economical time period. However, there is more likelihood of disturbance and the specimens may be less representative of mass effects.

The main advantage of the CU test compared to the CD test is that the shear stage is much faster, about eight times faster if side drains are provided and 16 times faster without side drains (Head, 1986). The stress paths from a CU test are also more informative.

Triaxial consolidated drained (CD) test
(Figure 7.40)

This test is carried out with the same saturation and consolidation stages as for the CU test and then ensuring that no pore pressures develop during the shearing stage by allowing full drainage. Thus total stresses applied to the specimen will also be the effective stresses inside the specimen. The rate of strain must be slow enough to ensure full dissipation throughout the specimen.

The main application for the test is to determine the effective stress parameters of the soil, c_d' and ϕ_d'. The main disadvantage is that the shear stage of the test takes much longer. However, for soils which are

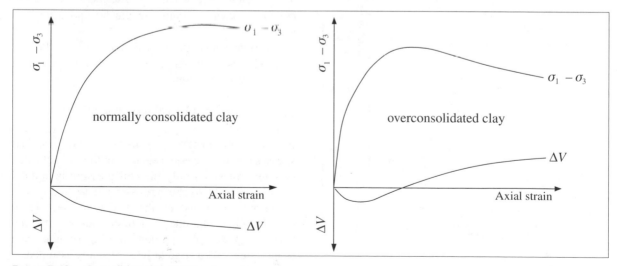

FIGURE 7.40 *Consolidated drained test on clay*

virtually fully saturated the saturation stage is not so important and drainage to an open burette can be permitted, eliminating the need for a back pressure system.

The axial stress σ_1 is increased while keeping the cell pressure σ_3 constant. This produces shear stresses in the specimen which give rise to volume changes which are measured. These volume changes will involve specimen reductions for a contractant soil and increases for a dilatant soil. The total stresses σ_1 and σ_3 at failure are also the effective stresses σ_1' and σ_3' so the Mohr circle can be plotted direct.

Critical state theory

This model of soil behaviour was developed at Cambridge University by Roscoe, Schofield and Wroth (Roscoe *et al.*, 1958). It has been described by Schofield and Wroth (1968) and simply explained by Atkinson and Bransby (1978). More recent coverage is given in Muir Wood (1990).

The theory provides a unified model for the behaviour of saturated remoulded clay by observing changes in:

● the mean stress or consolidation stress p'
● the shear stress or deviator stress q'
● the volume change v, specific volume.

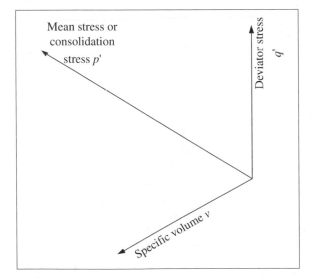

FIGURE 7.41   *$p' - q' - v$ plot*

These are observed on a three-dimensional plot, $p' - q' - v$, Figure 7.41, where:

$$p' = \tfrac{1}{3}(\sigma_1' + \sigma_2' + \sigma_3') \tag{7.27}$$

$$q' = \tfrac{1}{\sqrt{2}}[(\sigma_1' - \sigma_2')^2 + (\sigma_2' - \sigma_3')^2 + (\sigma_3' - \sigma_1')^2]^{1/2} \tag{7.28}$$

$$v = 1 + e \tag{7.29}$$

For the common case of axial symmetry (oedometer, triaxial tests) where $\sigma_2' = \sigma_3'$:

$$p' = \tfrac{1}{3}(\sigma_1' + 2\sigma_3') \tag{7.30}$$

$$q' = \sigma_1' - \sigma_3' \tag{7.31}$$

State boundary surface (Figure 7.42)

The theory postulates that there is a state boundary surface on the $p' - q' - v$ plot which separates states in which soil can exist and states in which soil could not exist. As a summary at this stage the complete state boundary surface is shown in Figure 7.42 and the various components which make up this surface are described separately below.

The surface is assumed to be a mirror image for negative values of q'. The state of a soil can only reach the state boundary surface by loading, unloading takes the state beneath the surface. Strains are not represented on the three-dimensional plot but it is assumed that deformations are elastic and recoverable while the state remains beneath the surface but when loading takes the soil to the boundary surface the soil yields and when the state moves across the boundary surface strains are plastic and irrecoverable.

The state boundary surface is a curved surface in all directions but when idealised it may be seen to be made up of the following.

Isotropic normal consolidation line (ICL) (Figure 7.43)

Isotropic compression means that the principal stresses applied are equal such as in the triaxial compression test when only the cell pressure is applied ($\sigma_1' = \sigma_2' = \sigma_3'$). No shear stresses are applied so $q' = 0$ and the state exists on the p'–v plane, see Figure 6.8 where p' is plotted at a natural scale and the line is curved. When p' is plotted on a logarithmic scale as in Figure 7.43 a straight line is usually assumed.

Normal compression or consolidation means that

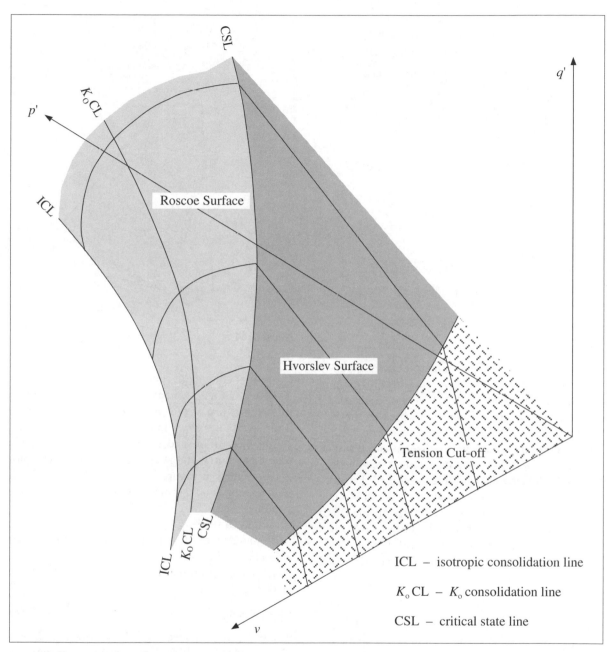

FIGURE 7.42 *Complete state boundary surface*

the consolidation stress p' is increasing and producing elastic and plastic strains and the plastic strains lead to a permanent reduction of the void space, therefore v decreases. Plastic strains are occurring by structural rearrangement of the soil skeleton with yielding and hardening taking place so the state of

the soil is moving along this part of the state boundary surface. This line is a yield curve. As shown in Chapter 6 when the line is plotted on a log scale a straight line is often obtained so the relationship for the ICL is given by:

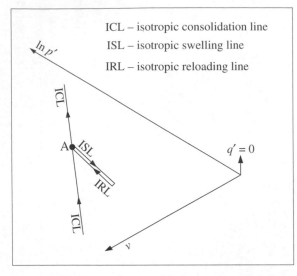

FIGURE 7.43 *Isotropic consolidation and swelling lines*

$$v = N - \lambda \ln p' \tag{7.32}$$

At the origin on a logarithmic plot the value of N is taken as the specific volume when p' is 1.0 kN/m^2.

If at any point (such as A in Figure 7.43) the stress is reduced the state will not return back up the ICL since there have been irrecoverable plastic strains. Only the elastic strains are recovered and the state then moves away from and inside the state boundary surface back along the isotropic swelling line (ISL).

If the soil is then loaded again the state will follow the swelling line along the recompression line (IRL) until reaching the state boundary surface at A when it will move down along the ICL with further elastic and plastic strains. The swelling and recompression lines are represented by:

$$v = v_\kappa - \kappa \ln p' \tag{7.33}$$

where v_κ is the specific volume when p' is 1.0 kN/m^2.

Soils whose state exists on the ICL part of the state boundary surface are described as normally consolidated and loading will immediately and continuously cause plastic strains while soils existing on a swelling line are described as overconsolidated and loading will produce only elastic strains until the state reaches the boundary surface.

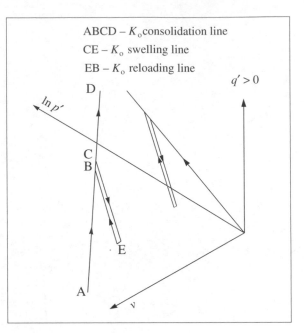

FIGURE 7.44 *K_0 consolidation and swelling lines*

K_0 normal consolidation line (K_0 CL) (Figure 7.44)
Although not an essential part of the state boundary surface this line is of fundamental importance since it represents the condition of the soil in the ground. During deposition and erosion the soil exists in a state of no horizontal strain (K_0 condition). The plot is represented in Figure 7.44 where the line ABCD (K_0CL) represents normal compression and CE is a swelling line (K_0SL) on which the soil is in an overconsolidated state.

Elastic and plastic strains occur on ABCD but only elastic strains occur on CE and EB. The equation for the K_0 normal compression line (K_0CL) is given by:

$$v = N_0 - \lambda \ln p' \tag{7.34}$$

where N_0 is the specific volume when p' is 1.0 kN/m^2 and the swelling line (K_0SL) is given by:

$$v = v_{\kappa 0} - \kappa \ln p' \tag{7.35}$$

where $v_{\kappa 0}$ is the specific volume when p' is 1.0 kN/m^2.

Critical state line (CSL) (Figure 7.45)
This line represents the state when the critical state strength (or ultimate strength) is being mobilised, when failure has occurred and with no further

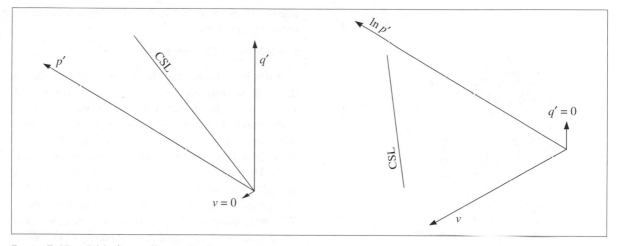

FIGURE 7.45   *Critical state line projections*

change in mean stress, deviator stress or volume (p', q' and v are constant) the soil continues to shear with plastic strains only. This state is akin to the constant volume state achieved by loose and dense sand when sheared, see Figure 7.28.

It must be emphasised that this condition only occurs with homogeneous shearing when all elements of the soil sample are shearing at the same rate. If shearing occurs on a thin zone or slip plane then rearrangement of clay particles semi-parallel to the slip surface may reduce the strength only on that surface to the residual strength, see below.

When projected onto the q'–p' plane (Figure 7.45) this line is represented by:

$$\pm q' = M p' \qquad (7.36)$$

and on the v–$\ln p'$ plane by:

$$v = \Gamma - \lambda \ln p' \qquad (7.37)$$

where Γ is the specific volume when p' is 1.0 kN/m².

The significance of this line is that irrespective of the starting point of a sample (normally consolidated or overconsolidated) and regardless of test path, undrained or drained, ultimate failure will occur once the state of the sample moves onto this line.

Roscoe surface (Figures 7.46, 7.47, 7.48, 7.49 and 7.50)

This is the part of the state boundary surface between the ICL and the CSL and the complete

surface is shown in Figure 7.42. Tests that commence on the ICL (isotropically normally consolidated) follow the Roscoe surface between the ICL and the CSL, irrespective of whether the test is undrained or drained. Soil states on this side of the critical state line are referred to as being on the wet side of critical.

No volumetric strains can occur during an undrained test so the path followed must exist on a

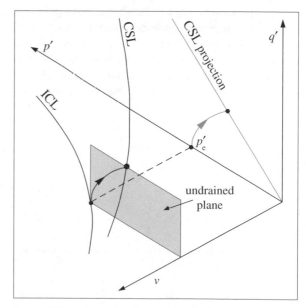

FIGURE 7.46   *Stress path for an undrained test*

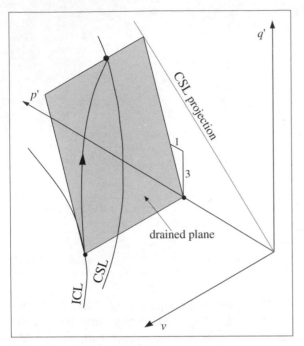

FIGURE 7.47 *Stress path for a drained test*

constant volume plane, Figure 7.46. There is an undrained plane for each value of v. The path followed by an undrained test starting from the ICL follows the Roscoe surface on the undrained plane up to the CSL.

For a drained test it can be shown that $dq/dp = 3$ so all drained planes must be parallel to the v axis and have a slope of 1 in 3 on the $q'-p'$ projection, Figure 7.47. Soil states starting from the ICL are normally consolidated and will have a tendency to contract or compress during shear so the test will

follow the Roscoe surface on the drained plane by reducing in volume.

SEE WORKED EXAMPLE 7.7

It has been found that the shape of the undrained test path starting from the ICL is the same for all values of p_e', the initial isotropic stress, see Figure 7.46. Normalising the $q'-p'$ plots by using q'/p_e' and p'/p_e' instead, one curve is obtained for undrained tests carried out at all values of v, as shown in Figure 7.48. The same curve is obtained for a drained test provided the value of p_e' associated with each value of specific volume as the sample compresses is used. This procedure is referred to as normalisation and can provide one relationship for a family of curves.

Soils which are lightly overconsolidated with different overconsolidation ratios can be obtained with the same moisture content and specific volume and these would lie on the same undrained plane, as shown in Figure 7.49. The paths for undrained tests will commence beneath the Roscoe surface and rise almost vertically until they reach the Roscoe surface where they move along the surface until reaching the critical state, Figure 7.50.

Tension cut-off (Figure 7.51)

When the mean effective stress in a soil is zero it is implicit that no shear stress can be applied, i.e. from equation 7.36 when $p' = 0$, $q' = 0$ so the state boundary surface must pass through this point.

The highest value of the ratio q'/p' corresponds to the case when $\sigma_3' = 0$. For a triaxial compression test

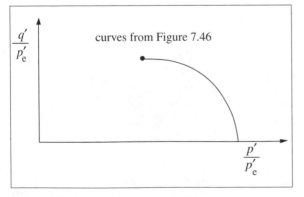

FIGURE 7.48 *Normalised q' – p' plot*

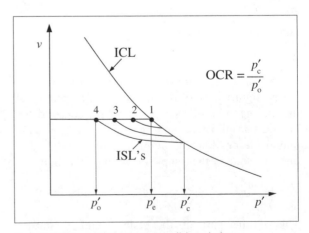

FIGURE 7.49 *Lightly overconsolidated clays*

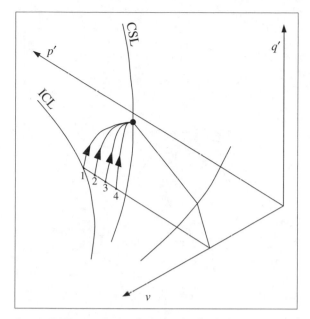

FIGURE 7.50 *Undrained test paths for lightly overconsolidated clays*

by increasing the vertical effective stress σ_1' the tensile strength will be mobilised. Then:

$$q' = \sigma_1' \quad p' = 1/3\,\sigma_1'$$

so

$$q' = 3\,p' \quad\quad\quad\quad (7.38)$$

giving an upper limit to the state boundary surface referred to as the tension cut-off, shown in Figure 7.51.

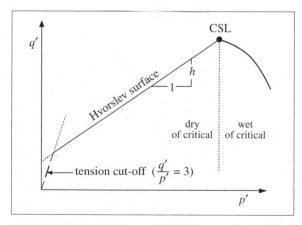

FIGURE 7.51 *Tension cut-off and the Hvorslev surface*

Hvorslev surface (Figures 7.51, 7.52 and 7.53)
This surface connects with the Roscoe surface along the critical state line, Figure 7.51 and is shown complete in Figure 7.42. It represents the state boundary surface for heavily overconsolidated samples. Soil states on this side of the CSL are referred to as being on the dry side of critical. For these soils, with different overconsolidation ratios following isotropic consolidation and swelling, Figure 7.52, the paths for undrained tests (all on an undrained, constant v plane) commence beneath the Hvorslev surface, rise near vertically until reaching the Hvorslev surface and then move along the surface until reaching the critical state.

The path for a drained test will lie on a drained plane described above, see Figure 7.53. The path will commence beneath the Hvorslev surface rising up the drained plane and then down the Hvorslev surface to the critical state line. Note that the maximum stress or peak strength q' occurs before the critical state is reached with volume expansion and work-softening occurring during the reduction in strength from the peak.

With overconsolidated clays mobilising the peak strength often produces thin slip zones or planes in the sample with only the material in this zone

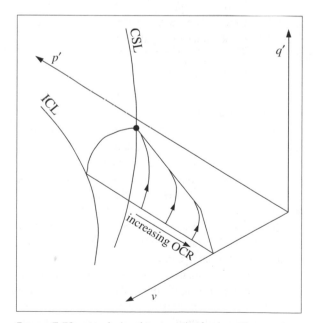

FIGURE 7.52 *Undrained test paths for heavily over-consolidated clays*

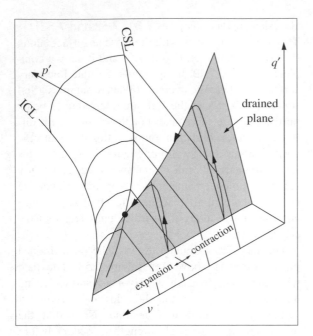

FIGURE 7.53 *Drained test paths for heavily over-consolidated clays*

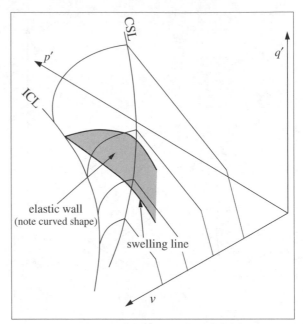

FIGURE 7.54 *The elastic wall*

moving down the Hvorslev surface and softening but with little softening of the remainder of the sample.

Assuming for one value of specific volume v the Hvorslev surface is a straight line with a slope h, Figure 7.51, the equation for this surface is:

$$q' = (M - h)\exp\frac{(\Gamma - v)}{\lambda} + hp' \qquad (7.39)$$

The elastic wall (Figure 7.54)

For an ideal isotropic elastic soil the volumetric strains are related only to the mean effective stress p' as seen in equations 7.5 and 7.6 and are not dependent on changes in q'. Along a swelling/recompression line the strains are also elastic so it can be postulated that on a surface vertically above a swelling line the strains will be elastic until this surface meets the state boundary surface when yielding and plastic strains can occur.

This surface defined as a vertical surface above a swelling line and beneath the state boundary surface is referred to as an elastic wall. As there are an infinite number of swelling lines so there will be an infinite number of elastic walls. The intersection of

the elastic wall and the state boundary surface will be a yield curve. Note that the swelling/recompression lines are curved so the elastic wall will be curved.

Assuming that below the state boundary surface the path of an overconsolidated soil remains on a particular elastic wall then the path followed during an undrained test will lie along the intersection of the constant volume undrained plane (Figure 7.46) and the elastic wall. Similarly for a drained test the path will follow the intersection of the drained plane (Figure 7.47) and the elastic wall.

Real soils

The above theory assumes that soils behave in an ideal manner, that they have isotropic structures and stress conditions, they are homogeneous throughout their mass and that they have no preferred structure within them, i.e. they are remoulded or reconstituted soils. Real soils in the ground do not behave in an ideal manner, they have anisotropic structures due to preferred particle orientations of the grains, they are subjected to anisotropic stress conditions and are usually non-homogeneous due to fabric effects such as layering and fissuring. As well as the fabric effects undisturbed soils have often developed some interparticle bonding which would be destroyed on remoulding and this is not included in the theories.

Nevertheless, the concepts of a state boundary surface, a critical state condition, the inter-relationships between mean stress, deviator stress and volume, the effects of drainage conditions, elastic and plastic straining, yielding and hardening provide a sound framework for the understanding of basic soil behaviour.

Residual strength

Although the critical state strength is often referred to as the ultimate strength this condition is achieved with homogeneous shearing, i.e. all of the sample is undergoing the same shear strain and these strains are not excessively large.

It has been observed, particularly from the study of old landslips (Skempton, 1964) where significant straining has occurred on thin shear surfaces that the operative shear strength on these surfaces was much lower than the critical state strength. For example, the residual ϕ_r' value for London Clay can be as low 10° whereas at the critical state ϕ'_{cv} is greater than 20°.

It is essential, therefore, to identify the presence, or otherwise, of pre-existing slip surfaces in a clay soil on a sloping site. Small changes in surface topography or pore pressure conditions can re-activate an ancient landslip with catastrophic consequences.

Residual strength is attained when large shear strains have occurred on a thin zone or plane of sliding in a clay soil where the clay particles have been rearranged to produce a strong preferred orientation in the direction of the slip surface. Lupini *et al.* (1981) recognised three modes of residual shear behaviour:

- *Turbulent*
  This occurs where behaviour is dominated by rotund particles. For soils dominated by platy particles with high interparticle friction this mode may also occur. In this mode energy is dissipated by particle rolling and translation. No preferred particle orientation occurs and residual strength still remains high so that ϕ_r' can be taken as ϕ'_{cv}.
- *Sliding*
  When behaviour is dominated by platy, low friction particles sliding occurs on a shear surface with strongly oriented particles and the strength is low. ϕ_r' depends mainly on the mineralogy, coefficient of interparticle friction, μ and pore water chemistry.
- *Transitional*
  This involves turbulent and sliding behaviour in different parts of a shear zone.

The residual shear strength can be obtained using a ring shear apparatus (Bishop *et al.*, 1971 and Bromhead, 1978). A ring-shaped thin sample of remoulded soil is sheared in a direct shear manner by rotating the upper half of the sample above the lower half with sufficient strain until a slip surface is formed on which the lowest strength is measured from the torsion applied. As illustrated in Figure 7.11 the residual strength τ_r is related to the normal stress σ_N' applied on the slip surface by:

$$\tau_r = \sigma_N' \tan \phi_r' \qquad (7.40)$$

although for many soils the plot of τ_r versus σ_N' shows a small cohesion intercept c_r' or a curvature of the plot so that the stress range applicable to the site conditions must be used for the determination of ϕ_r'.

In general, if the clay content is 40–50% or more or the plasticity index is 30–40% or more then the ϕ_r' value can be expected to be lower than 15° (Lupini *et al.*, 1981).

SUMMARY

Stresses and strains in soils in a two-dimensional plane can be analysed using the Mohr circle construction.

The shear strength of a soil varies with the amount of strain produced. After a yield stress has been reached plastic strains occur and the soil structure initially undergoes strain-hardening followed by strain-softening.

The Mohr–Coulomb relationship is the most commonly adopted failure criterion in soil mechanics. Stress paths are a useful way of demonstrating the changes of stress required to cause failure.

A soil can fail in an undrained or a drained manner depending on its permeability and the rate of applied loading.

Due to its high permeability, sand usually shears in a drained manner. The shear strength of a sand (or a non-cohesive or granular) soil is dependent on the confining stresses, the initial density, the particle sizes, angularity, uniformity coefficient and particle crushing strength. Loose sands undergo strain-hardening during shearing and do not achieve a peak strength while dense sands display dilatancy and undergo strain-softening after reaching a peak strength.

Due to the low permeability of a clay soil, the normal rates of loading in construction works often produce the undrained condition so the undrained strength in terms of the total stresses is required. For long-term stability of structures such as a slope the strength in terms of the effective stresses is required. The triaxial test method is the most commonly used to determine the undrained and drained strength of a clay soil. The shear box test is used to determine the drained strength of a sand. Reconstituted specimens are used since it is virtually impossible to obtain an undisturbed sample of sand.

The critical state theory involving the concepts of a state boundary surface, a critical state condition, the inter-relationships between mean stress, deviator stress and volume, the effects of drainage conditions, elastic and plastic straining, yielding and hardening provides a sound framework for the understanding of basic soil behaviour.

The residual strength of a clay soil is obtained when large shear strains have occurred on a thin zone or plane of sliding. The particles have been rearranged to produce a strong preferred orientation in the direction of the slip surface and the strength is the lowest available.

CASE STUDY

Quick clays

Case Objectives:

This case illustrates:

* that quick clays are a good example of how the soil structure affects the properties of a soil – they provided the impetus for the study of soil structure
* the significance of pore water chemistry in the formation of soil structures
* the effects of subsequent post-depositional changes to the pore water chemistry
* that clay minerals are not necessarily inert and will react to chemical and physical changes around them

They are post-glacial deposits found in areas of Scandinavia and Canada and have caused considerable problems particularly flow slides on slopes. There are a number of classification systems, the table below gives one of them.

The particles greater than 2 µm (clay size) are typically quartz and feldspar and the smaller clay minerals are predominantly illite and chlorite. Their index properties are similar to other glacial clays in that they are typically of low to intermediate plasticity with liquid limit between 20 and 50%, plasticity index between 5 and 15% and with high clay contents of 30 to 70% they give low activity values usually less than 0.5 so they are inactive. Their natural moisture content is greater than the liquid limit giving values of the liquidity index of between 1.3 and 2.8 (Gillott, 1979). Regardless of this, in the undisturbed state they can have undrained strengths in the very stiff range, >200 kN/m^2.

The soils were formed by sedimentation with a high water content and open fabric produced by flocculation in saline environments and the soil structure will remain in a metastable state until interference disturbs it to trigger a collapse mechanism. It has been found that the open fabrics are usually made up of granular particles and aggregations of smaller particles which are linked together by bridges of finer particles with a wide range of pore sizes. Clay coatings on granular particles and cemented junctions are

A quick clay is one of particularly high sensitivity when on remoulding a large decrease in strength can reduce the clay from a solid mass with a reasonable intact strength to a flowing liquid as illustrated.

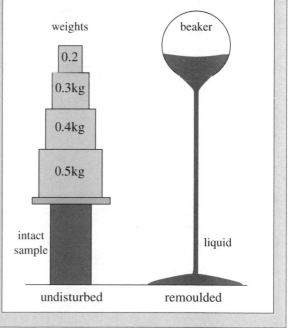

	sensitivity
insensitive	~1
slightly sensitive	1–2
medium sensitive	2–4
very sensitive	4–8
slightly quick	8–16
medium quick	16–32
very quick	32–64
extra quick	> 64

not common. The clay sizes contain a higher than normal proportion of platy primary minerals which may have resulted from glacial comminution and colloidal effects become more marked as particle sizes diminish.

The subsequent leaching by freshwater which can occur when the land rises or sea levels fall causes little change to the soil fabric but it can have a marked effect on the interparticle forces remaining, particularly the increase in repulsive forces over attraction forces in the double layer around the clay particles. One cause for this change to the interparticle forces is the depletion of divalent cations, Ca and Mg with the consequent higher percentage of monovalent cations, Na and K. Others are the formation of dispersing agents and thixotropic hardening which is described as a readjustment of the interparticle force balance which takes place over a period of time because of the viscous nature of the double layers.

The microfabric of quick clays can resemble that of similar much less sensitive clays but there must obviously be sufficient structural weaknesses in quick clay to permit the rapid destructuration that occurs. The current consensus appears to be that sensitive and quick behaviour may be induced by a number of mechanisms and a variety of small changes as summarised by Mitchell (1993) in the table below.

Mechanism	Approx. upper limit of sensitivity	Predominant soil types affected
Metastable fabric	slightly quick (8–16)	all soils
Cementation	extra quick (> 64)	soils containing iron oxide, aluminium oxide, calcium carbonate, free silica
Weathering	medium sensitive (2–4)	all soils
Thixotropic hardening	very sensitive	clays
Leaching, ion exchange, change in monovalent/divalent cation ratio	extra quick (>64)	glacial and postglacial marine clays
Formation or addition of dispersing agents	extra quick (> 64)	inorganic clays containing organic compounds in solution or on particle surfaces

Worked Example 7.1   Shear box test

The following results were obtained from a shear box test on a sample of dense sand. Determine the shear strength parameters for peak and ultimate strengths.

Normal load (N)	105	203	294
Shear load (N) at peak	95	183	265
Shear load (N) at ultimate	65	127	184

(N = Newtons)

Since the areas for both the normal stress and shear stress are the same it is not necessary to determine stress values, the shear loads can be plotted directly against the normal loads, as on Figure 7.55. Assuming $c' = 0$ the results give:

peak $\phi' = 42°$
ultimate $\phi' = 32°$

Drained conditions are assumed so total stresses equal effective stresses. The ultimate ϕ' could be taken as the value of ϕ_{cv}'.

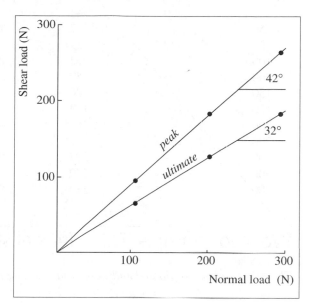

FIGURE 7.55   *Shear box test*

Worked Example 7.2   Triaxial saturation – coefficient B

The following are the results of the saturation stage of a triaxial test. Determine the value of B for each stage.

An explanation of the pore pressure parameters is given in Chapter 4. The parameter B is applied to the triaxial test to ensure the state of full saturation.

By increasing the cell pressure and the back pressure in a triaxial specimen of clay the voids are eventually filled with water by water entering the specimen from the back pressure system and air dissolving in the water.

The procedure is to first apply a cell pressure and to record the pore water pressure. Then the back pressure (in the pore water) is increased and the pore pressure recorded when it virtually equals the back pressure applied. The back pressure must be about 10 kN/m^2 less than the cell pressure to ensure positive effective stress in the soil.

B is determined from equation 4.8 and when it reaches about 0.97 the specimen may be assumed to be sufficiently fully saturated.

Cell pressure (kN/m^2)	Back pressure (kN/m^2)	Pore pressure (kN/m^2)	Δu (kN/m^2)	$\Delta\sigma_3$ (kN/m^2)	B
0	0	−4			
50	—	7	11	50	0.22
50	40	39	—	—	—
100	—	62	23	50	0.46
100	90	89	—	—	—
150	—	126	37	50	0.74
150	140	139	—	—	—
200	—	182	43	50	0.86
200	190	190	—	—	—
300	—	285	95	100	0.95
300	290	290	—	—	—
400	—	388	98	100	0.98

Worked Example 7.3   Triaxial shearing – coefficient A

The results of a consolidated undrained triaxial compression test with pore pressure measurements on a sample of saturated clay are given below. Determine the variation of the pore pressure parameter A during the test.

Cell pressure = 600 kN/m^2 throughout the test
Back pressure = 400 kN/m^2 at the start of the test
Effective cell pressure = 200 kN/m^2 = effective stress at the start of the test
The pore pressure parameter A can be obtained from Equation 4.13 assuming $B = 1$ and $\Delta\sigma_3 = 0$.

$\Delta\sigma_1-\Delta\sigma_3$	u	Δu	A
0	400	0	—
58	419	19	0.33
104	441	41	0.39
140	463	63	0.45
158	479	79	0.50
180	499	99	0.55
192	515	115	0.60

Worked Example 7.4   Consolidated undrained triaxial test

The results of a consolidated undrained triaxial compression test on a sample of fully saturated clay are given below. Each specimen has been consolidated to a back pressure of 200 kN/m^2.

Parameter (kN/m^2)	Specimen 1	Specimen 2	Specimen 3
Cell pressure	300	400	600
Deviator stress at failure	326	416	635
Pore pressure at failure	146	206	280

Calculate the required stresses and plot them on a Mohr cirle diagram to obtain the effective stress shear strength parameters for the clay.

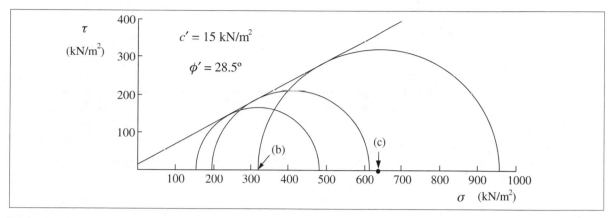

FIGURE 7.56 *Consolidated undrained triaxial test*

At failure:

Parameter	Specimen 1	Specimen 2	Specimen 3
σ_3	300	400	600
u_f	146	206	280
(b) σ_3'	154	194	320
$\sigma_1-\sigma_3$	326	416	635
(c) $\sigma_3' + \frac{1}{2}(\sigma_1' - \sigma_3')$	317	402	638

(c) represents the centre of the Mohr circle and (b) the point of circumscription, see Figure 7.56. The pore pressure parameter at failure, A_f for each specimen can be obtained from equation 4.13 assuming that $B = 1$ and $\Delta\sigma_3 = 0$.

Parameter	Specimen 1	Specimen 2	Specimen 3
Δu	−54	6	80
$\Delta\sigma_1-\Delta\sigma_3$	326	416	635
A_f	−0.17	0.01	0.13

Worked Example 7.5 Stress paths *t* and *s*

From the results of the consolidated undrained triaxial compression test given in Example 7.3 determine the values of t and s (total stresses) and t' and s' (effective stresses) and plot the stress paths.

$s = \sigma_3 + t \quad t' = t \quad s' = s - u$

$\sigma_1 - \sigma_3$	u	t	s	t'	s'
0	400	0	600	0	200
58	419	29	629	29	210
104	441	52	652	52	211
140	463	70	670	70	207
158	479	79	679	79	200
180	499	90	690	90	191
192	515	96	696	96	181

The total and effective stress paths are plotted on Figure 7.57.

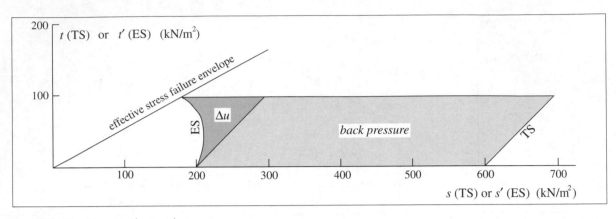

FIGURE 7.57   *Stress paths t and s*

Worked Example 7.6   Stress paths *q* and *p*

From the results of the consolidated undrained triaxial compression test given in Example 7.3 determine the values of q and p (total stresses) and q' and p' (effective stresses) and plot the stress paths.

$q = q' = \sigma_1' - \sigma_3'$
$p = \frac{1}{3}q + \sigma_3$
$p' = p - u$
$\sigma_3 = 600 \text{ kN/m}^2$

u	q	p	q'	p'
400	0	600	0	200
419	58	619	58	200
441	104	635	104	194
463	140	647	140	184
479	158	653	158	174
499	180	660	180	161
515	192	664	192	149

The total and effective stress paths are plotted on Figure 7.58.

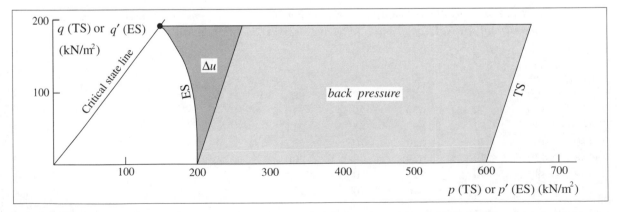

FIGURE 7.58   *Stress paths q and p*

Worked Example 7.7   Critical state approach to a drained test

A sample of clay has been isotropically normally consolidated to a stress $p_o' = 200$ kN/m^2 and a final void ratio of 0.92 and is sheared in a drained triaxial compression test. For the soil constants given below calculate q' and p' at failure and the final specific volume and volumetric strain at failure.

$\lambda = 0.16$ $\Gamma = 2.76$ $M = 0.89$

From Figure 7.47 the specimen state commences on the ICL and during shearing it rises up the Roscoe surface and over a drained plane to failure at the CSL.

The gradient of the drained plane in $p'-q'$ space is 3 so p' at failure is given by:

$$p_f' = p_o' + \tfrac{1}{3}q_f'$$

From Equation 7.36, $q_f' = M p_f'$

so $p_f' = p_o' + \tfrac{1}{3}M p_f'$

giving: $p_f' = \dfrac{p_0}{1 - \tfrac{1}{3}M} = \dfrac{200}{1 - \tfrac{1}{3} \times 0.89} = 284$ kN/m^2

$\therefore\ q_f' = 0.89 \times 284 = 253$ kN/m^2

Equation 7.37 gives the specific volume for the critical state line so the final specific volume is:

$v_f = 2.76 - 0.16 \ln 284 = 1.856$

The change in volume during shear is

$\Delta v = 1.920 - 1.856 = 0.064$

so the volumetric strain is

$$\dfrac{\Delta v}{v} = \dfrac{0.064}{1.920} \times 100 = 3.3\%$$

Exercises

7.1 The normal total stress acting on a plane within a soil is 185 kN/m^2 and the effective stress shear strength parameters are:

peak $c_p' = 5$ kN/m^2 $\phi_p' = 29°$ residual $c_r' = 0$ $\phi_r' = 12°$

When the pore water pressure in the soil is 90 kN/m^2 determine the maximum shear strength which can be obtained on the plane and the shear strength after considerable strain.

7.2 The vertical total stress acting on an element of soil in the ground is 125 kN/m^2 and the pore water pressure is 32 kN/m^2. The effective stress shear strength parameters of the soil are $c' = 8$ kN/m^2 and $\phi' = 28°$. The horizontal total stress is measured to be 97 kN/m^2. Confirm that the soil element is not at failure. (Hint: consider simply the apex of the Mohr circle for the *in situ* condition and compare this with the available shear strength, see Figure 7.18)

7.3 In Exercise 7.2, using equations 7.9 to 7.11, determine the lowest horizontal effective stress the soil can sustain before failure occurs. Assume the vertical stress remains constant. The horizontal stress at failure is referred to as the active pressure and a simpler method for its determination is given in Chapter 11.

7.4 From the results of the consolidated undrained test given in Exercise 4.10 for each reading determine the stresses:

(a) $\sigma_1, \sigma_1', \sigma_3'$
(b) the stress path values s' and t' and p' and q
(c) plot the stress path s' versus t' and determine the effective stress shear strength parameter ϕ'.
(d) plot the stress path p' versus q and determine the value of M.

7.5 A shear box test carried out at a slow rate of strain on 'undisturbed' specimens of a cemented sand gave the following results at failure:

Normal stress kN/m^2	40	80	120
Shear stress kN/m^2	47.5	80.0	112.0

Determine the effective stress shear strength parameters of the soil.

7.6 On a potential plane of sliding in a mass of this cemented sand the normal stress is estimated to be 100 kN/m^2 and the shear stress is 64 kN/m^2. Determine the factor of safety against failure.

7.7 The results of unconsolidated undrained (UU) tests on three similar specimens of a fully saturated clay soil at failure are given below. Determine the deviator stress and shear strength for each specimen.

	Cell pressure (kN/m^2)	Δl (mm)	Axial force (N)
(i)	100	11.4	331
(ii)	200	9.6	309
(iii)	400	8.5	312

Initial length = 76 mm   Initial diameter = 38 mm

7.8 The results of consolidated undrained (CU) tests on three similar specimens of a fully saturated clay soil at failure are given below.

	Cell pressure (kN/m²)	Deviator stress (kN/m²)	Pore pressure (kN/m²)
(i)	350	143	326
(ii)	400	208	354
(iii)	500	312	418

(a) Determine the effective stresses σ_1' and σ_3' for each specimen.
(b) Plot the Mohr circles and determine the effective stress shear strength parameters c' and ϕ'.
(c) Plot the stress path points s' and t', determine the values a and α and derive c' and ϕ'.
The back pressure used for all three specimens was 300 kN/m².

7.9 The results of consolidated drained (CD) tests on three similar specimens of a fully saturated clay soil at failure are given below.

	Cell pressure (kN/m²)	Axial force (N)	Δl (mm)	ΔV (ml)
(i)	100	300	7.75	2.8
(ii)	200	545	8.97	4.4
(iii)	300	787	10.34	5.9

$l_0 = 76$ mm $A_0 = 1140$ mm² $V_0 = 86.65$ ml

$\varepsilon_v = \Delta V/V_0$ $\varepsilon_a = \Delta l/l_0$

(i) Determine the effective stresses σ_1' and σ_3' for each specimen.
(ii) Plot the Mohr circles and determine the effective stress shear strength parameters c' and ϕ'.
(iii) Plot the stress path points s' and t', determine the values a and α and determine c' and ϕ'.

7.10 Specimens of fully saturated normally consolidated clay have been isotropically consolidated to 150 and 300 kN/m² with void ratios of 0.75 and 0.65, respectively. For the isotropic normal consolidation line (ICL) determine the values of N and λ for this clay.

7.11 The specimen consolidated to 300 kN/m² in Exercise 7.10 has been sheared to failure under drained conditions with the deviator stress at failure (at the critical state) of 368 kN/m². Determine the value of M for this clay.

7.12 What would be the deviator stress at failure in a drained test for the specimen consolidated to 150 kN/m² in Exercise 7.10?

7.13 In Exercises 7.11 and 7.12 determine the final void ratios for the specimens consolidated to 150 and 300 kN/m². $\Gamma = 2.37$.

7.14 Specimens of the clay in Exercise 7.10 are sheared to failure under undrained conditions. Determine the deviator stress and the pore pressure at failure for the specimens consolidated initially to 150 and 300 kN/m². $\Gamma = 2.37$.

Shallow foundations – stability

Shallow foundations

The term 'shallow foundations' refers to spread foundations as opposed to pile foundations. They are usually placed at shallow depths such as a depth less than their width, i.e. $D < B$. Alternatively, they may be considered as being constructed within the reach of normal excavation plant, i.e. D less than about 3 m where D/B may be up to 3. If they are open below ground level they would be treated as basement foundations or buoyant foundations.

Spread foundations

This description is often used to show that these foundations convert a localised line load or point load from the structure into a pressure by spreading it over the area of the foundation. The pressure then applied by the foundation must be supported by the soil. For line loading on a strip foundation:

$$\text{width } B = \frac{\text{wall line load (kN/m)}}{\text{applied bearing pressure}} \qquad (8.1)$$

For point loading on a pad foundation:

$$\begin{aligned}
\text{area } A &= \text{Length } L \times \text{width } B \\
&= \frac{\text{column load (kN)}}{\text{applied bearing pressure}}
\end{aligned} \qquad (8.2)$$

Raft foundations are of necessity the same or a similar size to the whole building itself and are designed to spread both line loads and point loads as much as possible over their whole area by virtue of their stiffness.

Design requirements (Figure 8.1)

The purposes of a foundation are to ensure:

● *Stability* of the structure supported.
  Failure could occur due to inadequate bearing capacity, overturning or sliding. There must be an adequate factor of safety against this occurrence. Conventional design is carried out by ensuring that the applied bearing pressure does not exceed the safe bearing capacity, q_s which is the ultimate bearing capacity, q_{ult} divided by an appropriate overall factor of safety F (Figure 8.1).

  In limit state design this is stated as verifying

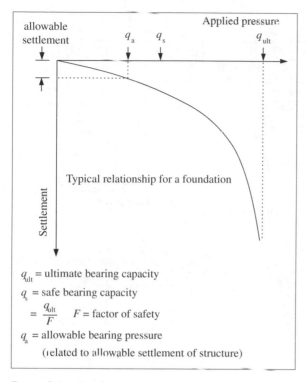

FIGURE 8.1 *Bearing pressures*

the inequality given in equation 8.30 (see later) such that the design action (load) does not exceed the design bearing resistance. It is then assumed that the foundation and the structure supported will not fail due to the ultimate limit state mechanisms considered.

● *Settlements* do not exceed the tolerable limits of the structure supported by the foundation.

The applied bearing pressure must not exceed the allowable bearing pressure, q_a otherwise settlements greater than the allowable or tolerable settlement of the structure will be produced (Figure 8.1). Settlement analyses are considered separately in Chapter 9. In limit state design this is stated as not exceeding a serviceability limit state.

If the settlements are greater than the allowable settlements then the foundation will have exceeded the serviceability limit state and can be considered to have 'failed'. Thus design of foundations should be more strictly based on the allowable bearing pressure rather than the safe bearing capacity.

● *Other ground movements* do not adversely affect the structure.

These are also serviceability limit states. Many of these ground movements are associated with the depth of the foundation, discussed below, while others such as mining subsidence, earthquakes, vibrations will require special measures.

Types of foundation

The type and shape of the loading generally determines the shape of a foundation, e.g. strip foundation beneath a wall; square pad foundation beneath a column; circular foundation beneath a circular tank; raft foundation beneath the whole of a building. Nearly all modern foundations will be constructed of concrete and all but the simplest will contain steel reinforcement, to provide bending and shear resistance and enable loads to be spread over wider areas.

Strip foundations (Figure 8.2)

Strip foundations are designed to provide sufficient width either as plain concrete for lightly loaded walls such as domestic properties or as reinforced concrete for heavier retaining walls.

For plain concrete foundations adequate stability in the concrete can be achieved if the thickness T is greater than the projection P from the face of the wall. However, equation 8.1 shows that for heavy loads or weak ground the width B and hence the thickness T must be large. To limit the depth of foundation and the amount of concrete required plain concrete foundations are restricted to narrow widths and small loads.

The efficiency of reinforcement in concrete means that wide foundations at shallow depths can be provided economically using reinforced concrete with transverse reinforcement at the base of the foundation resisting the bending stresses.

These foundations often traverse variable ground or weak spots, therefore it is advisable to provide at least nominal reinforcement longitudinally to increase bending resistance or stiffness along the length of the wall. This will reduce differential settlements and minimise the risk of cracking in the wall.

Deep strip foundations are difficult and expensive to construct since they require placing brickwork in a narrow confined trench, therefore the trench fill

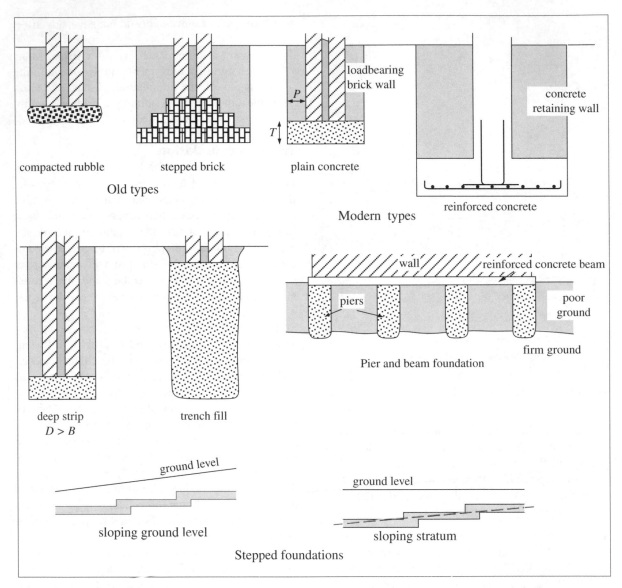

FIGURE 8.2 *Strip foundations*

foundation is often preferred. This entails placing concrete into a trench excavation immediately on reaching a suitable bearing stratum. It may be more expensive in concrete but it is quicker to construct, saves brickwork, does not rely on trench support and provides a good surface for the start of bricklaying. Alternatively, mass concrete piers could be constructed in a similar fashion to support a reinforced concrete beam spanning between the piers, referred to as a pier and beam foundation.

Pad foundations (Figure 8.3)
Unreinforced pad foundations are only used for small point loads otherwise excessive thicknesses T are necessary to provide the width required.

Pad foundations are often square to enable easier placing of reinforcement. However, where eccentric or inclined loading occurs rectangular pads with their long axis in line with the eccentricity will be more effective.

Where an external column has to be placed close

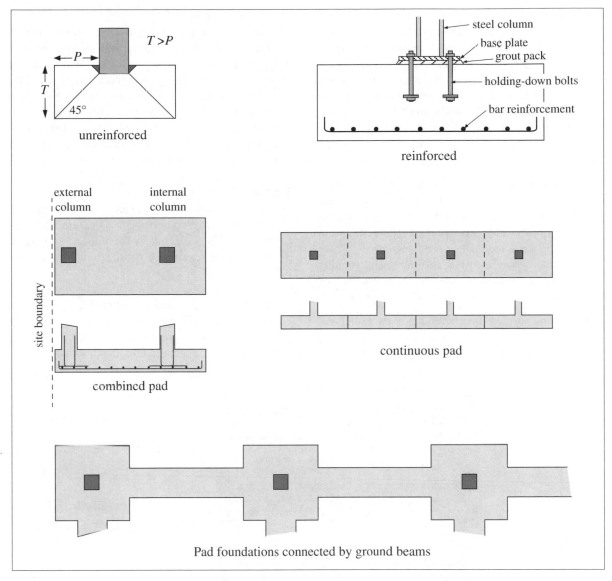

FIGURE 8.3 *Pad foundations*

to a site boundary it may be preferable to support it on a combined pad foundation where the balancing force of the internal column can be used.

When pad foundations are fairly close to each other it may be easier to construct them as continuous pads (by making them rectangular) to provide simpler excavation and easier formwork, reinforcement and concrete construction. They may be individually reinforced and, therefore, independent or they may be continuously reinforced to provide a strip foundation with greater overall longitudinal stiffness to minimise differential settlements. This foundation will then provide support for the infill walls.

Smaller pads which are further apart can be made to interact to transfer and share horizontal loading, moment loading and additional loading due to differential settlements by providing interconnecting ground beams.

Raft foundations (Figure 8.4)

Raft foundations which are very flexible with little stiffness will settle into a dish-shaped profile which may cause distress (distortion, cracking etc.) to the structure. The purpose of a raft foundation must be, therefore, to keep differential settlements to within the tolerable limits of the particular type of structure by providing sufficient stiffness, or resistance to bending deflections. This is achieved by interaction between:

1. A continuously reinforced concrete slab beneath the whole of the building. The reinforcement must be placed in two directions and at top and bottom of the slab, a convenient type being mesh or fabric reinforcement.
2. Concrete beams with bar reinforcement, running beneath the structural walls or beneath lines of columns.

Increasing stiffness is provided by the number and depth of beams and the amount of reinforcement in them. Some common types of raft foundations are illustrated in Figure 8.4.

If the differential settlements were reduced to zero the foundation would be described as rigid or infinitely stiff. This may seem a desirable goal but would mean designing for very large bending moments and since it has been established that most ordinary structures can tolerate some differential settlements a more economical solution is obtained by designing for the required stiffness to provide no more than the tolerable settlements.

Depth of foundations

Foundations must not be placed close to ground level (existing or proposed) because natural agencies cause ground movements in the upper horizons of the soil. Some examples of depth considerations follow.

Adequate bearing stratum (Figure 8.5)

Foundations should be taken below incompetent strata such as made ground or peat and founded in competent strata beneath. This depth should not be

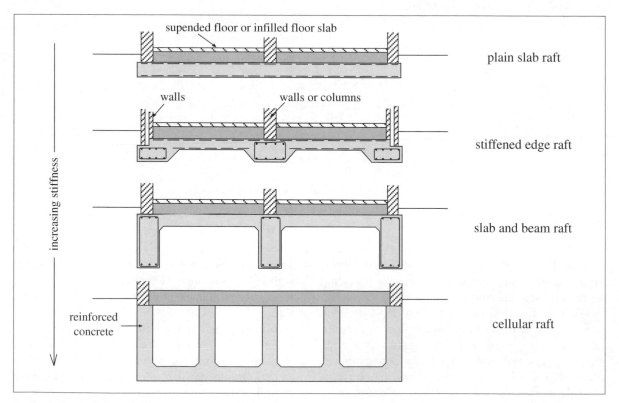

FIGURE 8.4 *Raft foundations*

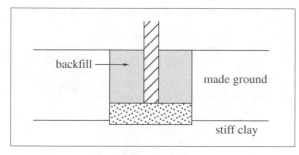

FIGURE 8.5 *Adequate bearing stratum*

considered for depth effects in settlement analyses or depth factors in bearing capacity calculations because it is not continuous with the ground supporting the foundation but it could be used to determine net bearing pressures.

Seasonal moisture variations (Figure 8.6)

In a clay soil the moisture content of the upper horizon can reduce considerably in dry periods and increase in wet periods producing volume changes and hence ground movements. These volume changes will be more severe during prolonged dry or wet periods (i.e. summer and winter) so the most noticeable movements are generally seasonal.

The amount of movement diminishes with depth (Figure 8.6). Thus there is a zone of seasonal moisture variations below which movements are insignificant. The depth to which movements occur will depend on the climate and the susceptibility of the soil to shrinkage or swelling – the shrinkage potential. An estimate of shrinkage potential based on classification tests is given in Table 8.1 (from *BRE Digest 240*).

The National House-Building Council (NHBC, 1992) has produced detailed recommendations on

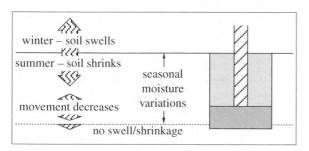

FIGURE 8.6 *Seasonal moisture variations*

TABLE 8.1 *Clay shrinkage potential*

Plasticity index %	Clay fraction %	Shrinkage potential
> 35	> 95	very high
22–48	60–95	high
12–32	30–60	medium
< 18	< 30	low

From BRE Digest 240, 1980

TABLE 8.2 *Foundation depths for clay soils unaffected by tree roots*

Shrinkage potential	Plasticity index %	Minimum depth (m)
high	> 40	1.0
medium	20–40	0.9
low	10–20	0.75

From NHBC 1992

depths of foundations in the UK. For a clay site unaffected by the presence of tree roots the depth of foundation (and zone of seasonal moisture variations) is given in Table 8.2, related to the shrinkage potential of the clay. These depths only apply outside the 'exclusion zone' around a tree, see below.

Effects of tree roots (Figures 8.7 and 8.8)

In 1976, 1985 and 1990 in the UK severe droughts occurred in the summer months forcing tree roots to extend in search of moisture beneath many brittle structures (load-bearing brickwork and plaster domestic properties). The resulting volume changes and ground movements caused varying degrees of damage (moderate to very severe) and resulted in insurance claims totalling many millions of pounds.

The radius of the exclusion zone D around a tree (Figure 8.7) has been related to the mature height of the tree H and the species of the tree. The Building Research Establishment (BRE) and the NHBC have recommended typical values of this exclusion zone, some of which are given in Tables 8.3 and 8.4, respectively. Providing foundations are outside the exclusion zone then no special precautions will be necessary.

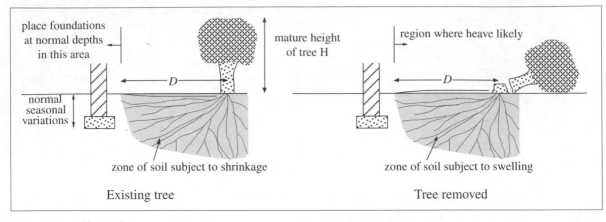

FIGURE 8.7 *Effects of tree roots*

If shallow foundations are to be sited within the exclusion zone then they must be placed down to considerable depths (*BRE Digest 298*) or, alternatively, a specially designed bored pile foundation should be provided. The latter is probably more economical and more effective (*BRE Digests 241 and 242*).

The most severe mode of deformation in which a structure can be placed is the hogging mode, as discussed in Chapter 9. This condition will be produced by shrinkage or swelling of a clay soil beneath part of a structure (Figure 8.8). Shrinkage will be caused by existing trees and planting new trees. Swelling or heave will be caused by removing trees prior to construction.

A clay soil will also shrink or swell in a horizontal direction, especially at shallow depths where the horizontal movements may far exceed the vertical movements. These horizontal movements can produce severe distortion of buried structures such as trench fill foundations, basements and services.

Frost action (Figure 8.9)

Permanently frozen ground (permafrost) exists in the cold regions of northern Canada and northern Russia, which understandably are areas of sparse population. In Europe, the Gulf Stream provides a more equable climate and frost action is less severe.

Nevertheless, during prolonged cold periods when the air temperature remains below freezing heat is conducted away from the earth's surface and a freezing front penetrates into the soil. As small pockets of water freeze and crystallise they give off latent heat. If this heat balances the heat lost by conduction to the earth's surface the freezing front becomes stationary and ice lens formation proceeds, fed by free

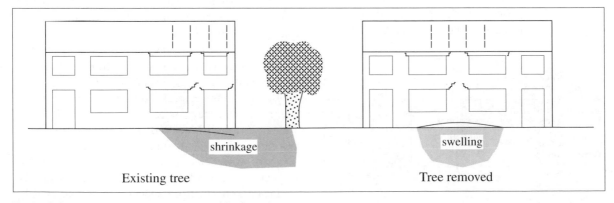

FIGURE 8.8 *Property movements caused by hogging*

TABLE 8.3 *Ranking of tree species known to have caused damage (from BRE Digest 298, 1987)*

Ranking [1]	Species	Maximum tree height H (m)	75 % of damage cases occurred within this distance (m)	Minimum recommended separation in very highly and highly shrinkable clays D (m)
1	oak	16–23	13	1 H*
2	poplar	24	15	1 H*
3	lime	16–24	8	0.5 H
4	common ash	23	10	0.5 H
5	plane	25–30	7.5	0.5 H
6	willow	15	11	1 H
7	elm	20–25	12	0.5 H
8	hawthorn	10	7	0.5 H
9	maple/sycamore	17–24	9	0.5 H
10	cherry/plum	8	6	1 H*
11	beech	20	9	0.5 H
12	birch	12–14	7	0.5 H
13	whitebeam/rowan	8–12	7	1 H*
14	cypress	18–25	3.5	0.5 H

1   In descending order of severity

*   Zone affected could be reduced to $0.5H$ for soils of low to medium shrinkage potential

TABLE 8.4 *Zone affected by tree roots – clay soils (from NHBC, 1992)*

Water demand	Maximum D/H		Typical tree species	Mature height H (m)
	coniferous	broad-leaved		
high	0.6	1.25	elm, oak, poplar, willow	16–24
moderate	0.35	0.75	ash, cherry, chestnut, lime, maple, plane, sycamore	10–20
low	–	0.5	beech, birch, holly	10–20

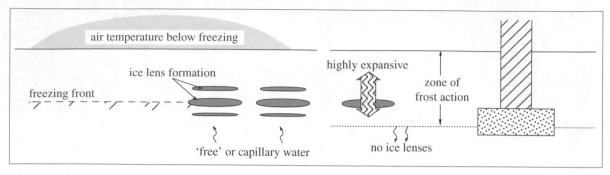

FIGURE 8.9 *Frost action*

water or capillary water from below. Ice formation is expansive so ground heave will result and when the ice thaws a very weak morass will be formed.

Some soils are more 'frost-susceptible' than others, allowing ice lens formation more easily. In particular, soils having a permeability comparable to a silt are prone to ice lens formation, e.g. silts, sands with moderate fines content and laminated or varved clays.

For highway pavements in the UK no frost-susceptible materials (including the subgrade) should be present within 450 mm of the ground surface. If damage does occur due to a particularly severe frost and especially at the edges of a carriageway then the road surface can be repaired relatively easily. However, for foundations of structures that are less easy to repair a minimum depth of 600 mm is recommended. For soils less susceptible to frost action such as clean and 'dry' sands or gravels shallower depths may be permitted.

River erosion (Figure 8.10)

Erosion of a river bed depends on the depth of water flowing and the erodibility of the soil forming the river bed. When river levels rise general scour of the river bed will occur, producing deeper channels, possibly in a different location to the normal deep channel. The deep flood channel could occur at different locations during subsequent floods.

When structures such as bridge piers and abutments obstruct the river flow additional local scour will occur which could extend below a shallow foundation causing an overturning failure and resulting in catastrophic collapse of the bridge. The case study following illustrates the importance of this problem.

The foundations must either be placed below the anticipated depth of scour or scour prevention measures such as sheet pile skirts, mattresses and rip-rap should be provided.

Water table (Figure 8.11)

In granular soils foundations should be kept as high above a water table as possible to avoid construction problems and to minimise settlements. However, water tables can fluctuate with higher levels during

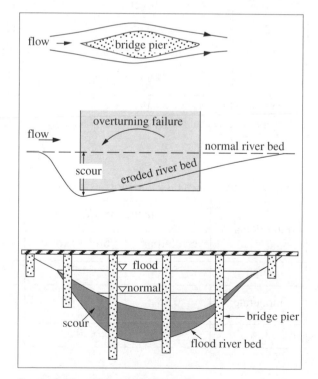

FIGURE 8.10 *River erosion*

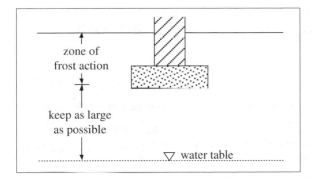

FIGURE 8.11 *Water table location*

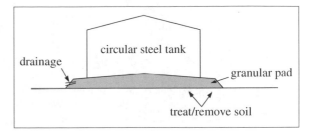

FIGURE 8.12 *Super-elevation*

prolonged wet periods and in the winter months. For design purposes the most severe level should be considered.

Super-elevation (Figure 8.12)
Steel storage tanks (for oil, liquids, chemicals) are usually placed above ground level to facilitate emptying and to prevent water collecting around and beneath the tank which could cause corrosion. The existing ground may require treatment or removal to eliminate the problems of shrinkage and swelling due to seasonal moisture movements or frost heave, particularly beneath the edges of the tanks.

Bearing capacity

Modes of failure (Figure 8.13)
Failure is defined as mobilising the full value of shear strength of the soil accompanied by large and excessive settlements. The mechanism causing failure for shallow foundations depends on the soil type, particularly its compressibility, and the type of loading.

If the soil is compressible, vertical displacements of the foundation are produced by volume reductions throughout the soil beneath the foundation with limited shear distortion and mass movement. This is described as punching shear.

If the soil is incompressible then volume reductions are not possible. Vertical displacements of the foundation can only be produced by mass movements with shear distortion occurring along a slip surface. This mode is described as general shear. The local shear mode is a transition between these two mechanisms. All of these mechanisms would be described as ultimate limit states since they involve the stability of the structure.

When a horizontal load as well as a vertical load is applied to a foundation there is a risk of failure by overturning due to the resultant inclined load, when a reduced bearing capacity is obtained. There must also be sufficient adhesion in the case of a clay or friction in the case of a sand at the underside of a foundation to prevent sliding due to the horizontal load component.

Bearing capacity – vertical load only
(Figure 8.14)
An expression for bearing capacity has only been derived for the general shear mode of failure (assuming the soil to behave as an incompressible solid) and is based on superposition of three components. Nevertheless, factors can be applied to compensate for the effects of compressibility and the superposition of the three components gives a conservative approach.

The expression is based on the failure mechanism illustrated in Figure 8.14 where the soil in zones I, II and III is in a state of plastic equilibrium. The active Rankine zone I is pushed downwards and in turn pushes the radial shear zones II sideways and the passive Rankine zones III sideways and upwards.

At failure the movement of these masses is mobilising the full shear strength of the soil which is obtained from the Mohr–Coulomb shear strength parameters c and ϕ and the total or effective stresses in the soil. These stresses are determined from the self-weight of the soil (due to gravity) and from the surcharge pressure acting at foundation level around the foundation (due to stress distribution).

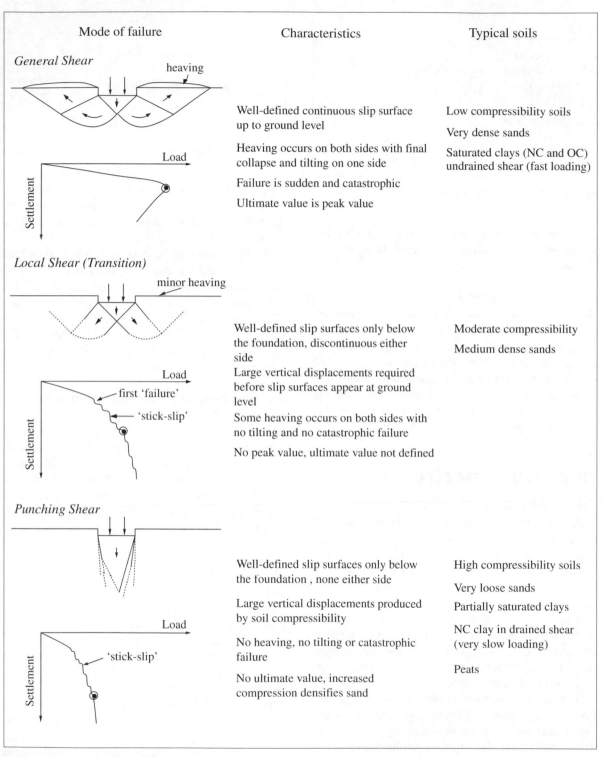

Mode of failure	Characteristics	Typical soils
General Shear	Well-defined continuous slip surface up to ground level	Low compressibility soils
	Heaving occurs on both sides with final collapse and tilting on one side	Very dense sands
	Failure is sudden and catastrophic	Saturated clays (NC and OC) undrained shear (fast loading)
	Ultimate value is peak value	
Local Shear (Transition)	Well-defined slip surfaces only below the foundation, discontinuous either side	Moderate compressibility
	Large vertical displacements required before slip surfaces appear at ground level	Medium dense sands
	Some heaving occurs on both sides with no tilting and no catastrophic failure	
	No peak value, ultimate value not defined	
Punching Shear	Well-defined slip surfaces only below the foundation , none either side	High compressibility soils
	Large vertical displacements produced by soil compressibility	Very loose sands
	No heaving, no tilting or catastrophic failure	Partially saturated clays
	No ultimate value, increased compression densifies sand	NC clay in drained shear (very slow loading)
		Peats

FIGURE 8.13 *Modes of failure*

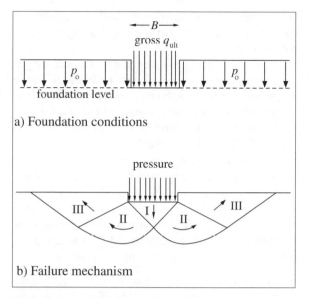

FIGURE 8.14   *Bearing capacity*

TABLE 8.5   *Bearing capacity factors*

ϕ	N_c	N_q	N_γ
0	5.14	1.0	0
1	5.4	1.1	0
2	5.6	1.2	0
3	5.9	1.3	0
4	6.2	1.4	0
5	6.5	1.6	0.1
6	6.8	1.7	0.1
7	7.2	1.9	0.2
8	7.5	2.1	0.2
9	7.9	2.3	0.3
10	8.4	2.5	0.4
11	8.8	2.7	0.5
12	9.3	3.0	0.6
13	9.8	3.3	0.8
14	10.4	3.6	1.0
15	11.0	3.9	1.2
16	11.6	4.3	1.4
17	12.3	4.8	1.7
18	13.1	5.3	2.1
19	13.9	5.8	2.5
20	14.8	6.4	3.0
21	15.8	7.1	3.5
22	16.9	7.8	4.1
23	18.1	8.7	4.9
24	19.3	9.6	5.7
25	20.7	10.7	6.8
26	22.3	11.9	7.9
27	23.9	13.2	9.3
28	25.8	14.7	10.9
29	27.9	16.4	12.8
30	30.1	18.4	15.1
31	32.7	20.6	17.7
32	35.5	23.2	20.8
33	38.6	26.1	24.4
34	42.2	29.4	28.8
35	46.1	33.3	33.9
36	50.6	37.8	40.0
37	55.6	42.9	47.4
38	61.4	48.9	56.2
39	67.9	56.0	66.8
40	75.3	64.2	79.5
41	83.9	73.9	95.1
42	93.7	85.4	114.0
43	105.1	99.0	137.1
44	118.4	115.3	165.6
45	133.9	134.9	200.8
46	152.1	158.5	244.7
47	173.6	187.2	299.5
48	199.3	222.3	368.7
49	229.9	265.5	456.4
50	266.9	319.1	568.5

The three components of the Terzaghi (1943) bearing capacity equation are:

$$gross\ q_{ult} = cN_c + p_oN_q + 1/2\gamma BN_\gamma \qquad (8.3)$$

where:

cN_c is due to cohesion and friction in the soil
p_oN_q is due to surcharge and friction in the soil
$1/2\gamma BN_\gamma$ is due to self-weight and friction in the soil
p_o = total overburden pressure at foundation level around the foundation (Figure 8.14)
γ = bulk unit weight of soil
B = width of foundation
N_c, N_q and N_γ are termed bearing capacity factors and are related to the ϕ value only.

Values of N_c and N_q, attributed to Prandtl and Reissner are given by:

$$N_c = (N_q - 1) \cot \phi \qquad (8.4)$$

$$N_q = \exp(\pi \tan \phi) \tan^2\left(45 + \frac{\phi}{2}\right) \qquad (8.5)$$

Values of N_γ have been obtained by several authors adopting different rupture figures to the ones used by Prandtl and Reissner, therefore superposition of the three components can only be an approximation, albeit a safe one. However, the contribution of the third term to bearing capacity is not signifi-

TABLE 8.6 *Shape factors (from Vesic, 1975)*

Shape of foundation	s_c	s_q	s_γ
strip	1.0	1.0	1.0
rectangle	$1 + \dfrac{B'}{L'} \dfrac{N_q}{N_c}$	$1 + \dfrac{B'}{L'} \tan\phi$	$1 - 0.4 \dfrac{B'}{L'}$
circle or square	$1 + \dfrac{N_q}{N_c}$	$1 + \tan\phi$	0.6

cant for narrow foundations so the more conservative expression given by Brinch Hansen (1970) is suggested:

$$N_\gamma = 1.5\,(N_q - 1)\,\tan\phi \qquad (8.6)$$

Values of N_c, N_q and N_γ from the above expressions are given in Table 8.5.

Shape and depth factors
(Tables 8.6 and 8.7)

The original Terzaghi equation was derived for a very long (strip) foundation where shearing in only two dimensions was assumed. However, for rectangular foundations shearing of the soil will also occur at the ends, producing an enhanced 'end effect'. For circular and square foundations a three-dimensional mass of soil will be sheared. These effects are catered for in a semi-empirical manner by modifying the Terzaghi equation (equation 8.3) using the shape factors s_c and s_q.

The shape factor s_γ associated with the self-weight term provides a reduction of bearing capacity due to shape because of the reduced confinement of the soil at the ends of rectangular, square or circular foundations. Values of the shape factors summarised from Vesic (1975) are given in Table 8.6.

Depth factors d_c, d_q and d_γ provide for the additional bearing capacity which could be obtained from shearing through the soil above foundation level. However, the original Terzaghi equation assumed that the soil above foundation level did not contribute to bearing capacity and there are good reasons for retaining this assumption such as:

- The soil above foundation level is usually inferior to the supporting soil. This is a major reason for taking foundations deep below ground level, e.g. made ground or shrinkable clay.
- Mobilising shearing resistance over this section depends on the soil below foundation level being virtually incompressible.

Expressions for depth factors, d_c, d_q and d_γ have been given by Brinch Hansen (1970) and Vesic (1975) and are reproduced in Table 8.7 but these should be used with caution.

See Worked Example 8.1

TABLE 8.7 *Depth factors (from Vesic, 1975)*

ϕ value	d_c		d_q		d_γ
$\phi = 0$ Clay undrained	$\dfrac{D}{B'} \le 1$ $\dfrac{D}{B'} > 1$	$1 + 0.4\dfrac{D}{B'}$ $1 + 0.4\tan^{-1}\dfrac{D}{B'}$ $\dfrac{D}{B'}$ in radians		1.0	1.0
$\phi > 0$ Clay drained Sand	$d_q - \dfrac{1 - d_q}{N_c \tan\phi}$		$\dfrac{D}{B'} \le 1$ $\dfrac{D}{B'} > 1$	$1 + 2\tan\phi\,(1 - \sin\phi)^2 \dfrac{D}{B'}$ $1 + 2\tan\phi\,(1 - \sin\phi)^2 \tan^{-1}\dfrac{D}{B'}$ $\dfrac{D}{B'}$ in radians	1.0

Use these factors with caution − see text

Bearing capacity – overturning
(Figure 8.15)

Many foundations have to be designed for the effects of horizontal loading or moment loading transferred via the structure to the foundation both of which could produce overturning. In order to proceed with the analysis it is necessary to resolve the loading into a single total vertical load V_T acting at the underside of the foundation (foundation level) with the moment and horizontal load effects catered for by placing the total load at an eccentricity e away from the centroid of the foundation.

If a horizontal load exists it is then applied at foundation level where it no longer contributes to overturning but it does produce a resultant inclined loading. Both eccentric and inclined loading produce a reduced bearing capacity and they are catered for in separate ways. Adequate resistance to sliding of the foundation subjected to the horizontal load must be provided.

Eccentric loading (Figure 8.16)

The location of V_T may be eccentric along both the long and short axes producing eccentricities e_L and e_B, respectively. A convenient and conservative approach attributed to Meyerhof (1953) is suggested

for the analysis of eccentric loading.

This assumes that V_T acts centrally on an effective foundation area (ignoring the area outside of this effective area) so that the bearing capacity equation for central vertical loading can be used. From then on only the effective width B' and effective length L' are used in the equations. Expressions for B' and L' are given in Figure 8.16 derived from:

$$\frac{B}{2} = \frac{B'}{2} + e_B \qquad (8.7)$$

SEE WORKED EXAMPLE 8.3

Inclined loading (Figure 8.17)

This produces a smaller failure zone and hence reduced bearing capacity. Following the work of several authors (summarised in Vesic, 1975) the general bearing capacity expression has been modified to:

$$gross\ q_{ult} = cN_c\,s_c\,d_c\,i_c + p_oN_q\,s_q\,d_q\,i_q + 1/2\gamma B'N_\gamma s_\gamma\,d_\gamma i_\gamma \qquad (8.8)$$

where i_c, i_q and i_γ are inclination factors, expressions for which are given in Table 8.8.

SEE WORKED EXAMPLE 8.4

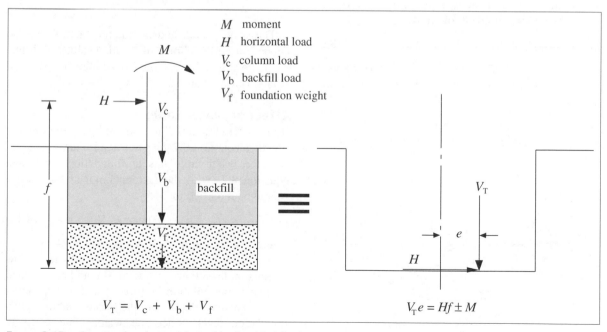

$$V_T = V_c + V_b + V_f \qquad\qquad V_T e = Hf \pm M$$

M   moment
H   horizontal load
V_c column load
V_b backfill load
V_f foundation weight

FIGURE 8.15   *Overturning (eccentric and inclined loading)*

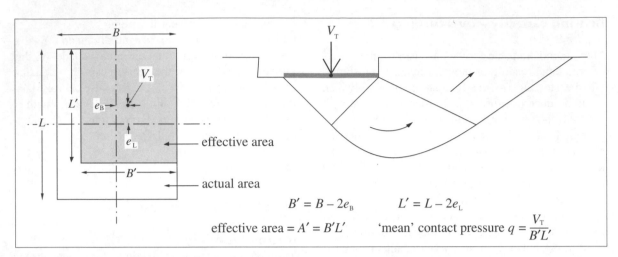

$$B' = B - 2e_B \qquad L' = L - 2e_L$$

$$\text{effective area} = A' = B'L' \qquad \text{'mean' contact pressure } q = \frac{V_T}{B'L'}$$

FIGURE 8.16 *Eccentric loading – effective area method*

Different soil strength cases

There are three cases usually considered in soil mechanics:

Case (a) is the *undrained* condition in clay (short-term case) and is represented by the shear strength parameters:

cohesion $c = c_u$ and $\phi = \phi_u = 0°$

giving the bearing capacity expression, equation 8.7 (with the factors removed for clarity) as:

$$gross \; q_{ult} = c_u N_c + p_o \qquad (8.9)$$

Case (b) is the *drained* condition in clay (long-term case) and is represented by the shear strength parameters:

cohesion $c = c'$ and $\phi = \phi' \; (> 0)$
giving the bearing capacity expression, equation 8.7 as:

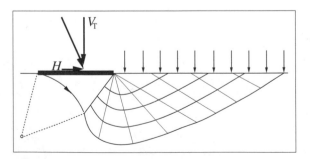

FIGURE 8.17 *Inclined loading*

$$gross \; q_{ult} = c'N_c + p_o N_q + 1/2 \; \gamma \; B' \; N_\gamma \qquad (8.10)$$

Case (c) is the *drained* condition in sand (short and long-term case) and is represented by the shear strength parameters:

cohesion $c = 0 \quad \phi = \phi' \; (> 0)$

giving the bearing capacity expression, equation 8.7 as:

$$gross \; q_{ult} = p_o N_q + 1/2 \; \gamma \; B' \; N_\gamma \qquad (8.11)$$

The shape, depth and inclination factors are also dependent on the choice of c and ϕ values so different factors will be obtained depending on the soil strength case assumed.

Effect of water table

If the water table lies B m or more below the foundation then equation 8.7 does not require modification.

If the water table lies at foundation level then equation 8.7 (with the factors removed for clarity) should be modified to:

$$gross \; q_{ult} = cN_c + p_o N_q + 1/2 \; \gamma' \; B' \; N_\gamma \qquad (8.12)$$

where γ' is the submerged unit weight $= \gamma - \gamma_w$

If the water table lies within B m below foundation level then a value of γ' could be obtained by linear interpolation between γ and γ' depending on the actual depth of the water table below foundation level. Since the water table is likely to fluctuate during the life of the structure it would

TABLE 8.8 *Inclination factors (from Vesic, 1975)*

ϕ value	i_c	i_q	i_γ
$\phi = 0$ Clay undrained	$1 - \dfrac{mH}{B'L'c_u N_c}$	1.0	1.0
$\phi > 0$ Clay drained Sand	$i_q - \dfrac{1 - i_q}{N_c \tan\phi}$	$\left[1 - \dfrac{H}{V_T + B'L'\,c'\cot\phi}\right]^m$	$\left[1 - \dfrac{H}{V_T + B'L'\,c'\cot\phi}\right]^{m+1}$

Exponent m

For horizontal load acting along the short axis (side B') use $\quad m = m_B = \dfrac{2 + \dfrac{B'}{L'}}{1 + \dfrac{B'}{L'}}$

For horizontal load acting along the long axis (side L') use $\quad m = m_L = \dfrac{2 + \dfrac{L'}{B'}}{1 + \dfrac{L'}{B'}}$

be prudent to adopt γ' irrespective of water table depth.

If the water table lies at a height h_w above foundation level then equation 8.7 should be modified to:

$$gross\ q_{ult} = cN_c + p_o'N_q + 1/2\gamma'B'N_\gamma + \gamma_w h_w \quad (8.13)$$

For a strip foundation when shape factors are unity and for vertical loading only when inclination factors are unity this expression may be modified to:

$$gross\ q_{ult} = cN_c + p_o'(N_q - 1) + 1/2\gamma'B'N_\gamma + p_o \quad (8.14)$$

where p_o' is the *effective* overburden pressure at foundation level.

SEE WORKED EXAMPLE 8.1

Net ultimate bearing capacity

In equations 8.7 to 8.12, q_{ult} is referred to as the gross ultimate bearing capacity. The net ultimate bearing capacity is the maximum *additional* pressure the soil can support in excess of the stress at foundation level which existed before placing the foundation. Thus:

$$net\ q_{ult} = gross\ q_{ult} - p_o \quad (8.15)$$

Overall factor of safety

This is used to obtain the net *safe* bearing capacity q_s:

$$net\ q_s = net\ q_{ult}\,/\,F \quad (8.16)$$

$$gross\ q_s = net\ q_{ult}\,/\,F + p_o \quad (8.17)$$

Alternatively, the overall factor of safety is obtained from:

$$F = \frac{gross\ q_{ult} - p_o}{gross\ q_{app} - p_o} = \frac{net\ q_{ult}}{net\ q_{app}} \quad (8.18)$$

where:

q_{app} = applied pressure
and the other values are defined above.

SEE WORKED EXAMPLE 8.3

A value of F is chosen to ensure that the risk of failure of the foundation and consequently the structure is minimal so a combination of structural factors and geotechnical factors must be considered. These include:

● *Uncertainty of loading*
With non-routine buildings and live loading, these effects are difficult to quantify, e.g. wind, water forces, moving loads, dynamic forces.

● *Likelihood of maximum design load*
For non-routine structures it is likely that unfavourable variations of loading will occur, whereas routine buildings are often designed on nominal loading which is unlikely to occur.

● *Consequences of failure*
The public will expect less risk to be taken with structures where failure could result in catastrophic consequences. More risk is taken with temporary works than permanent works.

● *Uncertainty of soil model*
Geological variations, inaccuracy of strength values, water table fluctuations, mode of failure, limitations of the analytical method all provide uncertainty.

● *Extent of investigation*
Sufficient depth of ground must be investigated to assess the layering of deposits, the uniformity of the ground conditions and a sufficient number of tests should be carried out to enable a reasonable choice of parameters. The more extensive the site investigation the more confidence there will be in the choice of the soil model.

Values for the overall factor of safety taking into consideration the above points have been suggested by Vesic (1975) and are reproduced in Table 8.9. Limit state design and partial factors are discussed later.

Effect of compressibility of soil

The above expressions assume that the soil is incompressible, that only mass movements of zones I, II and III (Figure 8.14) occur and the general shear failure case applies. Clays sheared in an undrained manner could be considered incompressible and to fail in the general shear mode so no reduction of bearing capacity need be imposed.

However, clays sheared slowly enough to attain drained conditions or sands which drain more rapidly will be compressible and the full amount of bearing capacity as given by equation 8.7 will not be achieved. Vesic (1975) has suggested the use of compressibility factors to modify this equation but these require knowledge of soil parameters which are not readily available and the method is somewhat tentative. The following simple approaches may be adopted:

1. For clays sheared in drained conditions the approach suggested by Terzaghi (1943) of using

TABLE 8.9 *Minimum safety factors for design of shallow foundations (after Vesic, 1975)*

Category	Characteristics of the category	Extent of soil investigation		Typical structures
		thorough, complete	limited	
A	Maximum design load likely to occur often; consequences of failure disastrous	3.0	4.0	railway bridges warehouses blast furnaces hydraulic structures retaining walls silos
B	Maximum design load may occur occasionally; consequences of failure serious	2.5	3.5	highway bridges light industrial and public buildings
C	Maximum design load unlikely to occur	2.0	3.0	apartment and office buildings

reduced strength parameters $c*$ and $\phi*$ in equation 8.7 where:

$$c* = 0.67\ c' \tag{8.19}$$

and

$$\tan \phi* = 0.67 \tan \phi' \tag{8.20}$$

2. For sands, Vesic (1975) proposed:

$$\tan \phi* = (0.67 + D_r - 0.75\ D_r^2) \tan\phi' \tag{8.21}$$

where D_r is the relative density of the sand, recorded as a fraction. Typical values of D_r are given in Table 1.6. This reduction factor only applies for loose and medium dense sands. For dense sands ($D_r > 0.67$) the strength parameters need not be reduced since the general shear mode of failure is likely to apply.

SEE WORKED EXAMPLE 8.2

Sliding

This topic is also covered for gravity retaining walls in Chapter 11. Failure by sliding will occur when the applied horizontal force H (or a component) exceeds the maximum resisting force, H_{max}. The overall factor of safety against sliding is then given as:

$$F = \frac{H_{max}}{H} \tag{8.22}$$

In some cases a proportion of the passive resistance of the ground, P_p, acting in front of the foundation may be included to give:

$$F = \frac{H_{max} + P_p}{H} \tag{8.23}$$

However, there may be good reasons to ignore this contribution such as shrinkage of the soil, poorly compacted backfill, erosion, possible future excavations and the large movements required to mobilise passive resistance.

For the undrained case in clay, H_{max} would be given by:

$$H_{max} = A'\ c_a \tag{8.24}$$

where:

A' = effective area = $B'\ L'$
c_a = adhesion = $\alpha\ c_u$

$\alpha = 1$ for soft and firm clays
$\alpha = 0.5$ for stiff clays

For a sand and the drained case in clay (c' assumed to be zero) H_{max} would be given by:

$$H_{max} = V_T'\ \tan\delta \tag{8.25}$$

where V_T' is the vertical effective load. If the water table lies above foundation level then V_T' will be given by:

$$V_T' = V_T - u\ A' \tag{8.26}$$

where u is the pore water pressure at foundation level.

SEE WORKED EXAMPLE 8.5

The skin friction angle δ can be taken as equal to ϕ' for cast *in situ* concrete foundations and $2/3\phi'$ for smooth precast concrete foundations (EC7, 1995). In limit state design the inequality:

$$H_s \leq S_d + E_{pd} \tag{8.27}$$

should be satisfied where:

H_d = horizontal component of the design load including the design active earth forces (= characteristic horizontal load multiplied by a partial factor).
S_d = design shear resistance between the base of the foundation and the ground (= H_{max} in the above equations).
E_{pd} = design resisting (passive) earth pressure force in front of the foundation (= P_p in the above equations).
S_d and E_{pd} should be obtained from the design strengths, i.e. the characteristic strength divided by a partial factor.

For an effective stress analysis in sands and clays, see equation 8.25, the sliding resistance is determined from the total vertical load V_T therefore the design value of the load in this respect should be the most unfavourable, i.e. the lowest load.

Allowable bearing pressure of sand

Settlement limit

A serviceability limit for a foundation is the allow-

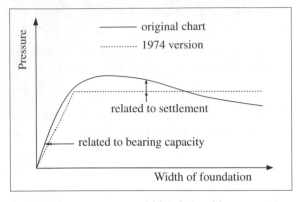

FIGURE 8.18 *Pressure – width relationship*

able (or tolerable, or permissible) settlement of the supported structure. For routine buildings on sands this is commonly taken as a total settlement of 25 mm on the understanding that differential settlements will then be within tolerable limits (Terzaghi and Peck, 1967).

The foundations must be designed to ensure that this settlement is not exceeded by using the allowable bearing pressure. For fairly wide foundations on sand it would be generally found that the overall factor of safety at this pressure will be far greater than the normal values of 3–4 so the foundation will certainly be 'safe' and within the ultimate limit state (see Figure 8.1).

Allowable bearing pressure
(Figures 8.18, 8.19 and 8.20)
Given the difficulties of obtaining undisturbed samples of sand the Standard Penetration Test (SPT) was developed in the United States as an empirical way of assessing the degree of compactness of a sand. The test procedure has since been standardised (BS 1377:1990 and ASTM D1586). However, given the wide variety of equipment still used to carry out the test and the many factors which can affect the result the repeatability (one operator obtaining the same result several times over) and reproducibility (several different operators obtaining the same result) are not good.

Terzaghi and Peck (1948) developed an empirical chart relating allowable bearing pressure q_a to the SPT 'N' value. However, this chart also related q_a to the width of foundation which complicated design

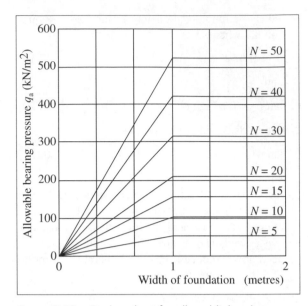

FIGURE 8.19 *Design chart for allowable bearing pressures (Adapted from Peck, Hanson and Thornburn, 1974)*

because of the iterative process required. Peck, Hanson and Thornburn (1974) simplified the chart into two regions, one where the design pressure was related to bearing capacity (for small foundation widths) and one where the design pressure was limited by settlement (Figure 8.18).

The chart is still very much empirical and any

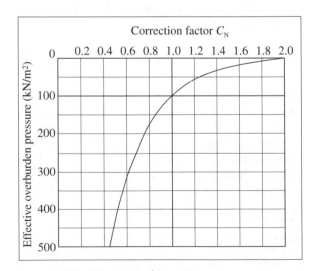

FIGURE 8.20 *Correction factor C_N*

TABLE 8.10 *Energy correction C_E (from Skempton, 1986)*

Country	Hammer	Release	C_E
Japan	Donut	Tombi	1.3
	Donut	2 turns of rope	1.1
China	Pilcon type	Trip	1.0
	Donut	Manual	0.9
USA	Safety	2 turns of rope	0.9
	Donut	2 turns of rope	0.75
UK	Pilcon, Dando	Trip	1.0
	Old standard	2 turns of rope	0.8

TABLE 8.11 *Correction for rod length C_R (from Skempton, 1986)*

Rod length, m	C_R
3–4	0.75
4–6	0.85
6–10	0.95
> 10	1.0

over-sophistication of approach will give an impression of accuracy which is unwarranted. A modified version of the Peck, Hanson, Thornburn chart is given in Figure 8.19 which has been simplified so that allowable bearing pressure can be reported in the form:

$$q_a = 10.5 \, N' \, B \quad \text{for width } B < 1 \text{ m} \tag{8.28}$$

and

$$q_a = 10.5 \, N' \quad \text{for width B} \geq 1 \text{ m} \tag{8.29}$$

where N' is a corrected value of the field value N given by:

$$N' = N \, C_E \, C_N \, C_R \, C_W \tag{8.30}$$

N is usually taken as the loosest or lowest value from any borehole on the site and within a depth equal to the width of the foundation B below foundation level. The depth of foundation does not seem to have a significant effect.

The chart was derived for square or rectangular foundations. Strip foundations can produce greater settlements for the same width and applied pressure so the allowable bearing pressure should be reduced by about 20% for strip foundations.

C_E is an energy ratio correction expressed by Skempton (1986) as:

$$\frac{\text{energy delivered to rod stem}}{\text{free-fall energy of hammer}} \cdot \frac{1}{60}$$

to allow for the different efficiencies of the various hammers used. Values of C_E are given in Table 8.10.

C_N is the correction for the effects of overburden pressure given by Peck *et al*, 1974:

$$C_N = 0.77 \log_{10}\left(\frac{20}{p'}\right) \tag{8.31}$$

where:
p' = effective stress at the depth of the test in ton/sq ft (1 ton/sq ft = 95.7 kN/m^2).

This expression normalises the N values to a standard stress level of 1 ton/sq ft or about 100 kN/m^2 for which the chart was derived.

Penetration resistance is related to confining stress so at shallow depths (p' less than 100 kN/m^2) the actual N value will underestimate the compactness or density of the sand so the values must be increased to obtain an N value as though it had been carried out at a confining stress of 100 kN/m^2. C_N from the above expression is plotted in Figure 8.20.

C_R is a correction for the effect of rod length on the energy delivered to the split spoon. Values of C_R are given in Table 8.11.

C_w is Peck's correction for the effect of a water table on the understanding that reduced effective stresses increase compressibility. However, as Skempton (1986) has shown, N is directly related to the effective stress existing in the ground so the effect of the water table will be reflected in the measured N value and further correction is unnecessary. However, if the water table may rise *after* the site investigation the correction C_w could be applied:

$$C_w = 0.5 + \frac{0.5 D_w}{D_f + B} \qquad (8.32)$$

where:

D_w = depth of the raised water table below ground level

D_f = depth of the foundation

Many other factors have been found to affect the SPT N value such as the diameter of the borehole, the use of liners in the split spoon, grain size, crushability of particles, overconsolidation, ageing of the deposit, fluctuating load and time after construction. Judgement must be exercised in the choice of an allowable bearing pressure.

Limit state design

Limit states

These are defined as states beyond which the structure no longer satisfies the design performance requirements. This could represent a range of limits from:

- cracking which gives an unsightly appearance
- unacceptable vibrations
- distortion or deflection that leads to loss of weathertightness or impaired durability or loss of function
- collapse of the whole or part of a structure due to excessive foundation movement such as subsidence, settlement or heave
- collapse of the whole or part of a structure due to foundation failure by exceeding the bearing capacity or sliding resistance or overall instability

It is necessary to distinguish between ultimate limit states and serviceability limit states. Ultimate limit states are defined as *states associated with collapse or with other similar forms of structural failure* and concern the safety of people and the structure. Serviceability limit states *correspond to conditions beyond which specified service requirements for a structure or structural element are no longer met* and concern the comfort of people, the functioning of the property and its appearance.

Limit state design is then carried out by setting up models for the structure and load cases in various design situations and verifying that the limit states are not exceeded when design values for actions, material properties and geometrical data are used in the (calculation) models. It is intended that ultimate limit states must be determined so that they may be possibilities but are unrealistic possibilities.

Serviceability limit states can be a little more flexible in that they depend on subjective views regarding people's comfort and their assessment of appearance and functioning of a building may be only temporarily impaired. Avoiding these limit states in all events may lead to poor economy of design so although the performance criteria must be deemed unacceptable possibilities they should not be unnecessarily severe. The approach in limit state design is to verify that:

- for ultimate limit states the effects of the design actions do not exceed the design resistance of the structure
- for serviceability limit states the effects of the design actions do not exceed the performance criteria of the structure

Verification of the limit state is provided by satisfying the following inequality:

$$V_d \leq R_d \qquad (8.33)$$

where V_d is the design action or effect of the design action (such as a moment) attempting to exceed the limit state and R_d is the design resistance (action or action effect) available to prevent the occurrence of the limit state and is obtained from the design values of the material properties.

Design values

The design value of an action (load) is obtained by multiplying the characteristic value of the action by partial factors:

$$G_d = G_k \, \gamma_G \quad \text{(permanent action)} \qquad (8.34)$$
$$Q_d = Q_k \, \gamma_Q \quad \text{(variable action)} \qquad (8.35)$$
$$A_d = A_k \, \gamma_A \quad \text{(accidental action)} \qquad (8.36)$$

where the subscript $_d$ represents the design value and subscript $_k$ represents the characteristic value and γ is the partial factor. The characteristic values of the actions can be obtained from relevant structural codes of practice.

The design value of a material property is obtained by dividing the characteristic values of the properties by a partial factor:

$$\tan \phi_d = \frac{\tan \phi_k}{\gamma_\phi} \qquad (8.37)$$

$$c_{ud} = \frac{c_{uk}}{\gamma_{cu}} \qquad (8.38)$$

where the subscripts $_d$ and $_k$ represent the design and characteristic value, respectively and γ is the partial factor.

Characteristic values

The determination of the characteristic values of the material properties causes the geotechnical engineer the biggest headache. EC7 requires that:

2.4.3 (5)P 'The characteristic value of a soil or rock parameter shall be selected as a cautious estimate of the value affecting the occurrence of the limit state'

and that selection of the value shall take account of:

- prior experience
- the mass characteristics of the ground, possibly requiring corrections to laboratory test results
- the variability of the property values in the field with depth, spatially across the site and statistically
- the extent of the zone of ground affected by the structure
- the changes that can be caused by construction activities including disturbance, weather deterioration, poor workmanship
- the type of limit state being considered
- for strength, the degree of mobilisation, i.e. peak, critical state or residual

Consideration should also be given to the consequences of a failure whether disastrous, serious or manageable both in the choice of characteristic values and partial factors.

In EC1 the choice of a characteristic value for both loads and material properties is directed towards a statistical approach and EC7 allows for these methods to be considered for ground properties. The characteristic value may be derived such that the calculated probability of a worse value governing the occurrence of a limit state is not greater than 5%, or alternatively the 5% fractile in 'a hypothetical unlimited test series'. This approach is successful with structural materials such as steel and concrete since their properties can be specified and their manufacture controlled to ensure compliance.

Take for example one cubic metre of concrete. Nearly 300 standard cube samples could be prepared and tested to give a representative series of data for statistical analysis and confidence in the value of the characteristic strength derived. Providing the ingredients and their proportions are maintained and the placement and construction controlled the structural engineer will have confidence in the properties of the concrete incorporated in the structure. Then consider one cubic metre of soil taken from the ground. Up to 500 undisturbed triaxial specimens could be prepared and tested to give a representative series of data for statistical analysis and give confidence in the value of characteristic strength for this cubic metre of soil.

However, this would be a laboratory result possibly requiring adjustment for mass effects such as fissures, differences in the field and laboratory behaviour such as plane strain and triaxial conditions, differences in stress effects produced by the curvature of the failure envelope, scale effects, time effects, temperature effects. In the geological context the ground model can vary dramatically due to stratification and weathering causing variations across a site and with depth. Further, the ground is sampled and tested relatively infrequently so only a very limited amount of information is usually obtained. Thus statistical approaches to characteristic values of soil properties are fraught with difficulties and potential danger.

For a man-made structure such as an earth dam where the material properties are specified then providing the source of soil material is constant, the placement and construction controlled and its remoulded or reconstituted properties are known and monitored throughout construction, then the geotechnical engineer will have more confidence in the properties of the soils incorporated into the structure. Each soil will have a characteristic strength and, by definition, the geotechnical engineer will accept that no more than 5% of the soils incorporated may not provide the characteristic strength.

Characteristic values may be lower values or upper values depending on the effects on the limit state considered. For example, lower values of density would be used where the soil acts favourably in support of the structure such as a passive pressure in front of a retaining wall. Higher values of density are used where the soil applies load to the structure such as active pressure behind the wall where it is unfavourable.

You will by now have realised that selecting soil parameters for design purposes is the most difficult yet most important task of the geotechnical engineer. Unfortunately, there are no clear guidelines that can be offered to address this problem. A useful discussion on this subject is given in Simpson and Driscoll, 1998. EC7 requires selection of a 'cautious estimate'. Other terms that appear in the literature include:

moderately conservative value
best estimate of the field value
more adverse than the most likely
pessimistic estimate
worst credible
midway between expected and worst credible

This task will continue to exercise all of the skills and judgement of even the most experienced geotechnical engineer. The accuracy adopted in calculation methods should be tempered in the light of the above and output values obtained for say bearing capacities or settlements should be rounded up or down and reported in approximate terms.

We should encourage engineers from other disciplines to be more sympathetic to the difficulties faced by the geotechnical engineer so they could accept some of the uncertainties inherent in the ground. Unfortunately all too often we are expected to provide accurate answers to problems where the ground conditions have been poorly investigated. I am not having a whinge!

Partial factors (Figure 8.21)

The concept of limit state design and partial factors is common in structural design but not in geotechnical design. The conventional approach in geotechnical design has been to follow an analytical method incorporating reasonable estimates of the load and material parameters to obtain a derived ultimate value such as bearing capacity. This has then been reduced by an overall factor to provide for safety and stability. This factor has without any real consideration also been deemed sufficient to allow for mobilisation of strength values, to provide acceptable deformations and possibly even cater for durability

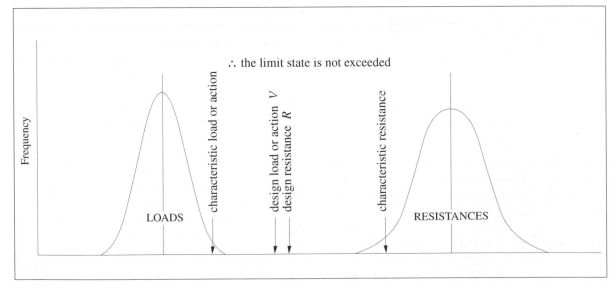

Figure 8.21 *Verification of the limit state*

TABLE 8.12 *Partial factors for persistent and transient situations (from Eurocode 7, 1995)*

Partial factors	Permanent actions		Variable actions	Ground properties			
	γ_G		γ_Q	γ_ϕ	$\gamma_{c'}$	γ_{cu}	γ_{qu}
Case	Unfavourable	Favourable	Unfavourable	$\tan\phi$	c'	c_u	q_u
Case A	[1.00]	[0.95]	[1.50]	[1.1]	[1.3]	[1.2]	[1.2]
Case B	[1.35]	[1.00]	[1.50]	[1.0]	[1.0]	[1.0]	[1.0]
Case C	[1.00]	[1.00]	[1.30]	[1.25]	[1.6]	[1.4]	[1.4]

and deterioration. Published values which provide some of these requirements are given in Table 8.9.

Partial factors were introduced into Danish geotechnical practice by Brinch Hansen in 1953 and will now form the basis for limit state design in EC7. Factors are applied to increase the loads or actions and to decrease the material strengths and are smaller than the overall factors. Applied to characteristic values they provide design values which are used to avoid the exceedance of a limit state. A statistical approach to their application is illustrated in Figure 8.21.

Three cases A, B and C have been introduced to ensure stability and adequate strength in the structure and the ground and the design must be verified for all three cases. The partial factors given for ultimate limit states for each case are reproduced in Table 8.12. Note that these are all boxed values.

These are recommended values for parameters. Each member country of the EC is required to review these and may substitute definitive values in their own National Application Document.

Case A considers stability and is only relevant to buoyancy problems where unfavourable water pressures beneath a buried structure may cause it to rise. Where permanent actions are favourable they are reduced by applying a partial factor less than unity. Where the failure of the structure or structural elements is governed by strength of the structural materials such as with the concrete elements of foundations, piles or retaining walls, the partial

factors for Case B are adopted. The actions are factored to give design values but the design values of the material properties are the characteristic values, i.e. they are not factored. The partial factors applied to strength of the structural materials are taken from the relevant Eurocode for that material.

Case C applies to failure in the ground, the strength of the structural elements is not involved. The permanent actions are not factored but the material properties are factored. Examples include slope stability problems, bearing capacity of the soil beneath foundations and overturning stability of retaining walls.

For the verification of the serviceability limit state the partial factor applied to the permanent and variable actions and to the ground properties is unity.

Annex B (informative) to Eurocode 7

A sample analytical method for bearing resistance calculation

This method provides the design bearing resistance R_d of a spread foundation. It is not given as a principle so alternatives are permissible during the ENV (experimental application) stage.

For all ultimate limit state load cases the inequality

TABLE 8.13 *Inclination factors (from Eurocode 7, 1995)*

Direction of horizontal load	i_c	i_q	i_γ
Horizontal load H parallel to L'	$\dfrac{i_q N_q - 1}{N_q - 1}$	$\left[1 - \dfrac{H}{V_T + B'L'c'\cot\phi}\right]$	$\left[1 - \dfrac{H}{V_T + B'L'c'\cot\phi}\right]$
Horizontal load H parallel to B'	$\dfrac{i_q N_q - 1}{N_q - 1}$	$\left[1 - \dfrac{0.7H}{V_T + B'L'c'\cot\phi}\right]^3$	$\left[1 - \dfrac{H}{V_T + B'L'c'\cot\phi}\right]^3$

$$V_d \leq R_d \tag{8.33}$$

must be satisfied where V_d and R_d are the design action and the design bearing resistance as defined above. Both short-term and long-term conditions must be checked.

Undrained conditions

The design bearing resistance is calculated from:

$$\frac{R_d}{A'} = (2 + \pi)c_u s_c i_c + q \tag{8.39}$$

where:

A' is the effective area $= B' \times L'$
$2 + \pi$ is the N_c value
s_c is the design value of the shape factor given by $1 + 0.2\ B'/L'$
i_c is the design value of inclination factor given by:

$$i_c = 0.5 + 0.5 \sqrt{\left(1 - \frac{H}{A'c_u}\right)} \tag{8.40}$$

q is the design total stress at foundation level, the same as p_o in equation 8.8. Note that depth factors are not included although it is stated that the effect of embedment depth should be included.

Drained conditions

The design bearing resistance is calculated from:

$$\frac{R_d}{A'} = c'\ N_c\ s_c\ i_c + q'\ N_q\ s_q\ i_q + 0.5\ \gamma\ B'\ N_\gamma\ s_\gamma\ i_\gamma \tag{8.41}$$

N_c and N_q are as given in equations 8.4 and 8.5 but N_γ is somewhat greater than in equation 8.6, given by:

$$N_\gamma = 2\ (N_q - 1)\ \tan\phi' \tag{8.42}$$

The shape factors are given as:

$$s_c = \frac{(s_q N_q - 1)}{(N_q - 1)} \tag{8.43}$$

$$s_q = 1 + \frac{B'}{L'}\sin\phi' \tag{8.44}$$

$$s_\gamma = 1 - 0.3\frac{B'}{L'} \tag{8.45}$$

Values of s_c and s_q are somewhat smaller than in Table 8.6 but values of s_γ are higher. The inclination factors are given in Table 8.13.

SEE WORKED EXAMPLES 8.6 AND 8.7.

SUMMARY

Shallow foundations comprise strip, pad and raft foundations depending on the load applied and the shape required. Their design is based on the need to achieve a stability requirement or an ultimate limit state and that they must satisfy settlement criteria and not be subjected to other ground movements referred to as serviceability limit states.

They are placed at shallow depths in the ground but not at ground level because of factors such as seasonal moisture variations, frost action, river erosion.

The stability condition is determined using bearing capacity theory applied to eccentric and inclined loading.

A method for the allowable bearing pressure of sand based on a limiting settlement is presented.

An introduction to geotechnical design as set out in the Eurocodes is given. These codes adopt limit state design and partial factors applied to characteristic values to give design values. The limit state is verified by ensuring that the design actions do not exceed the design resistance.

The determination of characteristic values of soil parameters provides the geotechnical engineer with the greatest challenge.

Foundation failure – Transcona Elevator, Canada

Case Objectives:

This case illustrates:

- the application of bearing capacity theory to a spread foundation – this theory was not well developed at the time
- that when most of the applied load is live or imposed load and it is placed quickly the soil behaves in an undrained manner and is given little opportunity for consolidation to occur and improve the shear strength
- that the stressed zone beneath a plate bearing test can bear little relationship to that beneath a large foundation
- the need to investigate the ground to at least within the stressed zone beneath the foundation

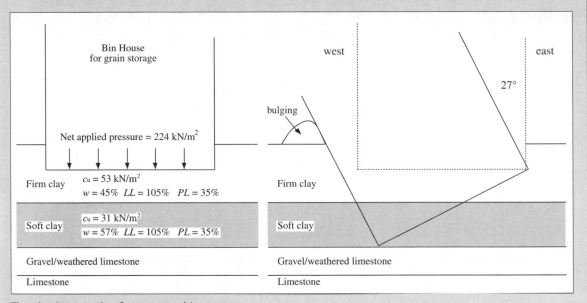

The classic example of a catastrophic failure of a shallow foundation due to exceeding the bearing capacity of the soil is the grain elevator at Transcona, Winnipeg, Manitoba, Canada. Descriptions of the structure and the bearing capacity analysis are given in Peck *et al.* (1953) and an eyewitness account is given in White (1953).

The structure comprised a reinforced concrete work house and adjoining bin house commenced in 1911 and completed in September 1913. The structure that underwent failure was the bin house. This comprised 65 bins in five rows of 13 bins

Do we learn from past mistakes?

In 1955 a very similar classic bearing capacity failure occurred in Fargo, North Dakota, USA involving the complete destruction of a grain elevator with similar size and weight and on a similar raft foundation on similar soils (Nordlund *et al.* 1970). By this time much more was known about soil mechanics and bearing capacity, in particular, so there was no excuse for this failure. To add insult, if the settlement measurements that were taken had been plotted and analysed they would have shown the imminence of a bearing capacity failure, the bins could have been unloaded and the structure would have been saved. It is important that we do not forget past failures and that we continue to learn, disseminate ideas and avoid the same mistakes.

adjacent to each other. The bins were 14 ft (4.27 m) diameter and 92 ft (28.05 m) high supported on a concrete raft 2ft (0.61 m) thick, 77 ft (23.47 m) wide and 195 ft (59.45 m) long at a depth of 12 ft (3.66 m). Filling with grain began in September 1913 and at lunchtime on 18 October, 1913 after 875,000 bushels of grain had been evenly stored in the bins settlement and tilting of the bin house was observed. Within an hour settlement was 1 ft (0.30 m) followed by tilting towards the west. The earth on the west side bulged up as the structure moved towards it and this slowed down the movement. The east side of the bin house moved away from the soil surrounding it leaving a gap to the depth of the raft foundation. The movement of the structure was gradual throughout this period and barely susceptible to the eye, diminishing throughout the day. The bin house eventually came to rest after about 24 hours but was at an angle of 26° 53' to the vertical.

The site investigation for the structure had comprised some small diameter plate bearing tests at foundation level which gave apparently satisfactory results. This is because these tests only stress the ground to a small depth and would only have tested the upper firmer clay. Subsequent boreholes showed the soils to comprise an upper layer of slickensided high plasticity (rich in montmorillonite) clay overlying a soft grey clay. Using an average weighted shear strength for the full thickness of both clay layers of 45 kN/m² Peck *et al.* (1953) determined a net ultimate bearing capacity of 246 kN/m² which was close to the estimated net applied pressure of 224 kN/m². The difference is probably associated mostly with the assumption of a weighted shear strength to give one value for the two clay layers and adoption of average values rather than more pessimistic values for each layer.

The bin house itself suffered little damage so it was subsequently emptied, underpinned with bored pile foundations taken to the limestone and relevelled although it ended up with a basement at about 10 m below ground level.

Apart from being a classic example of inadequate site investigation the case also showed that the full depth of a stressed zone must be investigated and that greater care should be exercised when the full live load can be realised unlike some structures such as office buildings where the live loading may be notional. Other relevant chapters are 5, Stress Distribution, 7, Shear Strength and 14, Site Investigation.

CASE STUDY

Bridge Scour - Schoharie Creek, New York State, USA

Case Objectives:

This case illustrates:

- that a multi-disciplinary approach to scour protection should be adopted with structural engineers, river engineers and geotechnical engineers appreciating the assumptions each one makes and the knock-on effects of any changes they make
- that regular inspections should be made and any changes to the river condition should be reviewed by the designers responsible for the bridge
- that the contractor should be made aware of the designer's assumptions for the ground conditions, material properties and the quality of work required.

In the United States nearly 500 bridges have failed since 1950 due to hydraulic conditions, mainly scour and this represents 60% of all bridge failures (Huber, 1991).

This bridge carried the New York State Thruway, Route 90 over Schoharie Creek, a major tributary of the Mohawk River, in central New York State. For most of the year the creek was a small river but it was known to be subject to severe floods. In 1955 the bridge had withstood a record-breaking flood soon after it had been constructed, but subsequent floods caused cumulative damage to the riprap protection surrounding the bridge piers.

In April, 1987 following heavy spring rains combined with snow melt from the mountains and the spill-over from a nearby dam the creek became a raging flood, the second largest the bridge had suffered. The strong currents eroded part of the river bed supporting a bridge pier which collapsed removing support from the bridge deck. Ten people died.

The bridge comprised five steel framed simple spans supported on four tall piers and two abutments. The structures were founded at shallow depths, in the river bed materials, on spread foundations. Piers 1 and 4 were at the banks of the creek while piers 2 and 3 stood within the creek. Drifting tree trunks caught on pier 3 increased the turbulence around the pier and with a curve in the river course directing a higher velocity flow towards this pier it failed due to scour and undermining, causing spans 3 and 4 to collapse into the river. The other piers failed within a few hours dropping further spans.

Other general factors contributed to the failure. The river bed comprised glacial till which was thought to be resistant to erosion but it contained layers of silt,

Bridge scour

Erosion occurs when the water velocity is sufficient to mobilise the river bed material. Most river beds are of a particulate nature and the increasing particle sizes of silt, sand, gravel and cobbles require increasing velocities to remove them. Scour may be general across a river bed or it may be local often with the river channel moving away from its previous location. Embankments across a flood plain will constrict flood waters into a narrower channel. River training works upstream can increase flows and erosive powers. Placing bridge piers in the path of a river provides obstructions to flow with damaging turbulence and vortices resulting in local scour around the bridge piers. Current velocities across a river are not constant so some piers can be more prone to scour than others. Redeposition of sediments only produces loose deposits which will be more easily removed later.

To minimise these effects the piers should be proportioned and streamlined to allow easier water flow around the piers. Other measures could include mattresses or stone layers placed on the river bed around a bridge pier to resist erosion and sheet piling driven into the river bed around the foundation to provide support to the soil beneath the pier in the event of erosion.

sand and gravel which were not protected against erosion. The construction excavations around the piers were supposed to be backfilled with riprap but in the event they were backfilled with easily eroded material with a riprap capping. The riprap stone sizes were too small for the current velocities experienced. The riprap layer was inadequately maintained even though bridge inspections had pointed out the necessity for repairs. Berms constructed upstream of the bridge in 1963 increased the flow of water through the bridge. Due to changes in the river downstream the hydraulic gradient at the bridge was increased. The structural detailing of the bridge had not considered the consequences and severity of a sudden, catastrophic collapse of the deck in the event of a pier collapse.

Other relevant chapters are 3, Permeability and 14, Site Investigation.

Worked Example 8.1   Vertical loading

A rectangular foundation 2.5 m wide and 3.5 m long is to be placed at a depth of 1.7 m below ground level in a thick deposit of firm saturated clay. The water table is at 1.2 m below ground level. Determine the gross and net ultimate bearing capacities for

(a) undrained and
(b) drained conditions.

Investigate the contribution made by including depth factors.

The soil parameters are $c_u = 65 \text{ kN/m}^2$ $\phi_u = 0°$
$\qquad\qquad\qquad\qquad\quad c' = 3 \text{ kN/m}^2$ $\phi_u = 27°$
$\qquad\qquad\qquad\qquad\quad \gamma_{sat} = 21.5 \text{ kN/m}^3$

(a) Undrained condition
Without depth factors
for $\phi = 0°$ $\qquad N_c = 5.14$ $\qquad N_q = 1$ $\qquad N_\gamma = 0$ $\qquad$ From Table 8.6, s_c $1 + 0.2 \times 2.5/3.5 = 1.14$
$p_0 = 21.5 \times 1.7 = 36.6 \text{ kN/m}^2$
From equation 8.9,   gross $q_{ult} = c_u N_c s_c + p_0$
$= 65 \times 5.14 \times 1.14 + 36.6 = 417.5 \text{ kN/m}^2$
net $q_{ult} = 417.5 - 36.6 = 380.9 \text{ kN/m}^2$

Assuming a factor of safety of 3 the net safe bearing capacity is $\frac{1}{3} \times 380.9 = 127 \text{ kN/m}^2$
With depth factors

$\dfrac{D}{B} = \dfrac{1.7}{2.5} = 0.68 \ (<1)$ From Table 8.7, $d_c = 1 + 0.4 \times 0.68 = 1.27$

gross $q_{ult} = 380.9 \times 1.27 + 36.6 = 520.3 \text{ kN/m}^2$ $\qquad$ net $q_{ult} = 520.3 - 36.6 = 483.7 \text{ kN/m}^2$

(b) Drained condition
Without depth factors
Assume the soil to be incompressible $\therefore$ no reduction of strength parameters is considered.
Using equations 8.8 and 8.13:
From Table 8.5, for $\phi' = 27°$ $\qquad N_c = 23.9$ $\qquad N_q = 13.2$ $\qquad N_t = 9.3$
From Table 8.6,

$s_c = 1 + \dfrac{2.5}{3.5} \times \dfrac{13.2}{23.9} = 1.40$ $\quad s_q = 1 + \dfrac{2.5}{3.5} \tan 27° = 1.36$ $\quad s_\gamma = 1 - 0.4 \times \dfrac{2.5}{3.5} = 0.71$

$p_0' = 36.6 - 0.5 \times 9.8 = 31.7 \text{ kN/m}^2$
$\gamma = 21.5 - 9.8 = 11.7 \text{ kN/m}^3$
gross $q_{ult} = 3 \times 23.9 \times 1.40 + 31.7 \times 13.2 \times 1.36 + 0.5 \times 11.7 \times 2.5 \times 9.3 \times 0.71 + 9.8 \times 0.5$
$= 100.4 + 569.1 + 96.6 + 4.9 = 771.0 \text{ kN/m}^2$
net $q_{ult} = 771.0 - 36.6 = 734.4 \text{ kN/m}^2$

If the foundation supports a net applied pressure of 127 kN/m² the factor of safety in the long term would be

$F = \dfrac{734}{127} = 5.8$

With depth factors

$$\frac{D}{B} = 0.68 \ (<1) \quad \text{From Table 8.7}$$

$$d_q = 1 + 2 \tan 27°(1 - \sin 27°)^2 \times 0.68 = 1.21 \qquad d_c = 1.21 - \frac{1 - 121}{23.9 \tan 27°} = 1.23$$

$$d_\gamma = 1.00$$
gross $q_{ult} = 100.4 \times 1.23 + 569.1 \times 1.21 + 96.6 \times 1.00 + 4.9 = 913.6 \ \text{kN/m}^2$

net $q_{ult} = 913.6 - 36.6 = 877.0 \ \text{kN/m}^2$

Worked Example 8.2   Effect of compressibility

For the foundation in Example 8.1 assess the effect of compressibility on the bearing capacity values. Ignore depth factors.

(a) Undrained condition
Assume the clay is incompressible for the undrained case so the bearing capacity is unchanged.

(b) Drained condition
The soil is assumed to be compressible so reduced strength parameters are used.
From equations 8.19 and 8.20
$c^* = 0.67 \times 3 = 2 \ \text{kN/m}^2$
$\phi^* = \tan^{-1}(0.67 \times \tan 27°) = 18.8°$
Consider the case without depth factors only.
for $\phi = 18.8°$ $\qquad N_c = 13.7 \qquad N_q = 5.7 \qquad N_\gamma = 2.4$

$$s_c = 1 + \frac{2.5}{3.5} \times \frac{5.7}{13.7} = 1.30 \quad s_q = 1 + \frac{2.5}{3.5} \tan 18.8° = 1.24 \quad s_\gamma = 0.71$$

gross $q_{ult} = 2 \times 13.7 \times 1.30 + 31.7 \times 5.7 \times 1.24 + 0.5 \times 11.7 \times 2.5 \times 2.4 \times 0.71 + 4.9$
$= 35.6 + 224.1 + 24.9 + 4.9 = 289.5 \ \text{kN/m}^2$
net $q_{ult} = 289.5 - 36.6 = 252.9 \ \text{kN/m}^2$

The factor of safety in the long term would be $F = \dfrac{252.9}{127} = 2.0$

Since the clay at shallow depths is of a firm consistency it is likely to be overconsolidated so the effects of compressibility are probably overstated.

Worked Example 8.3   Eccentric loading

The rectangular foundation in Example 8.1 is 0.5 m thick and backfilled with soil of unit weight 19.5 kN/m³. A column transmitting a vertical load of 850 kN is placed 0.6 m from the centre of the foundation on the long axis.
Determine the bearing capacity and factor of safety against general shear failure.
Unit weight of concrete is 25 kN/m³

Weight of backfill = $2.5 \times 3.5 \times 1.2 \times 19.5 = 204.8 \ \text{kN}$

Weight of foundation = $2.5 \times 3.5 \times 0.5 \times 25 = 109.4$ kN
Total vertical load = $204.8 + 109.4 + 850 = 1164.2$ kN

eccentricity $e_L = \dfrac{850 \times 0.6}{1164.2} = 0.44$ m $B' = B = 2.5$ m $L' = 3.5 - 2 \times 0.44 = 2.62$ m

(a) Undrained condition
Without depth factors

$$s_c = 1 + 0.2 \times \frac{2.5}{2.62} = 1.19$$

∴ gross $q_{ult} = 65 \times 5.14 \times 1.19 + 36.6 = 434.2$ kN/m^2 net $q_{ult} = 434.2 - 36.6 = 397.6$ kN/m^2

Gross applied pressure $= \dfrac{1164.2}{2.5 \times 2.62} = 177.7$ kN/m^2

Net applied pressure = $177.7 - 36.6 = 141.1$ kN/m$^{2'}$

Factor of safety $= \dfrac{397.6}{141.1} = 2.8$

(b) Drained condition
Without depth factors
for $\phi' = 27°$ $N_c = 23.9$ $N_q = 13.2$ $N_\gamma = 9.3$

$s_c = 1 + \dfrac{2.5}{2.62} \times \dfrac{13.2}{23.9} = 1.53$ $s_q = 1 + \dfrac{2.5}{2.62} \tan 27° = 1.49$ $s_\gamma = 1 - 0.4 \times \dfrac{2.5}{2.62} = 0.62$

gross $q_{ult} = 3 \times 23.9 \times 1.53 + 31.7 \times 13.2 \times 1.49 + 0.5 \times 11.7 \times 2.5 \times 9.3 \times 0.62 + 4.9$
$= 109.7 + 623.5 + 84.3 + 4.9 = 822.4$ kN/m^2
net $q_{ult} = 822.4 - 36.6 = 785.8$ kN/m^2

Factor of safety $= \dfrac{785.8}{141.1} = 5.6$

Compared with Example 8.1 the net bearing capacity has increased slightly but the net applied pressure has also increased. If the column load had been placed at the centre of this foundation the factor of safety would be given by:

Net applied pressure $= \dfrac{1164.2}{2.5 \times 3.5} - 36.6 = 133.1 - 36.6 = 96.5$ kN/m^2

(a) Undrained condition
Factor of safety $= \dfrac{380.9}{96.5} = 4.0$

(b) Drained condition
Factor of safety $= \dfrac{734.4}{96.5} = 7.6$

Worked Example 8.4   Eccentric and inclined loading

A horizontal load of 80 kN is applied 1.5 m above ground level to the column in Example 8.3 in the direction of the short side. Determine the bearing capacity and factor of safety against general shear failure.

eccentricity $e_B = \dfrac{80 \times 3.2}{1164.2} = 0.22$ m $B' = 2.5 - 2 \times 0.22 = 2.06$ m

(a) undrained condition

Without depth factors

$s_c = 1 + 0.2 \times \dfrac{2.06}{2.62} = 1.16$

The inclination of the load is in the direction of the short side so

$m_B = \dfrac{2 + \dfrac{2.06}{2.62}}{1 + \dfrac{2.06}{2.62}} = 1.56$ $i_c = 1 - \dfrac{1.56 \times 80}{2.06 \times 2.62 \times 65 \times 5.14} = 0.93$

gross $q_{ult} = 65 \times 5.14 \times 1.16 \times 0.93 + 36.6 = 397$ kN/m^2

net $q_{ult} = 397 - 36.6 = 360.4$ kN/m^2

Gross applied pressure $= \dfrac{1164.2}{2.06 \times 2.62} = 215.7$ kN/m^2

(b) Drained condition

$s_c = 1 + \dfrac{2.06}{2.62} \times \dfrac{13.2}{23.9} = 1.43$ $s_q = 1 + \dfrac{2.06}{2.62} \tan 27° = 1.40$ $s_\gamma = 1 - 0.4 \times \dfrac{2.06}{2.62} = 0.69$ $m_B = 1.56$

$i_q = \left[1 - \dfrac{80}{1164.2 + 2.06 \times 2.62 \times 3 \times \cot 27°} \right]^{1.56} = 0.90$ $i_\gamma = 0.933^{2.56} = 0.84$

$i_c = 0.90 - \dfrac{1 - 0.90}{23.9 \tan 27°} = 0.89$

From equations 8.8 and 8.13

gross $q_{ult} = 3 \times 23.9 \times 1.43 \times 0.89 + 31.7 \times 13.2 \times 1.40 \times 0.90 + 0.5 \times 11.7 \times 2.06 \times 9.3 \times 0.69 \times 0.84 + 4.9$

$= 91.3 + 527.2 + 65.0 + 4.9 = 688.4$ kN/m^2

net $q_{ult} = 688.4 - 36.6 = 651.8$ kN/m^2

Factor of safety $= \dfrac{651.8}{179.1} = 3.6$

Factors of Safety – Summary

Load case	Short term	Long term
Vertical, central load	4.0	7.6
Vertical, eccentric load	2.8	5.6
Eccentric and inclined load	2.0	3.6

Worked Example 8.5   Sliding

For the foundation detailed in Example 8.4 determine the factor of safety against sliding for the undrained and drained condition.

(a) Undrained condition
Assume $\alpha = 1$
From equation 8.24
$H_{max} = 2.06 \times 2.62 \times 65 \times 1 = 350.8$ kN

Overall factor of safety $= \dfrac{350.8}{80} = 4.39$

The factor of safety would still be adequate (2.2) if $\alpha = 0.5$ was used.

(b) Drained condition
Assume $c' = 3$ kN/m^2 and $\delta = \phi'$.
From equation 8.25
$V_T' = 1164.2 - 0.5 \times 9.8 \times 2.06 \times 2.62 = 1137.8$ kN
$H_{max} = 3 \times 2.06 \times 2.62 + 1137.8 \times \tan 27° = 16.2 + 579.7 = 595.9$ kN

Overall factor of safety $= \dfrac{595.9}{80} = 7.45$

Worked Example 8.6   Eccentric loading – Annex B Method (Eurocode 7)

Determine the design net bearing capacity and design bearing resistance and verify the appropriate limit state for the foundation described in Example 8.1.

This example will consider the ultimate limit state of the spread foundation in terms of bearing capacity. Firstly, a consideration of the load applied. In Example 8.1 the load is given without explanation. In limit state design both cases B and C must be considered, see Table 8.12. The characteristic loads would be the same for both cases but the partial factors are different. Thus both cases need to be compared with design loads or actions. Assume the soil parameter values given in Example 8.1 to be characteristic values.

Take the actions to be:
characteristic permanent action $G_k = 1400$ kN
characteristic variable action $Q_k = 750$ kN
Depth factors are not included as they have not been given in Annex B.

(a) Undrained condition – Case B
The design bearing resistance is given in equation 8.39.
The characteristic strength $c_{uk} = 65$ kN/m^2 = design strength ($\gamma_{cu} = 1.0$)
s_c is the design value of the shape factor so should include for some dimensional variation to give the lowest value of bearing resistance. Say the tolerance on the construction of the foundation is ±25 mm then the design width of 2.5 ± 0.025 m and design length of 3.5 ± 0.025 m should be used:

$s_c = 1 + 0.2 \times \dfrac{2.475}{3.525} = 1.140$ or $s_c = 1 + 0.2 \times \dfrac{2.475}{3.475} = 1.142$

Assume that the bulk unit weight is not factored i.e. the partial factor is unity

The design gross bearing capacity is

$$\frac{R_d}{A'} = 5.14 \times 65 \times 1.140 + 21.5 \times 1.7 = 380.9 + 36.6 = 417.5 \text{ kN/m}^2$$

or $= 5.14 \times 65 \times 1.142 + 21.5 \times 1.7 = 381.5 + 36.6 = 418.1 \text{ kN/m}^2$

The design net bearing capacity is 380.9 kN/m^2 or 381.5 kN/m^2

The design vertical bearing resistance R_d (action) is the lowest of

$380.9 \times 2.475 \times 3.525 = 3323.1$ kN

or $381.5 \times 2.475 \times 3.475 = 3281.1$ kN $\Leftarrow$

The design action is

$V_d = 1400 \times 1.35 + 750 \times 1.5 = 3015$ kN

so this ultimate limit state is not exceeded.

(b) Undrained condition – Case C

The design strength is $\dfrac{65}{1.4} = 46.4 \text{ kN/m}^2$

The shape factor is as before. The design gross bearing capacity is

$R_d = 5.14 \times 46.4 \times 1.140 + 21.5 \times 1.7 = 271.9 + 36.6 = 308.5 \text{ kN/m}^2$

or $5.14 \times 46.4 \times 1.142 + 21.5 \times 1.7 = 272.4 + 36.6 = 309.0 \text{ kN/m}^2$

The design net bearing capacity is 271.9 or 272.4 kN/m^2

The design vertical bearing resistance R_d is the lowest of

$271.9 \times 2.475 \times 3.525 = 2372.2$ kN

or $272.4 \times 2.475 \times 3.475 = 2342.8$ kN $\Leftarrow$

The design action is

$V_d = 1400 \times 1.00 + 750 \times 1.30 = 2375$ kN

so this ultimate limit state is exceeded and case C is critical. The foundation will have to be increased in size.

(c) Drained condition – Case B

The design bearing resistance is given by equation 8.41.

The design value of $\phi' = 27°$ ($\gamma_\phi = 1.0$)

The design value of $c' = 3$ kN/m^2 ($\gamma_{c'} = 1.0$)

As before, $N_c = 23.9$ $N_q = 13.2$

$N_\gamma = 2(13.2 - 1) \tan 27° = 12.4$

Expressions for the shape factors are given in Annex B of Eurocode 7.

$$s_q = 1 + \frac{2.475}{3.475} \times \sin 27° = 1.323 \qquad \text{or} \qquad 1 + \frac{2.475}{3.525} \times \sin 27° = 1.319$$

$$s_\gamma = 1 - 0.3 \times \frac{2.475}{3.475} = 0.786 \qquad \text{or} \qquad 1 - 0.3 \times \frac{2.475}{3.525} = 0.789$$

$$s_c = \frac{(1.323 \times 13.2 - 1)}{(13.2 - 1)} = 1.350 \qquad \text{or} \qquad \frac{(1.319 \times 13.2 - 1)}{(13.2 - 1)} = 1.345$$

For the determination of p_0' it is assumed that the value given for the water table level in Example 8.1 is a pessimistic value.

The design gross bearing capacity is

$3 \times 23.9 \times 1.350 + 31.7 \times 13.2 \times 1.323 + 0.5 \times 11.7 \times 2.475 \times 12.4 \times 0.786 = 791.5 \text{ kN/m}^2$

or $3 \times 23.9 \times 1.345 + 31.7 \times 13.2 \times 1.319 + 0.5 \times 11.7 \times 2.475 \times 12.4 \times 0.789 = 790.0 \text{ kN/m}^2$

The design net bearing capacity is $791.5 - 36.6 = 754.9 \text{ kN/m}^2$ or $790.0 - 36.6 = 753.4 \text{ kN/m}^2$

The design vertical bearing resistance is the lowest of

$754.9 \times 2.475 \times 3.475 = 6492.6$ kN $\Leftarrow$

or $753.4 \times 2.475 \times 3.525 = 6572.9$ kN

The design action is 3015 kN so the ultimate limit state is not exceeded and the undrained condition is critical.

(d) Drained condition – Case C

The design value of ϕ' is given by

$$\phi_d = \tan^{-1}\left(\frac{\tan 27°}{1.25}\right) = 22.2° \qquad (\gamma_\phi = 1.25)$$

The design value of c' is given by $c_d = \dfrac{3}{1.6} = 1.88$ kN/m² $\qquad (\gamma_{c'} = 1.6)$

From Table 8.5, for $\phi_d' = 22.2°$ $\qquad N_c = 17.1 \qquad N_q = 8.0$

$N_\gamma = 2(8.0 - 1) \tan 22.2° = 5.7$

$s_q = 1 + \dfrac{2.475}{3.475} \times \sin 22.2° = 1.269$ or $1 + \dfrac{2.475}{3.525} \times \sin 22.2° = 1.265$

$s_\gamma = 0.786$ or 0.789, as for Case B.

$s_c = \dfrac{(1.269 \times 8.0 - 1)}{(8.0 - 1)} = 1.307$ $\qquad$ or $\qquad$ $\dfrac{(1.265 \times 8.0 - 1)}{(8.0 - 1)} = 1.303$

The design gross bearing capacity is

$1.88 \times 17.1 \times 1.307 + 31.7 \times 8.0 \times 1.269 + 0.5 \times 11.7 \times 2.475 \times 5.7 \times 0.786 = 428.7$ kN/m²

or $1.88 \times 17.1 \times 1.303 + 31.7 \times 8.0 \times 1.265 + 0.5 \times 11.7 \times 2.475 \times 5.7 \times 0.789 = 427.8$ kN/m²

The design net bearing capacity is $428.7 - 36.6 = 392.1$ kN/m² or $427.8 - 36.6 = 391.2$ kN/m²

The design vertical bearing resistance is the lowest of

$392.1 \times 2.475 \times 3.475 = 3372.3$ kN $\Leftarrow$

or $391.2 \times 2.475 \times 3.525 = 3413.0$ kN

In all of the above situations the lowest bearing resistance is obtained from the smallest geometrical dimensions although these do not give the lowest shape factors.

The design action is 2375kN so this ultimate limit state (long term) is not exceeded. It can be seen that Case C is the most critical.

Worked Example 8.7   Eccentric and inclined loading – Annex B Method (Eurocode 7)

Determine the design net bearing capacity and design bearing resistance and verify the appropriate ultimate limit state for the foundation described in Example 8.3.

For simplicity, it is assumed that the critical condition is given by the smallest geometrical dimensions, depth factors are not included and that Case C is the most critical.

Take the actions acting on the column above the foundation to be:

characteristic permanent vertical action $G_{vk} = 800$ kN

characteristic permanent horizontal action $G_{hk} = 45$ kN

characteristic variable vertical action $Q_{vk} = 250$ kN

characteristic variable horizontal action $Q_{hk} = 20$ kN

The horizontal actions are applied to the column 1.2 m above ground level in the direction of the short side, parallel to B'.

The weights of the backfill and foundation are given in Example 8.3 as $204.8 + 109.4 = 314.2$ kN

The design vertical action is
$V_d = 1.0 \times (800 + 314.2) + 1.3 \times 250 = 1439.2$ kN

The design horizontal action is
$H_d = 1.0 \times 45 + 1.3 \times 20 = 71$ kN

The design moment action is
$M_d = 1.0 \times 45 \times 2.9 + 1.3 \times 20 \times 2.9 = 205.9$ kNm

The design eccentricities are

$$e_B = \frac{205.9}{1439.2} = 0.143 \text{ m}$$

$$e_L = \frac{800 + 1.3 \times 250}{1439.2} \times 0.6 = 0.469 \text{ m}$$

$B' = 2.5 - 2 \times 0.143 = 2.214$ m
$L' = 3.5 - 2 \times 0.469 = 2.562$ m

(a) Undrained condition

The design strength is $\dfrac{65}{1.4} = 46.4$ kN/m^2

$$s_c = 1 + 0.2 \times \frac{2.214}{2.562} = 1.173$$

$$i_c = 0.5 + 0.5 \sqrt{\left[1 - \frac{71}{2.214 \times 2.562 \times 46.4}\right]} = 0.927$$

The design gross bearing capacity is
$5.14 \times 46.4 \times 1.173 \times 0.927 + 36.6 = 259.3 + 36.6 = 295.9$ kN/m^2

The design net bearing capacity is 259.3 kN/m^2

The design vertical bearing resistance is
$R_d = 259.3 \times 2.214 \times 2.562 = 1470.8$ kN

The design vertical action V_d is 1439.2 kN so the ultimate limit state (short-term) is not exceeded.

(b) Drained condition
The design value of ϕ' is 22.2° and of c' is 1.88 kN/m^2

From Table 8.5, for $\phi_d' = 22.2°$ $N_c = 17.1$ $N_q = 8.0$ and $N_\gamma = 5.7$

$$s_q = 1 + \frac{2.214}{2.562} \times \sin 22.2° = 1.327$$

$$s_c = \frac{(1.327 \times 8.0 - 1)}{(8.0 - 1)} = 1.374$$

$$s_\gamma = 1 - 0.3 \times \frac{2.214}{2.562} = 0.741$$

The horizontal load is parallel to B' so the inclination factors from Table 8.13 are

$$i_q = \left[1 - \frac{0.7 \times 71}{1439.2 + 2.214 \times 2.562 \times 1.88 \times \cot 22.2} \right]^3 = 0.902$$

$$i_c = (0.902 \times 8.0 - 1) / (8.0 - 1) = 0.888$$

$$i_\gamma = \left[1 - \frac{71}{1439.2 + 2.214 \times 2.562 \times 1.88 \times \cot 22.2} \right]^3 = 0.862$$

The design gross bearing capacity is

$$\frac{R_d}{A'} = 1.88 \times 17.1 \times 1.374 \times 0.888 + 31.7 \times 8.0 \times 1.327 \times 0.902 + 0.5 \times 11.7 \times 2.214 \times 5.7 \times 0.741 \times 0.862$$

$$= 39.2 + 303.5 + 47.2 = 389.9 \text{ kN/m}^2$$

The design net bearing capacity is $389.9 - 36.6 = 353.3$ kN/m^2
The design vertical bearing resistance is
$353.3 \times 2.214 \times 2.562 = 2004.0$ kN
The ultimate limit state is not exceeded and the undrained case is the most critical.

EXERCISES

8.1 A square foundation, 3.5 m wide, is founded at 1.5 m below ground level in a stiff clay with undrained shear strength of 95 kN/m^2. Determine the net ultimate bearing capacity:
(a) with the depth factors included
(b) without the depth factors.

8.2 In Exercise 8.1, assuming the foundation and backfill have the same unit weight as the soil removed from the excavation determine the maximum central column load that can be applied. Assume an overall factor of safety of 3.

8.3 Determine the width required of a square foundation to support a central column load of 1450 kN placed at 1.2 m below ground level in a firm clay with undrained shear strength of 75 kN/m^2. Adopt a factor of safety of 3 and ignore depth factors. Assume that the foundation and backfill have the same unit weight as the soil removed.

8.4 For the foundation designed in Exercise 8.3 determine the factor of safety for long-term conditions given that the water table lies at 1.2 m below ground level.
$c' = 4$ kN/m^2, $\phi' = 32°$ and the unit weight of the clay is 21.5 kN/m^3.

8.5 A rectangular foundation, 3.0 m wide and 4.5 m long is to be placed at 2.0 m below ground level in a dense sand with $\phi' = 38°$ and unit weight of 19.5 kN/m^3. A water table exists at 0.5 m below ground level.
(a) determine the net ultimate bearing capacity for a central vertical load
(b) to keep the settlements within permissible limits the maximum column load that can be applied is estimated to be 3500 kN. Determine the factor of safety against bearing capacity failure.
Assume that the foundation and backfill have the same unit weight as the soil removed.

8.6 Repeat Exercise 8.5, assuming the water table to lie below the influence of the foundation.

8.7 A reinforced concrete retaining wall is to be founded at 1.5 m below ground level in a clay with the water table at 1.5 m below ground level as shown in Figure 8.22.
Properties of the materials are as follows:
clay: $c' = 4$ kN/m^2 $\phi' = 23°$ $\gamma = 22$ kN/m^3 $c_u = 40$ kN/m^2 $\phi_u = 0$
backfill: $\gamma = 21$ kN/m^3
concrete: $\gamma = 25$ kN/m^3.
The stem of the wall applies a line load of 90 kN/mrun to the base of the wall. Ignoring the effects of depth factors determine the factor of safety against bearing capacity failure for:
(a) short-term conditions and
(b) long-term conditions.

8.8 A reinforced concrete pad foundation, 3.5 m square and 0.6 m thick is to be founded at 1.8 m below ground level in a clay with the water table at 1.2 m below ground level. The properties of the materials are as follows:
clay: $c' = 2$ kN/m^2 $\phi' = 28°$ $\gamma = 21.5$ kN/m^3 $c_u = 85$ kN/m^2 $\phi_u = 0°$
backfill: $\gamma = 20$ kN/m^3
concrete: $\gamma = 25$ kN/m^3.

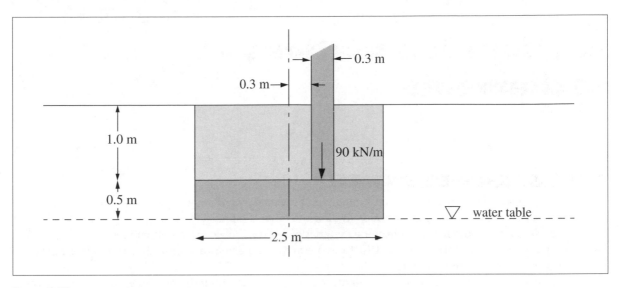

FIGURE 8.22

The foundation supports a vertical column load of 1600 kN at its centre and a horizontal load of 150 kN acts along one of the axes of the foundation at 1.5 m above ground level. Determine the factor of safety against bearing capacity failure for:
(a) short-term conditions
(b) long-term conditions.

8.9 For the foundation in Exercise 8.8 apply the method given in Annex B of Eurocode 7. The actions acting on the column are

$G_{vk} = 1000$ kN
$G_{hk} = 80$ kN
$Q_{vk} = 400$ kN
$Q_{hk} = 50$ kN

Assume that the soil parameters given in Exercise 8.8 are the characteristic values, the critical condition is given by the smallest geometrical dimensions, depth factors are not included and that case C is the most critical.

Shallow foundations – settlements

Foundations may move downwards due to shrinkage, erosion, subsidence, thawing, etc. but the foundation engineer prefers to avoid these situations rather than quantify them and accept them as inevitable. Settlements which must be quantified because they cannot be avoided are those caused by changes in stresses in the ground as a result of engineering works.

Foundations supported on sand are designed using the allowable bearing pressure as shown in Chapter 8 therefore it is expected that the foundation will not settle more than the allowable settlement. Semi-empirical methods to estimate the settlement of a foundation on sand are given in this chapter.

Foundations supported on clay are designed using the factored bearing capacity with the factor of safety chosen for reasons other than limiting settlements, described in Chapter 8. It is then necessary to estimate the settlement of a foundation on clay to determine whether it will lie within the permissible settlements of the structure.

Settlements produced by applied stresses originate from:

- immediate settlement, ρ_i

- consolidation settlement, ρ_c
- secondary compression, ρ_s

and amounts of each must be determined to give the total settlement, ρ_T as:

$$\rho_T = \rho_i + \rho_c + \rho_s \tag{9.1}$$

Simple methods have been established to determine settlements or vertical downward movement and these are covered in this chapter. However, methods to predict other movements are less well developed.

Some definitions of ground and foundation movement are given later in this chapter and these are useful in the assessment of the movement of a structure and the associated damage.

Immediate settlement

General method (Figure 9.1)

This is also termed undrained settlement since it occurs with no water entering or leaving the soil. No volume change means that the Poisson's ratio $v = 0.5$, see equation 7.5. The settlements are produced by shear strains within the soil, causing the soil

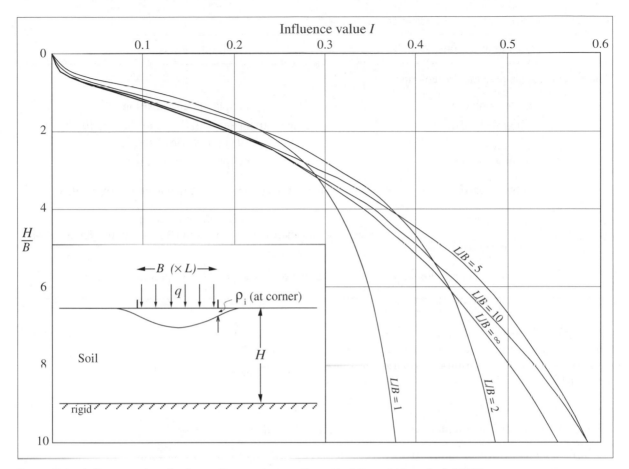

FIGURE 9.1 *Influence values for immediate settlement (from Ueshita and Meyerhof, 1968)*

surface to change shape. These strains are assumed to be elastic, so the settlements should be recovered on removal of the load.

Immediate settlements will occur virtually instantaneously on application of the loading therefore they will mostly occur during the construction period. For many structures these settlements are 'built in' during construction, before incorporation of the sensitive components such as the cladding or finishes. However, for those structures where the sensitive components are included at the beginning or throughout construction, such as with load-bearing walls immediate settlements can have a significant effect.

The foundation or loaded area is assumed to be flexible, producing a dish-shaped settlement profile with the maximum settlement at the centre of the foundation.

A rigorous solution for immediate settlements with the normal assumptions listed in Table 9.1 was given by Ueshita and Meyerhof (1968) for the conditions illustrated in Figure 9.1 using the expression:

$$\rho_i = \frac{qB}{E_u} I \tag{9.2}$$

where:

ρ_i = immediate settlement at the *corner* of the loaded area
q = uniform applied pressure
B = width of loaded area
I = influence factor, given in Figure 9.1
E_u = undrained modulus of soil

SEE WORKED EXAMPLE 9.1.

TABLE 9.1 *Immediate settlement – assumptions*

Normal assumptions	Modifications
Classical methods give the settlements:	
at the corner of	principle of superposition
a flexible loaded	average settlement (Christian and Carrier, 1978) or rigidity correction
rectangular area	principle of superposition
on the surface of	depth correction factor (Christian and Carrier, 1978)
a homogeneous	principle of layering modulus increasing with depth (Butler, 1976; Meigh 1974)
isotropic soil	
with linear stress-strain relationship	effect of plastic yield (D'Appolonia *et al.*, 1971)

Principle of superposition (Figure 9.2)

The above method gives the settlement at the corner of a loaded area. To determine settlements at other points beneath the foundation such as the maximum settlement at the centre, the principle of superposition must be used, illustrated in Figure 9.2.

SEE WORKED EXAMPLE 9.2.

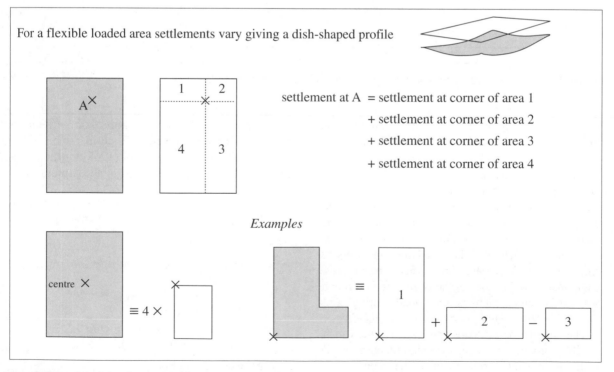

For a flexible loaded area settlements vary giving a dish-shaped profile

settlement at A = settlement at corner of area 1
+ settlement at corner of area 2
+ settlement at corner of area 3
+ settlement at corner of area 4

Examples

FIGURE 9.2 *Principle of superposition*

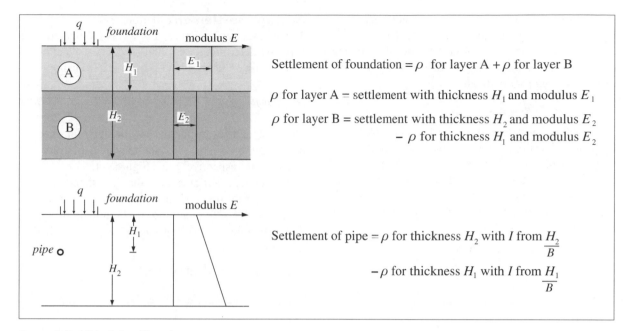

FIGURE 9.3   *Principle of layering*

Principle of layering (Figure 9.3)

Where there are two or more layers of soil with different modulus values the principle of layering can be used, as illustrated in Figure 9.3.

Where the settlement at a point within the soil layer is required for a buried structure such as a pipe, this principle may also be used. The settlement of the pipe will be due to the soil beneath it. A correction for the depth of embedment can be applied, see below.

SEE WORKED EXAMPLE 9.3.

Rigidity correction (Table 9.2)

A flexible foundation provides no resistance to deflection and will settle into a dish-shaped profile.

TABLE 9.2   *Rigidity correction factor μ_r*

L/B	$H/B = 1$	$H/B = \infty$
1	0.68	0.77
2	0.72	0.78
3, 4, 5	0.79	0.80

For a rigid foundation the settlement is the same at all points. Small reinforced concrete pad foundations will have enough stiffness to provide the rigid condition but larger raft type foundations are unlikely to be sufficiently stiff and will deflect to some extent. An approximate correction to the above method for the case of a rigid foundation is often quoted as:

$$\rho_{\text{rigid}} \approx 0.8 \times \rho_{\text{maximum flexible}} \qquad (9.3)$$

The general correction for rigidity can be represented as:

$$\rho_{\text{rigid}} = \mu_r \times \rho_{\text{maximum flexible}} \qquad (9.4)$$

where $\rho_{\text{maximum flexible}}$ is the settlement at the centre of the foundation assuming it to be flexible.

Fraser and Wardle (1976) demonstrated the effect of varying stiffness of a foundation on settlements and for a rigid foundation with infinite stiffness, the values of μ_r given in Table 9.2 have been derived from their results.

SEE WORKED EXAMPLE 9.5.

Depth correction (Figure 9.4)

Most foundations are placed at the base of an excavation. The effect of the depth of the foundation on settlement can be included using the factor μ_0 (from Burland, 1970) given in Figure 9.4. This method assumes that settlement is first determined by taking the loaded area to the surface of the soil layer and then correcting for depth using:

$$\rho_{\text{at depth}} = \mu_0 \times \rho_{\text{at surface}} \tag{9.5}$$

This method also assumes that the soil above the foundation has the same properties and is continuous with the soil beneath. This approach should not be confused with Fox's correction, see Chapter 10, which provides a correction for a loaded area which is embedded within the ground and not at the base of an excavation.

SEE WORKED EXAMPLE 9.4.

Average settlement (Figure 9.4)

The average settlement, ρ_{ave} of a flexible loaded area was first determined by Janbu *et al.* (1956) and modified later by Christian and Carrier (1978) using:

$$\rho_{\text{ave}} = \mu_0 \mu_1 \frac{qB}{E_u} \tag{9.6}$$

where μ_0 and μ_1 are factors for the depth of excavation and the thickness of the soil layer beneath the foundation, respectively, see Figure 9.4.

The average settlement for the dish-shaped profile of a flexible foundation is often assumed to be similar to the settlement of a rigid foundation.

The principle of layering can be applied with this method.

SEE WORKED EXAMPLE 9.5.

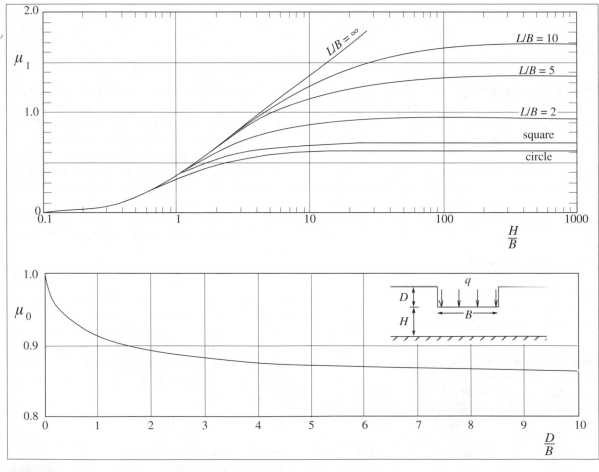

FIGURE 9.4 *Average settlement factors (from Christian and Carrier, 1978)*

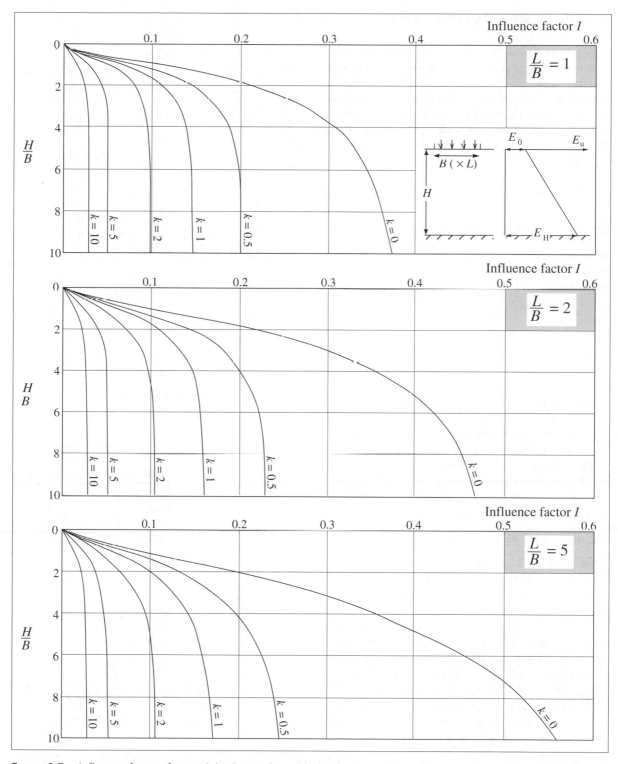

FIGURE 9.5 *Influence factors for modulus increasing with depth – immediate settlement (from Butler, 1974)*

Modulus increasing with depth
(Figure 9.5)

It has been found that for most soils the modulus increases with depth, thus assuming a constant modulus (the homogeneous case) will over-estimate settlements. Butler (1974) produced an approximate analysis based on Steinbrenner's influence values, for a soil with modulus increasing with depth, giving the immediate settlement at the corner of the loaded area as:

$$\rho_i = \frac{qB}{E_0} \, I \qquad (9.7)$$

where I is an influence factor related to:

● shape (L/B)
● thickness (H/B)
● a coefficient k given by:

$$k = \frac{E_H - E_0}{E_0} \, \frac{B}{H} \qquad (9.8)$$

Values of the influence factor I can be obtained from Figure 9.5. The curve for $k = 0$ represents the homogeneous or constant modulus case.

The method assumes that the foundation is placed on the surface of the compressible layer. A correction for a foundation at depth, μ_0, could be applied as described above.

The principle of superposition must be used for points other than the corner of the loaded area and the principle of layering can be used if layers with different modulus variation occur.

SEE WORKED EXAMPLE 9.6.

Effect of local yielding (Figures 9.6 and 9.7)

The above methods assume that the stress-strain relationship is linear which is a reasonable assumption provided the applied stress levels are low enough to prevent local plastic yielding of the soil. D'Appolonia et al. (1971) carried out finite element analyses of the problem assuming non-linear stress-strain properties after a yield condition had occurred (Figure 9.6). They incorporated the effect of the first local yielding beneath the foundation by modifying the elastic settlement ρ_i determined as above. This modification can be written in the form:

$$\rho_y = \rho_i \times f_y \qquad (9.9)$$

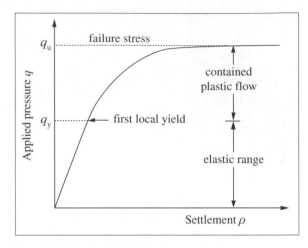

FIGURE 9.6 *Local yielding*

where:

ρ_y = immediate settlement including the effects of yield

ρ_i = immediate settlement from linear elastic theory

f_y = yield factor, given in Figure 9.7.

SEE WORKED EXAMPLE 9.7.

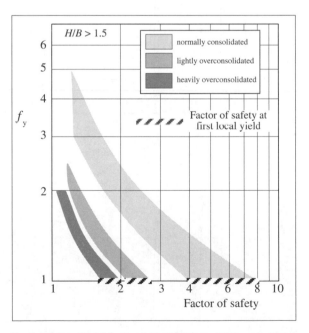

FIGURE 9.7 *Yield factor (adapted from D'Appolonia et al., 1971)*

Significantly, they found that yield and hence increased settlements can occur for normally consolidated clays if the factor of safety against bearing capacity failure is less than 4 to 8 whereas for heavily overconsolidated clays this does not occur unless the factor of safety is below 2.

Estimation of undrained modulus E_u
(Figure 9.8)
Due to sampling disturbance, scaling factors, the effects of stress relief and bedding errors laboratory tests are considered to give inaccurate values of the stress-strain relationship for soils. At best, they may be considered a lower bound estimate. *In situ* testing such as plate loading tests or pressuremeter tests may reduce some of these effects but these tests can still suffer from the effects of stress relief and there are uncertainties and assumptions made in their analysis.

For an estimate of settlement it is generally considered that the most accurate approach consists of correlations between the '*in situ*' modulus E_u and undrained shear strength c_u. The c_u values are measured on undisturbed samples in the laboratory triaxial test. The E_u values are derived from back-analysis of settlement observations on actual structures on a wide variety of soils. This should eliminate most of the effects of sampling disturbance, stress relief, scale effects and stress path.

A correlation presented by Jamiolkowski *et al.* (1979) is reproduced in Figure 9.8. The modulus values represent the secant modulus at 0.5 c_u (factor of safety = 2). The stress-strain relationship for soils is non-linear so higher modulus values should be expected at lower stress levels.

Consolidation settlement

General
The methods described below give the one-dimensional consolidation settlement, denoted as ρ_{od}, or more commonly, ρ_{oed} since the soil parameters are determined from the results of the oedometer test. These settlements will commence on application of the loading but a large proportion can be expected to occur after construction and thus affect the more sensitive parts of the structure.

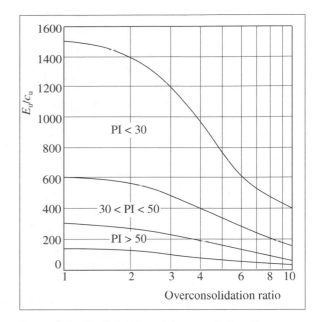

FIGURE 9.8 *Undrained modulus correlation (from Jamiolkowski et al., 1979)*

Compression index C_c method (Figure 9.9)
This method can be adopted for normally and lightly overconsolidated clays. The compression index C_c is the gradient of the void ratio–log effective stress plot for normally consolidated clay. For a lightly overconsolidated clay it is the gradient of the void ratio–log effective stress plot beyond the preconsolidation pressure p_c'.

For pressures lower than p_c' a smaller value C_s, referred to as the swelling index, is used for the overconsolidated portion of the plot. The accurate estimation of the preconsolidation pressure, p_c' is essential for lightly overconsolidated clays.

The method is illustrated in Figure 9.9. The soil deposit is split up into suitable sub-layers and the initial effective stress p_0' is determined at the mid-point of each sub-layer. The change in stress $\Delta\sigma$ caused by the structure or applied load is determined at the mid-points from the methods given in Chapter 5. The void ratio change Δe for each sub-layer is calculated from which the change in thickness ΔH is obtained. The consolidation settlement will be the summation of the changes in thickness. For a buried structure the settlements will be due to

$$\Delta e = C_c \log_{10}\left(\frac{p_0' + \Delta\sigma}{p_0'}\right) \qquad \therefore \text{ simplify to } \Delta e = C_c\,P \ \text{ where } P = \log_{10}\left(\frac{p_0' + \Delta\sigma}{p_0'}\right)$$

$$e_0 = w_0\,G_s \qquad\qquad \Delta H = \frac{\Delta e}{1 + e_0}\,H$$

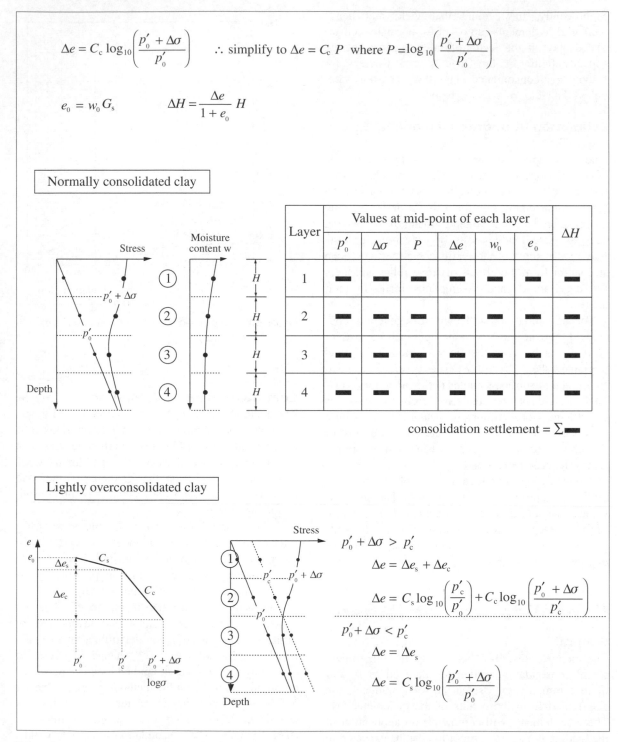

Normally consolidated clay

Lightly overconsolidated clay

FIGURE 9.9   Compression index method

changes in the thickness of the sub-layers beneath the structure. Greater accuracy is obtained with thinner sub-layers.

SEE WORKED EXAMPLE 9.8.

C_c values obtained from oedometer tests are likely to be under-estimated due to sampling disturbance. It is recommended that values of C_c and C_s are calculated from the *in situ* e–$\log \sigma$ curves which can be plotted using the methods described in Chapter 6.

Several correlations which relate C_c to a soil classification property have been published and two of these are given below:

$C_c \approx 0.009 \, (LL - 10)$ (Terzaghi and Peck, 1948)

$C_c \approx 0.5 \, \rho_s \, PI/100$ (Wroth, 1979)

where:

LL = liquid limit
PI = plasticity index
ρ_s = particle density

Oedometer or m_v method (Figure 9.10)

During an oedometer test values of the coefficient of compressibility, m_v are determined for each pressure increment applied above the vertical effective stress or overburden pressure p_o' at the depth from which the sample was taken. The change in thickness of a soil layer and hence the settlement of a foundation can be obtained from the expression:

$$\Delta H = \rho_{oed} = m_v \, H \, \Delta\sigma \qquad (9.10)$$

where $\Delta\sigma$ is the change in stress and H is the initial thickness of the soil layer.

Where the applied stress varies and possibly also the m_v values vary the deposits can be split up into layers and the change in thickness determined for each layer as illustrated in Figure 9.10. Values of the change in stress $\Delta\sigma$ can be obtained from the methods given in Chapter 5.

SEE WORKED EXAMPLE 9.9.

The coefficient of compressibility, m_v is not a true soil property, it depends on the pressure increments adopted. It is usually reported for a pressure range of p_o' + 100 kN/m² but where applied pressures are smaller than 100 kN/m², m_v values should be determined for the appropriate increments.

Typical values of m_v for different clay types are given in Table 9.3.

Total settlement

Skempton–Bjerrum method

The above methods determine the oedometer settlement, ρ_{oed} assuming that the pore pressure increase is produced by and is equal to the increase in vertical stress, i.e. $\Delta u = \Delta\sigma_v$, and that oedometer settlements are obtained from $m_v \, H \, \Delta\sigma_v$.

Skempton and Bjerrum (1957) realised that the

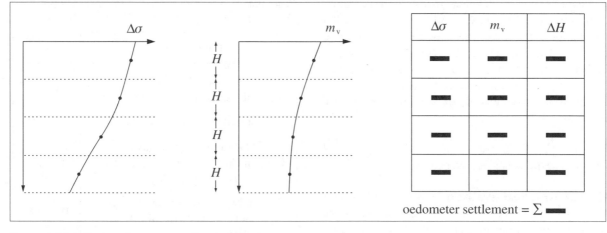

FIGURE 9.10 *Oedometer or m_v method*

TABLE 9.3 *Typical values of m_v*

Type of clay	m_v (m²/MN)
Very stiff heavily overconsolidated clay	< 0.05
Stiff overconsolidated clay	0.05–0.1
Firm overconsolidated clay, laminated clay, weathered clay	0.1–0.3
Soft normally consolidated clay	0.3–1.0
Soft organic clay, sensitive clay	0.5–2.0
Peat	> 1.5

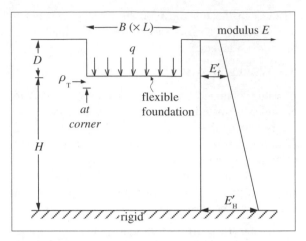

FIGURE 9.11 *Elastic drained method – definitions*

soil beneath structures is not laterally confined as in the oedometer and that generally $\Delta u < \Delta \sigma_v$ so that consolidation settlements will be less, given by $m_v H \Delta u$. They introduced a semi-empirical correction, μ to give the consolidation settlement ρ_c as:

$$\rho_c = \mu \, \rho_{oed} \tag{9.11}$$

with $\mu = A + \alpha (1 - A)$

A is Skempton's pore pressure parameter (equation 4.13) which is related to the overconsolidation ratio and the stress level. Lower values of μ would be obtained for the more heavily overconsolidated clays. α depends on the geometry of the loaded area, its shape, L/B and thickness, H/B with lower values of μ obtained for thicker soil layers and smaller foundations. However, the above expression for μ relies on A and m_v being constant with depth which does not occur. Some typical values of μ, suggested by Tomlinson (1986) are given in Table 9.4.

The total settlement ρ_T of a foundation is then given by:

$$\rho_T = \rho_i + \mu \, \rho_{oed} \tag{9.12}$$

The ratio ρ_i/ρ_T is considered below.

Elastic drained method (Figures 9.11–9.13)
For soils whose behaviour can be assumed to be linear elastic within the range of working stresses, with normal factors of safety, an elastic settlement procedure to determine total settlements is quite

simple and attractive. Many overconsolidated clays behave in this way so by using drained values of modulus E' and Poisson's ratio v' the total settlement (immediate and consolidation combined) can be obtained direct.

Using the same procedure as Butler (see above) Meigh (1976) gave the expression for the total settlement beneath the corner of a flexible rectangular area on a soil with modulus increasing with depth (see Figure 9.11) as:

$$\rho_T = \frac{qB}{E_f} \, IF_D F_B \tag{9.13}$$

TABLE 9.4 *Skempton–Bjerrum correction μ (from Tomlinson, 1986)*

Type of clay	μ
Very sensitive clays (Alluvial, estuarine and marine clays)	1.0–1.2
Normally consolidated clays	0.7–1.0
Overconsolidated clays (London Clay, Weald, Kimmeridge, Oxford, Lias clays)	0.5–0.7
Heavily overconsolidated clays (Glacial lodgement till, Mercia Mudstone)	0.2–0.5

The μ value for London Clay is usually taken as 0.5.

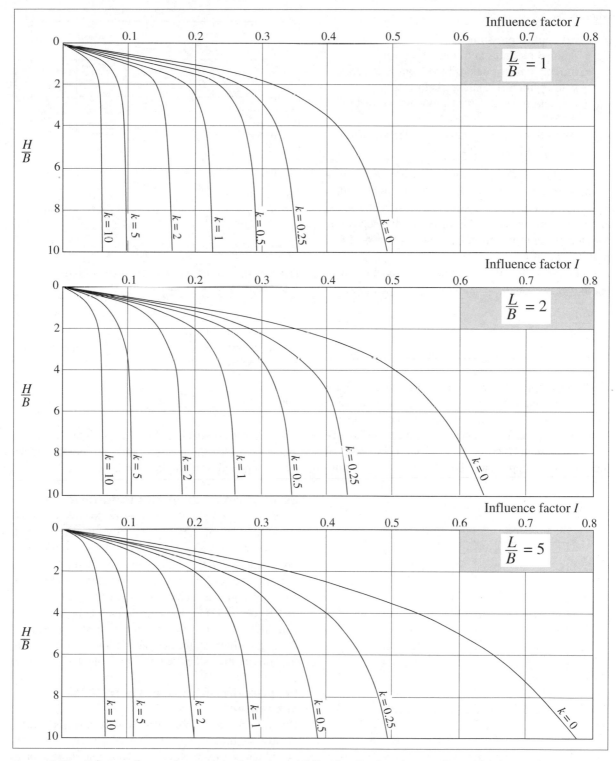

FIGURE 9.12   *Influence factors for modulus incresing with depth – Total settlement (from Meigh, 1976)*

where:

q = net applied pressure
E_f' = drained modulus at foundation level
I = influence factor determined from H/B, L/B and k, in Figure 9.12, but see below
F_D = correction for depth of embedment, in Figure 9.13
F_B = correction for roughness at underside of foundation, in Figure 9.14

The influence factor I is given assuming the base of the soil layer to be smooth. This tends to underestimate settlement so a simple adjustment for the effect of roughness at the base of the deposit is adopted by entering the charts at a value of $1.2 \times H/B$ instead of H/B.

k represents the rate of change of modulus with depth and is given by:

$$k = \frac{E_H - E_f}{E_f} \frac{B}{H} \quad (9.14)$$

v is assumed to be 0.2 and is incorporated into the values of I.

The principle of superposition must be used for points other than the corner of the loaded area and the principle of layering can be applied. Compared to the homogeneous, constant modulus case ($k = 0$) it can be seen that even a modest increase of modulus with depth can significantly reduce settlements.

SEE WORKED EXAMPLE 9.10.

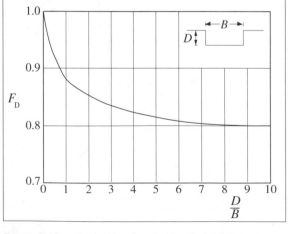

FIGURE 9.13   Correction for depth of embedment

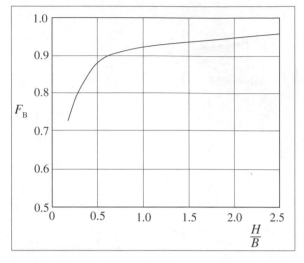

FIGURE 9.14   Correction for roughness of base of foundation

For the case of a rigid foundation, Meigh suggested using the mean settlement of the flexible foundation as:

$$\rho_{rigid} = 1/3 \, (\rho_{centre} + \rho_{corner} + \rho_{centre \, long \, edge})_{flexible} \quad (9.15)$$

A simpler approach and probably no less accurate would be to determine ρ_{centre} and multiply this value by μ_r, obtained from Table 9.2.

Estimation of drained modulus E'

As for the undrained modulus reasonable correlations have been developed between the drained modulus E' derived from back-analysis of settlement observations and the undrained shear strength, c_u.

Butler (1974) derived a correlation for London Clay (of high plasticity) as $E' = 130 \, c_u$ and Stroud and Butler (1975) gave correlations between E', c_u and SPT 'N' for soils of lower plasticity. Values relating to their 'suggested design line' which errs on the conservative side are given in Table 9.5.

Proportion of immediate to total settlement

Burland et al. (1978) have confirmed that sufficiently accurate values of the total settlement ρ_T can be obtained from the classical one-dimensional methods using equation 9.10 providing the soil can

TABLE 9.5 *Drained modulus values (from Stroud et al., 1975)*

Plasticity index %	E'/c_u
10 20	270
20–30	200
30–40	150
40–50	130
50–60	110

be considered to behave in an elastic manner. Thus for many overconsolidated soils with typical factors of safety greater than 2.5:

$$\rho_T = \rho_{oed} \qquad (9.16)$$

For soft yielding soils such as normally consolidated clays they found that the one-dimensional method gave values close to the consolidation settlement so that:

$$\rho_c = \rho_{oed} \qquad (9.17)$$

They also assessed the proportions of immediate to total settlement and their results are summarised in Table 9.6. Their analyses showed that the ratio $\dfrac{\rho_i}{\rho_T}$ decreases with:

- increasing non-homogeneity, of modulus increasing with depth
- increasing anisotropy, of the horizontal modulus E_H' compared with E_V'

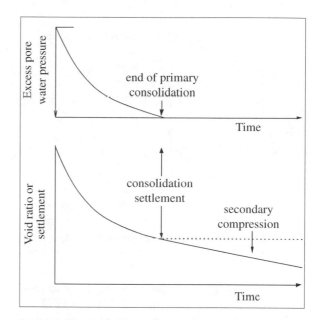

FIGURE 9.15 *Definition of secondary compression*

- decreasing drained Poisson's ratio v
- decreasing thickness of the soil layer

Secondary compression

Introduction (Figures 9.15 and 9.16)
With some soil types volume reductions and hence settlements have been found to continue even after primary consolidation has finished and when all pore water pressures have dissipated, see Figure 9.15. The settlements due to these volumetric strains are referred to as secondary compression. Drained creep can comprise both volumetric and shear strains.

Various mechanisms and models have been sug-

TABLE 9.6 *Typical ratios of immediate to total settlements (from Burland et al., 1978)*

Soil type	Immediate settlement	Consolidation settlement	Total settlement	ρ_i/ρ_T
	ρ_i	ρ_c	ρ_T	
soft yielding	$0.11\rho_{oed}$	ρ_{oed}	$1.1\rho_{oed}$	0.1
stiff elastic	$0.6\rho_{oed}$	$0.4\rho_{oed}$	ρ_{oed}	0.6

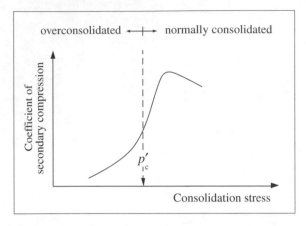

FIGURE 9.16 *Effect of preconsolidation pressure*

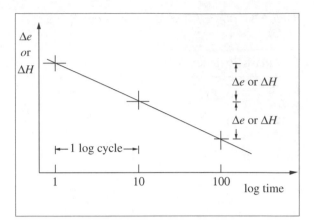

FIGURE 9.17 *Log time plot – secondary compression*

gested to explain the phenomenon but it seems most likely to be associated with a redistribution of the interactions (forces) between particles following the large structural rearrangements which occurred during the normal consolidation stage of loading. This is supported by the fact that secondary compressions are insignificant for stress levels below the preconsolidation pressure (when the soil is overconsolidated and mainly elastic strains occur) but can be large when stress levels exceed the preconsolidation pressure and plastic strains occur, see Figure 9.16.

It is also dependent on the type of particles, for example secondary compressions can be large when organic material is present, see below.

General method (Figure 9.17)

For volume changes to occur due to the expulsion of pore water there must be pore pressure gradients within the soil but these are considered to be so small and occur at such a slow rate that they are immeasurable. Quite commonly it is found that void ratio or thickness changes in the oedometer test plot linearly with the logarithm of time, as in Figure 9.17, with the gradient referred to as the coefficient of secondary compression, C_α where:

$$C_\alpha = \frac{\Delta e}{\Delta \log t} = \frac{\Delta e}{\log_{10} \frac{t_2}{t_1}} \qquad (9.18)$$

Secondary settlements, ρ_s can then be obtained from this laboratory test result as:

$$\rho_s = H \frac{C_\alpha}{1 + e_0} \log_{10} \frac{t_2}{t_1} \qquad (9.19)$$

where:

e_o = initial void ratio of soil
H = thickness of soil layer
t_2 = time after which settlement is required
t_1 = reference time

The amount of secondary settlement will depend on the time values chosen. For convenience, it will probably be sufficiently accurate to adopt $t_1 = 1$ year to allow for the construction period and primary consolidation and t_2 as the design life of the structure. The above expression presumes that equal settlements occur for each log cycle of time. e.g. settlement from 1 to 10 months (first year) = settlement from 10 to 100 months (first decade) = settlement from 100 to 1000 months (first century) and so on.

Estimation of C_α or ε_α values (Figure 9.18)

Values of C_α can be obtained from the laboratory oedometer test by continuing readings beyond the primary consolidation stage. However, as the times are plotted on a logarithmic scale, to establish the secondary compression line the test will be time consuming.

It may be sufficient to use an estimate from a correlation such as the one published by Mesri (1973), Figure 9.18, who reported the coefficient of secondary compression as ε_α:

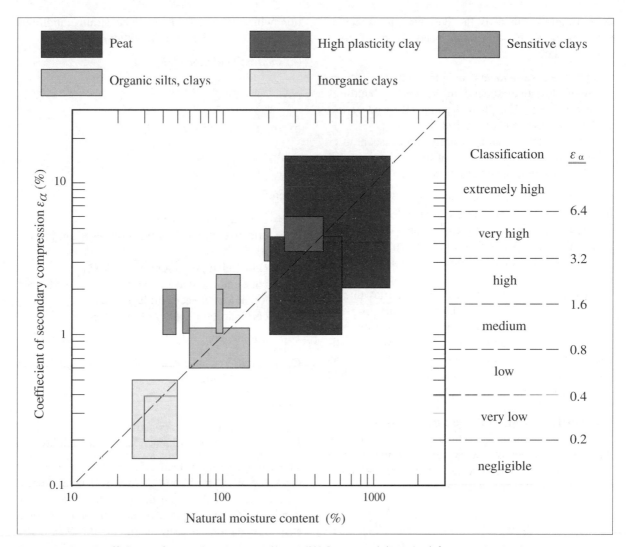

FIGURE 9.18 *Coefficients of secondary compression ε_α (%) for natural deposits (after Mesri, 1973)*

$$\varepsilon_\alpha = \frac{C_\alpha}{1 + e_0} \qquad (9.20)$$

This chart shows the significant effect on secondary compression of:

- organic content, especially peats
- clay mineralogy as represented by high plasticity
- metastable mineral grain structures as represented by sensitive clays.

Sands

There exists a large number of methods for estimating settlements of foundations on sands developed over the last 50 years. They are all based on empirical correlations between observed settlements and other appropriate parameters but probably because of the large range of variables involved not one of these methods can predict settlements particularly accurately. It should be accepted that only an approximate estimate is the best we can expect of these methods. However, this need not be of great concern since:

1. Observed settlements of pad-type foundations (width less than about 5 m) supporting buildings are typically less than 40 mm (Burland *et al.*, 1978 and Burland *et al.*, 1985).
2. Settlements in sands usually occur soon after applying the load so most of the settlements will take place during construction.

Greater settlements beneath larger raft foundations will be likely but these can also be affected by the presence of more compressible clay or silt layers within the sand deposit. Additional settlements can be expected due to vibrations from machinery and traffic or where there are large fluctuations in live loads or variable actions such as with silos due to filling and emptying, or chimneys, due to wind loading. Where there are compressible constituents present such as organic matter, clay or mica particles, shells or the sand itself is crushable such as coral sand, weathered sands, volcanic ash then greater settlements can be expected.

Methods of estimating settlements

Most of the methods for estimating settlements take the form of a quasi-elastic expression with empirical factors applied. The elastic expression for settlement in equation 9.13 is modified to the form:

$$\rho = \frac{q \times (\text{function of } B) \times \begin{pmatrix} \text{factors for shape,} \\ \text{thickness, depth,} \\ \text{water table, time} \end{pmatrix}}{\text{function of SPT' } N' \text{ or cone } q_c \text{ or modulus } E'} \tag{9.21}$$

Two methods are presented below, one based on the Dutch cone penetration test and the cone end resistance, q_c (Schmertmann's method) and the other adopting the standard penetration test 'N' value (Burland and Burbridge's method).

Schmertmann's method (Figure 9.19)

Proposed by Schmertmann (1970) and modified by Schmertmann *et al.* (1978) this method is based on a strain influence factor diagram for two cases, a square foundation with $L/B = 1$ where axi-symmetric stress and strain conditions occur and a strip founda-

tion with $L/B = 10$ where plane strain conditions exist. The expressions for the settlement are:

(a) *square foundation ($L/B = 1$)*

$$\rho = \frac{C_1 C_2}{2.5} \, \Delta p \sum_0^{2B} \frac{I_z \Delta z}{q_c} \tag{9.22}$$

b) *strip foundation ($L/B = 10$)*

$$\rho = \frac{C_1 C_2}{3.5} \, \Delta p \sum_0^{4B} \frac{I_z \Delta z}{q_c} \tag{9.23}$$

where:

p = gross applied pressure
p_o' = effective stress at foundation level around the foundation
Δp = net applied pressure = $p - p_o'$ kN/m^2
q_c = cone end resistance, kN/m^2, for each soil layer
Δz = thickness of each soil layer, metres
C_1 = correction for depth of foundation or confinement = $1 - 0.5 \dfrac{p_o'}{\Delta p}$ (C_1 should be ≥ 0.5)
C_2 = correction for creep or time related settlement = $1 + 0.2 \log_{10} 10t$

where:

t = time in years after construction
I_z = average strain influence factor for each layer

The average strain influence factor, I_z, is obtained as the value at the mid-point of each soil layer from a diagram drawn alongside the q_c–depth plot with a depth of $2B$ for a square foundation and $4B$ for a strip foundation as shown in Figure 9.19. The maximum value of I_z is given as:

$$I_{z\,max} = 0.5 + 0.1 \left(\frac{\Delta p}{\sigma_v'}\right)^{0.5} \tag{9.24}$$

where:

σ_v' = vertical effective stress at a depth of $B/2$ for a square foundation and B for a strip foundation, see Figure 9.19.

Values of Δz, average q_c and average I_z for each soil layer are required for the summation term and these can be conveniently presented in a table, as illustrated in Worked Example 9.11.

SEE WORKED EXAMPLE 9.11.

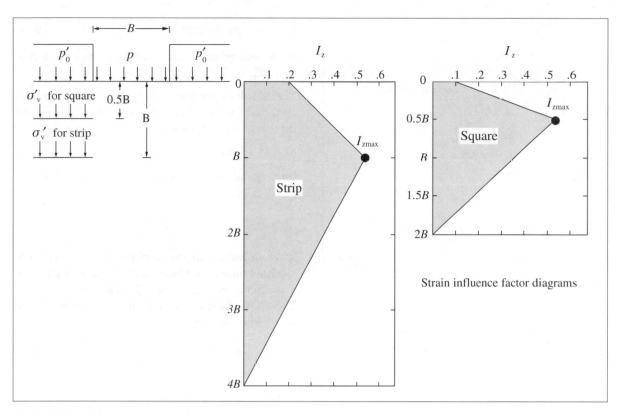

FIGURE 9.19 *Schmertmann's method*

Settlements for shapes intermediate between a square and strip can be obtained by interpolation.

Burland and Burbridge's method
(Figures 9.20 and 9.21)
Based on a statistical analysis of a large number of settlement observations from case studies, Burland and Burbridge (1985) proposed a method for normally consolidated and overconsolidated sand.

1. *Foundations on the surface of normally consolidated sand*
The average immediate settlement ρ_i at the end of construction is given by:

$$\rho_i = f_s f_1 q' B^{0.7} I_c \qquad (9.25)$$

where:

ρ_i = average immediate settlement, mm

f_s = shape factor = $\left[\dfrac{1.25 L/B}{L/B + 0.25} \right]^2 \qquad (9.26)$

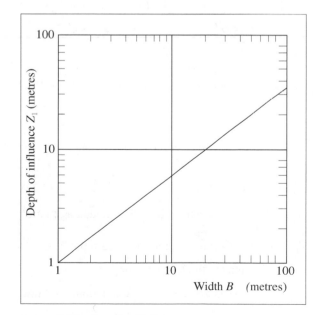

FIGURE 9.20 *Depth of influence*

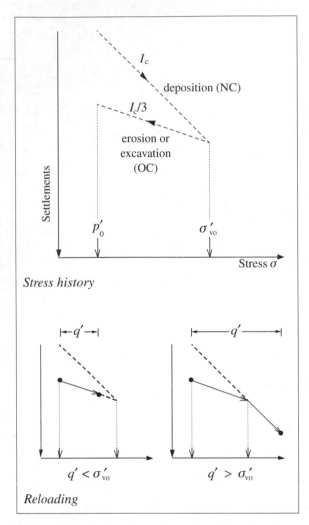

FIGURE 9.21 *Settlements of overconsolidated sand*

= 1.0 for square or circle ($L/B = 1$) and
= 1.56 for strip ($L/B = \infty$)

f_1 = thickness factor = $\dfrac{H_s}{Z_I}\left(2 - \dfrac{H_s}{Z_I}\right)$ (9.27)

H_s = thickness of sand below the foundation, metres (when $H_s < Z_I$)

Z_I = depth of influence, metres, see Figure 9.20

If $H_s > Z_I$ then the thickness factor $f_1 = 1$

q' = average gross effective foundation pressure, kN/m^2

B = width of foundation, metres

I_c = compressibility index = $\dfrac{1.71}{\overline{N}^{1.4}}$ (9.28)

$\overline{N}$ = average SPT value over the depth of influence Z_I

The N values obtained from the boreholes are not corrected for the effect of overburden pressure but they are corrected (using N') for grain size effects:

$N' = 15 + 0.5\,(N - 15)$ (9.29)

for very fine or silty sands below the water table and

$N' = 1.25N$ (9.30)

for gravel or sandy gravel

2. *Foundations on the surface of an overconsolidated sand or for loading at the base of an excavation in normally consolidated sand*

The average immediate settlement at the end of construction is given by:

$$\rho_i = f_s f_1 q' B^{0.7}\,\dfrac{I_c}{3}$$ (9.31)

when q' is less than σ_{vo}' and by:

$$\rho_i = f_s f_1 \left(q' - \dfrac{2}{3}\,\sigma_{vo}'\right) B^{0.7}\, I_c$$ (9.32)

when q' is greater than σ_{vo}'.

All of the terms are as described in (1) above. σ_{vo}' is the past maximum stress for an overconsolidated sand or the effective stress at foundation level or excavation level for a normally consolidated sand.

The method assumes that the compressibility of a sand is reduced by a factor of 3 when it has become overconsolidated, either by a geological removal process such as erosion or by a construction process such as excavation, Figure 9.21. There exists a past maximum stress σ_{vo}' below which the compressibility is given by $I_c/3$ and above which it is given by I_c. Settlements will be small if q' is less than σ_{vo}' but much larger if q' is greater than σ_{vo}'.

Unfortunately, there is no sure way of determining whether a sand is normally consolidated or overconsolidated nor of determining σ_{vo}' for an overconsolidated sand. In the absence of sufficient evidence it would be prudent, therefore, to assume that a sand is normally consolidated and to introduce the effect of

TABLE **9.7** *Time correction factors*

Loading condition	R_3	R_t
static loads	0.3	0.2
fluctuating loads	0.7	0.8

σ_{vo}' only when a foundation is placed at the base of an excavation.

Many of the cases studied displayed time-dependent settlement after the end of construction. A correction f_t should be applied to obtain the long-term settlement, thus:

$$\rho_t = \rho_i f_t \qquad (9.33)$$

where:

$$f_t = 1 + R_3 + R_t \log_{10} \frac{t}{3} \qquad (9.34)$$

t = time after the end of construction, years. Values of R_3 and R_t are given in Table 9.7.

SEE WORKED EXAMPLE 9.12.

The apparent accuracy of the above expressions for settlements should be tempered by the moderate accuracy of the statistical correlation and the inherent variability of all sand deposits. It is stated that the actual settlement could lie between ± 50% of the predicted value.

Permissible settlements

Having determined an amount of settlement for a foundation it is then necessary to know whether it will be acceptable or not. It should be less than the permissible (or tolerable, or allowable) settlement. However, there is no straightforward approach to this problem because the structure-foundation-soil interaction phenomenon is a complex and uncertain subject.

It must be remembered that structures may not move due to ground settlements alone. There may be movements and associated damage due to dimension changes in the structural materials such as due to moisture or temperature changes, creep or chemical reactions. Some examples of these are illustrated in

BRE Digest 361, 1991. There may be structural movements due to ground movements other than settlements as calculated above, such as mining subsidence, shrinkage or swelling of clay soils, erosion of sandy soils, slope instability, poorly compacted backfill, vibrations etc.

Definitions of ground and foundation movement (Figure 9.22)

Burland and Wroth (1974) have proposed some useful definitions of ground, foundation and structural movements to enable a detailed investigation of the various modes of movement of a structure. These are illustrated in Figure 9.22 where points A to D could represent points on a raft foundation, the locations of isolated foundations beneath the columns of a frame building or points along a loadbearing wall. The definitions are given as:

● *Settlement ρ*
  Downward movement at a point. This will vary across a non-rigid structure.
● *Heave ρ_h*
  Upward movement at a point. This will vary across an excavation or beneath a structure.
● *Differential settlement or differential heave – δρ or $\delta\rho_h$*
  The difference in settlement or heave between two points. Usually two adjacent points are chosen but the choice may be arbitrary, Figure 9.22a.
● *Horizontal displacement u*
  Extension or contraction of a building in the horizontal direction will result in tensile or compressive strains.
● *Rotation θ*
  The change in gradient of a line joining two reference points, such as between A and B in Figure 9.22a.
● *Tilt ω*
  The rigid body rotation of the whole of a structure or a well-defined part of it. This is obtained by drawing a straight line between the two edges of the settlement profile beneath a building or a well-defined part of it such as points A and D in Figure 9.22b. This is referred to as the tilt plane and will affect both the horizontal and vertical components of a building.

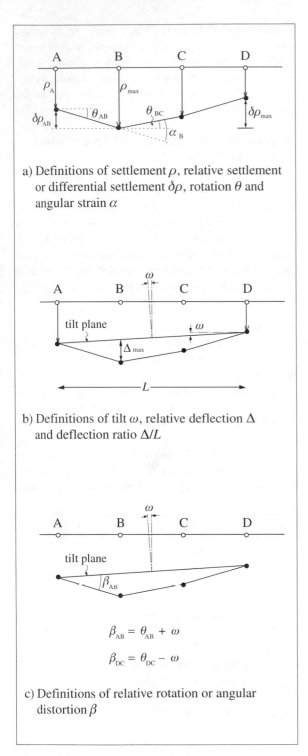

a) Definitions of settlement ρ, relative settlement or differential settlement $\delta\rho$, rotation θ and angular strain α

b) Definitions of tilt ω, relative deflection Δ and deflection ratio Δ/L

$$\beta_{AB} = \theta_{AB} + \omega$$

$$\beta_{DC} = \theta_{DC} - \omega$$

c) Definitions of relative rotation or angular distortion β

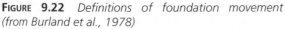

FIGURE 9.22 *Definitions of foundation movement (from Burland et al., 1978)*

- *Relative deflection Δ*
  The displacement relative to the 'tilt' plane, Figure 9.22b. Downward displacement is described as sagging and upward deflection as hogging.
- *Deflection ratio Δ/L*
  This denotes the degree of curvature to which the building or a part of it has been subjected and can represent a sagging ratio or a hogging ratio. The degree of curvature a building is subjected to will determine the amount of distortion and hence the degree of damage of the structure. The deflection ratio is the parameter adopted for Figure 9.22. Deflection ratio is preferred to angular distortion β since the latter is affected by the amount of tilt, see Figure 9.22c.
- *Angular strain α*
  This represents the total rotation at a point within a structure. For example, at B on Figure 9.22a it is given by:

$$\alpha_B = \theta_{AB} + \theta_{BC} \qquad (9.35)$$

  It is positive if it produces sag and negative if it produces hog. This parameter is particularly useful for assessing localised movements along a brick wall since cracks are often concentrated where angular strains are high. Cracks will occur at the base of the wall if it has sagged and at the top of the wall if it has undergone hogging.
- *Relative rotation (or angular distortion) β*
  The rotation of the line joining two reference points relative to the tilt line, Figure 9.22c.

Criteria for movements

From the work of Burland, Broms and de Mello (1978) there appear to be four criteria which must be satisfied:

1. *Visual appearance of the structure as a whole*
   Tilting of walls, floors and the whole building could be unpleasant or even alarming for the occupants and visitors. A deviation in excess of about 1 in 250 from the vertical or horizontal would probably be noticeable.
2. *Visual appearance of the architectural materials*
   Visible damage such as cracking or distortions of claddings can vary from being unsightly to

alarming. A classification for the assessment of damage has been suggested by the Building Research Establishment (*BRE Digest 251*, 1989). Although cracking is indicative of movements the classification is based on the ease of repair of the damage.

3. *Serviceability or function of the structure*
Movements can occur which affect the overall efficiency of the building such as reduced weathertightness, rain penetration, dampness, draughts, heat loss, reduced sound insulation, windows and doors sticking. Movements can also occur which affect the basic function or purpose of the structure such as with the operation of lifts or precision machinery, access ramps or steps, fracturing of service pipes.

4. *Stability*
Large movements and very severe damage to the cladding and fittings will have occurred before the structure itself fails due to instability so the above criteria are used to dictate permissible settlements.

Routine settlement limits (Figure 9.23)

Based on a large number of observations of structures Terzaghi and Peck (1948) and Skempton and MacDonald (1956) gave values of permissible settlements for framed structures and these are summarised in Table 9.8. It is emphasised that these refer to 'routine' buildings with fairly uniform distribution of loading and uniform ground conditions.

With regards to the interaction between a structure and its supporting soil a number of important points emerge:

● Permissible settlements for the same structure can be larger on clay soils than on sands. This is probably due to the longer period of time over which settlements occur on clay soils allowing the structure to gradually adjust to the settlements whereas if the structure was placed on sand it must respond immediately to settlements.

● Frame buildings and their claddings can tolerate more distortion than loadbearing walls. Many of the claddings in frame buildings are less sensitive to movement and they are installed at a time when much of the settlement has already elapsed whereas loadbearing walls are more brittle and

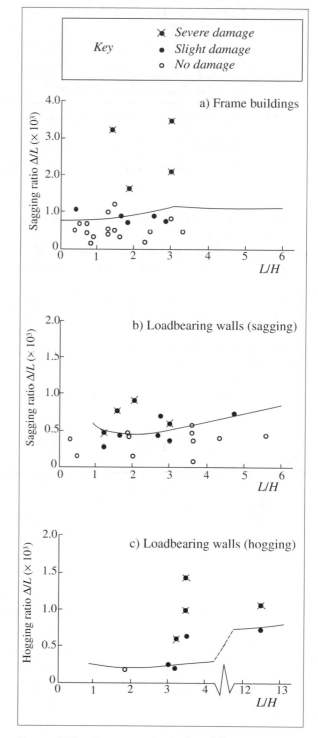

FIGURE 9.23 *Damage criteria for different types of structure (from Burland et al., 1978)*

TABLE 9.8 *Routine guides to permissible settlements*

Settlement	Sand		Clay
	Reference 1	Reference 2	Reference 2
maximum differential settlement $\delta\rho$	20	25	40
maximum settlement ρ (isolated foundations)	25	40	65
maximum settlement ρ (raft foundations)	50	40 – 65	65 – 100

Reference 1 – Terzaghi and Peck, 1948   Reference 2 – Skempton and MacDonald, 1956

are subjected to settlements from commencement of construction.

● Frame buildings without infill panels (open frames) can tolerate more settlement than infilled frames.

● Frame buildings on isolated foundations may distort differently from buildings on raft foundations.

● Load-bearing walls undergoing sagging can tolerate more distortion (nearly twice as much) than walls undergoing hogging.

● Longer structures can tolerate more relative deflection Δ.

● A stiff soil layer overlying the compressible soil causing the settlements will not prevent the total settlement occurring but will significantly reduce differential settlements.

Burland and Wroth (1974) proposed a simple criterion to relate the distortion of a structure (measured by the deflection ratio Δ/L) to the onset of visible cracking of the cladding or finishes and this relationship (Burland *et al.*, 1978) is reproduced in Figure 9.23 for frame buildings and load-bearing walls. If Δ/L lies above the criterion lines shown, it is likely that the buildings will suffer architectural damage.

SUMMARY

This chapter deals with the downward movement or settlement of a foundation, due to the stresses applied by a structure to the underlying ground. Settlements due to other phenomena must also be considered.

The total settlement of a foundation is the sum of the immediate, consolidation and secondary compression settlements. The immediate settlements will occur on application of the load so most should be completed by the end of construction. The consolidation settlement takes a longer time depending on the permeability of the soil and some of this will occur after the end of construction. The secondary compression occurs over a much longer period and can affect some structures for many years.

Immediate settlement is determined as the elastic undrained settlement at the corner of a flexible loaded area and methods are given for both homogeneous and non-homogeneous conditions. Corrections are available for the rigidity of the foundation, its depth and the possibility of local yielding of the soil. Using the principles of superposition and layering the settlement at any point near a loaded area and within the soil can be determined.

Consolidation settlement is determined using the compression index and swelling index method for normally consolidated and lightly overconsolidated clays. An accurate measure of the preconsolidation pressure is required and the *in situ* void ratio–log pressure curve should be constructed.

For more heavily overconsolidated clays the method using the coefficient of compressibility is often adopted with the Skempton–Bjerrum correction applied. The modulus of a soil increases with depth which will produce lower settlements. Methods are given which allow for these non-homogeneous conditions.

Settlements from secondary compression can be large for peat soils, normally consolidated highly plastic clays and sensitive clays.

Settlements of foundations on sand are obtained from empirical correlations. Two commonly used methods are given.

A criterion for the assessment of damage to a building is presented based on the onset of visible cracking. By determining the settlement profile beneath the structure, the deflection ratio, Δ/L that could occur will determine whether the structure will incur damage or not.

An existing structure may be affected by movements of varying types. Some useful definitions are given to aid in the assessment of the effects of these movements including tilt, relative deflection, deflection ratio and angular strain.

CASE STUDY

Leaning Tower of Pisa – Torre Pendente di Pisa, Italy

Case Objectives:

This case illustrates:

- careful and detailed measurements and thorough analysis are a prerequisite to the understanding of the development and causes of movements of structures
- the stability of tall structures can be impaired by leaning instability if the ground is compressible
- numerical modelling is a very useful tool in back-analysing the causes of movement, in predicting the nature and amounts of further movements and as a check on the monitoring of remedial measures
- a modern understanding of soil mechanics and structural stability has given insight into the mechanisms causing movement of the tower

History

The most famous bell tower (campanile) in the world was commenced in 1173 and is ascribed to the architect, Bonnano Pisano. In 1178 when the tower had reached a height of just over four storeys work was stopped, it is believed, because of construction difficulties and a war with Florence. By this time the tower was leaning to the north and must have been obvious. Nothing further was done for another 100 years, fortunately. Had work continued it is almost certain that the foundations would have undergone an undrained bearing capacity failure within the Upper Sand and the Pancone clay. During the 100-year break the strength of the soils increased due to consolidation beneath the weight of the tower.

During the 1270s the tower was continued to the seventh storey adopting the practice of attempting to straighten the tower by adding tapered masonry courses on the lower southern side. It is believed that the addition of stone on this side aggravated the southward lean and by 1278 it is estimated that the inclination was about 0.6° (Burland and Potts, 1995). Work ceased again in 1284 due to a war with Genoa and it was not until 1370 that the bell tower was completed when the inclination reached 1.6°. The significant amount of tilt at that time was recognised by the pronounced correction applied to the bell tower. However, the tilting continued unabated and in 1817, the first recorded measurement, the tilt was 4.9°.

The tower has a centre-line length of 58.4 m, a maximum base diameter of 19.58 m and has a total weight of 141,640 kN. The tower is constructed as a hollow cylinder with the 3 m thick walls comprising two layers (outer and inner) of marble with an annulus infilled with rubble and mortar which in places contains large voids. The axis of the tower is curved, in the shape of a banana or question mark.

In 1989 the 14th-century bell tower in Pavia, south of Milan, collapsed without warning killing four bystanders. This tower was not even leaning. The Tower of Pisa was closed to

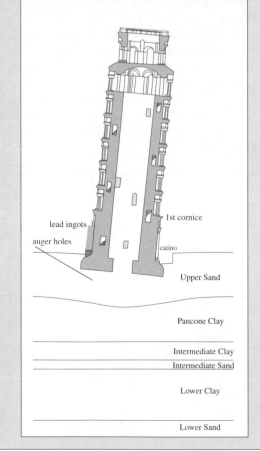

lead ingots
auger holes

1st cornice
catino

Upper Sand

Pancone Clay

Intermediate Clay
Intermediate Sand

Lower Clay

Lower Sand

Horizon	Geology	Strata	Nature	Thickness (m)
A	Estuarine deposits laid under tidal conditions	Upper Sand	Grey sand on N side yellow silty sand and silt on S side	10
B	Marine clay	Upper Clay (Pancone)	Soft gey clay of high and very high plasticity. Lightly overconsolidated and sensitive	11
		Intermediate Clay	Firm dark grey organic clay and stiff grey and yellow silty clay of intermediate plasticity Heavily overconsolidated	4
		Intermediate Sand	Grey clean sand and silty sand	2.5
		Lower Clay	Firm grey silty clay of high plasticity. Normally consolidated	13
C	Marine sand	Lower Sand	Dense silty sand	> 22

the public in 1990 and in the same year the Pisa committee was set up under the chairmanship of Professor Jamiolkowski and included Professor Burland of Imperial College to advise on remedial measures.

Soil Conditions
The soil conditions comprise three thick horizons, A, B and C. The deposit boundaries are essentially horizontal particularly in horizon B so the lean of the tower could not be ascribed to varying layer thicknesses. The upper surface of the Pancone clay is dished beneath the tower. The pore pressure distribution through horizon B is slightly below hydrostatic with downward seepage from Horizon A as a result of pumping from the Lower Sand.

Movements and Interventions
During its 800-year history the south side of the tower has settled 2.8 m and the north side 0.8 m, giving a differential settlement of 2m (Mitchell *et al.*, 1977). At the present time the tilt of the tower is due south with an overall tilt of 5.5°. It has been estimated that by 1300 the lean was 1.43 m and this has continued until today the tower is now 5.5 m out of plumb. The rate of movement in 1990 was estimated to be 1.2 mm per year.

There have been many interventions in the tower to correct the lean but all have failed. The most disastrous was in 1838. Due to the settlements the columns at ground level had become buried so a bright spark named Gherardesca suggested the digging of a catino or below ground walkway to expose the columns. This excavation was taken below the water table which is no more than about 1 to 2 m below ground level and the inflow of groundwater and loosening of the sands locally caused the tower to move dramatically by 0.5° or 0.5 m at the top of the tower. Since 1840 there have been 16 committees set up to address the problem and many interventions have taken place but all have failed. Significantly all interventions on the south side have resulted in adverse effects. Accurate measurements since 1911 have shown that when left untouched during 1935 to 1966 the inclination would increase at a rate of over 3.5 seconds of arc per year or just over 1 mm per year but by 1990 the rate of inclination had almost doubled. Significant interventions have accelerated the tilt. In 1934 361 holes were drilled into the foundation masonry and grouted causing a sudden increase in tilt of 31". Some soil and masonry drilling in 1966 and masonry drilling in 1985 caused small but distinct increases in tilt. In the early 1970s pumping from the Lower Sand produced a more regional subsidence and tilting of the whole Piazza towards the south-west inducing a tilt of 41" or almost 12 mm. The rate of tilting of the tower returned to its previous value when pumping was reduced. The tower is very sensitive to even the smallest interference.

Remediation
It has been shown that the marble masonry is under severe structural stress, especially at the level of the first

cornice on the south side. It is estimated that if the strain energy stored could be released in this area it would lead to a catastrophic buckling failure. In 1992 plastic coated steel tendons were wrapped around the second storey and tensioned to stabilise the masonry.

Recognising that leaning instability was the prime cause of the tilting, in 1993 as an emergency (and reversible) stabilisation measure lead ingots were placed in a gradual and controlled manner on a post-tensioned concrete ring base cast around the tower at plinth level. Continual monitoring showed that the tower tilted towards the north with each application of the load with ingots placed up to a maximum load of 700 tons. This reduced the southward inclination by about 1 minute of arc and reduced the overturning moment by 10%.

In 1995 a method of applying the downward force on the north side less obtrusively than the lead ingots was proposed using anchors drilled into the Lower Sand 45 m below ground level and tensioned to apply the required forces. These anchors were to be fixed into a reinforced concrete ring which would have to be constructed below the water table. To enable excavation in the dry the groundwater was excluded from the excavation by adopting ground freezing techniques. Unfortunately on 6 September 1995 the top of the tower moved 1.5 mm overnight, more than the movement in a whole year. The load of the lead ingots was increased to 900 tons to counteract the movements. This intervention was soon abandoned and in late 1996 the Committee was disbanded.

Causes of Movements

Since 1911 measurements have shown that the tower is no longer settling but tilting or rotating about a point near the axis of the tower at the first cornice level. From this it is concluded that the seat of the creep rotation is within Horizon A and not the Pancone clay. The most likely causes are considered to be (Burland and Potts, 1995):

1) fluctuations of the water table in Horizon A. It is known that during heavy rainstorms the water table rises quickly and more on the northern side than on the southern side causing a slight lifting of the tower on the northern side. These movements are mostly seasonal associated with the heavy rainstorms in the period September to December (Burland, 1997).

2) small racking movements of the foundations due to contraction and expansion effects in the tower as it heats and cools each day

3) seasonal fluctuations of the water pressures in the Lower Sands cause cyclic inclinations of the whole Piazza

The main cause of settlement of the tower is not due to the bearing capacity failure of the Pancone clay but is the result of the compressibility of the underlying soils with the deformation concentrated in Horizon A and the Pancone clay.

The tilt of the tower is a result of leaning instability which occurs in tall structures when the overturning moment produced by a small perturbation, or increase in the inclination $d\theta$, is equal to or greater than the resisting moment of the foundation soil produced by the same inclination. This will apply to tall structures when they reach a critical height. The seat of the continuous long-term tilting of the tower lies in Horizon A and not in the Pancone clay as was previously thought (Burland, 1997). Finite element analyses (Burland and Potts, 1995) have given very close agreement with the historical development of movements of the tower. They have shown that at the end of the first phase of construction (up to the fourth storey in 1178) there were extensive zones within both the Upper Sand and the Pancone clay where the shear strength was fully mobilised. Had loading continued then these zones would have extended and merged to produce the classical pattern of bearing capacity failure. The modelling of the 90-year break as a period of consolidation showed that there was a significant increase in the strength of the clay with the elimination of almost all of the zones of contained failure within the Pancone clay.

Soil Extraction

In 1997 a disastrous earthquake occurred near Assisi, east of Pisa with many historical buildings destroyed. This provided further impetus to the protection of historical buildings and in 1998 a soil extraction technique proposed by Professor Burland was approved. With the tilt of the tower at 5.44° it had been estimated that it was in a state of unstable equilibrium and was very close to falling down. With little margin for error a harness was wrapped around the tower to temporarily hold it in the event of unexpected movements. The soil extraction technique had undergone a successful trial in 1996. It comprises drilling continuous flight auger holes at an angle of 30° into the Upper Sand from the 12 boreholes spaced over a width of 5.5 m and taken to beneath the north side. Removal of small controlled volumes of soil has induced subsidence beneath the north side and is being continuously monitored for adverse effects but in the first three months the tower recovered 60 seconds of arc northwards and the top had moved back 18 mm with no settlement at the south side (Burland, 1999). To achieve a stable situation the tilt needs to recover another 0.5 m and this will take another 2 years.

Other relevant chapters are 5, Stress Distribution and 6, Consolidation.

CASE STUDY

Permissible! settlements – Palace of Fine Arts, Mexico City, Mexico

Case Objectives:

This case illustrates that:

- the nature of the ground was either completely ignored or totally misunderstood
- raft foundations even of considerable thickness can behave in a flexible manner, settling into a dish-shaped profile
- structures can undergo very large settlements but still provide their intended function (or most of it) although the serviceability is likely to be impaired
- cracked and heavily distorted structures are unsightly and undesirable whereas a titled structure can become a curiosity and tourist attraction

The Palacio de Bellas Artes, one of the finest architectural examples in the city, was commenced in 1904 and following several interruptions was completed in 1934. The structure settled considerably during and after construction and by 1950 it had settled 3 m below street level greatly distracting from its architectural merit. Regional settlements in Mexico City due to groundwater abstraction by pumping have taken place since the latter half of the 19th century and these have contributed to the structure's settlement, see the case study in Chapter 6.

The building comprises a structural steel frame supported on a raft (or mat) foundation 1.8 to 3 m thick. The building is approximately rectangular with overall dimensions of 81.4 m by 118.9 m. The raft foundation extends 1.5 m beyond the structure on all sides except on the south side where it extends 18m beyond the structure. The raft was constructed in at least two lifts with 1.2 m of unreinforced concrete placed first, then steel grillages followed by lighter weight concrete. The latter was made using porous volcanic rock as aggregate since settlements had already been observed following the initial pour. The grillages consisted of I-beams criss-crossing between the column bases but these were placed in the upper part of the raft.

A 1907 survey on top of the raft before the steel frame had been constructed showed a noticeable dish-shaped profile with a maximum settlement in the middle of 38 mm. By 1908 when the steel frame had been completed the partly built structure had settled 1.68 m and during 1910 a diagonal crack appeared across the raft. At this time the settlement was peaking at 43 mm per month (about 1.4 mm per day) and the north-west corner was settling faster, possibly as a result of more loading at this end.

In 1911–12 stabilisation measures were carried out consisting of encircling the structure with a row of steel sheet piling about 3 m away from the structure to resist the horizontal pressures or lateral squeezing of the soil and grout injection into the soils between the sheet piling and the structure was carried out to stiffen this soil and to (theoretically) increase the bearing area of the raft foundation. The grout was a 1:2 cement:sand mixture and about 20,000 bags of cement and 4000 m³ of sand were injected. This could have been a very early attempt at compensation grouting. The rate of settlement was reduced to 11 mm per month but it was still continuing. At this time cracks in the structure were so large that a person could crawl through them (Rossi, 1955).

Settlement measurements carried out in 1950 (Thornley et al., 1955) showed that the palace was settling at a rate of 38 mm per year more than the surrounding streets and that the Palace was in the centre of a settlement bowl extending beyond the structure. It was found that the Palace itself was settling at a rate of nearly 125 mm per year and that most of the settlement was occurring within the clay to 33 m below ground level.

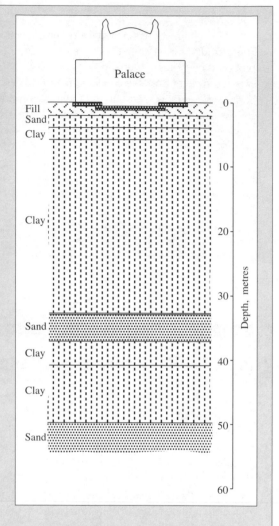

Pile foundations taken to the first sand stratum had been used successfully to support adjacent large buildings. Indeed, Rossi pointed out that several old Spanish buildings had to be removed to prepare the site. These were found to have been supported on pile foundations but the piles were removed and the voids filled with concrete. Rossi proposed demolishing the structure and starting again using a pile foundation but the architect refused.

It is easy to criticise in hindsight but foundation methods previously used on a site should not be ignored. Thornley's solution was to underpin the structure on piles taken to the first sand layer and allow the surrounding area to settle under the regional effects.

Other relevant chapters are 5, Stress Distribution and 6, Consolidation.

Worked Example 9.1   Immediate settlement - general method

A flexible foundation 10 m long, 5 m wide applies a uniform pressure of 75 kN/m² on the surface of a saturated clay, 20 m thick. The undrained modulus of the clay is 8 MN/m². Determine the immediate settlement at the corner of the foundation.

Using Figure 9.1, $B = 5$ m $\dfrac{H}{B} = \dfrac{20}{5} = 4$ $\dfrac{L}{B} = \dfrac{10}{5} = 2$ $I = 0.37$

From Equation 9.2

$$\rho_i = \frac{75 \times 5 \times 0.37 \times 1000}{8 \times 1000} = 17 \text{ mm}$$

Worked Example 9.2   Immediate settlement – principle of superposition

Determine the immediate settlement at the centre of the foundation in Example 9.1.

From Figure 9.2, the settlement at the corner of a quarter foundation is required.

$B = 2.5$ m $\dfrac{H}{B} = \dfrac{20}{2.5} = 8$ $\dfrac{L}{B} = \dfrac{5}{2.5} = 2$ From Figure 9.1, $I = 0.47$

$$\rho_i = \frac{75 \times 5 \times 0.47 \times 4 \times 1000}{8 \times 1000} = 44 \text{ mm}$$

Worked Example 9.3   Immediate settlement – principle of layering

The foundation in Example 9.1 is placed on two layers of clay, both 10 m thick. The modulus of the upper layer is 8 MN/m² and for the lower layer it is 16 MN/m². Determine the immediate settlement at the centre of the foundation.

From Figure 9.3
Settlement of upper layer

$B = 2.5$ m $H = 10$ m $\dfrac{H}{B} = \dfrac{10}{2.5} = 4$ $\dfrac{L}{B} = 2$ From Figure 9.1, $I = 0.37$

$$\rho_i = \frac{75 \times 2.5 \times 0.37 \times 4 \times 1000}{8 \times 1000} = 35 \text{ mm}$$

Settlement of lower layer

$B = 2.5$ m $H = 20$ m $\dfrac{H}{B} = \dfrac{20}{2.5} = 8$ $\dfrac{L}{B} = 2$ From Figure 9.1, $I = 0.47$

$B = 2.5$ m $H = 10$ m $\dfrac{H}{B} = \dfrac{10}{2.5} = 4$ $\dfrac{L}{B} = 2$ From Figure 9.1, $I = 0.37$

$$\rho_i = \frac{75 \times 2.5 \times (0.47 - 0.37) \times 4 \times 1000}{16 \times 1000} = 5 \text{ mm}$$

Settlement of foundation = 35 + 5 = 40 mm

Worked Example 9.4   Immediate settlement – correction for depth

The foundation in Example 9.2 is placed 2 m below ground level. Determine the immediate settlement at the centre of the foundation.

From Example 9.2 the settlement of the foundation on the surface of the clay is 44 mm.

Using Figure 9.4, $D = 2$ m $B = 5$ m $\frac{D}{B} = 0.4$ $\mu_0 = 0.94$

$\rho_{\text{at depth}} = 44 \times 0.94 = 41$ mm

Worked Example 9.5   Average settlement

Determine the average settlement for the foundation in Example 9.4.

From Figure 9.4, for $\frac{H}{B} = \frac{20}{5} = 4$ $\frac{L}{B} = 2$ $\mu_1 = 0.77$

From before, $\frac{D}{B} = 0.4$ $\mu_0 = 0.94$

$$\rho_{\text{ave}} = \frac{0.94 \times 0.77 \times 75 \times 5 \times 1000}{8 \times 1000} = 34 \text{ mm}$$

This compares reasonably with the settlement for a rigid foundation obtained by equation 9.4 and μ_r from Table 9.2:

$\mu_r \approx 0.77$

$\rho_{\text{rigid}} = 0.77 \times 41 = 32$ mm

Worked Example 9.6   Modulus increasing with depth

Determine the immediate settlement at the centre of the foundation in Example 9.1 when the undrained modulus of the clay increases from 4 MN/m² at ground level to 12 MN/m² at the base of the clay layer.

$B = 2.5$ m $H = 20$ m $E_0 = 4$ MN/m² $E_H = 12$ MN/m²

$k = \frac{12 - 4}{4} \times \frac{2.5}{20} = 0.25$ $\frac{H}{B} = \frac{20}{2.5} = 8$

From Figure 9.5, $\frac{L}{B} = 2$ $I = 0.30$ (Note: non-linear interpolation)

$$\rho_i = \frac{75 \times 2.5 \times 0.30 \times 4 \times 1000}{4 \times 1000} = 56 \text{ mm}$$

Comparing this answer with Example 9.2, it is not acceptable to take the 'average' modulus of 8 MN/m^2 and assume the homogeneous condition.

Worked Example 9.7   Effect of local yielding

The ultimate bearing capacity of the clay in the above examples is 150 kN/m^2. Determine the effect of yielding assuming the clay to be lightly overconsolidated.

The factor of safety is $\dfrac{150}{75} = 2$

From Figure 9.7 an average yield factor of about 1.15 is indicated so the above settlements should be increased by 15% to allow for the effect of yielding.

Worked Example 9.8   Consolidation settlement - compression index method

A flexible rectangular raft foundation, 4 m wide and 5 m long is to be constructed on the surface of a layer of soft normally consolidated clay, 8 m thick and will support a uniform pressure of 60 kN/m^2. The properties of the clay which are constant throughout the deposit are:
Bulk unit weight = 18.8 kN/m^3
Specific gravity = 2.72
Compression index C_c = 0.12
Assume γ_w = 9.8 kN/m^3
The water table is at ground level. The moisture content of the clay decreases linearly from 36% at ground level to 28% at the base of the clay layer. Determine the maximum consolidation settlement.

The maximum settlement will occur at the centre of the foundation where the maximum stress increases occur. The stress increases $\Delta\sigma$ have been determined using Figure 5.7 and are given in the table below. Splitting the clay into 4 no. 2 m thick sub-layers the values of p_0', $\Delta\sigma$ and w_0 are determined at the mid-point of each sub-layer and assumed to be the average values for each sub-layer, as shown on Figure 9.9.

Layer	Depth m	p_0' kN/m^2	$\Delta\sigma$ kN/m^2	P	Δe	w_0 %	e_0	ΔH m
1	1	9.0	56.4	0.861	0.103	35	0.952	0.106
2	3	27.0	32.4	0.342	0.041	33	0.898	0.043
3	5	45.0	16.8	0.138	0.017	31	0.843	0.018
4	7	63.0	9.6	0.062	0.007	29	0.789	0.008

$$\Sigma = 0.175 \text{ m}$$

The maximum consolidation settlement is 175 mm

Worked Example 9.9   Consoildation settlement – oedometer method

A flexible rectangular raft foundation, 4 m wide and 5 m long is to be constructed on the surface of a layer of firm overconsolidated clay, 8 m thick and will support a uniform pressure of 60 kN/m². The water table lies at 2 m below ground level. The bulk unit weight of the clay is 19.8 kN/m³, both above and below the water table. Using the oedometer test result in Example 6.1 determine the maximum consolidation settlement.

The stress increases in the soil due to the applied pressure at the mid-point of 2 m thick sub-layers are obtained from Example 9.8. Using the oedometer test result plotted as thickness versus log pressure values of m_v have been obtained for each sub-layer using the average values of p_0' and $\Delta\sigma$. The oedometer settlement for the foundation has been determined using equation 9.10. The oedometer test gave a value of $m_v =$ 0.135 m²/MN for the typical pressure increment of 50–100 kN/m² so the oedometer settlement has also been calculated with this value for all sub-layers. It can be seen that a sufficiently accurate estimate is obtained.

Layer	Depth m	p_0' kN/m²	$\Delta\sigma$ kN/m²	H_0 mm	ΔH mm	m_v m²/MN	$\rho_{oed}{}^1$ mm	$\rho_{oed}{}^2$ mm
1	1	19.8	56.4	19.840	0.202	0.18	20	15
2	3	49.6	32.4	19.700	0.072	0.11	7	9
3	5	69.6	16.8	19.652	0.032	0.10	3	5
4	7	89.6	9.6	19.615	0.014	0.07	1	3
							—	—
					Oedometer settlement $= \Sigma = $		31	32

[1] Using individual oedometer test m_v values
[2] Using one m_v value = 0.135 m²/MN for the pressure increment 50–100 kN/m²
Alternatively, the settlements could be determined from the pressure-void ratio plot using

$$\Delta H = \frac{\Delta e}{1 + e_0} H$$

where e_0 and Δe are the initial and change in void ratio, respectively, for each change in pressure p_0' to $p_0' + \Delta\sigma$ and H is the thickness of each sub-layer = 2000 mm.

Layer	Depth m	p_0' kN/m²	$\Delta\sigma$ kN/m²	e_0 mm	Δe mm	ΔH mm
1	1	19.8	56.4	0.635	0.0166	20
2	3	49.6	32.4	0.623	0.0059	7
3	5	69.6	16.8	0.619	0.0026	3
4	7	89.6	9.6	0.616	0.0012	1
						—

Oedometer settlement $= \Sigma = $ 31 mm

Worked Example 9.10   Total settlements - elastic drained method

A flexible rectangular foundation, 4 m wide and 8 m long is to be constructed 2 m below the surface of a layer of firm overconsolidated clay, 10 m thick and will support a uniform pressure of 60 kN/m². The drained modulus increases from 10 MN/m² at ground level to 20 MN/m² at the base of the clay. Determine the maximum total settlement.

The modulus at foundation level, $E_f' = 12$ MN/m²
From Figure 9.2 the settlement at the corner of a quarter foundation is obtained:

$$B = 2 \text{ m} \qquad \frac{H}{B} = \frac{8}{2} = 4 \qquad \frac{L}{B} = 2$$

From Equation 9.14,

$$k = \frac{20 - 12}{12} \times \frac{2}{8} = 0.167$$

Enter the charts, Figure 9.12, at $\dfrac{H}{B} = 1.2 \times 4 = 4.8 \quad \therefore \ I = 0.44$ (Note: non-linear interpolation)

$$\frac{D}{B} = \frac{2}{4} = 0.5$$
From Figure 9.13 $F_D = 0.92$

$$\frac{H}{B} = \frac{8}{4} = 2$$
From Figure 9.14 $F_B = 0.95$

$$\rho_T = \frac{60 \times 2 \times 0.44 \times 0.92 \times 0.95 \times 4 \times 1000}{12 \times 1000} = 15 \text{ mm}$$

Worked Example 9.11   Settlements on sand – Schmertmann's method

A square foundation, 4 m wide is placed at 2 m below ground level in a layered sand and applies a uniform gross pressure of 140 kN/m². The bulk unit weight of the sand is 18 kN/m³ and the water table lies at ground level. The cone end resistance of the layers is given on Figure 9.24. Determine the immediate and long-term settlement of the foundation.

Due to buoyancy from the water pressure at the underside of the foundation the resultant gross pressure will be
$140 - 2 \times 9.8 = p = 120.4$ kN/m²
$p_0' = (18 - 9.8) \times 2 = 16.4$ kN/m²
$\Delta p = 120.4 - 16.4 = 104.0$ kN/m²

$$C_1 = 1 - 0.5 \times \frac{16.4}{104.0} = 0.92$$

At $0.5B = 2$ m below foundation level $\sigma_v' = (18 - 9.8) \times 4 = 32.8$ kN/m²

From equation 9.24, $I_{zmax} = 0.5 + 0.1 \left(\dfrac{104.0}{32.8} \right)^{0.5} = 0.68$

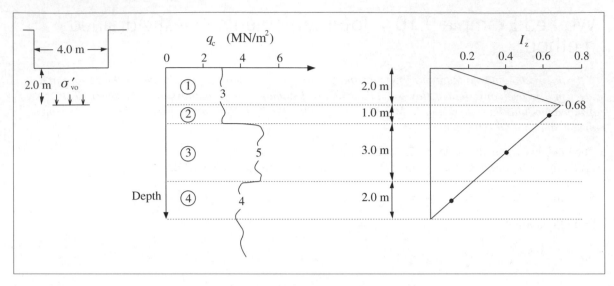

FIGURE 9.24

The variation of I_z has been plotted against depth for this value of I_{zmax} on Figure 9.24. Splitting the cross-section into suitable layers values of Δz and average values of q_c and I_z are obtained:

Layer	Δz m	q_c MN/m²	I_z	$\dfrac{I_z \Delta z}{q_c}$
1	2.0	3	0.39	0.26
2	1.0	3	0.62	0.21
3	3.0	5	0.40	0.24
4	2.0	4	0.11	0.06

$$\Sigma = 0.77$$

From equation 9.22, the immediate settlement will be

$$\rho_i = \frac{0.92 \times 104.0 \times 0.77 \times 1000}{2.5 \times 1000} = 29 \text{ mm}$$

For a design life of say, 30 years
$C_2 = 1 + 0.2 \log_{10} 10 \times 30 = 1.5$
$\therefore$ the long-term settlement will be
$\rho_t = 29 \times 1.5 = 44 \text{ mm}$

Worked Example 9.12   Settlements on sand - Burland and Burbridge method

A square foundation 8 m wide is placed at 2 m below ground level in a sand 6 m thick and applies a uniform gross pressure of 140 kN/m². The bulk unit weight of the sand is 18 kN/m³ and the water table lies at ground level. The average SPT 'N' value of the sand is 16. Determine the immediate and long-term settlements of the foundation.

Shape factor $f_s = 1$
From Figure 9.20, depth of influence, $Z_I = 5$ m
$H_s = 4$ m and is less than Z_I

$$\therefore \ f_i = \frac{4}{5}\left(2 - \frac{4}{5}\right) = 0.96$$

Assuming the sand to be normally consolidated
$\sigma_{v0}' = (18 - 9.8) \times 2 = 16.4$ kN/m²

$$I_c = \frac{1.71}{16^{1.4}} = 0.035$$

Due to buoyancy from the water pressure at the underside of the foundation the resultant gross pressure will be
$140 - 2 \times 9.8 = p = 120.4$ kN/m²
From equation 9.32

$$\rho_i = 0.96\left(120.4 - \frac{2}{3} \times 16.4\right) \times 8^{0.7} \times 0.035 = 16 \text{ mm}$$

Assuming fluctuating loads and a design life of 30 years, from Table 9.6 and Equation 9.34

$$f_t = 1 + 0.7 + 0.8 \log_{10} \frac{30}{3}$$

$\therefore$ the long-term settlement will be
$\rho_t = 16 \times 2.5 = 40$ mm

EXERCISES

9.1 A flexible rectangular foundation, 5 m wide and 10 m long is to be placed on the surface of a layer of clay, 20 m thick with undrained modulus of 20 MN/m^2. The foundation will support a uniform pressure of 120 kN/m^2. Determine the immediate settlement at the centre and corner of the foundation.

9.2 An existing pipe is buried at 5 m below ground level and will lie on the diagonal of the foundation in Exercise 9.1.
Determine the immediate settlement of the pipe:
(a) beneath the centre of the foundation
(b) beneath the corner of the foundation.

9.3 For the foundation in Exercise 9.1 determine:
(a) the immediate settlement of the foundation assuming it to be rigid
(b) the average immediate settlement of the foundation assuming it to be flexible.

9.4 A flexible rectangular foundation, 5 m wide and 10 m long, is to be placed on the surface of a layer of clay, 20 m thick with undrained modulus increasing from 10 MN/m^2 at ground level to 30 MN/m^2 at the base of the clay. The foundation will support a uniform pressure of 120 kN/m^2. Determine the immediate settlement at the centre and corner of the foundation.

9.5 A flexible rectangular foundation, 4 m wide and 8 m long, is to be placed on the surface of a layer of clay, 16 m thick and with undrained modulus of 8 MN/m^2. The foundation will support a uniform pressure of 40 kN/m^2. Determine the maximum immediate settlement assuming:
(a) linear elastic behaviour
(b) local yield occurs. With a factor of safety of 3 assume the yield factor f_y is 1.5.

9.6 A foundation is to be placed on the surface of a normally consolidated clay, 8 m thick, with the water table at 4 m below ground level. The following properties of the clay are constant with depth:
unit weight = 18.5 kN/m^3
specific gravity = 2.70
compression index = 0. 15
The increase in vertical stress provided by the foundation and the water content of the clay vary with depth:

Depth (m)	1	3	5	7
$\Delta\sigma_v$ (kN/m^2)	70	40	25	15
w_0 (%)	40	35	31	28

Using the compression index method determine the consolidated settlement of the foundation.

9.7 For the foundation described in Exercise 9.6 determine the consolidation settlement using the m_v method given that the coefficient of compressibility m_v decreases with depth in the form $m_v = 0.80 - 0.04_z$ where z is the depth below ground level and the Skempton–Bjerrum μ value is 0.8. Assume 2 m thick sub-layers.

9.8 A flexible rectangular foundation, 8 m wide and 16 m long is to be placed at 4 m below ground level in a layer of clay, 24 m thick and will impose a uniform pressure of 70 kN/m^2. The drained modulus of the clay increases from 4 MN/m^2 at ground level to 28 MN/m^2 at the base of the clay. Determine the total settlement at the centre and the corner of the foundation.

9.9 Determine the amount of secondary compression in the first 5 years and after a period of 30 years for a deposit of soft organic clay, 8 m thick with a coefficient of secondary compression, ε_α of 1.5%. Assume t_1 is 1 year.

9.10 A square foundation 10 m wide is to be placed at 2 m below ground level in a sand deposit 25 m thick and will impose a gross pressure of 90 kN/m^2. The water table exists at 4 m below ground level and the unit weight of the sand above and below the water table is 19.5 kN/m^3. The static cone resistance of the sand is constant with depth at 4MN/m^2. Determine the immediate settlement of the foundation and the long-term settlement after a period of 30 years using the Schmertmann method.

9.11 For the foundation described in Exercise 9.10 determine the immediate and long-term settlements using the Burland and Burbridge method. Assume the sand to be normally consolidated with a mean standard penetration resistance of $N = 10$ and the loads are static.

CHAPTER 10

Pile foundations

- To be aware of the variety of pile types available and their advantages and limitations.
- To calculate the load capacity of single driven and bored piles in both sand and clay soils.
- To appreciate the relative importance of shaft and base resistances, how they are mobilised and derived in sands and clays.
- To apply the limit state approach to the design of pile foundations.
- To assess the effects of negative skin friction acting on a pile foundation.
- To understand the effects produced by placing piles together in a group.

Pile foundations

If a structure cannot be satisfactorily supported on a shallow foundation for reasons such as incompetent ground, swelling/shrinking soil or open water then a pile foundation can provide an economic alternative. A single pile can be defined as a long, slender structural member used to transmit loads applied at its top to the ground at lower levels. This is achieved via:

- a shear stress mobilised on the surface of the shaft of the pile called skin friction in sands and adhesion in clays; and
- development of failure by bearing capacity at the base of the pile, called end bearing.

Loading conditions (Figure 10.1)

The loads applied by a structure are quite large and usually cannot be supported by a single pile so a number of piles are placed together to form a pile group with a substantial reinforced concrete pile cap placed on top to transfer and distribute the loadings from the structure to the piles. The various types of pile and loading conditions which piles may be subjected to are illustrated in Figure 10.1.

Types of pile

Piles can be classified according to:

- their method of installation, i.e. driven, bored, jacked
- their material type, i.e. timber, steel or concrete, precast or cast *in situ*, full length or segmental
- the plant used to install them such as driven, tripod, auger, under-reamed, continuous flight auger
- their size, i.e. small diameter bored, large diameter bored, under-reamed, mini-piling
- their effects during installation, i.e. displacement or replacement
- the way they provide load capacity, i.e. end bearing, friction piles, uplift piles, raking piles.

For further information on this subject the reader is referred to the books by Tomlinson (1987), Poulos (1980), Fleming *et al.* (1985) or Whitaker (1970).

Design of single piles

With so many different types of piles, different soil conditions and various effects of installation such as remoulding, softening, compaction and others, although the development of design processes has

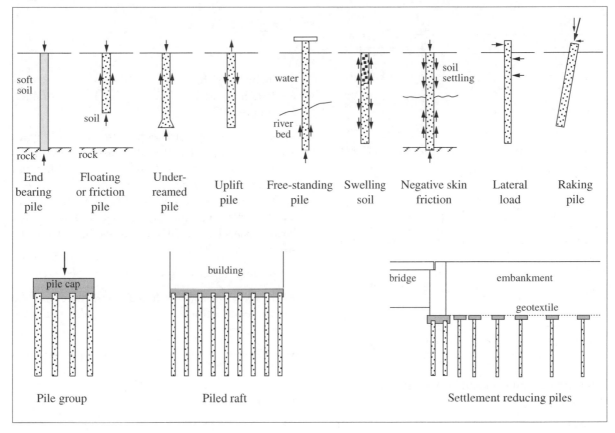

FIGURE 10.1 *Terminology and uses of piles*

evolved in a theoretical manner, there has always been a need for empiricism to obtain correlations and 'adjustment factors' for the various effects. Thus the predictions of load-carrying capacity of a pile will not be particularly accurate and should never be relied upon fully without some proof-load testing.

Pile load tests

A particular type of pile is chosen to suit the ground and other conditions and a preliminary estimate is made of its diameter and length to support a working load using the calculation methods given below. The working load is the load each pile would normally have to carry to support the design loads from the structure. Following this 'design' process a number of these piles should be installed on the site as trial piles and tested under load to confirm their adequacy and to assess the most economical solution. Back-

analysis of the load test results will provide useful information for predicting the performance of different sized piles on the site.

Eurocode 7 emphasises that the design of a pile must be based on static load tests or calculation methods that have been validated by these tests. If load testing is not carried out then the design would have to be based on calculation methods alone, using conservative values of soil parameters and with higher safety factors adopted so an economical design is less likely to be achieved.

Load capacity of single piles (Figure 10.2)

There are two resistances offered by a pile to the vertical load applied:

- shaft resistance
- base resistance.

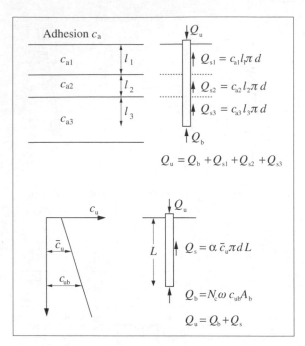

FIGURE 10.2 *Layered or non-homogeneous clays*

At failure the ultimate values of both these loads are mobilised and give:

$$Q_u = Q_s + Q_b \qquad (10.1)$$

where:

Q_u = ultimate applied load
Q_b = ultimate base load
Q_s = ultimate shaft load

Base resistance is obtained as a load by multiplying the cross-sectional area of the base of the pile by a bearing capacity stress:

$$Q_b = q_b A_b \qquad (10.2)$$

where:

q_b = end bearing resistance
A_b = base cross-sectional area

Shaft resistance is obtained as a load by multiplying the surface area of the shaft of the pile in contact with the supporting soil by a shear stress acting between the pile surface and the soil called skin friction in sands, f_s and adhesion in clays, c_a:

$$Q_s = \sum_0^L c_a \pi dl \quad \text{(clays)} \qquad (10.3)$$

$$Q_s = \sum_0^L f_s \pi dl \quad \text{(sands)} \qquad (10.4)$$

where:

d = diameter of the pile
l = length of the pile in contact with each stratum, (see Figure 10.2).

Piles usually penetrate several different soil types each providing different shaft resistances as shown in Figure 10.2, so the total shaft resistance is the sum of the individual values.

The weight of the pile W_p should be included in equation 10.1 but this can be ignored if the net bearing capacity is used to obtain the base resistance since:

net bearing capacity = gross bearing capacity – q

where q, as for shallow foundations, represents the surcharge or vertical total stress at the pile base level, σ_{vb} and:

$$W_p \approx A_b \, \sigma_{vb} \qquad (10.5)$$

or

$$W_p \approx W_s$$

where:

W_s = weight of soil removed/displaced
The full equation is then:

$$Q_u = Q_s + Q_b - W_p + W_s = Q_s + Q_b$$

Bored piles in clay

End bearing resistance q_b
This can be obtained from:

$$q_b = N_c \, \omega \, c_{ub} \quad \text{(kN/m}^2) \qquad (10.6)$$

N_c is a bearing capacity factor which for the undrained condition for a deep foundation can be shown to be theoretically between 8.0 and 9.8 (Whitaker and Cooke, 1966) and a value of 9 suggested by Skempton (1959) is widely adopted.

c_{ub} is the undrained shear strength at the base of the pile taken from the mean line of the shear strength/depth plot. These strengths used to be obtained from triaxial compression tests on 38 mm diameter specimens prepared in the conventional manner by jacking tubes into a U100 sample. If fissures exist within the clay these tests will be affected to varying degrees by fissuring within the specimens tested, as illustrated in Figure 7.35 and the mean results when plotted will not reflect the en masse fissure strength that would be operative beneath a pile. The factor ω is applied to convert these laboratory strengths into fissure strength and values suggested by Skempton (1966) are:

non-fissured clay $\omega = 1.0$
fissured clay, $d < 0.9$ m $\omega = 0.8$
fissured clay, $d > 0.9$ m $\omega = 0.75$

However, if the undrained shear strength is obtained from triaxial tests on 100 mm diameter samples which is more common nowadays then these tests will more likely represent the fissure strength and the ω factor need not be applied, especially if the lower bound of the data is used.

Before assuming the full value of end bearing it should be confirmed that the pile bore has been thoroughly cleaned out before concreting. If debris, slurry or any bentonite remain at the base of the pile then a reduced end-bearing resistance should be adopted.

Adhesion c_a

The shear stress developed between the pile surface and the soil will occur within a narrow zone adjacent to the pile. The adhesion c_a will generally be less than the initial undrained shear strength, c_u due to:

- remoulding caused by the action of the drilling tools
- softening caused by stress relief with moisture migrating from the nearby soil and from the wet concrete towards the annulus of soil around the pile. Any water present in the borehole will aggravate this problem.

The value of adhesion is likely, therefore, to be dependent on the drilling techniques adopted and, in particular, any delay before concrete is inserted.

A measure of adhesion is traditionally obtained from a total stress approach using an empirical modification of the undrained shear strength:

$$c_a = \alpha \, c_u \tag{10.7}$$

where α is an adhesion factor.

For soft and firm clays a value of $\alpha = 1.0$ can be used (Vesic, 1977). For stiffer London Clay, Skempton (1959) found values of α between 0.3 and 0.6 from back-analyses of a number of pile loading tests and recommended a value of 0.45 for piles extending beneath the highly fissured and weathered upper horizon of the London Clay. For short piles existing mostly within this upper zone and where delay before concreting is likely, such as for large-diameter under-reamed bored piles, a value of $\alpha = 0.3$ is recommended. These values are often used for other stiff, fissured clays. Values of α for piles installed into boulder clays and glacial tills are given in Figure 10.5.

When bentonite fluid or drilling mud is used to support the sides of the borehole then providing it is completely displaced by the concrete during the tremie process it should have no detrimental effects on the adhesion value (Fleming and Sliwinski, 1977). Tomlinson (1987) recommends that the adhesion values should be reduced by 20% to allow for drilling mud effects since it cannot be guaranteed that the mud will be removed entirely.

A maximum value of adhesion of 100 kN/m^2 is recommended for many soils (Skempton, 1959; Vesic, 1977) although for piles in glacial clays a maximum value of 70 kN/m^2 is recommended (Weltman and Healy, 1978).

SEE WORKED EXAMPLES 10.1 AND 10.2.

Driven piles in clay

End bearing resistance q_b

Because of the limiting diameter of conventional driven piles of a maximum 450 to 600 mm and the small cross-sectional area, the base load obtainable tends to be a small amount in relation to the shaft load. Nevertheless, it could be calculated from:

$$q_b = N_c \, c_{ub} \tag{10.8}$$

where N_c can be taken as 9 and c_{ub} is the undisturbed undrained shear strength at the base of the pile.

Adhesion c_a – installation effects

Driving a pile into clay requires considerable displacement and causes major changes in the clay. The effects of installation are different for soft clays and stiff clays.

Driving a pile into soft clay increases the total stresses which are transferred to a large rise in pore water pressure in the annulus of soil around the pile. This increase in pore pressure is larger for piles with a greater volumetric displacement, such as solid square piles compared to thinner H-section piles. Higher pore pressures can be generated in soils with a tendency for their mineral grain structures to contract on shearing, such as sensitive clays.

The time taken for this pore pressure to dissipate will depend on the initial excess pore pressure, the permeability of the soil, the permeability of the pile material and the number and spacing of the piles.

As the consolidation process occurs the effective stresses around the pile increase and the pile load capacity increases. Thus, the initial load-carrying capacity of a pile may be quite small but will increase with time. However, from measurements of pile load with time, illustrated in Tomlinson (1986), several weeks or months may elapse before the full load capacity is achieved.

Driving piles into stiff overconsolidated clays can produce three significant effects:

● Expansion of the soil surrounding the pile with associated radial cracking and opening of macro-fabric features, such as fissures. Any positive pore pressures set up during driving will rapidly dissipate into this open structure and expansion of the soil is more likely to produce negative pore pressures at least in the upper levels. Relatively short piles, therefore, may provide an initially high load carrying capacity but this could diminish with time. Longer piles are more likely to produce positive pore pressures in their lower regions.

● Ground heave comprising upward and outward displacement of the soil around a pile being driven. This effect can occur up to 10 pile diameters away from a pile (Cole, 1972) so driving piles in groups can magnify the effect. These ground displacements can cause damage to existing buried structures due to increased horizontal total stresses and to nearby driven piles, due to vertical displacements producing separation or

fracture of full length and segmental piles. Heave is particularly detrimental when the piles are intended to provide most of their load as base resistance since the piles may be lifted from their end bearing.

● 'Whippiness' or lateral vibrations set up in the pile once it has been partly driven into the clay. This produces a gap such as a 'post-hole' effect between the clay and the pile so no adhesion can exist over this length. Tomlinson (1970, 1971) also observed that any soil above the stiff clay was dragged down into this gap so soft clay overburden would produce a lower apparent adhesion but sand would produce a higher adhesion. The penetration of the pile into the stiff clay and the type of overburden, is, therefore, very important.

Adhesion c_a – values (Figures 10.3 –10.7)

The effects of the depth of penetration of the pile into the clay and the type of overburden are illustrated in Figure 10.3 which gives values of adhesion factor α for short penetration piles taken from Tomlinson (1987). For longer piles, ($L > 20$ or 40 diameters) the effect of the gap diminishes, as illustrated in Figure 10.4. It should be noted that the scatter of data points used to obtain these curves was considerable. A similar phenomenon has been observed for piles driven into glacial clays, see Figure 10.5.

The average adhesion on the shaft of a pile is then calculated from equation 10.7 using the adhesion factors from Figures 10.3, 10.4 or 10.5.

For very long piles driven into stiff clays it has also been found that the shaft capacity depends on the length of the pile but for probably different reasons. Vijayvergiya and Focht (1972) suggested a quasi-effective stress approach for the determination of average adhesion along long steel-pipe piles in the form:

$$\bar{c}_a = \lambda(\bar{\sigma}_m' + 2\bar{c}_m) \tag{10.9}$$

$$Q_s = \bar{c}_a A_s \tag{10.10}$$

where:

$\bar{c}_a$ = average adhesion along pile

$\bar{\sigma}_m'$ = mean effective vertical stress between ground level and the base of the pile

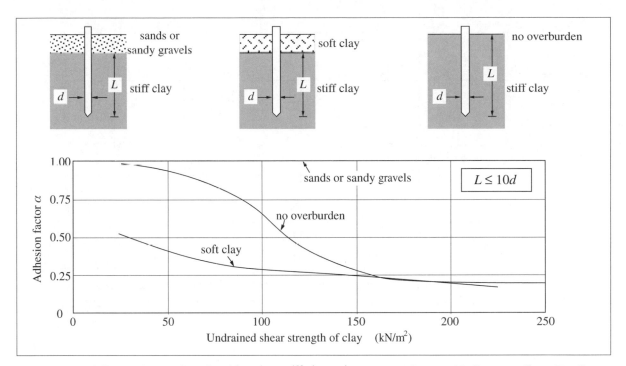

FIGURE 10.3 *Adhesion factors for piles driven into stiff clay – short penetration L < 10 diameters (from Tomlinson, 1987)*

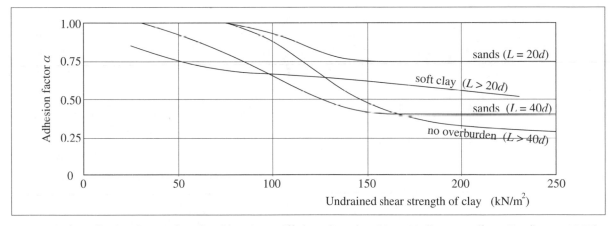

FIGURE 10.4 *Adhesion factors for piles driven into stiff clay – length > 20 to 40 diameters (from Tomlinson, 1987)*

$\bar{c}_m$ = mean undrained shear strength along the pile
λ = empirical factor, plotted in Figure 10.6.

Randolph and Wroth (1982) have shown that the average adhesion for long piles depends on the K_o

value, which tends to decrease with depth for an overconsolidated clay, see Chapter 4. They expressed this effect in terms of the c_u/σ_v' ratio which is related to the overconsolidation ratio, *OCR*.

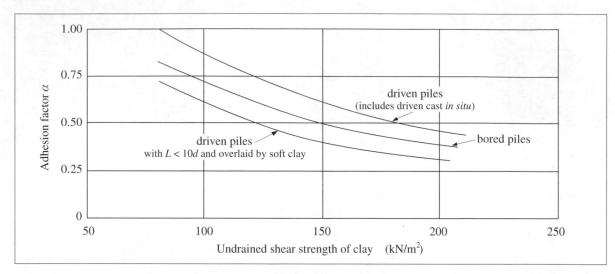

FIGURE 10.5 *Suggested adhesion factors for piles driven into boulder clay (from Weltman and Healy, 1978)*

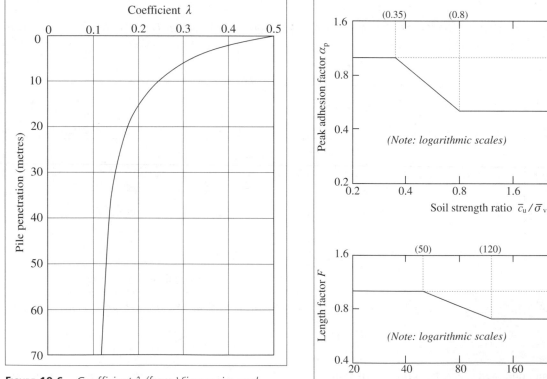

FIGURE 10.6 *Coefficient λ (from Vijayvergiya and Focht, 1972)*

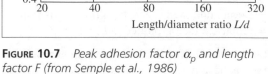

FIGURE 10.7 *Peak adhesion factor α_p and length factor F (from Semple et al., 1986)*

Design guidelines published by the American Petroleum Institute (API, 1993) give expressions for the adhesion factor in the form:

$$\alpha = 0.5\left(\frac{c_u}{\sigma_v'}\right)^{-0.5} \qquad \text{for } \frac{c_u}{\sigma_v'} \leq 1.0 \qquad (10.11)$$

$$\alpha = 0.5\left(\frac{c_u}{\sigma_v'}\right)^{-0.25} \qquad \text{for } \frac{c_u}{\sigma_v'} > 1.0 \qquad (10.12)$$

with the constraint that $\alpha \leq 1.0$.

Semple and Rigden (1986) proposed the relationships in Figure 10.7 giving the shaft load as:

$$Q_s = F\alpha_p \bar{c}_u \pi dl \qquad (10.13)$$

where $\bar{c}$ is the average undrained shear strength over the length of the pile and $\bar{\sigma}_v$ is the average vertical effective stress.

The value α_p is the peak adhesion factor for a rigid pile. The overall effect of increasing pile length is to reduce the average adhesion value. F is a length factor to account for flexibility and compressibility of the pile since:

- Lateral flexibility produces the whippiness or flutter effect as the pile is driven.
- The axial compressibility of the pile will permit greater displacements in the upper part of the pile than in the lower part, taking the skin friction beyond any peak value and towards a lower critical state or residual value.

SEE WORKED EXAMPLES 10.3 AND 10.4.

Effective stress approach for adhesion

On the basis that pile capacity increases with time after driving due to dissipation of pore pressures during consolidation and, hence, increases in effective stress, it is generally accepted that adhesion between the soil and the pile surface is frictional and is governed by the horizontal effective stress in the soil. Since the shear stress mobilised on the shaft surface is frictional it is referred to as skin friction, and is given by a general expression:

$$f_s = K_s p_0' \tan \phi' \qquad (10.14)$$

p_0' is the vertical effective stress at any depth z and K_s is a coefficient of horizontal earth pressure. Therefore $K_s p_0'$ represents the normal stress (horizontal stress) acting on the surface of the pile. K_s has a similar function to the value of K_0, the coefficient of earth pressure at rest. For normally consolidated clay it is given by $K_s = K_0 = 1 - \sin\phi'$ and can be expected to remain fairly constant with depth. Burland (1973) suggested a simpler form of equation 10.14 as:

$$f_s = \beta\bar{p}_0' \qquad (10.15)$$

where:

$\bar{p}_0'$ is the average effective vertical stress down the length of the pile and:

$$\beta = K_s\tan\phi'$$
$$= (1 - \sin\phi')\tan\phi' \qquad (10.16)$$

For a typical range of ϕ' values (15° to 30°) β varies between 0.2 and 0.29 and for bored piles in soft normally consolidated clay a value of about 0.3 is suggested. Meyerhof (1976) found that β varied from about 0.2–0.4 for short piles driven into soft clays (less than about 15 m long) to about 0.1–0.25 for very long piles, which may be due to some overconsolidation in the upper horizons and pile compressibility. Meyerhof also found higher values of β for tapered piles reflecting the higher horizontal stresses produced.

For piles bored or driven into stiff overconsolidated clays the K_0 and hence K_s value can be expected to vary with depth. Meyerhof (1976) stated that for bored piles K_s varies from about 0.7 K_0 to 1.2 K_0 but for driven piles it varies from about K_0 to more than 2 K_0. Randolph and Wroth (1982) suggest a relationship between β and the ratio $\bar{c}_u/\bar{\sigma}_v'$. However, although it is clear that shaft resistance is governed by friction and effective stresses the empirical correlations required to determine values of β and the scatter in the data available make this approach no better than the traditional total stress or α method, at the present time.

Driven piles in sand

Effects of installation

Driving piles into loose sands compacts them, increasing their density and angle of internal friction and increasing the horizontal stresses around the pile.

Driving piles into dense sands may not compact them. Instead, dilatancy and negative pore pressures may temporarily increase the pile load capacity, make driving difficult and possibly result in over-stressing and damage to the pile. Dissipation of this negative pore pressure after driving, referred to as relaxation will cause the pile load capacity to decrease so a false initial impression of load capacity may be derived from the driving records. This is often referred to as a 'false set'.

The extent to which driving may increase the density of the sand could be up to 4–6 diameters away from the pile and 3–5 diameters below the pile (Broms, 1966). This zone of influence is larger for loose sands than dense sands and will obviously affect the driving of piles in groups where piles are typically 2–3 diameters apart.

It is also presumed that the sands are hard, clean quartz grains that will not deteriorate under driving stresses. Softer crushable grains will produce lower angles of friction after driving and will be more compressible. It has also been found that the shaft friction at any point on the pile surface decreases as the pile is driven past this level. This is referred to as friction fatigue.

The design of a pile must consider the installation effects and the final state of the sand. It can only be considered as approximate and should be checked by *in situ* pile loading tests.

End bearing resistance q_b (Figure 10.8)

The conventional approach to end-bearing resistance is to use only the surcharge term of the bearing capacity equation (Equation 8.11) since $c' = 0$ and the width of a pile is small compared to its length:

$$q_b = N_q \, \sigma_v' \qquad (10.17)$$

where σ_v' is the vertical effective stress at the base of the pile and N_q is a bearing capacity factor. The values of N_q provided by Berezantsev *et al.* (1961) are commonly used, see Figure 10.8.

Critical depth (Figures 10.9 to 10.11)

Equation 10.17 suggests that as the pile penetrates deeper into the sand the end-bearing resistance will increase as vertical effective stress increases. However, field tests have shown that end-bearing resistance does not increase continually with depth.

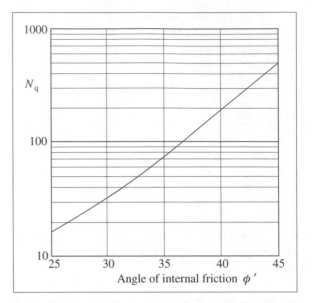

FIGURE 10.8 *Bearing capacity factor N_q for piles in sand (from Berezantsev et al., 1961)*

It seems more logical that end-bearing resistance depends on the mean effective stress at pile base level rather than just the vertical stress:

$$\sigma_m' = 1/3(\sigma_v' + 2\sigma_H') \qquad (10.18)$$

Since $\sigma_H' = K_o \, \sigma_v'$ end-bearing resistance will then be affected by the K_o value which for overconsolidated soils decreases with depth.

The Mohr–Coulomb criterion for soils at higher stress levels often shows some curvature rather than the straight line assumed. Thus as the stresses at pile

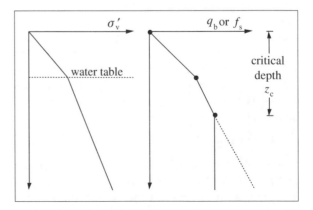

FIGURE 10.9 *Critical depth in sands*

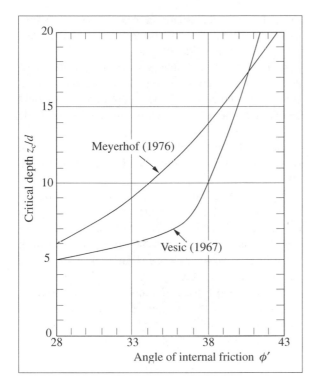

FIGURE 10.10 *Values of critical depth*

base level increase the ϕ' value and, hence, the bearing capacity factor N_q decreases. Arching is also considered to be a contributory factor.

The combined effect is to obtain decreasing end bearing resistance with depth. The simplest way of incorporating this effect into pile design is to adopt the concept of a critical depth z_c as shown in Figure 10.8, although this concept has been criticised, see below. Even though the vertical effective stress σ_v' increases with depth the end-bearing resistance q_b and the skin friction f_s are considered as constant below the critical depth, having the value at the depth z_c. The critical depth has been found to be shallow for loose sands and deeper for dense sands.

At the present time, values of this critical depth are somewhat tentative, the values suggested by Vesic (1967) and Meyerhof (1976) are given in Figure 10.10.

For the determination of N_q and z_c in Figures 10.8 and 10.10, respectively, the angle of internal friction ϕ' should relate to the state of the sand *after* pile installation. The values given in Table 10.1 are suggested (Poulos, 1980).

The initial angle of internal friction ϕ_1' before installation of the pile, is not an easy parameter to determine since sampling disturbance will largely

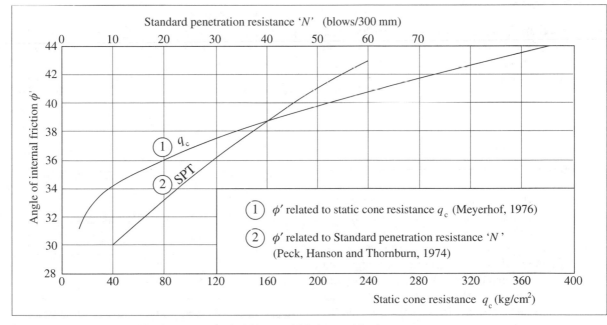

FIGURE 10.11 *Relationships between angle of internal friction and in situ tests*

TABLE 10.1 *Values of φ' after installation (from Poulos, 1980)*

Requirement	Values of ϕ after installation	
	Bored piles	Driven piles
N_q	$\phi_1' - 3$	$\dfrac{\phi_1' + 40}{2}$
z_c/d		
$K_s \tan\delta$	ϕ_1	$3/4\phi_1' + 10$

ϕ_1' is the ϕ value before installation

destroy the initial mineral grain structure making laboratory tests meaningless. The ϕ_1' value is usually obtained from correlations between the SPT 'N' value or the cone penetrometer q_c as illustrated in Figure 10.11.

There is some doubt about the existence of a true limiting value of end-bearing capacity at a certain critical depth (Randolph *et al.*, 1994). The more widely held view is that the end-bearing resistance increases with increasing depth but at a gradually decreasing rate.

Skin friction f_s (Figures 10.12 and 10.13)

Assuming effective stresses acting on the pile/soil surface the unit skin friction f_s at a depth z below the top of a pile is given by:

$$f_s = K_s\sigma_v' \tan \delta \qquad (10.19)$$

where:

K_s = coefficient of horizontal effective stress
δ = angle of friction between the pile surface and the soil.

Since both K_s and $\tan\delta$ will be governed by the method of installation and values of these factors may be difficult to assess separately a simple approach is to consider values of the lumped parameter $K_s\tan\delta$ (cf. β for the effective stress approach to piles in clay, Equation 10.14). Values of this parameter given in Figure 10.12 are related to the initial angle of internal friction. These are based on K_s values given by Meyerhof (1976), the ϕ values given

in Table 10.1 and assuming $\delta = 0.75\phi'$, for a normally consolidated sand. Higher values may be possible for overconsolidated sands.

It has been found that skin friction values also decrease with depth in a similar fashion to end-bearing resistance so a critical depth approach could be adopted. Values of z_c can be obtained from Table 10.1 and Figure 10.10.

The total shaft load is then summated from the shaft loads Q_{s1}, Q_{s2}, etc., as illustrated in Figure 10.13.

Again, there is considerable doubt concerning the existence of a limiting value of skin friction at a certain critical depth. It is accepted that the skin friction increases with depth down the pile, to a peak value just above the pile toe (Randolph *et al.*, 1994) and then decreases below this level. The observance of the so-called critical depth is now considered to be due largely to the presence of residual forces that were induced in a pile during installation and remain in the pile before loading. These authors also show

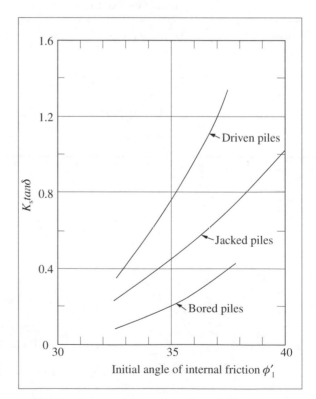

FIGURE 10.12 *Skin friction parameter $K_s \tan\delta$ (from Poulos, 1980)*

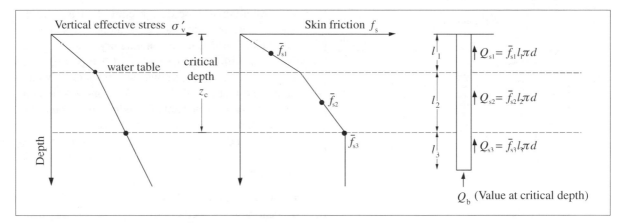

FIGURE 10.13 *Determination of shaft loads in sands*

that the estimation of separate shaft and base capacities is of poor accuracy.

SEE WORKED XAMPLE 10.5.

Bored piles in sand

Boring holes in sands loosens an annulus of soil around the hole and reduces horizontal stresses so bored piles constructed in initially dense sands can be expected to have low load capacity. If jetting techniques are used then the loosening can be even more severe. Casting concrete *in situ* will produce a rough surface but this effect is diminished by the loosening of the sand.

Poulos (1980) suggests using the methods given for driven piles but with reduced values of the final angle of internal friction as given in Table 10.1. Meyerhof (1976) suggests that for preliminary estimates the base resistance of a bored pile could be taken as one-third of the value determined for a driven pile with about one-half for the shaft resistance.

Factor of safety (Figure 10.14)

A factor of safety is applied to the ultimate load Q_u to safeguard against the uncertainties in the ground conditions and installation effects and to limit settlement to a permissible value.

Although piles are often designed by applying a factor of safety to the ultimate load to obtain a working load, the overriding performance criteria for a pile is that it must not settle more than a permissible amount. Tomlinson (1987) stated that from his experience for piles up to 600 mm diameter if an overall factor of safety of 2.5 is adopted to give:

$$\text{working load} = \frac{\text{ultimate load}}{\text{factor of safety}} \tag{10.20}$$

the settlement of the pile under the working load is unlikely to exceed 10 mm.

For piles larger than 600 mm diameter, it has been found that the two components: shaft resistance and base resistance, are mobilised at different amounts of settlement. Approximately, the full shaft resistance is mobilised at a pile head settlement of about 1–2% of the pile diameter, whereas to mobilise the full base resistance the pile must be pushed down about 10–20% of the diameter.

This is illustrated in Figure 10.14 for a clay soil. For a typical pile diameter and a permissible settlement of ρ_{all}, it can be seen that when the working load is applied a large proportion of the shaft resistance is mobilised at this settlement with only a small proportion of the base resistance acting. For piles in sands the shaft load component would tend to be less than the base resistance.

Thus, a more logical approach is to apply different factors (of safety) to the two components. These are referred to as partial factors. For bored piles in

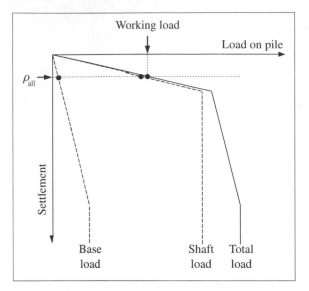

FIGURE 10.14 *Mobilisation of base and shaft loads*

London Clay, Burland *et al.* (1966) suggest that providing an overall factor of safety of 2 is obtained, partial factors on the shaft and base of 1 and 3, respectively, should also be applied so that the working load Q_a is the smaller of:

$$Q_a = \frac{Q_{ult}}{2} = \frac{Q_s + Q_b}{2} \qquad (10.21)$$

$$Q_a = \frac{Q_s}{1} + \frac{Q_b}{3} \qquad (10.22)$$

Burland *et al.* stated that the latter expression generally governs the design for large under-reamed piles and the former commonly governs the design for straight-shafted piles.

For soils other than London Clay, when there is less certainty about the ground conditions, applied loadings and installation effects then higher factors of safety should be applied, such as 2.5 overall, and 1.5 and 3.5 for the shaft and base load, respectively.

Limit state design

Eurocode 7 makes it clear that calculations for pile capacity are unreliable and that a pile must not be designed on the basis of calculations alone. A calculation method must be validated by static pile load

tests. Alternatively the design could be based on the results of static pile load tests alone. In some circumstances load tests are mandatory, such as where there is insufficient experience of a type of pile in the given ground conditions, where there is insufficient confidence in the effects of non-axial loading on a pile such as horizontal loading and when piles do not behave as they are expected to.

In Eurocode 7 the design requires that the following ultimate limit states should be sufficiently improbable:

- overall stability failure
- bearing resistance failure
- collapse or severe damage to a supported structure caused by displacement of the pile foundation.

Also that the serviceability limit states in the supported structure caused by displacements of the pile foundation are also sufficiently improbable.

Overall stability failure could occur when a pile foundation is situated on a slope and a slip surface passing beneath the foundation could provide a mechanism for failure.

Ultimate bearing resistance from pile load tests

The emphasis of the Code is that design is to be based on the use of static load tests. The ultimate characteristic bearing resistance R_{ck} of a pile including both shaft and base resistances is determined from measured values of the characteristic load R_{cm}, obtained from 1, 2 or more static pile load tests. There will be higher confidence in the results if more than one test is carried out. To allow for the variability of ground conditions and installation effects an allowance ξ is applied depending on the number of load tests carried out giving:

$$R_{ck} = \frac{R_{cm}}{\xi} \qquad (10.23)$$

Values of ξ are given in Table 10.2 and as both conditions *a* and *b* must be satisfied the lowest value of R_{ck} in each case is adopted.

The ultimate design bearing resistance, R_{cd} is obtained from the characteristic value, R_{ck} by dividing it into its constituent components, shaft and base resistances, R_{sk} and R_{bk}, such that:

TABLE 10.2 *Values of factor ξ (from Eurocode 7, 1995)*

Number of load tests	1	2	> 2
a Factor ξ on the mean value of R_{cm}	1.5	1.35	1.3
b Factor ξ on the lowest value of R_{cm}	1.5	1.25	1.1

$$R_{ck} = R_{sk} + R_{bk} \qquad (10.24)$$

These components can be distinguished either by setting up the load tests to measure them separately or by estimating their proportions from a calculation method. Then the ultimate design bearing resistance is obtained from:

$$R_{cd} = \frac{R_{sk}}{\gamma_s} + \frac{R_{bk}}{\gamma_b} \qquad (10.25)$$

where γ_s and γ_b are partial factors obtained from Table 10.3.

If it is not possible to distinguish between base and shaft resistances sufficiently accurately then a partial factor γ_t can be applied to the ultimate characteristic total pile resistance to give:

$$R_{cd} = \frac{R_{ck}}{\gamma_t} \qquad (10.26)$$

Simpson *et al.* (1998) recommend that although this gives a more conservative approach, providing the shaft resistance is not too low, the design value obtained from Equations 10.24 and 10.25 should be used.

TABLE 10.3 *Values of partial factors (from Eurocode 7, 1995)*

Partial factors	γ_b	γ_s	γ_t
Driven piles	1.3	1.3	1.3
Bored piles	1.6	1.3	1.5
CFA continuous flight auger piles	1.45	1.3	1.4

Ultimate bearing resistance from ground test results

The validity of any empirical or analytical calculation method must be demonstrated by static load tests carried out in comparable conditions to the site conditions. To provide adequate safety against bearing resistance failure the following inequality must be satisfied:

$$F_{cd} \leq R_{cd} \qquad (10.27)$$

where:

F_{cd} = ultimate limit state axial design compression load

R_{cd} = the sum of all of the ultimate limit state design bearing resistance components.

The design bearing resistance, R_{cd} of a pile is obtained from:

$$R_{cd} = R_{bd} + R_{sd} \qquad (10.28)$$

where:

$$R_{bd} = \text{design base resistance} = \frac{R_{bk}}{\gamma_b} \qquad (10.29)$$

$$R_{sd} = \text{design shaft resistance} = \frac{R_{sk}}{\gamma_s} \qquad (10.30)$$

where:

R_{bk} = characteristic base resistance
R_{sk} = characteristic shaft resistance
γ_b = partial factor on base resistance
γ_s = partial factor on shaft resistance

Values of γ_b and γ_s are given in Table 10.3.

The characteristic bearing resistance R_{bk} is obtained from:

$$R_{bk} = q_{bk} A_b \quad \text{(see Equation 10.2)} \qquad (10.31)$$

where q_{bk} is the characteristic end-bearing resistance given by expressions such as equations 10.6, 10.8 or 10.17.

The characteristic shaft resistance is obtained from:

$$R_{sk} = \sum_{i=1}^{n} q_{sik} A_{si} \qquad (10.32)$$

where q_{sik} is the characteristic shaft resistance per unit area in the *i*th layer (adhesion in clay, skin friction in sand) given by expressions such as Equations 10.7, 10.9, 10.14, 10.15 or 10.19.

It is assumed that the characteristic shaft and base resistances are calculated from methods that have been calibrated or validated by load tests and that there are established correlations for aspects such as the empirical 'adjustment' factors applied to the above equations, e.g. for α, K or δ. To allow for the variability that is likely within these correlations the calculated values of the shaft and base resistances are obtained from mean (not characteristic) values of the strength parameters and then are divided by [1.5] to obtain the characteristic values of adhesion c_{ak}, e.g.

$$c_{ak} = \frac{c_a \text{ calculated from mean } c_u \text{ values}}{1.5} \qquad (10.33)$$

The partial factor 1.5 is a boxed value, subject to review by each member country of the EU.

The ultimate bearing resistance of a pile can also be determined from the results of pile-driving formulae and from wave equation analysis but these methods must have been validated by static pile load tests in similar conditions.

Negative skin friction

Causes of negative skin friction
(Figure 10.15)

If a pile extends through a soft compressible soil and into more competent strata many designers choose to ignore the possible supporting capacity from positive skin friction in these soils, as a conservative approach. If there is a possibility that this soil may undergo settlement at some time, even many years after the building is completed, then an allowance for negative skin friction should be included.

Negative skin friction is the shear stress applied to the surface of a pile by a soil undergoing settlement. The load applied is an adverse action because it acts downwards on the pile (hence the term downdrag) and it is a permanent action that the pile must always support in addition to the structural load. To allow for this load the piles must have greater supporting capacity which usually means greater length in the supporting soils. If this load is not catered for then the pile foundation is likely to experience settlements in excess of those anticipated.

Settlements can occur due to:

- Self-weight compression of under-consolidated soils, such as recent deposits, hydraulic fills and made ground.
- Consolidation of soils under an applied stress such as the placement of a layer of fill over the site. Note that all of the strata above the consolidating layer will be undergoing settlement and so will be applying negative skin friction to the piles.
- Consolidation of soils due to a decrease in pore water pressures in the ground, caused by natural or artificial drainage or groundwater lowering for construction purposes. The increase in effective stress will affect all of the soils around and beneath the piles, so apart from overall settlements of the pile foundation the settlements in the compressible soils will also apply negative skin friction to the piles.
- Consolidation of soils around the piles during the dissipation due to the excess pore pressures set up as the piles were driven into the soil. This is especially significant in the more sensitive and normally consolidated soils.

Determination of negative skin friction
The frictional nature of the shear stress is determined for both clays and sands using expressions

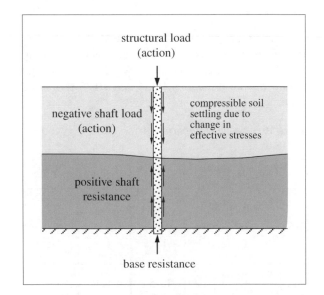

FIGURE 10.15 · *Negative skin friction*

similar to equations 10.14 and 10.15. From these equations it can be seen that the shear stress per unit area of pile shaft surface increases with depth and should be integrated to obtain the total negative skin friction load using expressions such as equations 10.3 and 10.4. This load is then added to the structural load and the pile must be designed to support both loads by increasing its length within the supporting soil where positive skin friction is obtained or by a more substantial socket into a bedrock.

Eurocode 7 requires that the downdrag force must be the maximum which could be generated and should be calculated from adverse estimates of the particular situation. The characteristic values of the soil parameters, i.e. δ or β, will be cautious upper values, not the lower values. EC7 states that for the ultimate limit states in which the soil strength acts in an unfavourable manner the value of the partial factor γ_m should be less than unity so that the design value is greater than the characteristic value. Alternatively, multiplying by the partial factors given in Table 10.3 is suggested (Simpson *et al*, 1998). The design of the pile should be checked for both cases B and C, see Chapter 8, using the methods suggested by these authors.

Other factors which should be included are the effects of the variation of negative skin friction with depth, the amount of soil displacement relative to the pile, the yielding of the pile toe and the effects of pile groups. These are discussed in more detail in Tomlinson (1987).

Pile groups

It is not common for a single pile to support a structural load on its own because the structural load is rarely a simple vertical load. Often there are overturning effects, moments and lateral loading and there is the difficulty of transferring the load axially down the pile. Large-diameter piles supporting a single column load would be the exception. Piles are grouped together usually on a square grid pattern with the structural load transferred to and shared between the piles by a thick reinforced concrete pile cap. A larger pile group supporting the whole of a building would be called a piled raft foundation.

Pile spacing

The centre-to-centre spacing between piles, s, is typically 3 diameters for friction piles and may be 2 diameters for end-bearing piles. For under-reamed piles the spacing should be no less than 2 pile base diameters.

If the piles are too far apart the bending stresses in the pile cap are large necessitating thicker concrete and more reinforcement to distribute the applied loads to the piles. If the piles are too close then the soil between can become excessively disrupted and disturbed leading to high pore pressure increases, ground heave and poor efficiency of loading. Interference due to misalignment may also be a problem.

Stressed zone (Figure 10.16)

The zone of the soil stressed around a single pile is much smaller than around and beneath a pile group. This has a number of consequences:

- The installation method has less effect on group behaviour than on single pile behaviour since around and beneath a group the zone affected by disturbance is relatively small and the stresses will be transferred to undisturbed soil.
- Compressible layers existing beneath the base of a pile group will produce settlement of the group while they may not affect the result of a single pile load test.
- Because of the above and other factors extrapola-

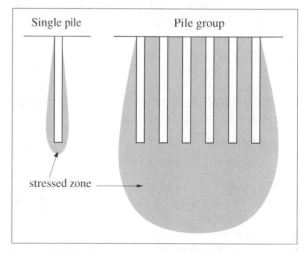

FIGURE 10.16 *Stressed zone around piles*

tion from the performance of a single pile load test to the behaviour of a group must be treated with caution.

Load variation (Figure 10.17)

It is a fallacy to presume that each pile in a pile group carries the same load. The loads in the piles differ and this is well illustrated by reference to Figure 5.1.

If a flexible pile cap is provided and a uniform load is applied then the contact pressures will be fairly uniform for both clays and sands and piles placed beneath the flexible pile cap will carry fairly similar loads. This could arise for piles placed beneath an embankment to reduce the settlement of the embankment. It is to be noted, however, that a dish-shaped profile of settlement is obtained which many structures cannot tolerate.

The purpose of a rigid pile cap is to even out the settlement profile and to produce similar settlements. From Figure 5.1 it can be seen that if the piles are embedded in a clay then the rigid cap will increase the loads on the outer piles and decrease loads on the centre piles. This was confirmed for model piles in a clay tested by Whitaker (1957) who showed that for typical pile spacings of less than 4 diameters the corner piles carry the largest load and the centre piles carry least load, see Figure 10.17.

From Figure 5.1 for pile groups in sands with rigid pile caps the distribution of load can be expected to be reversed with the centre piles carrying the largest loads and the corner piles the least load and this has been confirmed by Vesic (1969).

Efficiency (Figure 10.18)

It is usually incorrect to assume that the group failure load equals the sum of the pile failure loads with the piles acting individually. The difference is represented by an efficiency factor η:

$$\eta = \frac{\text{average load per pile at failure of group}}{\text{failure load of single isolated pile}} \quad (10.34)$$

For piles driven into loose or medium dense sands the effect of compaction will lead to an efficiency η greater than 1 with higher efficiencies for closer pile spacing and in looser sands. For piles driven into dense sands the efficiency is unlikely to exceed 1 and could be less if driving causes disturbance. For

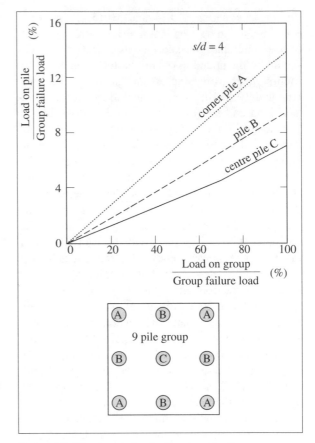

Figure 10.17 *Load distribution in a 9-pile group (from Whitaker, 1970)*

bored pile groups in sand the overlapping disturbance zones between piles is likely to reduce efficiency.

The model tests of Whitaker (1957) on piles in clay, see Figure 10.18, showed the significant effect of the spacing of the piles on efficiency. For a square group of piles the efficiency at a spacing of 8 diameters or more, even with a rigid pile cap was close to unity giving equal pile loads. As the spacing decreases the efficiency decreases but not too significantly.

With wider spacing the pile group reaches a state of failure referred to as individual pile penetration, see Figure 10.19, with the soil remaining static and the piles alone moving down. This, of course, assumes a 'free-standing' group. However, as the pile spacing decreases, at a critical pile spacing the

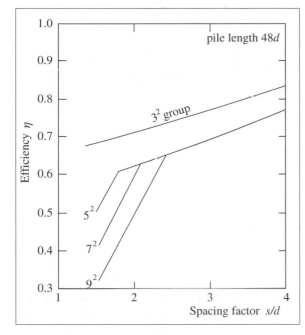

FIGURE 10.18 *Efficiency of freestanding pile groups (from Whitaker, 1970)*

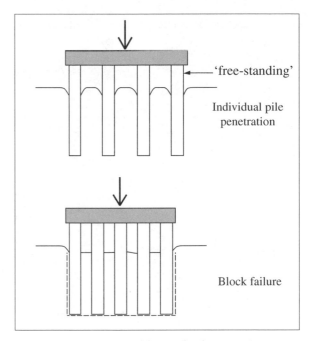

FIGURE 10.19 *Modes of failure for free-standing pile groups*

soil between the piles moves down with the piles causing 'block failure' when the efficiency diminishes dramatically, as shown in Figure 10.18. The efficiency was also found to decrease for longer piles and for larger pile groups.

Ultimate capacity (Figures 10.19 and 10.20)

The ultimate capacity of a free-standing pile group in clays should be taken as the lesser of:

a) the sum of the failure loads of the individual piles in the group; or

b) the bearing capacity of a block of soil bounded by the perimeter of the pile group, i.e.

$$P_{ult} = c_u N_c s_c d_c B_g L_g + 2(B_g + L_g)L_p \bar{c} \qquad (10.35)$$

where:

$c_u N_c s_c d_c$ is the bearing capacity of the clay beneath the pile group

B_g and L_g are the plan width and length of the pile group

L_p is the length of the piles

$\bar{c}$ is the average c_u value over the length of the piles.

The above refers to 'free-standing' pile groups

where the underside of the pile cap is not in contact with the supporting soil or the soil between the pile cap and the supporting soil is compressible. This latter situation is commonly found for pile foundations where the piles are taken through inferior superficial soils.

If the pile cap is constructed in contact with the supporting soil then this is referred to as a 'pile foundation' and the cap itself can support a proportion of the load applied. At failure, the effect of a contacting pile cap is to induce block failure, irrespective of pile spacing, as shown in Figure 10.20. The ultimate capacity of a pile foundation with its pile cap in contact with the supporting soil can then be obtained from the lesser of:

a) the block failure mode as given by Equation 10.35 above; or

b) the sum of the failure loads of the individual piles plus the failure capacity of the remainder of the pile cap contact surface

$$P_{ult} = n(Q_s + Q_b) + c_{uc} N_c s_c d_c (B_c L_c - nA_p) \qquad (10.36)$$

where:

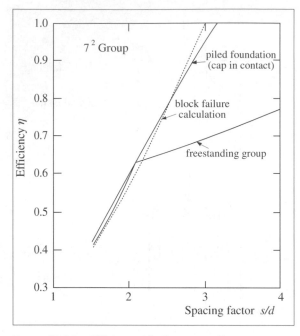

FIGURE 10.20 *Efficiency of pile groups (from Whitaker, 1970)*

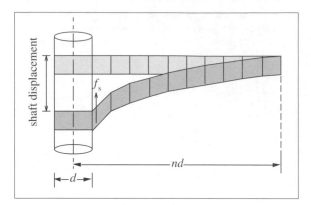

FIGURE 10.21 *Shear displacement distribution around a pile (from Whitaker, 1970)*

n = number of piles
Q_s = shaft load of a single pile obtained from the average shear strength along the pile
Q_b = base load of a single pile obtained from the shear strength at the base of the pile
c_{uc} = undrained shear strength at pile cap level
$N_c s_c d_c$ = factors obtained from bearing capacity theory, Chapter 8.
B_c and L_c = plan width and length of the pile cap
A_p = cross-sectional area of a single pile.

Settlement ratio (Figures 10.21 – 10.23)

As a pile is pushed downwards shear stresses and strains will develop in the soil around the pile. Cooke (1974) assumed that these shear stresses are transferred radially between successive annular soil elements and diminish with distance from the shaft of the pile as shown in Figure 10.21. The point where shear strains were insignificant was found theoretically to be at $n = 22$ assuming a continuous medium and by experiment in a discontinuous fissured clay using sensitive instruments to be at $n = 10$.

Normal pile spacings are much less than these values so the piles within a pile group will effect a complex pile-to-pile interaction when they are loaded, producing additional settlements on surrounding piles and vice versa. This additional settlement has been computed by Poulos (1968) using an interaction factor α_F given by:

$$\alpha_F = \frac{\text{additional settlement caused by adjacent pile}}{\text{settlement of pile under its own load}}$$

(10.37)

Values of α_F are given in Figure 10.22 for rigid piles, infinite soil thickness and homogeneous undrained soil ($v = 0.5$). They are smaller for flexible piles, finite layer depths and for soil modulus increasing with depth but slightly higher for under-reamed piles and drained soil ($v < 0.5$). Adopting a super-position principle the interaction factors can be used to determine the cumulative settlement of a group of piles and give a settlement ratio R_s:

$$R_s = \frac{\text{settlement of group}}{\text{settlement of single isolated pile}}$$

(10.38)

when the loads in the group and the single pile are at the same proportion of their failure load (and same factor of safety).

Settlement ratios have been computed by Cooke *et al.* (1980) and are illustrated in Figure 10.23. They also stated that for pile spacings greater than 2 diameters settlement ratios are unlikely to exceed 6 for rectangular pile groups with any number of piles, in soil increasing in stiffness with depth.

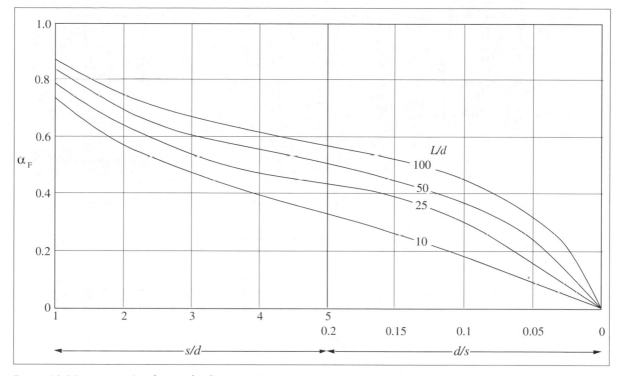

FIGURE 10.22 *Interaction factors for floating piles (from Poulos, 1980)*

Settlement of pile groups (Figures 10.24 and 10.25)

A convenient method of estimating settlement of a pile group is to assume that the group is represented by an 'equivalent raft' at some depth below ground level and then to determine settlements using the conventional methods given in Chapter 9.

Uniform loading over this raft is assumed. This is obtained by dividing the total load (net) by the area of the equivalent raft. The size and depth of the equivalent raft are determined as shown in Figure 10.24 for different soil conditions.

The assumed raft is embedded within the compressible deposit so the depth correction factor μ given by Fox (1948) may be used, Figure 10.25. However, this would entail moving the equivalent raft to the surface of the deposit to determine the settlements for a surface loaded foundation. This may be appropriate for immediate settlement but not for consolidation settlement. The consolidation

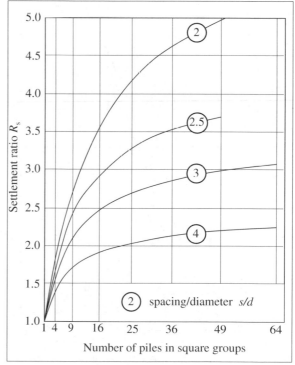

FIGURE 10.23 *Settlement ratios for pile groups (from Cooke et al., 1980)*

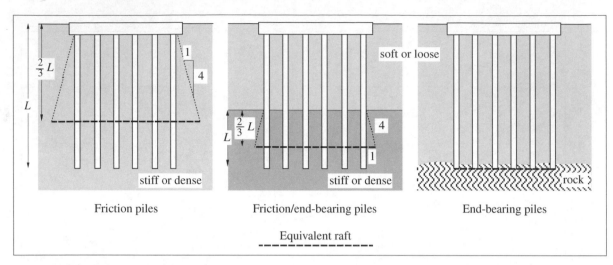

FIGURE 10.24 *Equivalent raft foundation*

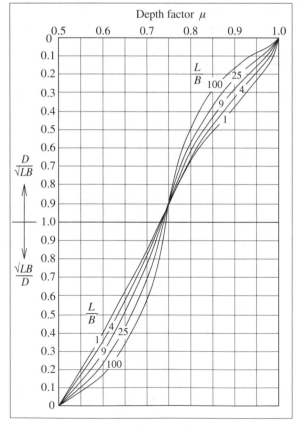

FIGURE 10.25 *Depth correction factor μ (from Fox, 1948)*

settlement should be determined from the stresses applied by the equivalent raft where it exists in the ground and for the thickness of soil beneath this location.

SUMMARY

The first requirement in the design of a pile foundation is to choose the most appropriate and economical type of pile for the site conditions.

Although the student must understand the fundamental factors affecting a pile foundation it must be appreciated that the reliability of calculations based on analytical methods is often poor. Eurocode 7 makes it clear that pile designs must be validated by static load tests carried out with the chosen pile type and in the range of ground conditions anticipated.

Bored and driven piles in clay obtain most of their load capacity from shaft resistance whereas in sands driven piles achieve much higher base resistances. Bored piles in sand are not recommended due to the effects of disturbance.

The limit state design approach can be carried out from pile load test results directly, or indirectly by validation of calculation methods based on ground test results, pile-driving formulae or wave equation analysis.

If negative skin friction is likely but ignored then a pile foundation will be subjected to settlements greater than anticipated. If it is allowed for then a more costly pile will result, with greater lengths required.

When piles are grouped together interaction effects produce load variations within the group and reduced efficiency of loading, especially for clays. The settlement of a pile group is greater than an individual pile and may bear no relationship to the load-settlement behaviour of a single pile.

Pile Group Effects – Charity Hospital, New Orleans, USA

Case Objectives:

This case illustrates:

- the significance of the settlement ratio for pile groups
- that extrapolation from individual pile load tests can be dangerous
- that site investigations should be extended well below the depth of a pile foundation

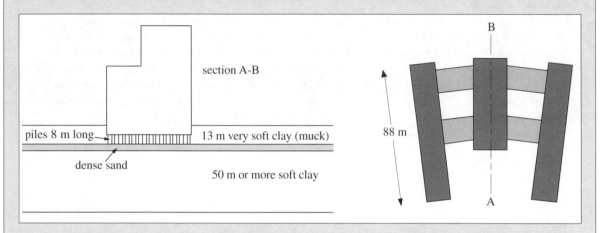

This case was described by Terzaghi *et al.*, 1996. The ground conditions comprised 13 m of very soft clay (muck) overlying a 1.8 m layer of dense sand. This was underlaid by 50 m or more of soft compressible clay. Because of the very soft clay a pile foundation was proposed. A test on an individual pile driven into the dense sand gave a settlement of only 6 mm under a load of 270 kN so the settlement of the entire structure was also expected to be small. The load per pile was chosen as 135 kN. The pile foundation was then constructed from a basement structure with about 10,000 timber piles 7.9 m long driven through the very soft clay to achieve end bearing on the dense sand.

By the time the structure was completed it had settled up to 100 mm and two years after construction the settlement had reached 300 mm. The final maximum consolidation settlement was estimated to be 500 mm but with secondary compression this settlement was to be far exceeded. The structure underwent significant differential settlements, settling into the classic bowl-shaped profile, and the steel frame and stone cladding suffered severe damage.

Referring to Figure 10.16 the stressed zone around a single pile is very small and in this case would have been concentrated in the dense sand. From the diagram above the stressed zone would have extended well into the soft clay beneath resulting in the settlements experienced. It is suspected that apart from inadequate knowledge of pile group effects at the time there was insufficient ground investigation carried out.

Other relevant chapters are 5, Stress Distribution and 9, Settlements. Worked example 5.4 shows that the stresses applied by the building to the soft clay were considerable.

CASE STUDY

Negative Skin Friction

Case Objectives:

This case illustrates:
- that negative skin friction load can exceed the design load of the pile
- that negative skin friction develops with time as pore water pressures dissipate
- that negative skin friction is determined by the effective stress in the soil

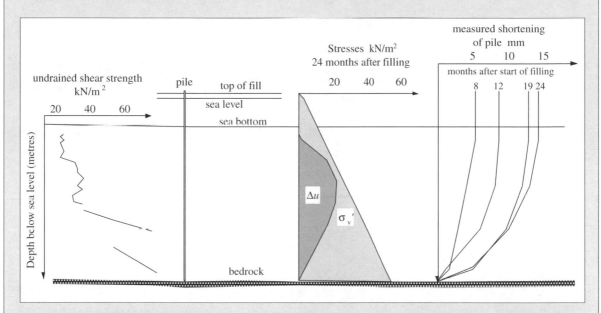

The following is a description of the pile testing reported by Johannessen and Bjerrum (1965). The test site was in an area reclaimed from the sea where the water depth was about 8 m and 10 m fill would be placed. The ground conditions comprised about 45 m of late-glacial marine clay which was soft becoming firm with depth, with moisture contents typically midway between the liquid and plastic limits. Beneath the clay was bedrock consisting of Ordovician schist. During the two-year monitoring period the sea bottom settled 1.2 m under the weight of the fill.

The piles were 470 mm cruciform-shaped hollow steel piles with a closed base and were driven in April 1962 and the fill was placed in the following 12 months. Mechanical strain measurements were obtained down the inside of the pile. Piezometers to measure pore water pressure were placed in the clay and settlement plates were installed on the sea bed. Anchors installed in the clay and connected to the surface with steel wires were used to measure settlements at different depths. As the fill was placed the excess pore water pressure in the soft clay increased and after completion of the fill these pressures dissipated. Both the increase and decrease were influenced by major drainage boundaries at the top and bottom of the clay.

The negative skin friction was assumed to be determined by Equation 10.14 with $K\tan\phi'$ constant with depth. The shortening of the pile was calculated down the length of the pile and compared with the observed shortening to back-analyse a value of the factor $K\tan\phi'$. For the ultimate condition when all the excess pore pressures had dissipated the value of this factor was estimated to be 0.20.

The time-dependent nature of negative skin friction is illustrated in the diagram of shortening of the pile with depth. After two years the pile shortening was 14 mm and the final shortening after all excess pore pressures had dissipated was estimated to be 25 mm.

It was estimated that the stress in the steel just above the pile point was nearly 200 N/mm^2 and the load at the pile point caused by the negative skin friction could be 250 tons. It is not surprising that the pile point punched into the rock and settled about 60 mm. Yielding of the pile toe into the rock allowed at least the lower part of the pile to elongate and caused a temporary relief of stresses in the pile.

Other relevant chapters include 6, Consolidation and 9, Settlements.

Worked Example 10.1   Bored piles in clay

Determine the length of a 600 mm diameter pile to support a working load of 1200 kN at a site where two layers of clay exist. The upper layer is 8 m thick and has an undrained shear strength of 80 kN/m². The lower layer is 20 m thick and has an undrained shear strength of 120 kN/m².

Assume

(i)   the top 1 m of the pile does not support load due to clay/concrete shrinkage

(ii)   an adhesion factor of 0.5 $N_c = 9$ $\omega = 1.0$

(iii) Factors of safety are 1.5 and 3.0 on the shaft load and base load, respectively

Shaft load in upper layer Q_{s1}
Length of pile in contact with upper layer is 7.0 m

$$\frac{0.5 \times 80 \times \pi \times 0.6 \times 7.0}{1.5} = 352 \text{ kN}$$

remaining load of 1200 − 352 = 848 kN must be supported in the lower layer

Base load in lower layer Q_b

$$\frac{9 \times 1.0 \times \pi \times 120 \times 0.6^2}{3 \times 4} = 102 \text{ kN}$$

remaining load of 848 − 102 = 746 kN must be supported by the shaft in the lower layer. Assuming the length of pile in the lower layer to be L.

$$\frac{0.5 \times 120 \times \pi \times 0.6 \times L}{1.5} = 746 \text{ kN}$$

$L = 9.9$ m
Length of pile required is 8.0 + 9.9 = 17.9 m, say 18 m
If end bearing is ignored or cannot be relied on then

$$\frac{0.5 \times 120 \times \pi \times 0.6 \times L}{1.5} = 848 \text{ kN}$$

$L = 11.25$ m
Length of pile is 8.0 + 11.25 = 19.25 m

Worked Example 10.2   Bored piles in fissured clay

Determine the length of a pile, 1200 mm diameter, to support a working load of 4500 kN in a thick deposit of fissured clay with undrained shear strength increasing linearly with depth from 55 kN/m² at ground level and at 5 kN/m² per metre depth. The same assumptions as in Example 10.1 are used but with $\omega = 0.75$ for a large diameter pile.

At 1 m below ground level $c_u = 55 + 5 = 60$ kN/m²
Let L be the length of pile below 1 m
At the base of the pile $c_{ub} = 60 + 5L$ kN/m²

Mean shear strength over effective length of shaft $L = 60 + \dfrac{5L}{2} = 60 + 2.5L$

$$4500 = \frac{9 \times 0.75 \times \pi \times (60 + 2.5L) \times 1.2^2}{3.0 \times 4} + \frac{0.50 \times \pi \times (60 + 2.5L) \times L \times 1.2}{1.5}$$

giving $L^2 + 28.05L - 1383.6 = 0$ and $L = 25.73$ m

Check: $\dfrac{9 \times 0.75 \times \pi \times 188.7 \times 1.2^2}{3.0 \times 4} + \dfrac{0.50 \times \pi \times 124.3 \times 25.73 \times 1.2}{1.5} = 480 + 4019 = 4499$ kN

∴ length of pile is 1.0 + 25.7 = 26.7 m, say 27 m

Worked Example 10.3   Short driven pile in clay

A 250 mm square concrete pile is driven 2.5 m into a very stiff clay with undrained shear strength of 150 kN/m². Determine the working load of the pile assuming an overall factor of safety of 2.0 for:
(i)   soft clay overlying
(ii) sand overlying

For both cases the base resistance can be calculated from
$Q_b = 9 \times 150 \times 0.25^2 = 84.4$ kN

i) Soft clay overlying
From Figure 10.3, $\alpha = 0.25$
$Q_s = 0.25 \times 150 \times 4 \times 0.25 \times 2.5 = 93.8$ kN

Working load = $\dfrac{84.4 + 93.8}{2} = 89$ kN

ii) Sand overlying
From Figure 10.3, $\alpha = 1.0$
adhesion = $1.0 \times 150 = 150$ kN/m²
Assuming a maximum adhesion of 100 kN/m²
$Q_s = 100 \times 4 \times 0.25 \times 2.5 = 250$ kN

Working load = $\dfrac{84.4 + 250}{2} = 167$ kN

Worked Example 10.4   Long driven pile in clay

A closed end steel tubular pile, 0.9 m diameter, is driven into a stiff clay with a penetration of 50 m. The undrained shear strength of the clay is 120 kN/m² and the submerged unit weight is 11 kN/m³. Determine the working load assuming an overall factor of safety of 2.0.

Base load is $9 \times 120 \times \pi \times 0.9^2/4 = 687$ kN
i) Using the λ method
Assuming a water table at ground level or submerged conditions
mean effective stress = $25 \times 11 = 275$ kN/m²
From Figure 10.6, $\lambda = 0.13$
$c_a = 0.13 (275 + 2 \times 120) = 67$ kN/m²
Shaft load = $67 \times \pi \times 0.9 \times 50 = 9472$ kN

Working load = $\dfrac{687 + 9472}{2.0} = 5080$ kN

ii) Using the $\dfrac{c_u}{\sigma_v}$ ratio method

$$\dfrac{c_u}{\sigma_v} = \dfrac{120}{275} = 0.44 \qquad \dfrac{L}{d} = \dfrac{50}{0.9} = 55.6$$

From Figure 10.7, $\alpha_p = 0.82$ and $F = 0.95$

Shaft load $= 0.95 \times 0.82 \times 120 \times \pi \times 0.9 \times 50 = 13215$ kN

Working load $= \dfrac{687 + 13215}{2.0} = 6951$ kN

Worked Example 10.5   Driven pile in sand

A concrete pile, 350 mm square, is to be driven into a thick deposit of medium dense sand with a SPT 'N' value of 20 and a bulk unit weight of 19 kN/m³. The water table lies at 2.0 m below ground level. Estimate the length of pile required to support a working load of 450 kN assuming an overall factor of safety of 2.5.

From Figure 10.11 an estimate of the initial angle of friction is $33°$

From Table 10.1 the angle of friction after driving can be obtained

$$\phi' = \dfrac{33 + 40}{2} = 36.5° \quad \text{or} \quad \phi' = {}^3/_4 \times 33 + 10 = 34.75°$$

From Figure 10.10, using Meyerhof's relationship a critical depth at $\dfrac{z_c}{d} = 10.5$ is indicated

$z_c = 10.5 \times 0.35 = 3.7$ m

depth, m	σ_v' kN/m²
0	0
2.0	38.0
3.7	53.6

From Figure 10.8, $N_q = 95$

$q_b = 95 \times 53.6 = 5092$ kN/m²

Base load (factored) $= \dfrac{5092 \times 0.35^2}{2.5} = 249.5$ kN

$\therefore$ pile must be longer than the critical depth

From Figure 10.12, $K_s \tan \delta = 0.70$

Shaft load from ground level to 2.0 m $= Q_{s1}$

$$Q_{s1} = \dfrac{38.0 \times 0.43 \times 2.0 \times 0.5 \times 0.35 \times 4}{2.5} = 9.2 \text{ kN}$$

Shaft load from 2.0 to 3.7 m $= Q_{s2}$

$$Q_{s2} = \dfrac{(38.0 + 53.6) \times 0.43 \times 1.7 \times 0.5 \times 0.35 \times 4}{2.5} = 18.7 \text{ kN}$$

Shaft load required below 3.7 m $= Q_{s3}$

$Q_{s3} = 450 - (249.5 + 9.2 + 18.7) = 172.6$ kN

Length of pile below 3.7 m = L

$$L = \frac{172.6 \times 2.5}{53.6 \times 0.43 \times 0.35 \times 4} = 13.4 \text{ m}$$

Pile length = 3.7 + 13.4 = 17.1 m, say 17.5 m

EXERCISES

10.1 A bored pile, 750 mm diameter and 12 m long is to be installed on a site where two layers of clay exist. The upper firm clay layer is 8 m thick and has an undrained shear strength of 50 kN/m². The lower stiff clay layer is 12 m thick and has an undrained shear strength of 120 kN/m². Determine the working load the pile could support assuming the following:
(a) $\alpha = 1.0$ for firm clay and 0.5 for stiff clay $\qquad N_c = 9 \qquad \omega = 1.0$.
(b) Factors of safety of 1.5 and 3.0 are applied to the shaft load and base load, respectively.
(c) The top 1 m of the firm clay is ignored due to shrinkage.

10.2 For the ground conditions and assumptions described in Exercise 10.1 determine the length of pile required to support a working load of 1500 kN.

10.3 A bored pile, 900 mm diameter and 15 m long, is to be installed in a fissured clay whose strength increases with depth from 75 kN/m² at ground level and at 4 kN/m² per metre of depth. Determine the working load the pile could support assuming the following:
(a) $\alpha = 0.45$ and $\omega = 0.8$
(b) $F = 1.5$ for the shaft load and $F = 3$ for the base load
(c) The top 1 m is ignored.

10.4 A large diameter under-reamed bored pile is to be installed in stiff clay with undrained shear strength of 125 kN/m². The main shaft of the pile is 1.5 m diameter and the base of the under-ream is 4.5 m diameter with a height of 3.0 m and the total length of the pile from ground level to the base of the under-ream is 27 m. Determine the working load the pile could support assuming the following:
(a) $\alpha = 0.3 \quad N_c = 9 \quad \omega = 1.0$.
(b) Adhesion should be ignored over the height of the under-ream, over the main shaft of the pile up to 2 shaft diameters above the top of the under-ream and over the top 1 m of the pile.
(c) A factor of safety of 3.0 should be applied to the base load but full mobilisation of shaft adhesion can be assumed.

10.5 A closed-end pipe pile, 1.2 m diameter is to be installed offshore with a penetration of 40 m into the stiff clay below the water. The undrained shear strength of the clay is 140 kN/m² and its submerged unit weight is 11.5 kN/m². Determine the working load the pile could support with an overall factor of safety of 2.5 using:
(a) the λ method and
(b) the c_u/σ_v' method.

10.6 A jacked pile, 0.35 m diameter and 9 m long is to be installed into a dense 'dry' sand with an initial angle of friction of 40° and bulk unit weight of 20.5 kN/m³. The water table lies well below the base of the pile. Assuming the critical depth as given by Meyerhof's curve in Figure 10.10 estimate the working load the pile could support with an overall factor of safety of 2.5.

10.7 A cylindrical pile, 450 mm diameter and 15 m long is to be driven into a medium dense sand with a water table at 1.5 m below ground level. The initial angle of internal friction of the sand is 34° and its bulk unit weight is 19.5 kN/m³. Assuming the critical depth as given by Meyerhof's curve in Figure 10.10 estimate the working load the pile could support with an overall factor of safety of 2.5.

CHAPTER 11

Lateral earth pressures and retaining structures

- To appreciate that the horizontal pressure acting on a wall is determined by the amount and type of movement, the flexibility of the wall and any propping or anchoring applied.
- To understand the Rankine theory of active and passive pressure and how it is modified by the presence of wall friction, inclined walls and backfill surfaces.
- To construct active and passive pressure diagrams appreciating the difference between undrained and drained conditions, the effects of the soil cohesion component, the presence of a water table and surcharge loads.
- To be aware of the range and styles of walls in various applications.
- To carry out a check on the stability of a gravity retaining wall including rotational failure, overturning, bearing pressure and sliding.
- To carry out a check on the stability of a cantilever and propped embedded wall.
- To appreciate that the design of a multi-propped strutted excavation is based on an empirical approach developed from experience.
- To carry out a check on the internal and external stability of a reinforced soil structure.

Lateral earth pressures

The assessment of earth pressures is based generally on an effective stress analysis. The pressure in the pore water in a fully saturated soil is hydrostatic, i.e. $u_H = u_V = u$ so there is only one value. The pressure within the mineral grain structure (the effective stress) is not the same in all directions. The vertical effective stress in a soil can be obtained from simple considerations of depth multiplied by the bulk or submerged unit weight and is treated as a principal stress, σ_V'. For the design of vertical walls the horizontal effective stress acting, σ_H' is required and a coefficient of earth pressure, K is used to relate the two stresses:

$$K = \frac{\sigma_H'}{\sigma_V'} \tag{11.1}$$

Various factors determine the horizontal stress acting on a wall but initially the assumption of a smooth wall is considered. The amount and type of movement of the wall has a major effect on the horizontal stresses developed, as described below.

Effect of horizontal movement

1 None – 'at rest' condition
Consider an element of soil in the ground which is at equilibrium, with no movement of any kind. There will be a vertical effective stress σ_V' and a different horizontal effective stress σ_H' which are both principal stresses and can, therefore, be represented on the Mohr circle diagram (see Figures 4.10 and 4.11).

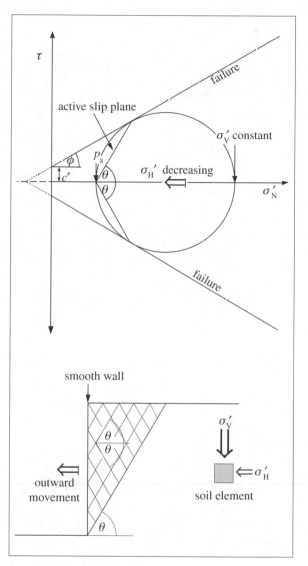

FIGURE 11.1 *Active Rankine state*

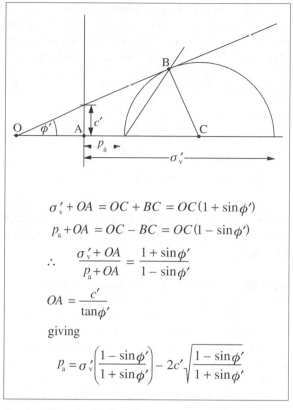

$$\sigma_v' + OA = OC + BC = OC(1 + \sin\phi')$$

$$p_a + OA = OC - BC = OC(1 - \sin\phi')$$

$$\therefore \quad \frac{\sigma_v' + OA}{p_a + OA} = \frac{1 + \sin\phi'}{1 - \sin\phi'}$$

$$OA = \frac{c'}{\tan\phi'}$$

giving

$$p_a = \sigma_v'\left(\frac{1 - \sin\phi'}{1 + \sin\phi'}\right) - 2c'\sqrt{\frac{1 - \sin\phi'}{1 + \sin\phi'}}$$

FIGURE 11.2 *Active Rankine pressure*

The vertical effective stress σ_v' in the ground will remain constant and since it will be the larger value it will be the major principal stress. As horizontal expansion of the soil occurs when the wall moves away from the soil and more of the strength of the soil is mobilised the horizontal stress on the wall decreases. When the failure strength of the soil is mobilised the minimum horizontal stress is termed the active pressure p_a and will be the minor principal stress.

This state can be represented by a Mohr circle (Figure 11.1) which touches the failure envelope. Shear failure will occur at angles θ to the major principal plane so that a network of shear planes will form at angles θ to the horizontal behind the wall where:

$$\pm\theta = 45 + \frac{\phi'}{2} \qquad (11.2)$$

The soil is obviously not in a state of failure and the ratio of the stresses is given by the coefficient of earth pressure 'at rest' K_0 (Equation 4.2). The values and variation of K_0 are discussed in Chapter 4.

2 Horizontal expansion – active pressure (Rankine theory) (Figures 11.1 and 11.2)

This theory considers the ratio of the two principal stresses when the soil is brought to a state of shear failure throughout its mass (plastic equilibrium).

The horizontal stress or active pressure p_a can be obtained in terms of the vertical stress from the geometry of the Mohr–Coulomb failure envelope as explained in Figure 11.2 and is usually given by:

$$\text{minimum } \sigma_H' = p_a = \sigma_v' K_a - 2c' \sqrt{K_a} \qquad (11.3)$$

where:

$$K_a = \frac{1 - \sin \phi'}{1 + \sin \phi'} = \tan^2 \left(45 - \frac{\phi'}{2} \right) \qquad (11.4)$$

SEE WORKED EXAMPLE 11.1.

This 'local' Rankine state of stress will only occur within a wedge defined by θ to the horizontal. The soil outside this wedge is considered to be unstrained.

3 Horizontal compression – passive pressure (Rankine theory) (Figure 11.3)

Consider a wall moving or being pushed towards the soil behind. The vertical effective stress σ_V' in the ground will remain constant but the horizontal stress σ_H' must increase until the soil is brought to the state of plastic equilibrium. σ_H' will be greater than σ_V' so σ_V' will be the minor principal stress. The maximum horizontal stress required to produce failure of the soil is termed the passive pressure p_p and will be the major principal stress.

This state can be represented by another Mohr circle (Figure 11.3) which touches the failure envelope. Shear failure will occur at angles θ to the major principal plane so that a network of shear planes will form at angles θ to the vertical, or $90-\theta$ to the horizontal.

The horizontal stress or passive pressure p_p can be obtained in terms of the vertical stress from the geometry of the Mohr–Coulomb failure envelope using a similar procedure as given in Figure 11.2. It is then represented by:

$$\text{maximum } \sigma_H' = p_p = \sigma_v' K_p + 2c' \sqrt{K_p} \qquad (11.5)$$

where:

$$K_p = \frac{1 + \sin \phi'}{1 - \sin \phi'} = \tan^2 \left(45 + \frac{\phi'}{2} \right) \qquad (11.6)$$

SEE WORKED EXAMPLE 11.6.

This 'local' Rankine state of stress will only occur

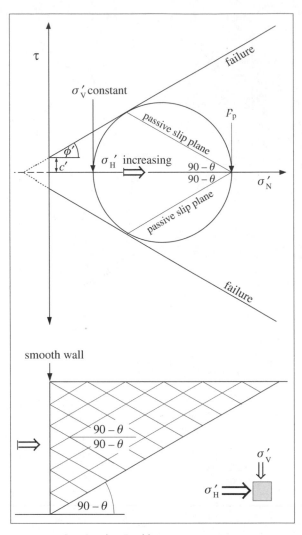

FIGURE 11.3 *Passive Rankine state*

within a wedge defined by $90-\theta$ to the horizontal. The soil outside this wedge is considered to be unstrained.

4 Amount of movement required (Figure 11.4)

Generally, much greater movement of the wall is required to mobilise the full value of the passive pressure compared to the small movements required to mobilise the full value of the active pressure. A typical relationship for sands is illustrated in Figure 11.4 where x represents the inward or outward movement of the wall. It is observed that loose sands

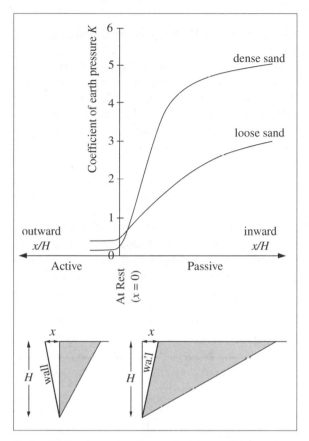

FIGURE 11.4 *Movements required to mobilise earth pressure*

provide greater active pressures and overturning forces and lower passive pressures and restraining forces than dense sands.

Gravity walls, cantilever walls, sheet-pile walls and timbered walls could be considered to yield sufficiently so that the full active pressure is mobilised. The strain required to mobilise the full active pressure behind a wall will mobilise only a portion of the passive pressure in front of the wall so the full amount of passive pressure should never be relied on.

With some structures where yielding is restricted, such as bridge abutments, propped or anchored basement walls and rectangular culverts the horizontal pressure acting could be greater than the active pressure and nearer the 'at rest' condition.

When higher values of K_0 already exist in the ground the amount of movement required to

mobilise active conditions increases since greater expansion is required. On the other hand, less movement is required to mobilise passive conditions since the soil is already highly compressed by the high horizontal stresses. In these circumstances, a greater proportion of the passive resistance can be assumed.

5 Type of movement (Figure 11.5)

Equations 11.3 and 11.5 suggest that the active pressure p_a and passive pressure p_p increase linearly with

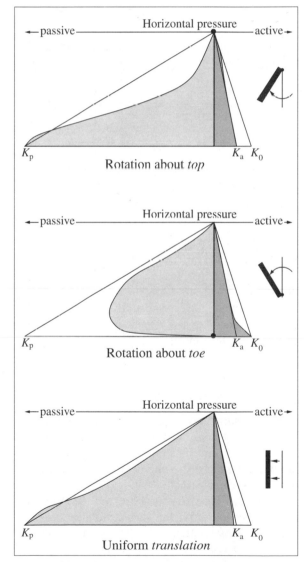

FIGURE 11.5 *Pressure distributions alongside rigid walls*

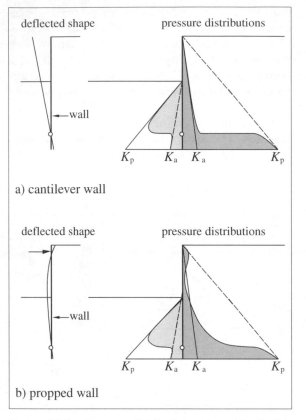

FIGURE 11.6 *Pressure distributions alongside flexible walls*

depth as vertical stress σ_V' increases uniformly. It has been found, however, that different pressure distributions are obtained depending on whether the wall movement comprises:

- rotation about the top
- rotation about the toe
- uniform lateral translation.

Typical variations of pressure developed behind a rigid wall in a dense sand due to each of these types of movements are illustrated in Figure 11.5. These are based on the distributions given in Padfield and Mair (1984) and IStructE (1989).

It has also been shown (IStructE, 1989) that the full passive and active thrusts are mobilised at small movements for rotation about the top and translation, with much larger rotations required (about 2–3 times as much) to fully mobilise these thrusts when rotation about the toe occurs.

Effect of wall flexibility and propping
(Figure 11.6)

Steel sheet pile walls are more flexible than reinforced concrete cantilever walls, embedded diaphragm or contiguous bored pile walls. If a wall deflects due to lateral stress there is a redistribution of stresses due to stress transfer and arching. This redistribution of stress is greater as the wall deflects more.

If the top of the wall is restrained by a prop, strut or anchor then load is attracted to this area with an increased pressure behind the wall which may reach the passive pressure. As the depth of embedment of the wall increases the bottom of the wall becomes more fixed and rotation is restricted. This fixity is provided by the passive resistance behind the wall at its toe.

Effect of wall friction (Figure 11.7)

The Rankine theory assumes that the surface of a wall is smooth but usually in practice it is rough. If the soil moves downwards or upwards against the

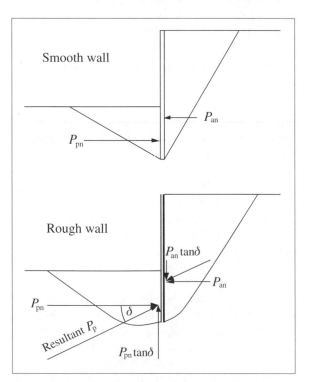

FIGURE 11.7 *Effect of wall friction*

wall a shear stress is transmitted producing wall friction denoted by $f_s = \sigma_H'\tan\delta$ if it is frictional and adhesion denoted by $c_a = c_w$ if it is cohesive.

If the settlement of the wall is negligible but it rotates or moves sideways the active wedge will settle relative to the wall and the passive wedge will rise relative to the wall. The forces applied are then forces P_{an} and P_{pn} normal to the wall and $P_{an}\tan\delta$ acting downwards on the active side and $P_{pn}\tan\delta$ acting upwards on the passive side.

SEE WORKED EXAMPLE 11.4.

Coulomb theory – active thrust
(Figure 11.8)

To a certain extent the effects of wall friction, sloping wall and sloping ground surface can be included using the method proposed by Coulomb (1776). A straight trial surface bounding a wedge of soil of weight W is considered as shown in Figure 11.8.

As the wedge moves downwards due to gravity the shear strength of the soil is assumed to be fully mobilised on the presumed failure plane and wall friction or adhesion is mobilised on the back of the wall. The shear strength and the wall friction act in support of the wedge of soil so the active thrust transmitted to the wall will be smaller for stronger soil and greater wall friction.

From the area of the wedge and the unit weight of the soil the weight W is known. The directions of the resultant forces acting on the wedge, R and P_a, are known, therefore assuming $c' = 0$ the triangle of forces can be completed to obtain a value of P_a for the trial surface chosen. The method is repeated for a number of trial failure planes to obtain the maximum value of P_a.

By considering the trigonometry of the wedge values of P_a and W can be determined as functions of α, β, θ and δ. The maximum value of the resultant P_a is then given by:

$$P_a = \frac{1}{2} K_a \gamma H^2 \tag{11.7}$$

where K_a is determined by the Coulomb equation assuming:

$$\frac{\partial P}{\partial \theta} = 0 \tag{11.7}$$

$$K_a = \left(\frac{\sin(\alpha - \phi)/\sin\alpha}{\sqrt{[\sin(\alpha + \delta)]} + \sqrt{\left[\dfrac{\sin(\phi + \delta)\sin(\phi - \beta)}{\sin(\alpha - \beta)}\right]}} \right)^2 \tag{11.8}$$

The point of application of the thrust P_a (or P_{an}) can be taken as $1/3H$ vertically above the base of the wall assuming a uniform ground slope β. If the

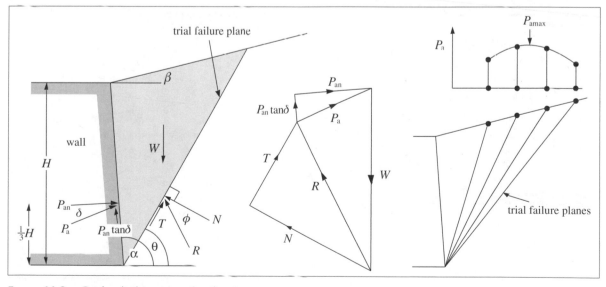

FIGURE 11.8 *Coulomb theory – active thrust*

ground surface is irregular then the centre of gravity of the critical failure wedge (giving the maximum thrust) must be determined. The point of application of the active thrust is assumed to be the point where a line through the centre of gravity of the wedge and parallel to the failure plane cuts the back of the wall.

See Worked Example 11.7.

For the case of a smooth, vertical wall ($\delta = 0$, $\alpha = 90°$) and a horizontal soil surface ($\beta = 0°$) equation 11.8 reduces to the Rankine condition, given by equation 11.4.

If a water table exists behind the wall then it is likely that seepage will be occurring to a vertical wall drain or a toe drain. To assess the effects of this a flow net should be constructed and the pore pressure distribution determined to obtain the variation of effective stresses. Alternatively, a simple approach not requiring a flow net is to assume that the difference in total head either side of the structure is distributed evenly around the structure. The pore pressures are then obtained from the expression, pressure head = total head – elevation head, as described in Chapter 3 and illustrated later.

Coulomb theory – passive thrust
(Figure 11.9)

Passive thrust is produced on the back of the wall as it is pushed towards the wedge of soil of weight W. Assuming a plane trial surface the shear strength of the soil on this 'failure' plane is fully mobilised as the wedge is forced upwards and wall friction or adhesion is mobilised on the back of the wall acting downwards. The shear strength and the wall friction resist upward movement of the wedge so the passive thrust transmitted to the wall will be larger for stronger soil and greater wall friction.

The directions of the resultant forces acting on the wedge, R and the passive thrust P_p are known so assuming $c' = 0$ the triangle of forces can be completed to obtain a value of P_p for the trial surface chosen. The method is repeated for a number of trial failure planes to obtain the minimum value of P_p.

By considering the trigonometry of the wedge values of P_p and W can be determined as functions of α, β, θ and δ. The minimum value of the resultant P_p is then given by:

$$P_p = \frac{1}{2} K_p \gamma H^2 \qquad (11.9)$$

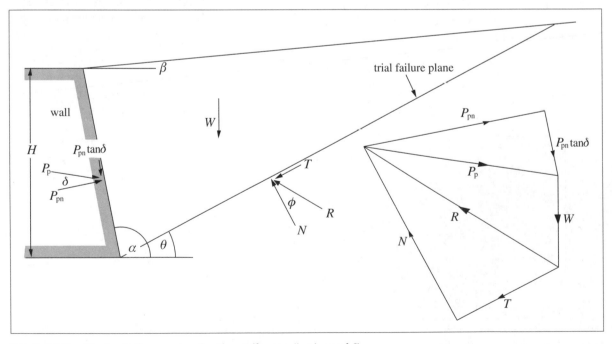

Figure 11.9 *Coulomb theory – passive thrust (for small values of δ)*

SEE WORKED EXAMPLE 11.4.

where K_p is:

$$K_p = \left(\frac{\sin(\alpha + \phi)/\sin \alpha}{\sqrt{[\sin(\alpha - \delta)]} - \sqrt{\left[\dfrac{\sin(\phi + \delta)\sin(\phi + \beta)}{\sin(\alpha - \beta)} \right]}} \right)^2$$

(11.10)

Limitations of the Coulomb theory
(Figure 11.10)

The trial failure surfaces are assumed to be planes for both the active and passive cases whereas in practice the actual failure surfaces have curved lower portions due to wall friction. For the active case the error in assuming a plane surface is small and K_a is under-estimated slightly.

For the passive case the error is also small providing wall friction is low, but for values of $\delta > \phi'/3$ the error becomes large with K_p significantly over-estimated. Because of this the approach usually adopted is to use earth pressure coefficients, see below.

Earth pressure coefficients (Figure 11.11)

To take account of the effects of wall friction Equations 11.3 and 11.5 have been generalised to:

$$p_{an}' = K_a \sigma_V' - K_{ac} c'$$ (11.11)

$$p_{pn}' = K_p \sigma_V' + K_{pc} c'$$ (11.12)

The coefficients of earth pressure K_a and K_p are given for the *horizontal* component of pressure p_{an} or force P_{an} so that for active conditions:

pressure $\quad p_{an} = p_a \cos\delta$ (11.13)

force $\quad P_{an} = P_a \cos\delta$ (11.14)

SEE WORKED EXAMPLES 11.4 AND 11.6.

where p_a and P_a are the resultant values of pressure and force acting at an angle δ to the back of the wall. In the case of an inclined wall the values p_{an} and P_{an} would be those normal to the back of the wall. The shear force acting on the wall is given by:

$$P_{an} \tan\delta$$ (11.15)

The coefficients K_a and K_p have been determined by Kerisel and Absi (1990) assuming the curved failure surface to be a logarithmic spiral. Values of K_a and K_p for a horizontal backfill and vertical wall are given in Figure 11.11. They give the horizontal components of active and passive pressures. Wall friction forces $P_{an} \tan\delta$ or $P_{pn} \tan\delta$ may occur on the back of the wall.

Values of K_{ac} and K_{pc} can be obtained with sufficient accuracy from the expressions:

$$K_{ac} = 2 \sqrt{\left[K_a \left(1 + \frac{c_w'}{c'} \right) \right]}$$ (11.16)

$$K_{pc} = 2 \sqrt{\left[K_p \left(1 + \frac{c_w'}{c'} \right) \right]}$$ (11.16)

where c_w' is the drained or effective stress wall adhesion. Since low values of c' are usually adopted it is reasonable to assume that c_w' is also low, if not zero.

The above expressions will be appropriate for granular soils and overconsolidated clays where the

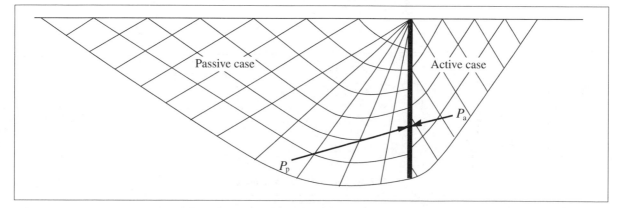

FIGURE 11.10 *Curved failure surfaces due to wall friction*

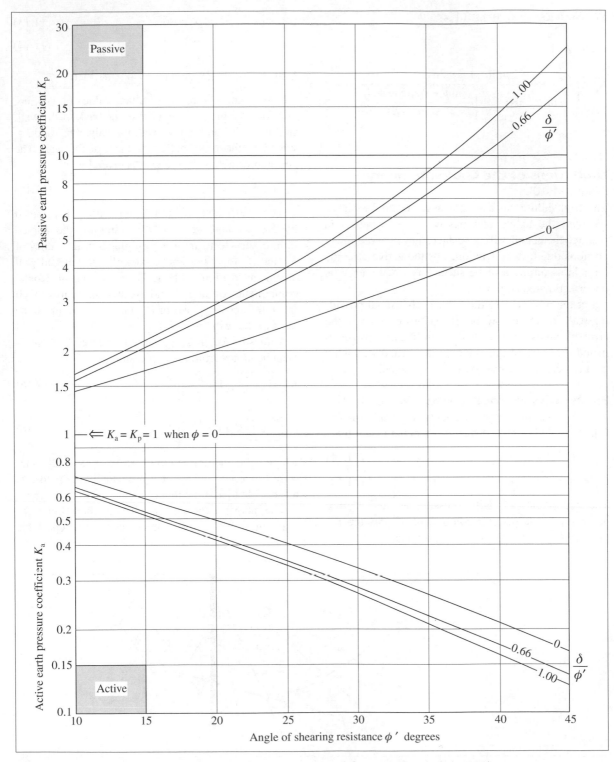

FIGURE 11.11   *Horizontal earth pressure coefficients K_a and K_p (after Kerisel and Absi, 1990)*

critical condition will be the drained case and effective stress parameters are applicable.

The angle of wall friction δ will depend on the frictional characteristics of the soil and the roughness of the wall and is usually given as a proportion of ϕ', the value δ never exceeding ϕ'.

The relative movement of the wall and soil must also be considered. For active conditions wall friction should only be considered if the soil moves downwards relative to the wall. If the wall also has a tendency to settle then it is safer to ignore wall friction. For passive conditions wall friction can be considered where the wall settles relative to the soil such as a load-bearing basement wall. Values of wall friction and wall adhesion from EC7 and BS8002 are given in the annex to this chapter. Commonly adopted values for temporary works (Williams and Waite, 1993) are given in Table 11.1.

Effect of cohesion intercept c'
(Figure 11.12)

Although the value of c' is typically small for overconsolidated clays it may have a marked effect on the pressures produced with lower active thrusts and greater passive thrusts. On the active side a depth of theoretical negative pressure is obtained from equation 11.11. This pressure cannot act in support of the wall so it is presumed to be zero over this depth.

On the passive side the amount of movement required to fully mobilise passive thrust will be large and the shear strength may have dropped to the critical state value so the use of $c' = 0$ will give the safest, albeit conservative approach. For a normally consolidated uncemented clay and for a compacted clay the cohesion intercept could also be expected to be zero.

The effect of a cohesion intercept is shown in Figure 11.12. For the active case there is a depth of soil z_0 over which the active pressure is theoretically negative. This depth is given when $p_{an} = 0$ in equation 11.11:

$$z_0 = \frac{K_{ac}c'}{\gamma K_a} \tag{11.18}$$

If wall friction and adhesion are ignored this reverts to the Rankine case when:

TABLE 11.1 *Values of skin friction and wall adhesion (from Williams and Waite, 1993)*

Analysis	Angle of wall friction δ		Wall adhesion	
	active	passive	active	passive
Effective stress	$0.67\phi'$	$0.5\phi'$	$0.5c'$	$0.5c'$
Total stress	–	–	$0.5c_u$	$0.5c_u$
			$50kN/m^2$ maximum	$25kN/m^2$

Note: c' is normally taken as zero

$$z_0 = \frac{2c'}{\gamma\sqrt{K_a}} \tag{11.19}$$

SEE WORKED EXAMPLE 11.3.

Values of c' in excess of about 5 to 10 kN/m² should be treated with caution. However, it is considered that the wall should not be assumed to be subjected to *no* pressure, therefore if a water table is not present the minimum equivalent fluid pressure should be adopted, see below.

Minimum equivalent fluid pressure
(Figure 11.12)

When a cohesion intercept c' or c_u is assumed a depth of negative active pressure z_0 is obtained which may result in the soil theoretically supporting itself and applying no active pressure on the wall.

To ensure that there is always some positive total pressure on the wall, CP2 (1951) and Padfield and Mair (1984) recommend the use of a minimum equivalent fluid pressure given by an 'equivalent fluid' acting behind the wall with a density of 5 kN/m³ (or 30 lb/ft³). The equivalent fluid pressure at a depth z m behind the wall must not be less than $5z$. This minimum pressure must be greater than the total active pressure (effective soil pressure and water pressure), otherwise it need not be used.

Effect of water table (Figure 11.13)

The presence of a water table has two effects, as illustrated in Figure 11.13a:

1 Effective vertical stresses are reduced below the water table so horizontal active and passive pressures (which are effective stresses) are reduced.

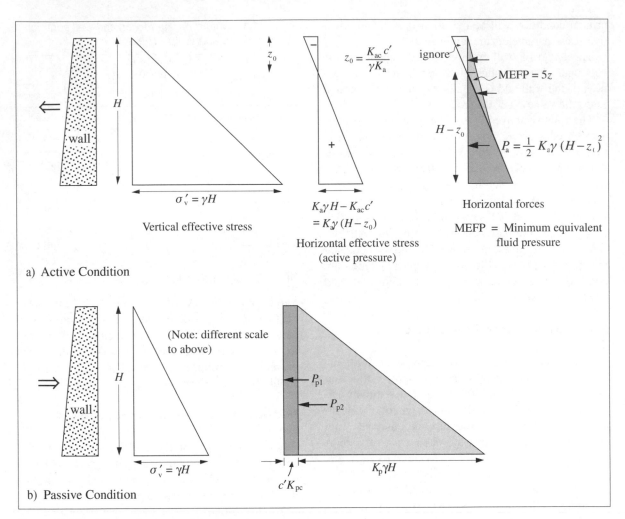

$$z_0 = \frac{K_{ac} c'}{\gamma K_a}$$

$$\sigma'_v = \gamma H$$

Vertical effective stress

$$K_a \gamma H - K_{ac} c'$$
$$= K_a \gamma (H - z_0)$$

Horizontal effective stress
(active pressure)

MEFP = 5z

$$P_a = \frac{1}{2} K_a \gamma (H - z_0)^2$$

Horizontal forces

MEFP = Minimum equivalent
fluid pressure

a) Active Condition

(Note: different scale
to above)

$$\sigma'_v = \gamma H$$

P_{p1}

P_{p2}

$c' K_{pc}$

$K_p \gamma H$

b) Passive Condition

Figure 11.12 *Effect of cohesion intercept c'*

2 The pressure in the pore water below the water table is hydrostatic so a horizontal water thrust P_w must be added to the horizontal soil thrust to give the total thrust.

SEE WORKED EXAMPLE 11.2.

BS 8002:1994 recommends that the water pressure regime should be the most onerous that is considered reasonably possible.

If there is a difference in water level on each side of the wall and seepage occurs beneath and around the wall the pore pressure should be determined from a flow net, as described in Chapter 3. However,

a reasonable simplifying approach (Burland *et al.*, 1981) is to assume that the hydraulic head varies linearly down the back and up the front of the wall as shown in Figure 11.13b.

Taking the water level behind the wall as the datum the elevation head at the toe of the wall is:

$$\text{elevation head} = -(h + d - j) \tag{11.20}$$

The total head varies linearly along the flow path length $(h + d - i - j)$ so that the total head at any depth x below the datum is:

$$\text{Total head} = \frac{-x(h + i - j)}{(h + d + j) + (d - i)} \tag{11.21}$$

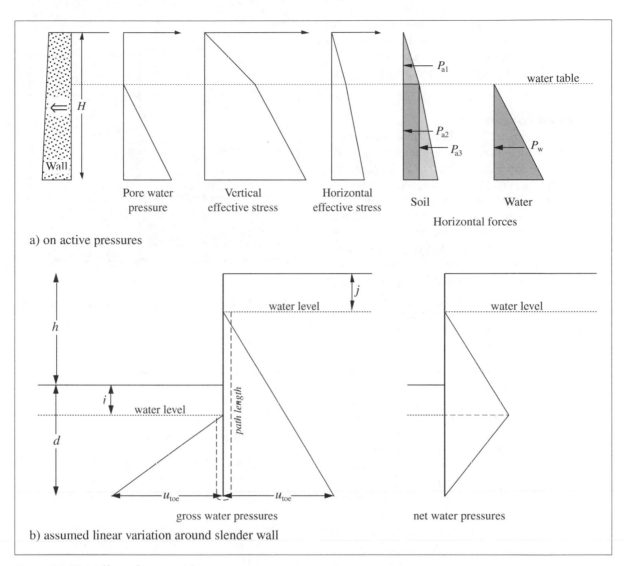

FIGURE 11.13   *Effect of water table*

At the toe of the wall the value of x is:

$$x = h + d - j \qquad (11.22)$$

The pressure head is given by:

Pressure head = total head − elevation head   (11.23)

and the pore water pressure at the toe of the wall is:

$$u_{toe} = \text{pressure head at toe} \times \gamma_w \qquad (11.24)$$

Combining Equations 11.20 to 11.24 the pore pressure at the toe of the wall is:

$$u_{toe} = \frac{2\gamma_w (h + d - j)(d - i)}{(h + d + j) + (d - i)} \qquad (11.25)$$

The simplified diagram for pore pressure shown in Figure 11.13b can then be used with this value to obtain the forces from the water. The net water pressure diagram simplifies the design.

Given the significant effects of water pressures, they should be mitigated as much as possible by the provision of drains. With backfilled gravity walls this can be achieved during construction by incorporating granular or geotextile layers. However, for embedded walls the installation of drains is not feasible and dewatering may be necessary to reduce the water forces, at least during the temporary construction period.

Undrained conditions

When a low permeability clay exists behind or in front of a wall the shear stresses induced by movement of the wall will cause changes in the pore pressures within the clay. If the permeability is very low these pore pressures will dissipate only slowly so the clay will behave in an undrained manner and total stress theory can be applied for design of the wall. BS 8002:1994 suggests that this could apply when the mass permeability is less than 10^{-8} m/s.

This condition may apply for temporary works design where the soil requires support for a short period of time and when the consequences of failure are not severe. In the event, however, the time required for dissipation may be very short due to:

- The presence of fabric within the clay making its mass permeability much greater than its intrinsic permeability. If fissures, joints, bedding, silt or sand partings, silt-filled fissures or a higher porosity due to weathering exist then the pore pressures can dissipate rapidly and the 'long-term' condition will soon be obtained when the effective stress approach must be used.
- Expansion of the soil in the active state behind a wall is likely to open up any fabric present, accelerating the softening process and providing the 'long-term' condition very quickly whereas compression of the soil on the passive side may slow down this process and provide undrained conditions during the period of loading.
- The development of vertical tension cracks which may fill with water, see below.

Earth pressures – undrained condition
(Figure 11.14)

The undrained condition will occur in the short term for a homogeneous intact clay so this condition is only appropriate for temporary works. It is generally considered that the long-term condition will soon apply so it is safest to assume this latter condition.

Nevertheless, if the appropriate soil parameters, c_u

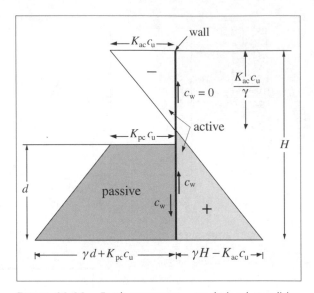

FIGURE 11.14 *Earth pressures – undrained condition* $(\phi = 0°)$

(> 0) and $\phi_u = 0$, are inserted in equations 11.4 and 11.6 then:

$$K_a = K_p = 1 \tag{11.26}$$

and in equations 11.16 and 11.17:

$$K_{ac} = K_{pc} = 2\sqrt{\left(1 + \frac{c_w}{c_u}\right)} \tag{11.27}$$

where c_w is the undrained or total stress wall adhesion value.

The vertical stress becomes the total stress, $\sigma_V = \gamma z$ so from equation 11.11 the active normal pressure is:

$$p_{an} = \sigma_V - K_{ac}c_u \tag{11.28}$$

and from equation 11.12 the passive normal pressure is:

$$p_{pn} = \sigma_V + K_{pc}c_u \tag{11.29}$$

The variation with depth of these pressures is shown in Figure 11.14. On the passive side wall adhesion may be assumed to act but where the pressures are negative on the active side wall adhesion cannot be assumed since the soil in this region is effectively supporting itself in tension as the wall deflects outwards.

Tension cracks (Figure 11.15)

On the active side the theoretical pressure is negative down to a depth where $p_{an} = 0$. From equation 11.28 this depth is given as:

$$z_0 = \frac{K_{ac}c_u}{\gamma} \qquad (11.30)$$

where K_{ac} is given by Equation 11.27.

As the soil cannot readily support tension, vertical tension cracks may occur down to this theoretical depth. The actual depth to which tension cracks develop is likely to be affected by the support provided to the wall. They are unlikely to extend below the excavation level and may be limited to the level of a strut or anchor.

It is commonly assumed that the tension crack will fill completely with water so the hydrostatic pressure must be considered, as shown in Figure 11.15.

Loads applied on soil surface

1) Uniform surcharge (Figure 11.16)

BS 8002:1994 recommends that all walls should be designed with a minimum surcharge of 10 kN/m² on the active side and with a minimum depth of unplanned excavation on the passive side in front of the wall. This depth should be:

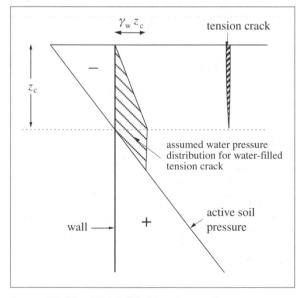

FIGURE 11.15 *Water-filled tension crack*

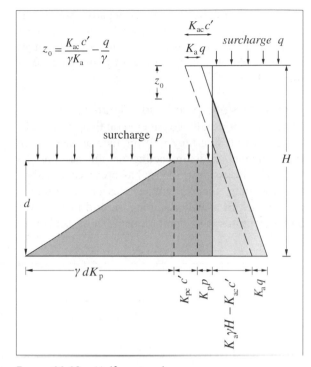

FIGURE 11.16 *Uniform surcharge*

- not less than 0.5 m and
- not less than 10% of the retained height.

EC7 states that the depth of unplanned excavation should be no more than 0.5 m.

Where anchored sheet piling is used for walls in harbours, river banks and canals, erosion or excessive dredging could reduce the depth of soil on the passive side of the wall. The designed depth of embedment is often increased by say 20% to allow for this, as well as the depth allowed for unplanned excavation.

If a surcharge is applied uniformly over the soil surface on the active or passive side the vertical stresses in equations 11.11, 11.12 and 11.28 and 11.29 are increased by the surcharge pressure q so equation 11.11 becomes:

$$p_{an}' = K_a(\gamma z + q) - K_{ac}c' \qquad (11.31)$$

SEE WORKED EXAMPLE 11.5.

and equation 11.28 becomes:

$$p_{an} = (\gamma z + q) - K_{ac}c_u \qquad (11.32)$$

The depth of the theoretical negative pressure is then altered, as shown in Figure 11.16. If wall friction is assumed then the values of the shear forces acting on the wall are increased by the presence of the surcharge.

2) Line loads and point loads (Figure 11.17)

These are not usually considered on the passive side. On the active side they will produce an increase in the horizontal pressure acting on the back of the wall. The Boussinesq theory has provided a method for obtaining the horizontal pressure distribution on the back of a wall assuming the soil to be elastic and incompressible. Unfortunately it is neither of these.

The modifications suggested by Terzaghi (1954) have been adopted in the *NAVFAC Design Manual* (1982) and these are reproduced in Figure 11.17. It is likely that these horizontal pressures are underestimated (Padfield *et al.*, 1984) therefore a conservative approach is suggested.

Earth pressures due to compaction
(Figures 11.18 and 11.19)

When a gravity wall is backfilled it is necessary to compact the soil in layers using fairly heavy compaction plant, as described in Chapter 13. As each layer is compacted close to the wall horizontal stresses are induced which remain locked into the soil when the plant is removed. These can be signifi-

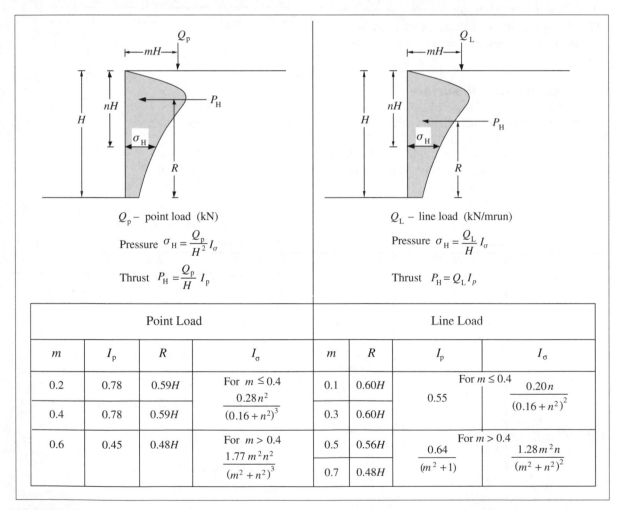

Point Load				Line Load			
m	I_p	R	I_σ	m	R	I_p	I_σ
0.2	0.78	0.59H	For $m \le 0.4$ $\dfrac{0.28n^2}{(0.16+n^2)^3}$	0.1	0.60H	For $m \le 0.4$ 0.55	For $m \le 0.4$ $\dfrac{0.20n}{(0.16+n^2)^2}$
0.4	0.78	0.59H		0.3	0.60H		
0.6	0.45	0.48H	For $m > 0.4$ $\dfrac{1.77\,m^2n^2}{(m^2+n^2)^3}$	0.5	0.56H	For $m > 0.4$ $\dfrac{0.64}{(m^2+1)}$	For $m > 0.4$ $\dfrac{1.28\,m^2n}{(m^2+n^2)^2}$
				0.7	0.48H		

Figure 11.17 *Horizontal pressures and thrusts on rigid walls due to surface loads (From Navfac, 1982)*

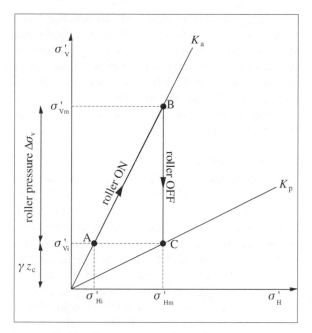

FIGURE 11.18 *Stresses beneath a roller*

cant especially in the higher levels of the backfill where the normal theory would suggest that only the active pressures are acting.

The effects of compaction pressures can be visualised by reference to Figure 4.11 in Chapter 4. Under the compaction plant the vertical total stress

will be increased. As free-draining backfill is normally adopted no pore pressures will develop so the effective stresses will be equal to the total stresses. This could be at point B in Figure 4.11, with a high horizontal stress required to maintain the stability of the compacted fill. When the plant is removed the vertical stress decreases but most of the horizontal stress remains.

Assuming that K_a conditions exist when the compaction roller is applied and K_p conditions occur when the roller is removed, Ingold (1979) showed that there is a critical depth z_c at which the maximum horizontal stress σ_{Hm}' is retained in the soil on unloading. This is illustrated in Figure 11.18. The initial vertical stress is:

$$\sigma_{Vi}' = \gamma \, z_c \tag{11.33}$$

With an increase in vertical stress from the roller of $\Delta\sigma_v$ the maximum vertical effective stress is:

$$\sigma_{Vm}' = \gamma \, z + \Delta\sigma_v \tag{11.34}$$

and from Figure 11.18 it can be seen that:

$$\sigma_{Hm}' = K_a\sigma_{Vm}' = K_p\sigma_{Vi}' \tag{11.35}$$

Assuming that at shallow depths γz in Equation 11.34 is negligible compared to $\Delta\sigma_v$ and that the increase in vertical stress from the roller pressure is given by:

$$\Delta\sigma_v = \frac{2p}{\pi z} \tag{11.36}$$

where p is the line load per unit length then:

$$z_c = K_a \sqrt{\frac{2p}{\pi\gamma}} \tag{11.37}$$

The maximum horizontal stress is:

$$\sigma_{Hm}' = \gamma z_c K_p = \sqrt{\frac{2p\gamma}{\pi}} \tag{11.38}$$

and the depth h_c to which the maximum pressure extends can be obtained:

$$h_c = \frac{1}{K_a} \sqrt{\frac{2p}{\pi\gamma}} \tag{11.39}$$

Using these values the diagram of earth pressure variation including the compaction pressures is

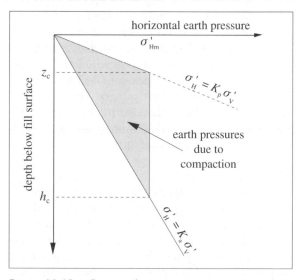

FIGURE 11.19 *Compaction pressures*

plotted, see Figure 11.19, and the forces and over-turning moments may be calculated.

In view of the large overturning moment produced by compaction pressures it would seem advisable to restrict the use of plant close to the wall although settlements could be excessive with poorly com-pacted fill and construction control may be difficult. If the wall can deflect then the movement may be sufficient to reduce these pressures but they could affect the structural design (Simpson *et al.*, 1998).

Retaining structures

Introduction (Figure 11.20)

There is a wide variety of structures used to retain soil and/or water for both temporary works and per-manent works. Some of the more common types of retaining structures for different purposes are illus-trated in Figure 11.20.

Mass concrete or masonry walls rely largely on their weight for stability against overturning and sliding. They are unreinforced so their height must be limited to ensure internal stability of the wall in bending and shear when subjected to the lateral stresses. They are typically no more than about 3 m high. Providing a minimum slope of 1:50 (horizon-tal:vertical) on the front face avoids the illusion of a vertical wall tilting forwards.

Reinforced concrete walls are more economical with the reinforcement enabling the stem and base sections to be designed as cantilevered structural ele-ments. Overall stability is provided by an adequate base width and the weight of backfill resting on the base slab behind the stem.

Basement walls

Unlike the above wall types which are freestanding, basement walls are restrained by embedment in the ground, a base slab, suspended basement floors and possibly external ground anchors. The latter are more commonly adopted for temporary support during construction with permanent propping pro-vided by the subsequent basement slab and floor construction. Alternatively, temporary internal prop-ping in the form of struts may be provided and replaced eventually with the permanent base slab and floors.

Ground movements produced around deep base-ments by the removal of vertical and horizontal stresses must be minimised, particularly if there are existing structures nearby. The top-down method of construction (IStructE, 1975) has been developed to ensure minimal ground movements.

There are two basic approaches to the construc-tion of a basement:

1 Backfilled basements

Construction takes place in an open excavation with either unsupported sloping sides or vertical sides supported by temporary shoring or sheet piling. Sloping sides occupy a large space around the base-ment and may require dewatering to ensure their sta-bility but they can be the most economical method for shallow basements. If space is limited then verti-cal faces could be cut and supported by timbering, steel trench sheeting or H-section steel soldier piles and timber lagging. These methods rely on the ground having some self-supporting ability for a short time so that the supports can be installed.

If the ground has poor self-supporting capabilities then steel sheet piling driven into the ground as ver-tical support before excavation commences will retain and exclude both soil, groundwater and open water. The sheet piling then either acts as a can-tilever or is supported internally by a system of walings and horizontal or raking struts, or externally by a system of stressed ground anchors.

The basement walls and base slab are constructed with conventional in situ reinforced concrete. This should be of good quality and well compacted to provide a dense, impermeable structure and to max-imise resistance to water penetration into the base-ment. To ensure water-tightness this form of construction can be surrounded by an impermeable membrane such as a layer of asphalt tanking or card-board panels filled with bentonite.

2 Embedded walls

Excavations are supported by reinforced concrete diaphragm walls, contiguous bored pile walls or secant bored pile walls. These are constructed around the basement perimeter before excavation commences, occupying minimal space but providing support to the soil and groundwater both in the tem-porary condition during excavation and construction

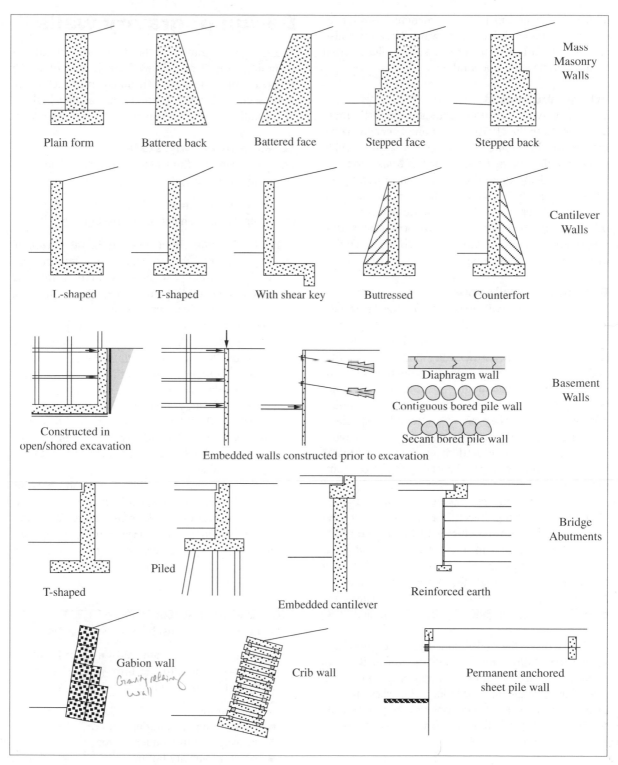

FIGURE 11.20 *Typical retaining structures*

of the basement and in the permanent condition as the final structural basement wall. They may also provide support to vertical loads such as the external columns and walls of a building.

Bridge abutments

The many types of bridge abutment are well illustrated in Hambly (1979), the more common forms are shown in Figure 11.20. These walls provide support to the retained soil and act as foundations for the bridge deck, therefore apart from providing the normal stability conditions they must also be designed to ensure tolerable settlements for the bridge deck. Horizontal outward movement and/or rotation must be minimal to ensure correct operation of the bridge deck bearings.

Gabions and cribwork

Even when faced with masonry or other materials, concrete walls can appear hard and uncompromising. Gabions can blend sympathetically with the environment as they resemble open stone walling and cribwork can be 'softened' by using timber for construction and encouraging plant growth. They are both highly permeable so additional drainage should not be required. They are very flexible, especially gabions, and only nominal foundations are usually required. Since large settlements can be tolerated without apparent distress they are suitable on compressible soils.

They both rely for their strength on the interaction from the tensile properties of the gabion wire or steel mesh cages and the stretcher and header bond of the cribwork combined with the compressive and shear strength properties of the contained stone. The main disadvantages are that the wire mesh cages of gabions are prone to corrosion and abrasion although their life can be extended by galvanising and PVC coating.

They require the soil retained to have some self-supporting abilities during construction so they are commonly used to provide additional support to steep cuttings and natural slopes and where the toe of a slope is cut back. Gabions are frequently used for river bank protection where they provide useful erosion protection.

Design of gravity walls

The design of gravity walls should be carried out according to EC7 and BS8002:1994, both of which adopt a limit state approach and partial factors. In some sections below, the traditional methods adopting overall or lumped factors of safety are also given.

Serviceability limit states

The serviceability limit states to consider are that there should not be:

a) substantial deformation of the structure or
b) substantial movement of the ground.

The serviceability limit state will be the governing criterion for equilibrium since the deformations required to fully mobilise the shear strength are large compared to the deformations acceptable in practice. The fully active pressure is a minimum value obtained at the maximum strength so the active earth pressure under working conditions will be more severe than under the ultimate limit state.

Without complex analyses and simplifying assumptions the displacements of a gravity wall cannot be confidently predicted. However, it is considered (BS 8002:1994) that the serviceability limit state can be assured by limiting the design strength to a value that is mobilised and not the fully active or fully passive values.

For the structural design the forces and bending moments derived from the earth pressures decrease as the structure deforms. The most severe earth pressures determined for the serviceability limit state should be used as these occur during working conditions. Partial factors to increase the loads derived need not be applied.

Ultimate limit states (Figure 11.21)

Retaining structures must be designed to prevent:

a) Collapse or serious damage by instability of the earth mass. The ultimate limit state of a gravity wall must be checked for:

- slip failure
- overturning or rotational failure
- bearing pressure under the toe
- bearing capacity failure
- translational failure or sliding.

It is carried out by ensuring equilibrium conditions using the design actions and design strength of the soils. Design values of strength are given in the annex to this chapter.

b) Failure of structural members in bending and shear, referred to as internal stability.

c) Excessive deformation of the wall or ground that could cause adjacent structures or services to reach their ultimate limit state.

Only the collapse mechanisms of the earth mass are considered in this chapter.

Slip failure

The factor of safety against overall failure along a deep-seated slip surface extending beneath the wall can be obtained using the methods of analysis given in Chapter 12, Slope Stability. If the wall is associated with loading applied to the ground, such as a wall at the toe of an embankment, then the short-term conditions (for clays, the undrained case) will be the more critical. If the wall is constructed within an excavation then the long term drained condition will be the more critical case.

Adequate drainage measures (permeable blankets, pipes etc.) behind the wall and within the backfill can provide a lower equilibrium phreatic surface. However, the long-term effectiveness of this drainage must not be in doubt.

Considerations of the value of the overall factor of safety to adopt are given in Chapter 12. However, the consequences of failure of a retaining wall are likely to be much more serious than a slope. In

Hambly (1979) overall factors of safety are given on the basis of confidence in the accuracy of soil strength values, i.e.

$F \geq 1.25$ – for soil strengths based on back-analysis of failure of the same type of soil.

$F \geq 1.5$ – for soil strengths based on laboratory or *in situ* tests.

An approach to limit state design in slope stability analysis is described in Chapter 12.

Overturning or rotational failure

For limit state design the approach would be to ensure that equilibrium is achieved with:

$$\Sigma \text{overturning moments} \leq \Sigma \text{ resisting moments}$$
(11.40)

Design soil parameters are used to determine the design earth pressures and with factored values of any loads applied.

The overall factor of safety against overturning about the toe can be obtained from:

$$F_{\text{overall}} = \frac{\Sigma \text{ resisting moments}}{\Sigma \text{ overturning moments}} \qquad (11.41)$$

It is recommended (CP2:1951) that a minimum overall factor of safety of 2 be obtained.

Passive resistance in front of the wall is usually ignored because considerable rotation is required before it is fully mobilised and this mode of movement may not achieve the maximum value expected, see Figure 11.5. If a wall is supported at a higher level by a prop, tie or anchor then the reaction force provided at this level may be added to the restraining moments.

SEE WORKED EXAMPLE 11.7.

Bearing pressure under the toe
(Figure 11.22)

If it is assumed that soil can sustain a linear stress distribution and that it remains elastic, without plastic yielding, a trapezoidal distribution of pressure can be analysed as in Figure 11.22 to give the maximum and minimum pressures applied.

BS 8002:1994 requires that the maximum pressure beneath the toe of the wall does not exceed the allowable bearing pressure of the soil. This is not an

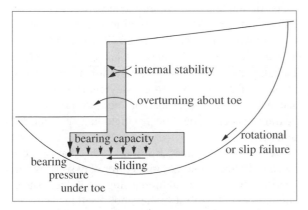

FIGURE 11.21 *Stability of gravity walls*

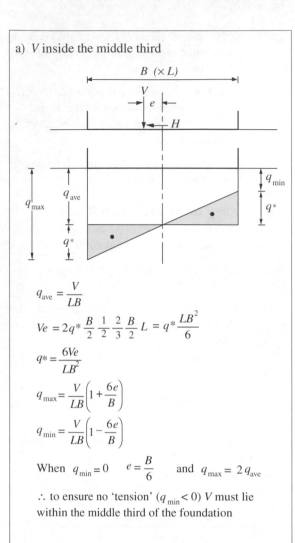

a) *V* inside the middle third

$$q_{ave} = \frac{V}{LB}$$

$$Ve = 2q^* \frac{B}{2} \frac{1}{2} \frac{2}{3} \frac{B}{2} L = q^* \frac{LB^2}{6}$$

$$q^* = \frac{6Ve}{LB^2}$$

$$q_{max} = \frac{V}{LB}\left(1 + \frac{6e}{B}\right)$$

$$q_{min} = \frac{V}{LB}\left(1 - \frac{6e}{B}\right)$$

When $q_{min} = 0$ $\quad e = \frac{B}{6}$ $\quad$ and $\quad q_{max} = 2q_{ave}$

$\therefore$ to ensure no 'tension' ($q_{min} < 0$) *V* must lie within the middle third of the foundation

b) *V* outside the middle third

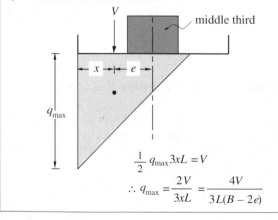

$$\frac{1}{2} q_{max} 3xL = V$$

$$\therefore q_{max} = \frac{2V}{3xL} = \frac{4V}{3L(B - 2e)}$$

FIGURE 11.22 *Middle third rule*

entirely satisfactory criterion since the maximum pressure occurs over a very small area just beneath the toe and is likely to stress the soil sufficiently to develop plastic yielding, a state beyond the criterion for the allowable bearing pressure.

The allowable bearing pressure of a soil is limited by the amount of settlement that can be tolerated. This settlement lies within the elastic range of the soil and depends on the pressure applied across the *full* width of a foundation. A direct comparison of the two values is unrealistic and could lead to onerous conditions, particularly if only small settlements can be tolerated.

This criterion may be satisfied by:

1) Designing the wall with an overall factor of safety against overturning, of say 2 or more.
2) Designing the wall so that the resultant vertical thrust V lies within the middle third of the base of the wall. In this case, q_{max} will be no more than twice q_{ave}. Hambly (1979) summarises the recommendations of Huntington (1957) that overturning stability should be controlled by keeping the vertical thrust:

   a) within the middle third for walls on firm soils
   b) within the middle half for walls on rock
   c) at or behind the centre of the base for walls on very compressible soils to avoid forward tilting.

SEE WORKED EXAMPLE 11.7.

Bearing capacity

The above methods only consider the effects of eccentric loading and ignore the significant effect of inclined loading. The horizontal load combined with the vertical load produces an inclined resultant applied to the soil. From Chapter 8 (Shallow Foundations – Stability) a more rational approach is to adopt the effective area (Meyerhof) method to account for eccentric loading and to modify the bearing capacity equation by including the inclination factors.

Experience indicates (BS 8002:1994) that the serviceability limit state is satisfied for clay soils by the bearing capacity calculation using undrained shear strength values with a higher mobilisation factor in the range of 2.0 to 3.0.

Sliding

This topic is also covered for foundations in Chapter 8. Excessive horizontal movement of a gravity wall could occur if there is an insufficient safety margin against sliding. A general discussion on failure by sliding is given in Chapter 8, considering both overall factors of safety and limit state design with partial factors.

Values of wall friction and adhesion are given in the annex at the end of this chapter. Base friction should be treated separately from wall friction as the former is dependent on the natural foundation strata and wall friction is determined by the nature of the backfill.

The ϕ' value is the triaxial angle of shearing resistance. The ϕ' value appropriate beneath a wall would be the plane strain value which is somewhat higher than the triaxial value (see Chapter 7). However, using a lower value will compensate for the mobilisation of shear strengths beyond the peak values as the wall moves forwards. Disturbance of the soil formation level is also likely to reduce the ϕ' values.

If there is a likelihood that the overturning produces small minimum bearing pressures on the underside of the wall near the heel then it is suggested that the effective width B' obtained from the effective area method (Figure 8.16) be used in the equations, instead of B.

SEE WORKED EXAMPLE 11.7.

Internal stability

This is concerned with the structural integrity of the wall elements. Brickwork or masonry walls should be proportioned so that they are not in tension at any point, otherwise, buckling or bursting failures could occur. Mass concrete walls should be proportioned so that the permissible compressive, tensile and shear stresses are not exceeded. Reinforced concrete walls are designed as cantilevered structural elements.

Embedded walls

These walls may be distinguished from gravity walls in that they are constructed *in situ* prior to excavation so they support *in situ* soils whereas gravity walls are constructed first and then support backfill. Embedded walls are slender structures which means:

1 Their own self-weight is ignored and they do not normally interact vertically with any soil beneath, compared with cantilever gravity walls.
2 They do not require a check for sliding or bearing capacity failure, overturning is the main overall stability consideration.
3 They rely on mobilisation of passive resistance in front of the wall for support below excavation level.
4 They must be expected to deflect, at least below excavation level.
5 They are commonly propped or anchored over the excavation depth so the pressures which may develop behind a wall will depend on the flexibility of the wall, the amount of support provided and the stage at which it is applied. Actual pressure distributions are, therefore, complex and to some extent dependent on the method of construction.

Embedded walls should be designed to prevent:

- overall deep-seated rotational or slip failures
- lack of vertical equilibrium
- structural failure due to the maximum bending moment or shear force
- excessive deformation
- overturning instability/moment equilibrium
- translational failure.

Only the latter two are considered in this chapter, in order to determine the depth of penetration required below excavation level and the force in a prop or anchor.

Cantilever embedded walls – general

Steel sheet piling driven into the ground for temporary works is commonly used to support the vertical sides of excavations during construction. To avoid internal propping or external anchoring it is preferable if the wall can be designed to act in the cantilever mode. Following completion of the below ground structure and backfilling the sheet piles are usually removed. This type of wall should be limited to a maximum height of 3–5 m depending on the soil types and the presence of water. Deflections and outward movement at the top of the wall may be significant.

Contiguous or secant bored pile walls and diaphragm walls are also frequently used in cantilever mode for permanent applications, such as for retaining structures alongside urban highways, bridge abutments and for basement walls. Due to the minimal vibrations produced during boring these methods can be adopted for walls close to existing structures. Diaphragm walls would be more suitable where a high water table must be retained and where a greater bending resistance or section modulus is required when a substantial reinforcement cage can be incorporated.

Cantilever embedded walls – design
(Figure 11.23)

For construction in sands an effective stress design approach with full pore pressure conditions must be used.

For piling used to support temporary works in clays, the effective stress condition should normally be assumed since the equilibration of pore pressures can be rapidly achieved. This is because of the expansion of the soil on the active side of the wall (tension cracks in the extreme), swelling of the soil on the passive side due to unloading and the presence of macro-fabric, such as fissures and laminations which are present in most clays. Notwithstanding these considerations a greater risk is usually taken with temporary works so lower factors of safety are adopted. For permanent works the long-term effective stress condition is assumed.

The stability of a cantilever wall is derived from the passive resistance obtained in the embedded

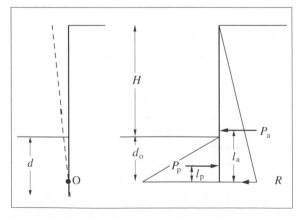

FIGURE 11.23 *Cantilever embedded wall*

portion below excavation level (see Figure 11.6). If the wall rotates about the point O in Figure 11.23 then passive resistance is mobilised in the soil above O on the excavation side and below O on the retained side. Because of the restraint below O this is referred to as the fixed earth condition. Without this a cantilever wall would not be stable. To simplify the design the passive resistance below O is assumed to be a force R acting at O and moments about O are taken for the active and passive thrusts P_a and P_p.

Single anchor or propped embedded walls – general

A single anchor or prop near the top of the wall prevents outward deflection at this location and modifies the pressures mobilised behind the wall. The flexibility of the wall also allows it to deflect, further modifying the pressures due to an arching action within the retained soil. The pressure distributions behind such a wall are likely to be complex. The types of anchorages are described below.

Single anchor or propped embedded walls – design (Figure 11.24)

For design purposes a simplified distribution of earth pressures is assumed, see Figure 11.24. The overall stability of the wall is considered by taking moments about the prop level Q of the active and passive thrusts P_a and P_p assuming a free-earth support condition. In this condition the depth of embedment is not sufficient to prevent rotation at the toe of the wall, as in the fixed-earth condition. However, rotation of the toe is necessary for the consideration of overall stability in terms of moments about the prop level.

Caution should be exercised when assuming values of wall friction δ and wall adhesion c_w because of the uncertainties of relative vertical movements between the wall and the soil.

Design methods

The guidance given in EC7, BS8002 and CIRIA 104 is not entirely compatible. There are a number of alternative ways of modifying the active and passive thrusts and moments to ensure stability (CIRIA 104; Padfield and Mair, 1984). BS 8002:1994 offers five traditional design methods as informative references. The designer must be aware of the limitations of

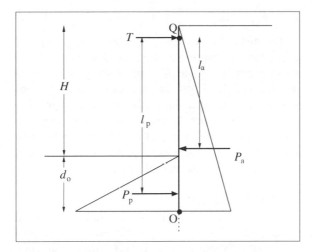

FIGURE 11.24 *Anchored or propped embedded wall*

each method and the sensitivity to changes in parameter values. Further, each method has its own set of overall factors of safety to apply. Personal preference plays a large part in the choice of which method to use (CIRIA 104; Williams *et al.*, 1993). Four of the methods are described below.

Gross pressure method (Figure 11.23)

This method was first adopted in CP2:1951. The depth of embedment d is obtained by equating moments about the point O of the full value of the active thrust P_a but balanced by a reduced value of P_p. Moments are taken about the point O near the toe of the wall for the cantilever mode and about the prop level Q for the propped wall. From Figure 11.23:

$$P_a = \frac{1}{2} K_a \gamma (H + d)^2 \tag{11.42}$$

$$P_p = \frac{1}{2} K_p \gamma d^2 \tag{11.43}$$

$$M_p = P_p l_p \tag{11.44}$$

$$M_a = P_a l_a \tag{11.45}$$

$$F_p = \frac{M_p}{M_a} \tag{11.46}$$

A cubic equation in d is obtained which is solved by substituting trial values of d. This value of d is

then increased by 20% to give the full depth of embedment and to ensure that the passive resistance below O represented by the reaction R will be obtained. This increase is not an additional factor of safety.

SEE WORKED EXAMPLE 11.8.

Recommended values of the overall factor of safety F_p are given in CIRIA 104. For an effective stress analysis, F_p can vary from 1.0 to 2.0. It depends on the value of ϕ', whether the works are temporary or permanent and whether moderately conservative or worst credible parameters are used. Water pressures and their resultant thrusts are not factored.

There is an anomaly with this method for undrained conditions when $\phi_u = 0°$ and $K_a = K_p = 1$ since the factor of safety decreases with increasing depth of penetration.

Net available passive resistance method (Figure 11.25)

This method was developed by Burland, Potts and Walsh (1981). It assumes a modified distribution of pressures obtained by subtracting the hatched area in Figure 11.25 from both the active and passive pressures and applies moment equilibrium to the remaining forces. The net water pressure diagram is obtained by subtracting the water pressures on the passive side of the wall from the active side.

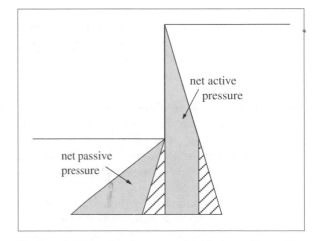

FIGURE 11.25 *Cantilever embedded wall – Net available passive resistance method*

Moments are taken about the point O near the toe of the wall for the cantilever mode and about the prop level Q for the propped wall. The factor of safety F_r is determined from:

$$F_r = \frac{\text{moments from net available passive resistance}}{\underset{\text{active forces}}{\text{moments from net}} + \underset{\text{forces}}{\text{water}} + \underset{\text{forces}}{\text{surcharge}}}$$

$$(11.47)$$

Values of F_r are given in CIRIA 104. For an effective stress analysis F_r can vary between 1.0 and 2.0 depending on whether temporary or permanent works are involved and whether moderately conservative or worst credible parameters are used.

SEE WORKED EXAMPLE 11.8.

Factor on strength method

In this method the values of active and passive pressures are determined from soil strength parameters reduced by a single factor of safety F_s to give mobilised values:

$$c_m' = \frac{c'}{F_{sc}} \tag{11.48}$$

$$\tan\phi_m' = \frac{\tan\phi'}{F_{s\phi}} \tag{11.49}$$

$$c_{um} = \frac{c_u}{F_{ss}} \tag{11.50}$$

Moment equilibrium is then carried out in the conventional way by equating the mobilised moments on the active and passive sides. The pressure diagrams from the methods of analysis above can be used since the limiting equilibrium condition (with an overall factor of safety of 1) will give the same result.

Recommended values of the factor of safety F_s for stiff clays are given in CIRIA 104. Values lie between 1.0 and 1.5, depending on the ϕ' value, whether temporary or permanent works are involved and whether moderately conservative or worst credible parameters are used.

This method is similar to a limit state or partial factor approach although both strength parameters are reduced by the same factor. For an effective stress analysis of permanent works with ϕ' greater than 30° (CIRIA 104) the factors are:

$$F_{sc} = F_{s\phi} = 1.2 \tag{11.51}$$

The method should only be used to determine the depth of penetration. With reduced strength values the active forces are increased and the passive forces are reduced. This distorts the calculations for the structural bending moment in the wall stem. The depth of penetration determined using the method is also sensitive to the value of F_s. The design bending moment can be calculated using unfactored soil parameters at limiting equilibrium.

BS 8002 method

The factor on strength approach is similar to a partial factor method as adopted in EC7 and recommended by BS 8002. The procedure is to determine the depth of penetration that ensures overall equilibrium of the wall using design earth pressures derived from design soil strengths. These are described in the annex to this chapter.

SEE WORKED EXAMPLE 11.8.

Anchorages for embedded walls
(Figure 11.26)

Anchorages are essential in water-front structures. Props or struts inside an excavation provide severe restrictions to the safe and efficient construction operations while anchorages permit an unrestricted excavation. Anchorages, however, affect and occupy the ground around the excavation and behind the water-front structure so adjacent buildings, services and other works and the rights of adjoining owners must be considered.

The most common form of anchorages for sheet piling are shown in Figure 11.26. Deadman anchors rely on the mobilisation of passive resistance in front of them so adequate compaction and prevention of disturbance to the backfill in front of the anchorages is essential. They must be placed beyond the lines AB and BC in Figure 11.26 so that the passive restraint mobilised in front of them as they compress the soil is not affected by expansion of the soil in the active wedge DEF behind the wall.

If smaller individual anchor blocks are used for each tie rod then increased passive restraint can be expected due to the three-dimensional shear zone in front of the anchor block and shearing resistance on the sides of the block. Deflections at the top of the

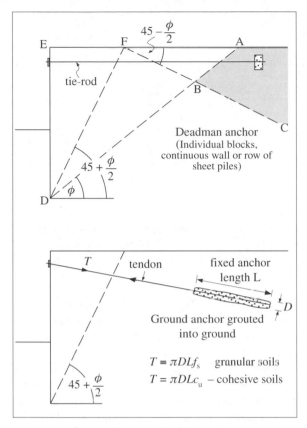

FIGURE 11.26 *Anchorages for sheet piling*

wall are to be expected before sufficient passive restraint in front of the anchor block can be mobilised.

Ground anchors consisting of corrosion-protected tendons of tie-bars or wire strands are inserted in boreholes drilled from the front face of the wall and bonded into the ground by various grouting techniques depending on the soil type to form a fixed anchor length. The design, construction and testing of ground anchors is described in BS 8081:1989. Their main advantage could be in restricting deflection of the wall and hence minimising both horizontal and vertical ground movements around the excavation. This is achieved by excavating a small depth to the level of the anchor position, installing the anchor, stressing the tendon and locking this force against the wall before continuing further excavation.

The anchor or prop force T is given by a consider-

ation of horizontal equilibrium. The force T is determined per metre of wall so the force in the anchor or prop will be given by T multiplied by the spacing between the anchors or props.

Strutted excavations and cofferdams

Introduction (Figure 11.27)

For excavations up to about 6 m deep the earth pressures acting on the supports can be affected by many factors such as shrinkage and swelling of both the soil and the supports, temperature changes, the procedures adopted, materials used and quality of workmanship employed. The design of the supports, therefore, cannot be based on any reliable theory and they are empirically chosen based on experience. Guidance on the design and construction of strutted excavations or cofferdams is given in BS8002:1994; Irvine and Smith (1983) and Williams and Waite (1993).

A strutted excavation is constructed by first driving two rows of steel sheet piling to the full depth required. A small amount of excavation is then carried out and the first frame of walings and struts are fixed, at level 1 in Figure 11.27. As excavation proceeds the level 1 strut restricts inward yielding at this level. Struts are progressively fixed as the excavation deepens so the mode of deformation is similar to a wall rotating about its top. For comparison, the likely pressure variation for this condition for a rigid wall is also shown.

Strut loads (Figure 11.28)

The design of strutted walls is based on a semi-empirical procedure proposed by Terzaghi and Peck (1967). They determined 'apparent pressure diagrams' which were back-analysed from the strut load measurements taken from various sites. A typical apparent pressure diagram is shown in Figure 11.27. They suggested that trapezoidal pressure envelopes could be used to determine the strut loads. These envelopes embraced all of the distributions obtained from the field measurements to ensure that the maximum likely loads in the struts are catered for and that progressive failure should not occur due to one strut failing and shedding excess load onto other struts.

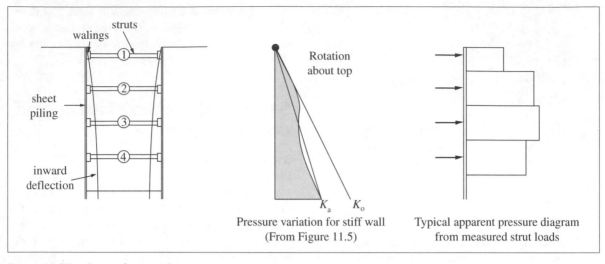

FIGURE 11.27 *Strutted excavations*

For deep excavations in sands, up to about 12 m deep the pressure envelope given in Figure 11.28, case a) will give the maximum strut loads.

For deep excavations in clays, Terzaghi and Peck showed that considerable variations in strut loads can be obtained, up to ± 60% from the average load so the methods proposed should be used with caution. The behaviour of a strutted excavation in clay was found to be dependent on a stability number N:

$$N = \frac{\gamma H}{c_u} \qquad (11.52)$$

which is related to the stability of the clay beneath and around the excavation.

When N is less than about 4 the soil around the excavation is still mostly in a state of elastic equilibrium and the pressure envelope in Figure 11.28, case b) can be used. When N exceeds about 6 movements of the sheet piling and ground movements can become significant because plastic zones are beginning to form near the base of the excavation and as N increases these plastic zones and the associated movements increase. In this case higher pressures on the sheeting will occur and the pressure envelope given in Figure 11.28, case c) should be used.

Terzaghi and Peck found that the reduction coefficient m appears to be 1.0 for most clays providing a lowest average shear strength from the site investiga-

tion results is used. However, they showed that for a truly normally consolidated clay or a soft sensitive clay the value of m can be as low as 0.4.

When N exceeds about 7 for a long excavation or about 8 for a circular or square excavation then complete shear failure, base heaving and extensive collapse of the excavation will be imminent.

When there is a surcharge and a water pressure acting on the back of the sheet piling the total design pressure envelope should include the horizontal pressures due to the surcharge and water as well as the Terzaghi and Peck envelope (Williams *et al.*, 1993).

Reinforced soil

Earth structures on their own are quite weak in tension, relying instead on their compression and shear strength properties for their stability. Inserting tensile reinforcement into soil in the direction of the maximum tensile strains which are usually in the horizontal direction will enable vertical faced masses of soil to remain stable.

The inclusion of reinforcements to improve the stability of soil structures has been practised in simple forms for centuries, e.g. by incorporating fibrous plant and wood materials. More recently

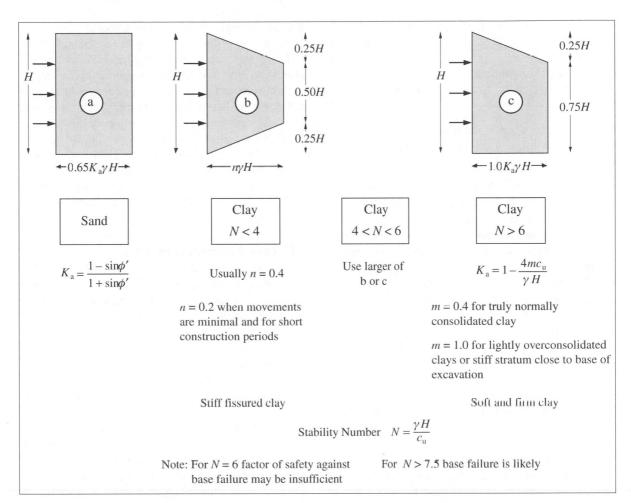

FIGURE 11.28 *Apparent pressure diagrams for strutted excavations (from Terzaghi and Peck, 1967)*

metallic or plastic strips, bars, sheets and grids have been used to good effect.

Reinforced soil provides a relatively cheap form of construction for retaining walls, bridge abutments, marine structures, reinforced slopes and embankments because of the speed and simplicity of construction and they can provide an aesthetic appearance. The space required for its construction is minimal so it is useful where land-take is a problem. It is a flexible form of construction in that it can follow curved lines and it can tolerate some settlements so it can be placed on poorer ground. This chapter describes the use of reinforced soil in vertical walls.

Reinforced soil walls – construction

A reinforced soil wall is constructed using layers of compacted frictional fill material with the reinforcement placed horizontally at suitable vertical intervals and tied to interlocking precast reinforced concrete facing units with a joint filler between the units. The fill material must be frictional to provide an adequate bond or pull-out resistance between the reinforcement surface and the soil. As well as sufficient strength and bond the reinforcement must have sufficient tensile stiffness. If large extensions were required in the reinforcement before sufficient tensile force could be mobilised then the allowable deformations of the soil structure could be exceeded.

The integrity of the structure is dependent on the

long-term durability of the reinforcement so the fill must not have an aggressive nature. Fills with high resistivity, high redox potential, low water content and neutral pH are preferable.

The reinforcement types used in vertical walls include:

● metallic strips, grids and meshes
● polymeric strips, sheets, grids and meshes
● anchors.

Soil nailing is also a reinforcing technique but is applied to the reinforcement of *in situ* soils, e.g. for stabilising steepened cutting slopes.

The facing units are only intended to provide local support for the backfill to prevent spillage or erosion of the front face. In the original Vidal method a half-round aluminium, steel or galvanised steel skin section was tied to the reinforcements. The most commonly used facing units consist of discrete panels of precast reinforced concrete with overlapping and interlocking joints filled with inert, compressible filler. Other types include full height panels usually of precast concrete, or where large settlements are expected flexible wrap-around facings of metal sheeting or polymeric material. If open grid types are used they can sustain the development of vegetation which will enhance the stability of the structure and improve its appearance.

It is important that water does not enter the structure, to avoid the build-up of pore pressures which will aggravate the stability condition and increase the deformations of the structure and to prevent the deterioration of the reinforcement and facings, especially if the water contains aggressive chemicals. Water can enter the structure from the top surface so an adequate surface drainage system should be provided. Groundwater in the retained ground may also enter the structure especially on sloping hillsides. As the frictional fill used is granular it will also be free-draining and should provide sufficient permeability for discharging groundwater. If the fill is not of sufficient permeability a positive system of pipes, coarse drainage blankets, filters and weep-holes may be required.

Effects of reinforcement (Figure 11.29)

The reinforcement acts to improve stability within the soil by reducing the forces causing failure and increasing the overall shear force resisting failure, as illustrated in Figure 11.29. The effect of horizontal reinforcement is to reduce the magnitude of the horizontal deformations by providing greater horizontal confinement, or stiffness. This, in turn, will reduce the vertical deformations or settlements.

The majority of reinforcing elements are flexible with little bending stiffness and are manufactured to withstand only tensile forces. The magnitude of the tension developed in the reinforcement as the fill deforms depends on the tensile stiffness of the reinforcement. For a given soil deformation, with stiff reinforcement a higher tensile force can be sustained but for more extensible reinforcement greater soil

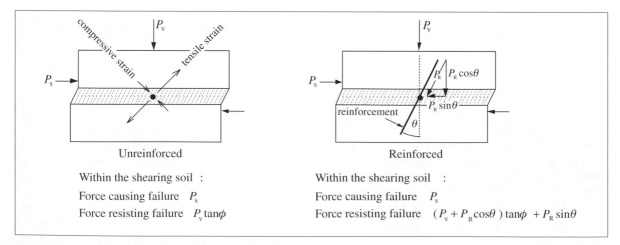

FIGURE 11.29 *Influence of reinforcement on a shear plane (from Jewell and Wroth, 1987)*

deformations are required to develop the design strength of the reinforcement.

BS 8006:1995 classifies reinforcement as inextensible where the design load can be sustained at a total axial strain less than or equal to 1% and as extensible when strains greater than 1% are required to support the design load. Steel and some polymeric materials may be classified as inextensible whereas most polymeric materials would be classified as extensible.

Reinforced soil walls – design

Design of reinforced soil walls in the UK is carried out according to BS 8006:1995, Code of practice for the use of strengthened/reinforced soils and other fills. A limit state approach is adopted based on the principle that the design strength should be greater than or equal to the design load. The ultimate limit state is associated with collapse, structural failure or major damage. Serviceability limit states may occur when:

- the magnitudes of deformations exceed acceptable limits
- other types of damage affect the appearance of the structure, increase its maintenance requirements or reduce the service life of the structure.

As with reinforced concrete retaining walls both the external and internal stability must be considered.

External stability (Figure 11.30)

The external stability of a reinforced soil wall is checked in a similar way to concrete gravity retaining walls. In order to perform these checks the size of the reinforced block is required and BS8006 gives guidance on the choice of initial dimensions for the block. For walls with a horizontal backfill surface the reinforcement length can be taken as $0.7H$ where, in this case, H is the height of the wall from its toe to the top of the facing units. For walls with sloping backfill the mechanical height, H of the structure is defined by a line at an angle of $\tan^{-1}(0.3)$ from the toe of the wall to its intersection with the ground surface, Figure 11.30a.

For horizontal ground in front of the wall the depth of embedment D_m at the toe of the wall should be at least $H/20$ and for ground sloping away the depth should be greater.

It is assumed that the reinforced block of soil acts as a rigid structure for this purpose with an active thrust from Rankine theory acting on the back of the block so that rotational or slip failure, bearing capacity and sliding or translational failure can be assessed in the same manner as for gravity walls. No

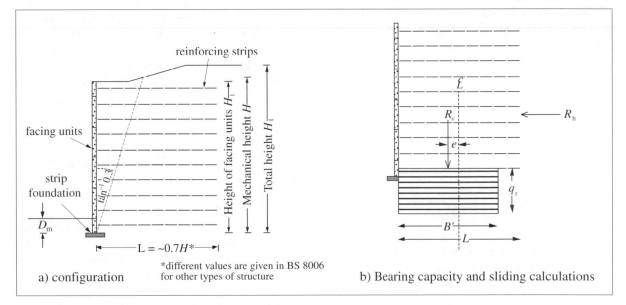

a) configuration

*different values are given in BS 8006 for other types of structure

b) Bearing capacity and sliding calculations

Figure 11.30 *Reinforced soil wall – external stability*

consideration is given to moment equilibrium or overturning or the maximum bearing pressure beneath the toe. The bearing capacity calculation is referred to as bearing and tilt failure.

For the bearing capacity analysis BS8006 recommends the equivalent area or Meyerhof method to determine the gross applied bearing pressure q_r acting over the equivalent width B', Figure 11.30b. This is determined from the eccentricity e, as described in Chapter 8. The mass of the reinforced soil block, any backfill above and the earth pressure behind the block are increased by a load factor of 1.5 to provide the worst combination of loading for this case. The ultimate limit state is confirmed if:

$$q_r \leq \frac{q_{ult}}{f_{ms}} + \gamma D_m \qquad (11.53)$$

where:

q_{ult} = net ultimate bearing capacity of the foundation soil
f_{ms} = partial material factor, = 1.35 in BS 8006
D_m = depth of embedment of wall
γ = bulk unit weight of foundation soil

The net ultimate bearing capacity is determined using characteristic values of the soil strength and other parameters and should include the effects of the inclined loading.

The ultimate limit state for base sliding is checked by factoring the resultant horizontal load R_h and the design strengths. For the long-term stability case where there is soil-soil contact at the base of the structure:

$$f_s R_h \leq R_v \frac{\tan\phi_p'}{f_{ms}} + \frac{c'}{f_{ms}} L \qquad (11.54)$$

and where there is soil-reinforcement contact at the base:

$$f_s R_h \leq R_v \frac{\alpha' \tan\phi_p'}{f_{ms}} + \frac{\alpha_{bc}' c'}{f_{ms}} L \qquad (11.55)$$

where:

R_v = resultant factored vertical load. For the worst combination, the load factor of 1.0 is applied to the mass of the reinforced soil block and any overlying fill.
R_h = resultant factored horizontal load. For the worst combination, the load factor of 1.5 is applied to the earth pressure behind the structure.
ϕ_p' = peak angle of shearing resistance
c' = peak value of cohesion
L = effective base width for sliding, in BS 8006 this is the full width of the reinforced block
f_s = partial load factor for base sliding
= 1.2 where the base comprises soil-soil contact
= 1.3 where the base comprises reinforcement-soil contact
f_{ms} = partial material factor
= 1.0 applied to $\tan\phi_p'$
= 1.6 applied to c'
α' = interaction coefficient for the soil-reinforcement interface on $\tan\phi_p'$
α_{bc}' = adhesion coefficient for the soil-reinforcement interface on c'.

The serviceability limit state is checked by calculating the settlement of the reinforced soil block. The methods given in Chapters 6 and 9 are applied to determine the settlement from the underlying ground due to the imposed loading. Load factors of 1.0 are adopted for the mass of the soil block, backfill above and the earth pressure behind. The compressibility or modulus of the foundation soil is taken as the characteristic value with a partial material factor of 1.0.

Providing the soil backfill is of sufficient permeability and well compacted, internal settlements due to self-weight compressions should be small. If internal settlements are significant, particularly with higher walls, then provision should be made for vertical movement of the reinforcement relative to the facing panels.

Reinforced soil walls can tolerate total and differential settlements without significant distress within themselves. BS8006 suggests a maximum differential settlement of 1 in 100 as a normal safe limit for discrete concrete panels.

If the actual dimensions of the reinforced soil block are not sufficient to provide the limit states then it may be necessary to increase the length of the reinforcement to provide a wider reinforced block and reduce the effect of the active thrust.

Internal stability – general

The integrity of the reinforced soil block must be maintained throughout the life of the structure. This is ensured by:

- tensile capacity in the reinforcement elements
- the shear stress (interaction or adhesion) transferred between the reinforcement and the fill
- the compression of the fill due to the transferred shear stress
- consideration of the economic ramifications of failure
- allowance for the creep strains produced by some polymeric reinforcements in the long term

Failure mechanisms may occur due to:

- tensile rupture of the reinforcement elements
- loss of bond between the reinforcement and the soil, known as pull-out resistance or adherence
- forward sliding on a horizontal plane – this plane could lie within the fill, on the surface of a sheet reinforcement in contact with the fill or at the level of an open geogrid and soil combination
- instability of wedges within the reinforced soil block.

In the analysis for the ultimate limit state it is assumed that:

a) The characteristic value of the peak angle of shearing resistance ϕ_p' is used, divided by a partial material factor

b) The design loads are obtained from the characteristic loads multiplied by the partial load factors. For tensile rupture factors of 1.5 are adopted on the mass of the reinforced block, any backfill above the block and the earth pressures behind the block, to give the maximum tension in the elements. For pull-out resistance load factors of 1.0 are applied to the mass of the block and any backfill above to provide the least shear resistance at the reinforcement-soil interface.

c) For extensible reinforcements the tie-back wedge method is used and for inextensible reinforcements the coherent gravity method is used.

Tensile rupture

The design tensile strength of reinforcement, T_D is obtained from an ultimate value T_U (or T_B) divided

by a partial material factor, f_m. For metallic reinforcements:

$$T_D = \frac{T_U}{f_m} \tag{11.56}$$

where:

T_U = ultimate tensile strength (or base strength) of the reinforcement.

Polymeric reinforcements can exhibit long-term creep behaviour and the ultimate or base strength is determined on the following basis:

- The reinforcement should not fail in tension during the life of the structure. This strength is given by the peak tensile creep rupture strength, T_{CR}.
- The strains in the reinforcement should not be excessive at the end of the structure's design life. This is given by the average tensile strength, T_{CS} based on creep strain considerations.

Both strengths are obtained for the appropriate in-service temperature. For polymeric reinforcements the design tensile strength is the lesser of:

$$T_D = \frac{T_{CR}}{f_m} \text{ or } \frac{T_{CS}}{f_m} \tag{11.57}$$

The values of the partial material factor, f_m are governed by consideration of several factors including:

a) The intrinsic properties of the material, f_{m1}. This is made up of a factor to account for the consistency (or inconsistency) of manufacture and is related to the quality control and variability of the material and the tolerances of its shape and dimensions. Another factor considers the extrapolation of the test data obtained from short duration tests to the long-term in-service capacity

b) Construction and environmental effects, f_{m2}. This is made up of a factor to account for short-term and long-term effects of damage during installation and construction. Another factor considers the detrimental effects of the environment, especially aggressive chemicals or elevated temperatures.

Guidance on appropriate values is given in BS8006. The partial factor f_m is determined from:

$$f_m = f_{m1} \times f_{m2} \qquad (11.58)$$

See Worked Example 11.9.

Pull-out resistance or adherence

This is a measure of the shear force, T_{Aj} that can be developed at each reinforcement/soil interface along the length of reinforcement, L_{ej}, in the resistant zone:

$$T_{Aj} = \frac{L_{ej}P_j}{f_p}\left[\frac{\alpha' \tan \phi_p'}{f_{ms}}(f_{fs}\gamma_1 h_j + f_f w_s) + \frac{\alpha_{bc}' c'}{f_{ms}}\right] \qquad (11.59)$$

where:

P_j = total horizontal width of reinforcement per metre run on top and bottom faces

f_{ms} = partial material factor on $\tan\phi_p'$ and c'

f_p = partial factor for pull-out resistance, = 1.3 for the ultimate limit state

f_{fs} = partial load factor applied to the soil self-weight, = 1.0 or 1.5 according to the load combination applied. 1.0 is probably the most critical value.

γ_1 = soil backfill unit weight

h_j = depth of jth layer below top of structure

f_f = partial load factor applied to surcharge loads; only applied to dead loads

w_s = surcharge due to dead loads only

See Worked Example 11.9.

Internal stability – tie-back wedge method

1) Local stability (Figure 11.31)

This method applies the basic principles for a classical approach to retaining wall design. It is used for walls containing extensible reinforcements such as polymeric types and assumes that active earth pressures are mobilised. The simple case of a vertical wall supporting soil with a horizontal backfill and without surcharges is considered here. For the ultimate limit state the maximum tensile force per metre run, T_j within the jth layer at the depth h_j from the top of the wall is given by:

$$T_j = K_a \sigma_{vj} S_{vj} \qquad (11.60)$$

where:

K_a = coefficient of active earth pressure given by

the Rankine earth pressure theory, see equation 11.4

σ_{vj} = factored vertical effective stress at the level of the jth layer, given by the Meyerhof distribution, Figure 11.31:

$$\sigma_{vj} = \frac{R_{vj}}{L_j - 2e_j} \qquad (11.61)$$

S_{vj} = vertical spacing between the reinforcement at the jth level

R_{vj} and e_j are the vertical total load and eccentricity at the jth level.

Additional tensile forces are developed when there are surcharges or imposed loads placed on the top of the wall. Expressions for these cases are given in BS8006. The tensile force per metre run is checked against the design strength T_{Dj} of the reinforcement and the design pull-out resistance T_{Aj}:

$$T_j \le \frac{T_{Dj}}{f_n} \quad \text{and} \quad T_j \le \frac{T_{Aj}}{f_n} \qquad (11.62)$$

where:

f_n = partial factor for the economic ramification of failure.

The partial factor f_n allows for the long-term risk associated with different classes of structure. BS8006 includes three classes, each related to the damage that could be caused if failure occurred. A partial factor is not assigned to category 1 which includes walls of low height and where failure would result in minimal damage. A partial factor of $f_n = 1.0$

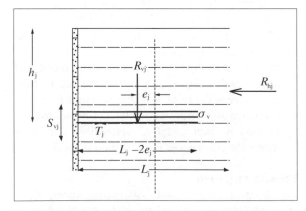

Figure 11.31 *Reinforced soil wall – local stability*

is applied to category 2 structures, where failure would result in moderate damage and loss of services. Category 3 applies to important or high risk structures such as those supporting roads, railways, buildings, river and coastal works, dams and slopes. A partial factor of 1.1 is applied.

SEE WORKED EXAMPLE 11.9.

A force per metre run is a convenient parameter for sheet or mesh reinforcements but for strip reinforcements at a horizontal spacing S_H the tensile force in each strip would be:

$$T_j S_H \qquad (11.63)$$

2) Wedge stability (Figure 11.32)

Behind the vertical face of the wall it is assumed that there is a zone or wedge of soil deforming horizontally which produces strains sufficient to develop the full active condition in the soil. The stability of a wedge of soil at any location within the reinforced soil block is checked by considering a polygon of forces to give the gross tensile force T required to balance the weight of the wedge W and the shear strength F mobilised on the base of the wedge. Sufficient trial wedges are checked to determine the maximum value of T. Any surface loads must be included.

For stability the gross tensile force T must be less than or equal to the sum of the resistances provided by the reinforcement layers in the resistant zone:

$$T \leq \sum_{j=1}^{m} \left(\frac{T_{Dj}}{f_n} \right) \quad \text{or} \quad T \leq \sum_{j=1}^{m} \left(\frac{T_{Aj}}{f_n} \right) \qquad (11.64)$$

The lesser value of the design tensile rupture strength T_{Dj} or the design pull-out resistance T_{Aj} is used for each layer.

The serviceability limit state is checked to ensure that under working conditions the structure will maintain its function. Checks for post-construction movements are made considering both external and internal factors such as:

- foundation settlements
- differential settlements
- internal compression of the fill
- internal creep strain if soil backfill contains a high fines content
- internal creep strain of the reinforcement.

For polymeric reinforcements the latter can be assessed from creep test curves of the reinforcement, as shown in Figure 11.33. The post-construction strain for the jth layer under an average tensile force T_{avj} can be obtained as the difference between the isochrone for the end of the design life and the end of construction. Maximum values of post-construction strain given in BS8006 are 0.5% for bridge abutments and 1.0% for retaining walls.

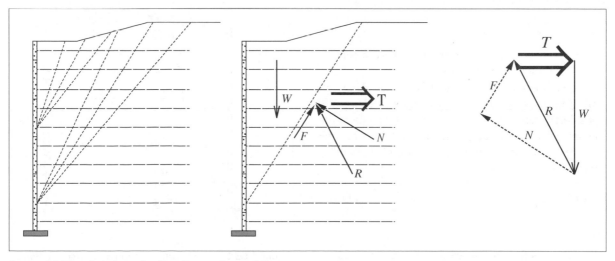

FIGURE 11.32 *Reinforced soil wall – wedge stability*

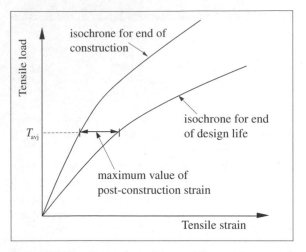

FIGURE 11.33 *Post-construction strain*

Coherent gravity method

It has been observed that with inextensible reinforcements the horizontal earth pressures in the upper part of the wall tend towards the K_0 condition rather than active conditions. In BS8006 this method assumes that the horizontal pressures are given by a coefficient K, varying from K_0 at the top of the wall and decreasing linearly to K_a at a depth of 6 m. Thus:

$$K = K_0 - \frac{z}{z_0}(K_0 - K_a) \quad \text{for } z \le z_0 \ (= 6 \text{ m}) \quad (11.65)$$

$$K = K_a \quad \text{for } z > z_0 \ (= 6\text{m}) \quad (11.66)$$

where:

z = depth measured from the top of the mechanical height H of the wall.

The simple case of a vertical wall supporting soil with a horizontal backfill and without surcharges is considered here. For the ultimate limit state the maximum tensile force per metre run, T_j within the jth layer at the depth h_j from the top of the wall is given by:

$$T_j = K \sigma_{vj} S_{vj} \quad (11.67)$$

where:

K = the coefficient of earth pressure derived as above.

The value of T_j must be less than or equal to the design tensile strength of the reinforcement, according to equation 11.62.

As with the tie-back wedge method the length of the reinforcement within the resistant zone is required and, therefore, it is necessary to determine the boundary of this zone. A line of maximum tension following a log spiral shape is assumed but for calculation purposes BS8006 recommends the use of a 'maximum tension line 2' as shown in Figure 11.34. This applies where there is no superimposed strip load on top of the structure. A further tension line is considered for this load case, and is detailed in BS 8006.

For the pull-out resistance T_{Aj} the shear stress along the length of reinforcement in the resistant zone, L_{ej} is integrated to give:

$$T_{Aj} = \frac{2BL_{ej}}{f_p}\left[\frac{\alpha' \tan \phi_p'}{f_{ms}}f_{fs}\gamma_1 h_j + \frac{\alpha_{bc}' c'}{f_{ms}}\right] \quad (11.68)$$

where:

B = width of the reinforcement and the other parameters are defined above.

The value of T_{Aj} must be greater than or equal to the value of T_j according to equation 11.62. Further analyses are required where surface loads such as vertical and shear strip loads from bridge abutments are applied and where the structure has an unusual

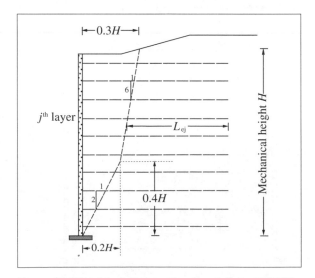

FIGURE 11.34 *Maximum tension line 2 (from BS8006: 1995)*

geometrical form (BS 8006:1995).

The deformation of this type of structure will be determined by the construction and post-construction strains in the metallic elements. These can be calculated from:

$$\varepsilon_j = \frac{T_{avj}L}{EA_j} \qquad (11.69)$$

where, for the reinforcement:

T_{avj} = average tensile load along its length
L = length
E = elastic modulus
A_j = cross-sectional area of the jth layer

SUMMARY

Pore pressures in a soil are hydrostatic. Earth pressures are seated in the soil structure and are dependent on the effective stresses. The magnitude of the earth pressures is determined by the amount and type of movement. The minimum pressures are the active pressures, when the retained soil is allowed to expand and the maximum pressures are the passive pressures when the soil is compressed. In both cases the full shear strength of the soil is mobilised.

The Rankine theory is frequently used to determine earth pressures and applies when the wall friction is zero, i.e. for a smooth wall. The Coulomb wedge theory is used for rough walls when friction is operative and for general cases of inclined walls and inclined backfill. Earth pressure coefficients are used to overcome the limitations of the Coulomb theory.

The effects of a cohesion intercept, the presence of a water table, surcharge loadings and compaction pressures are described.

There is a wide variety of retaining walls available for different circumstances. They tend to group into gravity walls that are constructed and then backfilled behind and embedded walls that are driven, bored or excavated into the ground before any excavation takes place.

The design of a gravity wall must prevent the occurrence of deep-seated slip failure, overturning or rotational failure, excessive bearing pressure beneath the toe, bearing capacity failure and base sliding.

An embedded wall may act as a cantilever, when the fixed-earth condition within its depth of embedment maintains its stability. When an anchor or prop below the top of the wall provides a supporting force the free-earth condition is assumed in the embedment depth. The ultimate limit state is determined for the moment equilibrium of the forces acting and BS 8002 recommends the use of partial factors on loads and strengths.

Due to the many factors outside the province of soil mechanics the loads on struts in a braced excavation are calculated using an empirical method.

The incorporation of reinforcement in an earth structure provides significant economies of soil materials and space occupied and has a wide range of applications. Stability and serviceability of a reinforced soil structure are ensured by considering the external factors such as bearing capacity and settlements and internal factors such as the capacity of the reinforcement to withstand rupture or pull-out.

Annex – limit state design

Design values of shear strength

A careful study of the requirements of BS8002:1994 and EC7:1994 should be carried out before embarking on a serious design of a retaining structure.

Eurocode 7

Characteristic values are obtained for each soil parameter in the manner described in Chapter 8.

The design value is:

$$\text{design value} = \frac{\text{characteristic value}}{\text{partial factor}}$$

The values of the partial factors are given in Table 8.13.

The design values of K_a and K_p are determined from the design values of the soil parameters.

BS8002:1994

Representative values of shear strength

These are a conservative estimate of the mass strength of the soil. A conservative value is defined as one which is more adverse than the most likely value, tending towards the limit of the credible range. They are obtained from an assessment of reliable site investigation data. In the absence of laboratory test data the following values are given:

Cohesive soils

Plasticity index %	ϕ_{cv}' *degrees*
15	30
30	25
50	20
80	15

Cohesionless soils

for siliceous sands and gravels:

ϕ_{peak}' or $\phi_{max}' = 30 + A + B + C$

$\phi_{cv}' = 30 + A + B$

Angularity	*A degrees*	*Grading of sand*	*U*	*B degrees*	*SPT N' value*	*C degrees*
rounded	0	uniform	< 2	0	< 10	0
subangular	2	moderate	2 to 6	2	20	2
angular	4	well graded	> 6	4	40	6
					60	9
estimated visually		based on uniformity coefficient			N' is corrected for effect of overburden pressure	

Design values of shear strength

1) Total stress analysis

$$\text{design } c_u = \frac{\text{representative } c_u}{M}$$

M ≥ 1.5 if wall displacements are not to exceed 0.5% of wall height
M > 1.5 if large strains are required to mobilise the peak strength

2) Effective stress analysis
Design strength is the lesser of:

a)
$$\text{design} \tan \phi' = \frac{\text{representative peak} \tan \phi'_{max}}{M}$$

$$\text{design } c' = \frac{\text{representative } c_{peak}}{M}$$

$M = 1.2$ for both values.
or
b) the representative critical state strength, i.e.
design $\tan \phi'$ = representative $\tan \phi'_{cv}$
$c'_{cv} = 0$
These values should ensure that the wall displacement will not exceed 0.5% of the wall height.

Design values of K_a and K_p
These are determined from the design value of the soil parameters, i.e. design ϕ' and design δ.

Values of wall friction δ and adhesion *a*
Eurocode 7

	δ	$a\ (= c_w)$
smooth	0	0
completely rough	ϕ_{cv}	c'
sand and gravel		
precast concrete/steel sheet pile	$2/3\phi_{cv}$	$a = c_w = 0$
cast *in situ* concrete	ϕ_{cv}	$a = c_w = 0$
clay, undrained	0	0
steel sheet pile		

BS8002:1994
Representative values
Obtain values from shear box tests, large size preferable, with soil in one half and wall material in the other half. If no test results are available use the following:
granular soils
$c' = 0$
representative strength $\delta = \phi_{cv}$, for rough surface texture, coarser than that of the median particle size
representative strength $\delta = 20°$ for smooth surface texture, finer than that of the median particle size
clay, undrained – representative value = remoulded strength c_u
clay, effective stress – representative value = δ

Design values
The design value is the lesser of:
a) the representative strength, see above
or

b) design value = 0.75 × design strength i.e.

effective stress analysis
design $\tan\delta$ = 0.75 × design $\tan\phi$ = 0.625 × representative $\tan\phi_{max}$ (M = 1.2)
c'_w = 0

total stress analysis
design c_w = 0.75 × design c_u = 0.5 × representative c_u (M = 1.5)

Basement Excavation – Tokyo, Japan

Case Objectives:

This case is a good example of the use of the observational method, described by Peck (1969).

Observational Method

Where there is complex soil-structure interaction and uncertainties in predicting performance the observational method can save time and money and give better assurances for the safety of the works providing the design can be modified during construction and appropriate steps can be taken. The first step is to develop a design based on the most probable conditions of the ground model, soil parameters, calculation model and boundary conditions and then make predictions of behaviour.

The parameters derived from the output such as deformation must be capable of measurement on site so that a monitoring system can regularly check performance. A further set of calculations based on the most unfavourable conditions and for every foreseeable eventuality will give the performance values when contingency plans should be triggered. These would involve the need to modify some part of the construction either by adopting different materials, methods or reprogramming the work. With constant monitoring and review and checks on the current performance, if necessary, modification of the design and/or construction methods can be implemented.

The method relies on sufficient knowledge of the ground model and how it will behave, the reliability of monitoring (both the instruments and the analysis of the readings) and greater cooperation between the designer and constructor.

Top-down basement construction

The observational method was applied to the design and construction of a deep basement in Tokyo, Japan by Ikuta *et al.*, 1994. The basement to the multi-storey office block was constructed using diaphragm walls 0.8 m thick and 25.5 m deep to support the side walls. These were supported by the permanent basement floor concrete slabs with an excavation depth of 18.3 m. A raft foundation was placed at the bottom of the basement.

To do this the top-down construction method was adopted in stages as shown in the diagram below.

The alluvium was very soft clay and the diluvium was stiff clay. The Kasuza layer is a mudstone. Sub-artesian groundwater conditions existed in the Tokyo gravel and mudstone. The instruments installed were lateral earth pressure cells on the back and front of the wall, pore water pressure cells, stress transducers and inclinometers.

Due to the limited site investigation information available at the initial design stage more conservative (most unfavourable) design values had to be adopted and two levels of temporary diagonal steel struts were proposed. Additional shear strength tests on the upper part of the diluvium showed it to be stronger and stiffer so most probable values could then be adopted. These values had a significant effect on the lateral pressures and wall displacements. It was also found that the total lateral earth pressure on the retained side was

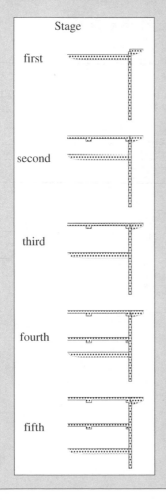

Stage

first

second

third

fourth

fifth

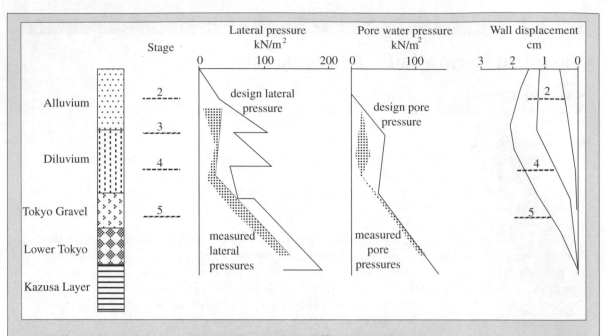

very similar to the pore water pressure so that the active (effective stress) pressure must have been very small. With these beneficial factors and back-analysis of the monitoring results a review of the design meant that it was possible to dispense with the struts. This removed a costly and time-consuming part of the temporary works.

The measured bending moments were found to be comparable to the predicted values and the measured stresses in the diaphragm wall reinforcing bars were smaller than the allowable stress for long-term loading. On the other hand, if the conditions had been found to be worse than the most probable the contingency measures already determined to deal with the situations would have been triggered.

Other relevant chapters are 7, Shear Strength and 14, Site Investigation.

Worked Example 11.1   Active thrust

Determine the total active thrust on the back of a smooth vertical wall, 6 m high, which is supporting granular backfill with bulk unit weight 19 kN/m³ and angle of friction 33°. The soil surface is horizontal and the water table lies below the base of the wall.

From equation 11.4,

$$K_a = \frac{1 - \sin 33°}{1 + \sin 33°} = 0.295$$

The backfill is cohesionless, equation 11.3 gives the pressure at the base of the wall
$p_a = 6.0 \times 19 \times 0.295 = 33.6$ kN/m²
The pressure distribution is triangular so the active thrust is

$$\frac{1}{2} \times 33.6 \times 6.0 = 100.8 \text{ kN/m run of wall}$$

This thrust acts at $\frac{1}{3} \times 6.0 = 2.0$ m above the base of the wall

Worked Example 11.2   Active thrust ÷ with water table present

For the same conditions as in Example 11. 1 but with a water table at 1.8 m below ground level determine the total thrust on the back of the wall. Assume the saturated unit weight of the backfill is 20 kN/m³ below the water table. The vertical effective stress and active pressure are plotted in Figure 11.35.

At 1.8 m below ground level
$\sigma_v' = 1.80 \times 19 = 34.2$ kN/m² $\sigma_H' = 34.2 \times 0.295 = 10. 1$ kN/m²
At 6.0 m below ground level
$\sigma_v' = 34.2 + 4.20 \times (20 - 9.8) = 77.0$ kN/m² $\sigma_H' = 77.0 \times 0.295 = 22.7$ kN/m²
The active thrusts are then given by the areas of the active pressure diagram

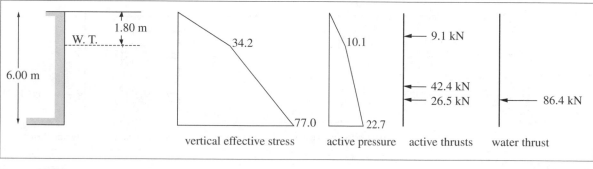

FIGURE **11.35**

$\frac{1}{2}(10 + 1 + 22.7) 4.2$

10.1×4.2

68.98

i) $\frac{1}{2} \times 10.1 \times 1.8 = 9.1$ kN/m run acting at 4.80m above the base of the wall

ii) 10. $1 \times 4.2 = 42.4$ kN/m run at 2. 10 m

iii) $(22.7 - 10.1) \times \frac{1}{2} \times 4.2 = 26.5$ kN/m run at 1.40 m

The force from the water is

$\frac{1}{2} \times 4.2 \times 9.8 \times 4.2 = 86.4$ kN/m run at 1.40m above the base of the wall

The total force acting on the back of the wall is 164.4 kN/m run

---

Worked Example 11.3   Active thrust – two layers and effect of c'

For the same conditions as in Example 11. 1 but with a clay deposit 3.0 m below the top of the wall, determine the total thrust on the back of the wall.

The vertical effective stress and active pressures are plotted in Figure 11.36.
In the sand, the active pressures are
i)   at 1.8 m $34.2 \times 0.295 = 10.1$ kN/m run
ii) at 3.0 m $46.4 \times 0.295 = 13.7$ kN/m run in the granular soil
     In the clay

$$K_a = \frac{1 - \sin 25°}{1 + \sin 25°} = 0.406$$

iii) at the top of the clay, from equation 11.3 $p_a = 46.4 \times 0.406 - 2 \times 10 \times \sqrt{0.406} = 6.1$ kN/m²
iv) at the bottom of the clay $p_a = 80.0 \times 0.406 - 2 \times 10 \times \sqrt{0.406} = 19.7$ kN/m²
The active thrusts are then

i) $\frac{1}{2} \times 10. 1 \times 1.8 = 9.1$ kN/m run acting at 4.80m above the base of the wall

ii) $10.1 \times 1.2 = 12.1$ kN/m run at 3.60 m

iii) $\frac{1}{2} \times 1.2 \times 3.6 = 2.2$ kN/m run acting at 3.40m

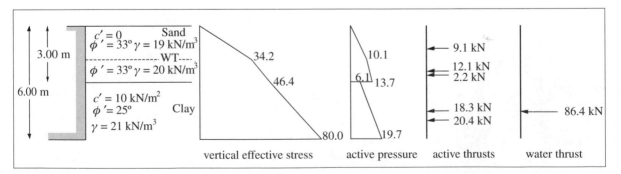

FIGURE **11.36**

iv) $6.1 \times 3.0 = 18.3$ kN/m run at 1.50 m

v) $\frac{1}{2} \times 13.6 \times 3.0 = 20.4$ kN/m run acting at 1.00m

The water force is 86.4 kN/m run at 1.40 m
The total force acting on the back of the wall is 148.5 kN/m run

Worked Example 11.4   Active thrust – wall friction

For the same conditions as in Example 11. 1 but with a rough wall with $\delta = 0.75 \, \phi'$ determine the total thrust on the back of the wall.

From Figure 11.11, $K_a = 0.24$
From equation 11.11 the horizontal (normal) component of the active thrust P_{an} is

$$P_{an} = \frac{1}{2} \times 0.24 \times 19 \times 6.0^2 = 82.1 \text{ kN/m run}$$

From equation 11.15 the shear force acting downwards on the back of the wall is
$82.1 \times \tan(0.75 \times 33) = 37.8$ kN/m run
From equation 11.15 the resultant force acting at $^3/_4 \times 33° = 24.75°$ to the horizontal will be

$$P_a = \frac{82.1}{\cos(0.75 \times 33°)} = 90.4 \text{ kN/m run}$$

Worked Example 11.5   Active thrust – surcharge

Determine the total active thrust on the back of a rough vertical wall, 6 m high, which is supporting a saturated clay with unit weight 21 kN/m³, $c' = 10$ kN/m² and $\phi° = 25°$ and with a uniform surcharge of 10 kN/m² acting on the soil surface. The water table lies below the base of the wall.
Assume $c_w \, '/c' = 0.5$ and $\delta = {}^3/_4 \, \phi'$

The vertical effective stresses and active pressures are plotted in Figure 11.37.

From Figure 11. 11, for $\dfrac{\delta}{\phi} = 0.75$ and $\phi' = 25°$ $K_a = 0.34$

From equation 11.16 $K_{ac} = 2\sqrt{(0.34 \times 1.50)} = 1.43$

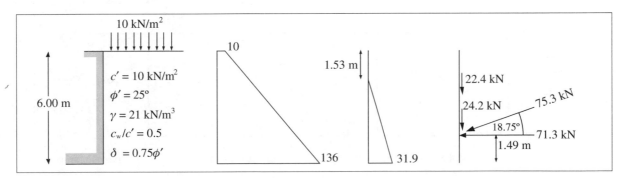

FIGURE 11.37

From equation 11.31 the depth to zero active pressure is given by

$0 = 0.34 (21 \times z_0 + 10) - 1.43 \times 10$

giving $z_0 = 1.53$ m

Active pressure at the base of the wall is

$p_{an}' = 0.34 (21 \times 6.0 + 10) - 1.43 \times 10 = 31.9$ kN/m²

Active thrust $P_{an} = 31.9 \times (6.0 - 1.53) \times \dfrac{1}{2} = 71.3$ kN/m run

acting at $\dfrac{1}{3} (6.0 - 1.53) = 1.49$ m above the base of the wall

From equation 11.15 the shear force acting on the back of the wall will be

$71.3 \times \tan (\dfrac{3}{4} \times 25) = 24.2$ kN/m run

From equation 11.14 the resultant force P_a acting at $\dfrac{3}{4} \times 25° = 18.75°$ to the horizontal will be

$P_a = \dfrac{71.3}{\cos(0.75 \times 25°)} = 75.3$ kN/m run

The adhesion on the back of the wall is $0.5 \times 10 \times (6.0 - 1.53) = 22.4$ kN/m run

Worked Example 11.6   Passive thrust

For the sheet pile wall shown in Figure 11.38 determine the total passive thrust acting on the left hand side of the wall. Assume the water table lies below the base of the wall.

Assume $\delta = 0.67 \, \phi' \, c_w = 0.5 \, c'$

From Figure 11.11, $K_p = 3.3$

From equation 11.17, $K_{pc} = 2\sqrt{(3.3 \times 1.5)} = 4.45$

At the top of the clay, from equation 11.12

$p_{pn}' = 3.3 \times 0 \times 21 + 4.45 \times 10 = 44.5$ kN/m²

At the bottom of the clay

$p_{pn}' = 3.3 \times 6.0 \times 21 + 4.45 \times 10 = 460.3$ kN/m²

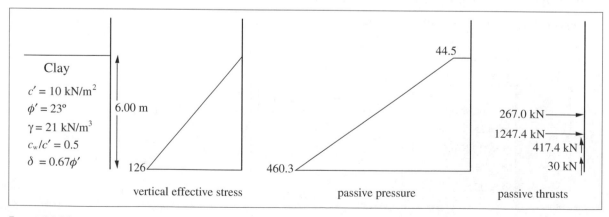

FIGURE 11.38

The passive thrusts are then

i) $44.5 \times 6.0 = 267.0$ kN/m run acting 3.0 m above the base of the wall

ii) $(460.3 - 44.5) \times 6.0 \times \dfrac{1}{2} = 1247.4$ kN/m run acting at 2.0 m

These thrusts act normal to the wall
A shear force acts upwards of
$(267.0 + 1247.4) \tan(0.67 \times 23°) = 417.4$ kN/m run
Adhesion also acts upwards
$0.5 \times 10 \times 6.0 = 30$ kN/m run

Worked Example 11.7   Gravity wall

The gravity wall shown in Figure 11.39 is to support backfill with unit weight 19.5 kN/m³ and shear strength parameters c'= 0 and ϕ'= 36°. The unit weight of the wall material is 24 kN/m³ and the angle of friction δ between the wall and the backfill is 27° and at the base of the wall it is 25°. Determine the factors of safety against overturning and sliding and the maximum and minimun bearing pressures.

The weight of the wall is
$1.62 \times 5.0 \times 24 = 194.4$ kN/m run acting at 0.81 m from the toe
$0.88 \times 5.0 \times 0.5 \times 24 = 52.8$ kN/m run acting at 1.91 m from the toe
From equation 11.8

$$\alpha = 100° \qquad \beta = 15° \qquad \delta = 27° \qquad \phi = 36°$$

$$K_a = \left[\frac{\sin 64°/\sin 100°}{\sqrt{\sin 127°} + \sqrt{(\sin 63° \, \sin 21°/\sin 85°)}} \right]^2 = 0.39$$

From equation 11.7

$$p_a = \frac{1}{2} \times 0.39 \times 19.5 \times 5.0^2 = 95.1 \text{ kN/m run}$$

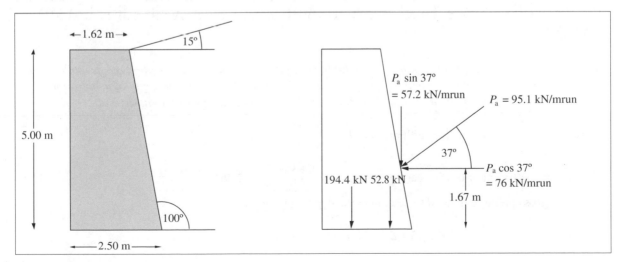

FIGURE 11.39

The point of application of this thrust is taken as $\frac{1}{3}H$ vertically, i.e. 1.67 m. This thrust acts at δ (=27°) to the normal to the base of the wall, i.e. 37° to the horizontal. Resolving this thrust vertically and horizontally gives horizontal component = 95.1 cos 37° = 76.0 kN/m run
vertical component = 95.1 sin 37° = 57.2 kN/m run
Taking moments about the toe the factor of safety against overturning is given.

$$F = \frac{194.4 \times 0.\,81 + 52.8 \times 1.91 + 57.2 \times 2.21}{76.0 \times 1.67} = 3.0$$

From equations 8.22 and 8.25, the factor of safety against sliding is

$$F = \frac{(194.4 + 52.8 + 57.2)\tan 25°}{76.0} = 1.87$$

Taking moments about the heel for the eccentricity e
$\Sigma M = 194.4 \times 1.69 + 52.8 \times 0.59 + 57.2 \times 0.29 + 76.0 \times 1.67 = 503.2$ kN/m run
The total vertical load = 194.4 + 52.8 + 57.2 = 304.4 kN/m run
and acts at $\frac{503.2}{304.4} = 1.65$ m from the heel
The eccentricity e is 1.65 –– 1.25 = 0.40 m on the left hand side of the wall centre-line.
$\frac{2.5}{6} = 0.417$ m so this is just within the middle third,

From Figure 11.22

$$q_{max} = \frac{304.4}{2.5}\left(1 + \frac{6.0 \times 0.40}{2.5}\right) = 238.7 \text{ kN/m}^2$$

$$q_{min} = \frac{304.4}{2.5}\left(1 - \frac{6.0 \times 0.40}{2.5}\right) = 4.9 \text{ kN/m}^2$$

Worked Example 11.8   Cantilever sheet pile wall – limit state design

A cantilever sheet pile wall is to retain soil 5 m high. Using the factor on strength method the design ϕ' value of 25° is determined. The design unit weight of the soil is 20 kN/m³. Determine the depth of penetration required. Using another method show that the limit equilibrium condition gives the same result.

The design coefficient of active earth pressure $K_a = \dfrac{1 - \sin 25°}{1 + \sin 25°} = 0.406$

The design coefficient of passive earth pressure $K_p = \dfrac{1 + \sin 25°}{1 - \sin 25°} = 2.464$

Using the gross pressure method to determine the depth of penetration

the active thrust $P_a = \dfrac{1}{2} \times 0.406 \times (5 + d)^2 \times 20 = 4.06\,(5 + d)^2$

acting at $\dfrac{5+d}{3}$ from the toe of the wall

the passive thrust $P_p = \dfrac{1}{2} \times 2.464 \times d^2 \times 20 = 24.64\, d^2$

acting at $\dfrac{d}{3}$

Equating the moments of the active and passive thrusts gives
$6.86d^3 - 20.3d^2 - 101.5d - 169.17 = 0$
$d = 6.07$ m
check:

$P_a = \dfrac{1}{2} \times 0.406 \times 11.07^2 \times 20 = 497.5$ kN $l_a = \dfrac{11.07}{3}$ $M_a = 497.5 \times \dfrac{11.07}{3} = 1835.9$ kNm

$P_p = \dfrac{1}{2} \times 2.464 \times 6.07^2 \times 20 = 907.9$ kN $l_p = \dfrac{6.07}{3}$ $M_p = 1837.0$ kNm $\therefore$ OK

Using the Burland-Potts method the active pressure at the bottom of the wall is $0.406 \times 20 \times 11.07 = 89.89$ kN/m^2. The active pressure at excavation level is $0.406 \times 20 \times 5 = 40.6$ kN/m^2

From Figure 11.25 the width of the hatched area at the bottom of the wall is $89.89 - 40.6 = 49.29$ kN/m^2. The passive pressure at the bottom of the wall is $2.464 \times 20 \times 6.07 = 299.13$ kN/m^2
deducting the hatched area gives $299.13 - 49.29 = 249.84$ kN/m^2

the overturning moment is $\dfrac{1}{2} \times 40.6 \times 5 \times \left(6.07 + \dfrac{5}{3}\right) + \dfrac{1}{2} \times 40.6 \times 6.07^2 = 1533.2$ kNm

the resisting moment is $\dfrac{1}{2} \times 249.84 \times 6.07^2 \times \dfrac{1}{3} = 1534.2$ kNm $\therefore$ OK

Worked Example 11.9   Reinforced soil

An abutment wall 6.4 m high is reinforced with polymeric sheet reinforcement at vertical spacings of 0.8m. The angle of friction between the sheet and the fill is 22°, the creep limited tensile strength of the reinforcement is 33kN/m and the interaction coefficient for the soil-reinforcement interface is 0.90. The unit weight of the frictional fill is 19.5 kN/m³ and its peak angle of friction is 35° with c' assumed to be zero. Carry out a calculation to check on the local stability of the reinforcement layer at 2.4 m below the top of the reinforced soil block.

The length of the reinforcement is taken initially as $0.7H = 0.7 \times 6.4 = 4.5$ m

a) *Tensile rupture*
The partial factor f_m is given by
$f_m = f_{m1} \times f_{m2}$
where $f_{m1} = f_{m11} \times f_{m12}$
and $f_{m2} = f_{m21} \times f_{m22}$

Values of these factors are obtained from Annex A in BS8006. The characteristic base strength is specified so $f_{m11} = 1.0$. The test data has accrued from appropriate creep tests but to allow for extrapolation of the data to cover the design life of the reinforcement $f_{m12} = 1.05$. To allow for installation damage $f_{m21} = 1.1$ and as the fill and the general environment are non-aggressive $f_{m22} = 1.0$.
$f_{m1} = 1.0 \times 1.05 = 1.05$
$f_{m2} = 1.1 \times 1.0 = 1.1$
$f_m = 1.05 \times 1.1 = 1.155$

The design tensile strength

$$T_D = \frac{33.0}{1.155} = 28.57 \text{ kN/m}$$

As the structure is an abutment the partial factor for the economic ramifications of failure is 1.1.

$$\frac{T_{Dj}}{f_n} = \frac{28.57}{1.1} = 25.97 \text{ kN/m}$$

b) *Pull-out resistance*

The design shear strength of the fill is

$$\tan \phi_d = \frac{\tan \phi_p}{f_{ms}} \text{ since } f_{ms} = 1.0 \; \phi_d = \phi_p = 35° \text{ assuming this value to be a characteristic or worst credible value.}$$

The length of reinforcement in the active zone

$$= 4.0 \times \tan \left(45 - \frac{35}{2} \right) = 2.08 \text{ m}$$

the length of reinforcement L_{ej} in the resistant zone is $4.5 - 2.08 = 2.42$ m

From equation 11.59 the pull-out resistance will be

$$T_{Aj} = \frac{2.42 \times 2.0}{1.3} \left(\frac{0.9 \times \tan 35°}{1.0} \times 1.0 \times 19.5 \times 2.4 \right) = 109.8 \text{ kN/m}$$

From equation 11.62

$$\frac{T_{Aj}}{f_n} = \frac{109.8}{1.1} = 99.8 \text{ kN/m}$$

For the tensile force within the reinforcing layer using equation 11.60:

$$K_a = \frac{1 - \sin 35°}{1 + \sin 35°} = 0.27$$

$$R_{vj} = 2.4 \times 4.5 \times 19.5 \times 1.0 = 210.6 \text{ kN/m}$$

The load factor on earth pressure behind the structure for this case is $f_{fs} = 1.5$.
$$R_{hj} = 0.5 \times 0.27 \times 19.5 \times 2.4^2 \times 1.5 = 22.74 \text{ kN/m}$$

$$\text{Eccentricity } e_j = \frac{22.74 \times \frac{2.4}{3}}{210.6} = 0.086 \text{ m}$$

the vertical stress on the jth layer

$$\sigma_{vj} = \frac{210.6}{4.5 - 2 \times 0.086} = 48.66 \text{ kN/m}^2$$

$$T_j = 0.27 \times 48.66 \times 0.8 = 10.51 \text{ kN/m}$$

For the design tensile strength
$10.51 < 25.97 \therefore$ OK
For the design pull-out resistance
$10.51 < 99.8 \therefore$ OK

EXERCISES

11.1 A retaining wall, 5 m high supports backfill with a horizontal surface and a water table at 2 m below ground level. The unit weight of the backfill soil is 19 kN/m³ above the water table and 20 kN/m³ below and the angle of internal friction of the backfill is $\phi' = 36°$ with $c' = 0$. Assuming the wall to be smooth determine the total thrust (from the soil and the water) and its point of application as the wall moves away from the soil.

11.2 For the wall and soil conditions described in Exercise 11.1 determine the total passive thrust and its point of application if the wall moves towards the soil.

11.3 For the wall and soil conditions in Exercise 11.1 but with a layer of clay below 3 m below ground level determine the total active thrust on the back of the wall and its point of application assuming drained conditions in the clay. Properties of the clay are:
$c' = 12$ kN/m² $\phi' = 26°$ $\gamma = 21.5$ kN/m³

11.4 For the wall and soil conditions in Exercise 11.1 but with wall friction acting on the back of the wall with $\delta = 0.67 \phi'$ determine
a) the total horizontal thrust (active and hydrostatic) on the back of the wall and its point of application
b) the shear force acting on the back of the wall

11.5 The cantilever retaining wall shown in Figure 11.40 supports a free draining backfill with the following properties:
$\phi' = 37°$ $c' = 0$ $\gamma = 21$ kN/m³   concrete $\gamma = 25$ kN/m³
Assuming that earth pressures are calculated on a vertical line above the heel of the wall and that soil friction acts along this line ($\delta = \phi'$) determine
a) the factor of safety against overturning
b) the factor of safety against sliding, assuming the angle of friction on the base $\delta = 25°$
c) the maximum and minimum bearing pressures beneath the base of the wall.

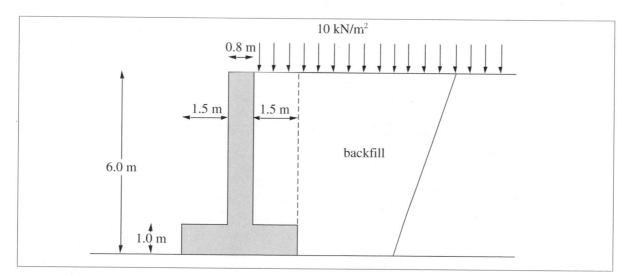

FIGURE 11.40

11.6 The anchored sheet pile wall shown in Figure 11.41 supports the bank of a canal with a water level normally inside the canal at the same level as the water table in the ground. This water table and water level is initially at 2 m below the upper ground level. However, the water level in the canal has been lowered rapidly to its base level while the water table behind the wall has remained unchanged. Assuming hydrostatic conditions on both sides of the wall determine
1) the depth of embedment required for stability and
2) the anchor force
using the BS 8002 method.

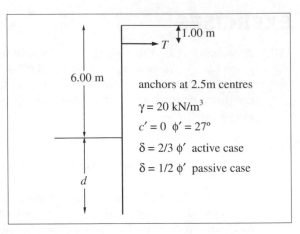

Figure 11.41

11.7 A reinforced soil wall acting as an abutment is 7 m high and is reinforced with strip elements 8.0 m long, 60 mm wide and 3 mm thick with horizontal spacings of 1.0 m and vertical spacings of 0.6 m. The angle of friction between the elements and the fill is 25° and the ultimate tensile strength of the strips is 300 N/mm². The unit weight of the fill is 19 kN/m³ and its angle of friction is 37° with $c' = 0$. Assume the partial factor $f_m = 1.1$. The highest strip lies at 0.7 m below the top of the wall and the lowest strip is at 6.7 m below the top of the wall. Carry out a check on the local stability of the reinforcement and determine the length required.

Slope stability

Natural and artificial slopes

Sloping ground can become unstable if the gravity forces acting on a mass of soil exceed the shear strength available at the base of the mass and within it. Movement of the mass of soil down the slope will then occur.

This can have catastrophic consequences to life and property if buildings exist on, above or below the slope. However, in remote, unpopulated areas mass movements may have minimal effect, merely being part of the natural degradation of the land surface. On some coastal cliffs instability involving the destruction of property is often accepted since the costs of resisting natural erosion processes with cliff stabilisation measures are prohibitive.

Types of mass movement (Figure 12.1)

Mass movements, generally referred to as landslides, can take many forms. Skempton and Hutchinson (1969) have classified types of mass movement, as illustrated in Figure 12.1.

Flows are distinguished from slides in that slides comprise the movement of large, generally continu-ous masses of soil on one or more slip surfaces whereas flows consist of slow movement of softened or weathered debris such as from the base of a broken-up slide mass (earth flows) or somewhat faster movement of clay debris softened and lubricated by water (mudflows) with slip surfaces less evident.

Natural slopes

These slopes have evolved as a result of natural processes, largely erosion, over a long period of time. If the soils comprising these slopes have sufficient strength to support the gravity forces then the slope will remain stable without any mass movement.

Where they have been affected by instability in the past then the previous mass movements will have produced a 'first-time slide' resulting in considerable strains usually along a slip surface. The maximum strength mobilised on parts of the slip surface will be initially the peak strength but after a small amount of strain the critical state strength will pertain. As the strength available is reduced on these parts the peak strength will be mobilised on other portions of the slip surface until the strength along the whole surface is mostly at its critical state value. This

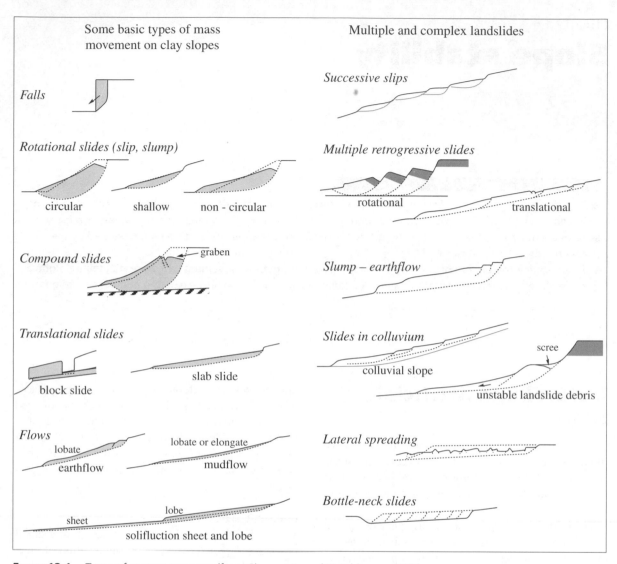

FIGURE 12.1 *Types of mass movement (from Skempton and Hutchinson, 1969)*

process is referred to as progressive failure. It can be seen that relying on the peak strength is a dangerous assumption and assuming the critical state strength is a cautious approach.

If the soil is prone to further reduction in strength following large strains, progressive failure will take the strength from the peak to the critical state and eventually to the residual value. The slope will then be left with a slip surface along which the lowest possible shear strength exists, the residual strength, see Chapter 7.

The slope will remain in a state of incipient failure (only just stable) until it is affected by:

● A change in the pore pressure condition within the slope. The most common causes are exceptional rainfall and changes in the drainage pattern. Placing soil over the toe of a slope could permit a build-up of pore water pressures within the slope as will curtailment of pumping from aquifers close to the slope.

● A change in the geometry of the slope. This can be caused by undercutting due to erosion at the toe of the slope, or accumulation of soil at the top of the slope.

- Engineering works. The construction of a cutting at the toe of a slope or an embankment at the top could precipitate failure.

Artificial slopes or earthworks

Soil has been used for many construction purposes ever since human activity commenced. Major earthworks can be classified as:

- excavations
- cuttings
- embankments
- earth dams
- spoil heaps, tailings dams.

Excavations are usually made for temporary works to enable construction of a structure below ground level or for extraction and quarrying purposes. Since the ground is only open for a short period of time more risk is often taken on the sides of the excavation remaining stable and for economic reasons as steep a slope as possible is cut. This stability relies on the soil remaining in the undrained condition for the period of exposure. It is advisable to cover the slope surfaces to prevent infiltration of water and where necessary control the flow of groundwater towards the excavation face. Once the structure is complete the excavation is backfilled and the stability is restored. Quarry slopes are usually left to degrade with time. The most common forms of failure would be falls, rotational slides and translational slides.

Cuttings are formed from sloping sided excavations below ground level. The soil and groundwater conditions comprising the slopes will be determined by the natural geology at the site so a slope may have to be designed for a range of these conditions. They must be stable in the short and long term. The long-term state is the most critical, when pore pressures have risen to equilibrium and the drained condition applies. Thus most cutting slopes are stable in the short-term but become less stable with time and may fail, sometimes many years later. Modes of failure could include falls, rotational, composite and slab slides, and slumping.

A first-time slide mobilising the critical state strength can be assumed if pre-existing slip surfaces

are not present. If pre-existing slip surfaces are present with only the residual strength available then cuttings are likely to re-activate the slipping, often with significant effects.

Embankments are formed by placing a mound of fill material above ground level with sloping sides. The strength and groundwater conditions within an embankment can be controlled to a certain extent by specifying good quality fill and drainage measures so the stability of the embankment slopes themselves will be more certain. However, the soil and groundwater conditions beneath the embankment will be determined by the natural geology at the site so the possibility of foundation failure must be considered.

Modes of failure would usually be rotational or composite often with circular slip surfaces, but sometimes with non-circular slip surfaces depending on the stratigraphy. Foundation failures beneath the embankment would be more likely during construction with slope failures within the embankment more likely in the long-term.

Earth dams require much greater care in their design. They are higher structures with the materials forming them subjected to higher stress levels and seepage forces. They are of composite construction, formed from materials of differing permeabilities to restrict flow through. However, these materials also have different stress–strain properties so that some zones may be more highly stressed than imagined.

The pore pressure condition will be affected by seepage of water through the various zones so the stability must be assessed for the steady seepage condition on the downstream face with the rapid drawdown of reservoir levels producing a potentially critical condition on the upstream face. Due to the zoning of materials potential slip surfaces will tend to be non-circular. The consequences of failure are catastrophic so these structures must be designed with a negligible risk of failure.

Placing and compacting fill materials will tend to produce a soil mass with its critical state strength available and with no pre-existing slip surfaces present where the residual strength could pertain. However, if soil containing pre-existing slip surfaces is left in place beneath an embankment or earth dam then re-activation of slipping along these planes will mean that the earth structure will be less stable than if the soil had been removed and replaced with compacted fill. The failure of Carsington Dam is an example of this.

Spoil heaps and tailings dams are mounds of waste products from industrial processes, mining residues etc. They may be placed by end tipping or loose dumping in a solid form or by pumping in a hydraulic form and can result in a variable and potentially weak cross-section. For economic reasons they are constructed as steep as possible and if little concern is given to the control of surface water, groundwater or process water then they can be a disaster waiting to happen.

Short-term and long-term conditions
(Figures 12.2, 12.3 and 12.4)

When a cutting or embankment is constructed the total stresses in the ground are changed. This results in a change in pore water pressure (see equation 4.13) and since the factor of safety of a slope decreases as the pore pressure increases the most critical condition will occur when pore pressures are greatest.

Cuttings (Figure 12.2)

In Figure 12.2 the total stress and pore pressure variations are shown for a point P within a cutting, for different types of soil.

The reduction in total stresses during construction will lead to a decrease in the pore pressure as the soil structure attempts to expand. If the excavation is carried out quickly then there will be no time available for redistribution of pore pressures. The pore pressure reduction will be greatest at the end of construction and its amount will depend on the type of soil present, particularly the pore pressure parameter value A.

Following construction these out-of-balance pore pressures adjust by increasing gradually towards the steady state seepage flow pattern appropriate to the new slope profile, when the long-term condition applies. Depending on the mass permeability of the soil this redistribution will require varying amounts of time. With sands the redistribution will be rapid whereas with intact clays it may take several years. For a cutting the lowest factor of safety is associated with this long-term condition.

Embankments (Figure 12.3)

For an embankment the increase in total stresses during construction will lead to an increase in the pore pressure within the foundation soil (as at the point P in Figure 12.3) as the soil structure attempts to contract. If the construction is carried out rapidly the pore pressure increase will be greatest at the end of construction, assuming the soil behaves in an undrained manner. The lowest factor of safety will be associated with this short-term undrained condition. Most embankment failures occur at or near the end of construction as foundation failures. After construction the pore pressures dissipate vertically and horizontally through the foundation soil until they return to equilibrium with the original water table level.

Embankments on soft normally consolidated clays are at the greatest risk of failure associated with the largest rise in pore pressures. Their stability can be improved by allowing pore pressure dissipation to occur as construction proceeds but this will require a slower rate of construction, a good knowledge of the mass permeability and possibly measures to accelerate dissipation such as vertical drains.

Earth dams (Figure 12.4)

For an earth dam the shear stresses and the pore pressures will vary both during and after construction depending on the state of the dam and the level of the water impounded in the reservoir, see Figure 12.4. The following conditions should be checked:

1. During and shortly after construction – both upstream and downstream faces.
2. With the reservoir full (steady seepage) – downstream face.
3. Following rapid drawdown of the impounded water – upstream face.

These conditions are shown on the lower diagram in Figure 12.4.

Methods of analysis

Plane translational slide – general
(Figure 12.5)

This method assumes movement of a mass of soil above a single planar slip surface parallel to the ground surface with end and side effects ignored. This type of analysis is most applicable to granular soils, soils with no cohesion ($c' = 0$), soils with

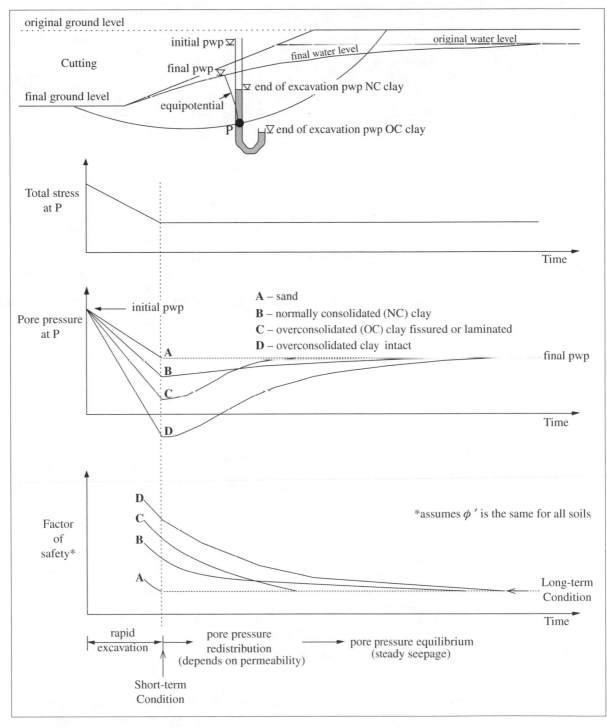

FIGURE 12.2 *Cuttings – short- and long-term conditions*

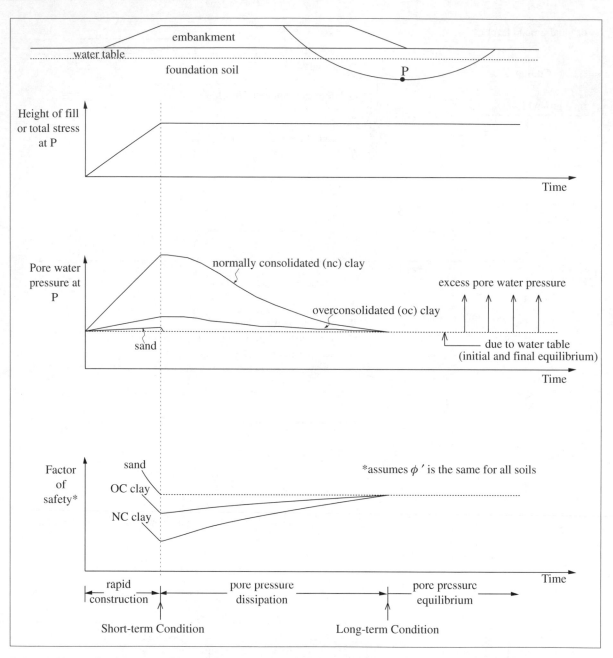

FIGURE 12.3 *Embankments – short and long-term conditions*

bedding or laminations dipping parallel to the slope, soils with weathering profiles producing upper weaker horizons and slopes where there has already been a shallow slab slide such that the shear strength on the slip surface has reduced to its residual value.

Consider a vertical segment of soil of width b and unit thickness in a slope inclined at an angle β, see Figure 12.5. The water table is parallel to the ground surface at a depth h_w with steady seepage assumed to be taking place parallel to the ground surface. The

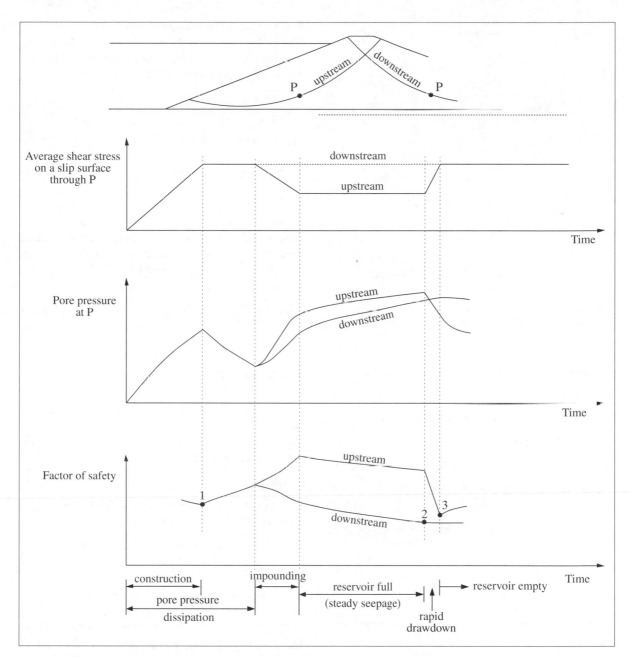

FIGURE 12.4 *Earth dams – stability conditions (from Lambe and Whitman, 1979)*

pore water pressure u on the slip surface will be given by:

$$u = (z - h_w)\gamma_w \cos^2\beta \qquad (12.1)$$

Assuming the unit weights of the soil above and below the water table (bulk and saturated unit weights) to be the same, γ, the weight of the segment W will be:

$$W = \gamma z b \qquad (12.2)$$

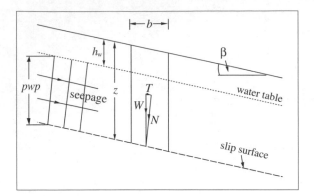

FIGURE 12.5 *Plane translational slide*

The tangential force down the slope T will be:

$$T = W \sin\beta = \gamma z \, b \, \sin\beta \qquad (12.3)$$

The tangential stress down the slope τ will be:

$$\tau = \gamma z \, \sin\beta \cos\beta \qquad (12.4)$$

The normal total force on the slip surface will be:

$$N = W \cos\beta \qquad (12.5)$$

The normal total stress on the segment σ will be:

$$\sigma = W \cos\beta \cos\beta / b = \gamma z \cos^2\beta \qquad (12.6)$$

The normal effective stress in the segment σ' will be:

$$\sigma' = \sigma - u = \gamma z \cos^2\beta - (z - h_w)\gamma_w \cos^2\beta$$
$$= \cos^2\beta \, (\gamma z - \gamma_w z + \gamma_w h_w) \qquad (12.7)$$

The shearing resistance at the base of the segment τ_f will be:

$$\tau_f = c' + \sigma' \tan\phi' \qquad (12.8)$$

The factor of safety is given by:

$$F = \tau_f / \tau \qquad (12.9)$$

For the general case:

$$F = \frac{c' + \tan\phi' \cos^2\beta(\gamma z - \gamma_w z + \gamma_w h_w)}{\gamma z \sin\beta \cos\beta} \qquad (12.10)$$

SEE WORKED EXAMPLE 12.1C.

Plane translational slide – special cases

Several special cases can be considered, as described below.

1 Dry cohesionless slope $c' = 0$, $h_w = z$
This is the case for a deep water table with no pore pressures (positive or negative) assumed on the slip surface. From equation 12.10:

$$F = \frac{\tan\phi'}{\tan\beta} \qquad (12.11)$$

SEE WORKED EXAMPLE 12.1B.

For the critical case (when $F = 1$)

$$\beta = \phi'$$

2 Wet cohesionless slope $c' = 0$ $h_w = 0$
This is the most critical condition, when the water table is at ground level and seepage is occurring down-slope. From Equation 12.10:

$$F = \frac{\gamma_{sub} \tan\phi'}{\gamma \tan\beta} \qquad (12.12)$$

since $\gamma - \gamma_w = \gamma_{sub}$

SEE WORKED EXAMPLE 12.1A.

3 Cohesionless slope ($c' = 0$) with pore pressure ratio r_u
The value of the pore pressure ratio on the slip surface is given by:

$$r_u = u/\gamma z \qquad (12.13)$$

Note that this is not the average value for the slope. Equation 12.10 can be written as:

$$F = (1 - r_u \sec^2\beta)\frac{\tan\phi'}{\tan\beta} \qquad (12.14)$$

4 Cohesionless slope with a water table below the slip surface producing suction at the slip surface
The capillary zone above the water table in a fine sand, silt or clay can be significant (see Chapter 4) and will produce negative pore pressures within this zone given by:

$$u_s = -\gamma_w h_s \qquad (12.15)$$

where h_s is the vertical distance between the water table and the slip surface.

The factor of safety can then be obtained as:

$$F = \frac{\tan\phi' \; (\gamma z \cos^2\beta + \gamma_w h_s)}{\gamma z \sin\beta \cos\beta} \qquad (12.16)$$

A simpler expression can be obtained if seepage through the suction zone parallel to the surface is assumed when:

$$u_s = -\gamma_w \, h_s \cos^2\beta \qquad (12.17)$$

and

$$F = \left(1 + \frac{\gamma_w h_s}{\gamma z}\right) \frac{\tan\phi'}{\tan\beta} \qquad (12.18)$$

SEE WORKED EXAMPLE 12.1.

Caution must be exercised when assuming suctions to act in support of a slope. Percolation following a rainstorm can quickly cancel the suctions, lowering the factor of safety below unity and instigating slope failure.

5 Submerged slope with no seepage
With water above the slope the effective stresses are given by the submerged unit weight of the soil, γ_{sub} and equations 12.4 and 12.7 can be written as:

$$\tau = \gamma_{sub} z \sin\beta \cos\beta \qquad (12.19)$$

and

$$\sigma' = \gamma_{sub} z \cos^2\beta \qquad (12.20)$$

Inserting these into Equation 12.10 gives:

$$F = \frac{c' + \gamma_{sub} \, z \cos^2\beta \tan\phi'}{\gamma_{sub} \, z \sin\beta \cos\beta} \qquad (12.21)$$

$$= \frac{c'}{\gamma_{sub} \, z \sin\beta \cos\beta} + \frac{\tan\phi'}{\tan\beta} \qquad (12.22)$$

Compared to Equation 12.12 it can be seen that the effect of submergence is to enhance the stability of a wet slope. In fact a cohesionless submerged slope has the same factor of safety as a dry slope. When drawdown of the water level occurs the exposed slope will develop into case 2 and the factor of safety in this region will drop significantly. Thus, maintaining water levels can be a slope stabilisation expedient.

6 Layered soil profile (Figure 12.6)
Apart from the completely dry or completely wet cases, 1 and 2, the other cases require some information about, or an assumption concerning, the depth to the presumed slip surface, z. Worked Example 12.1 shows that for a homogeneous soil the factor of safety decreases as the presumed slip surface moves deeper so the method has obvious limitations for a homogeneous slope. With a layered soil profile the tangential shear stress down the slope could be plotted with depth, from equation 12.4, with the available shearing resistance τ_f superimposed, from equations 12.7 and 12.8, as:

$$\tau_f = c' + \tan\phi' \cos^2\beta \; (\gamma z - \gamma_w z + \gamma_w h_w) \qquad (12.23)$$

A typical layered profile is illustrated in Figure 12.6, where the likely horizons for failure are identified.

7 Undrained conditions
All of the above cases assume the long-term condition to be critical using effective stress soil parameters. If the slope behaves in an undrained manner (with $\phi_u = 0°$) the factor of safety is given by:

$$F = \frac{c_u}{\gamma z \sin\beta \cos\beta} \qquad (12.24)$$

Plane translational slide – partial factors
Treating the disturbing action as the tangential stress τ down the slope, equation 12.4, and the resisting action as the shearing resistance on the slip surface, τ_f, equations 12.7 and 12.8, the inequality:

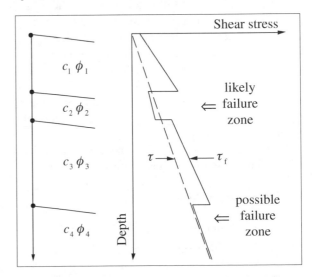

FIGURE 12.6 *Translational slide – layered soil profile*

$$V_d \leq R_d \qquad (12.25)$$

where:

V_d = design load acting down the slope
R_d = design resistance acting up the slope

can be written as:

$$\tau \leq \tau_f \qquad (12.26)$$

Adopting partial factors the disturbing stress τ acting down the slope will be:

$$\tau = \gamma_k \, \gamma_G \, z \sin\beta \cos\beta \qquad (12.27)$$

and the resisting stress τ_f acting up the slope will be:

$$\tau_f = \frac{c_k'}{\gamma_c} + \frac{\tan\phi_k'}{\gamma_\phi} \cos^2\beta \left(\frac{\gamma_k}{\gamma_m} z - \gamma_w z + \gamma_w h_w \right) \qquad (12.28)$$

where:

γ_k = characteristic bulk unit weight
c_k' = characteristic cohesion intercept
ϕ_k' = characteristic angle of internal friction (critical state or residual)
γ_c = partial factor on cohesion intercept
γ_ϕ = partial factor on angle of internal friction
γ_m = partial factor on bulk unit weight
γ_G = partial factor on the permanent actions

SEE WORKED EXAMPLE 12.2.

Variable actions would occur when there are dynamic loads such as wave action or seismic effects. The material factor γ_m is not applied to the bulk unit weight in the disturbing stress unless it is applied in an adverse manner. If it can vary significantly then the effect of its variation should be established. The parameter h_w is not factored but a reasonable worst-case scenario or most unfavourable value should be adopted.

From Eurocode 7, Case C is the most appropriate and values of the partial factors are given below.

Partial factors for Case C
Permanent actions $\gamma_G = 1.0$
Unfavourable and favourable
variable actions $\gamma_Q = 1.3$ (unfavourable)
$\gamma_\phi = 1.25$
$\gamma_c = 1.6$
$\gamma_{cu} = 1.4$

Circular arc analysis – undrained condition or $\phi_u = 0°$ analysis
(Figures 12.7 and 12.8)
This is an analysis in terms of total stresses and applies to the short-term condition for a cutting or embankment assuming the soil profile to comprise fully saturated clay.

A segment of the slope of unit thickness bounded by a circular arc (a presumed slip surface) is considered, see Figure 12.7. The tendency for gravity forces to rotate the circular segment about its circle centre is resisted by the shear strength mobilised along the slip surface so moment equilibrium is applied. The factor of safety against instability is given by:

$$F = \frac{\text{shear resistance moment}}{\text{overturning moment}}$$

The shear resistance moment is equal to the force along the length L_{AB} of the circular arc multiplied by the radius of the circle, R:

$$\text{shear resistance moment} = c_u \, L_{AB} \, R = c_u \, R^2 \theta \qquad (12.29)$$

where θ is in radians.

The overturning moment is Wd. W is the total weight of the segment, determined from γA where γ is the bulk unit weight of the soil and A is the area of the segment. d is the horizontal distance from the circle centre to the centroid of the segment. The factor of safety is then:

$$F = \frac{c_u R^2 \theta}{Wd} \qquad (12.30)$$

The determination of W and d can entail lengthy calculations so it will be more convenient to split the segment into a number of slices as shown in Figure 12.8 where the factor of safety is then given by:

$$F = \frac{\Sigma c_u b \sec\alpha}{\Sigma W \sin\alpha} \qquad (12.31)$$

The area of each slice can be obtained from the mid-height multiplied by the width of the slice and the angle α can be scaled off or calculated. Note how the sign of the overturning forces changes to the left and right of the circle centre. Inter-slice forces are ignored.

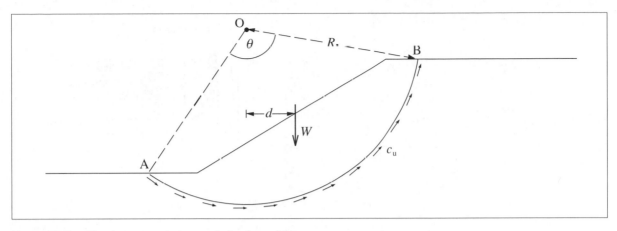

FIGURE 12.7 *Circular arc analysis – undrained condition*

SEE WORKED EXAMPLE 12.3.

It is necessary to find the circular arc which gives the lowest factor of safety, the critical circle. A number of trial circles must then be analysed in the same way but with different circle centres and different points where the circle cuts the slope. This process can be time-consuming so the use of charts for homogeneous slopes and computer programs for heterogeneous slopes is recommended.

The slices approach is also useful where the soil profile consists of different layers and where there are other forces to consider such as a foundation at the top of the slope or a water-filled tension crack.

Tension crack (Figure 12.9)

As the condition of limiting equilibrium develops with the factor of safety close to 1 a tension crack may form near the top of the slope through which no shear strength can be developed and if it fills with water a horizontal hydrostatic force P_w will increase the disturbing moment by $P_w y_c$. The factor of safety will be further reduced because of the shorter length of circular arc along which shearing resistance can be mobilised.

The depth of a tension crack can be taken as:

$$z_c = \frac{2c_u}{\gamma} \qquad (12.32)$$

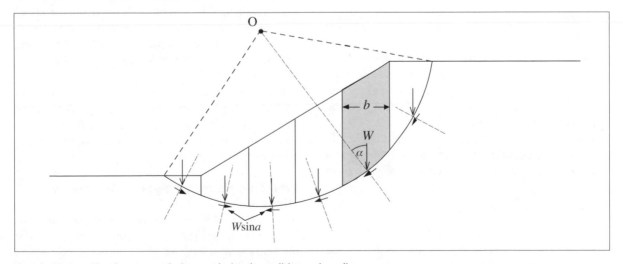

FIGURE 12.8 *Circular arc analysis – undrained condition using slices*

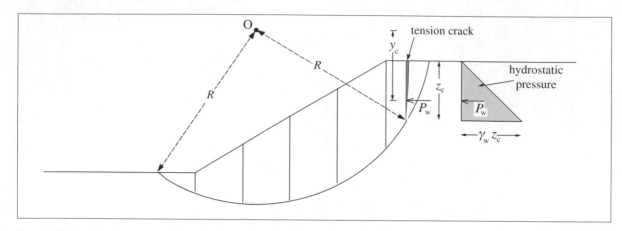

FIGURE 12.9 *Circular arc analysis – undrained condition with tension crack present*

and the hydrostatic force P_w is:

$$P_w = 1/2 \; \gamma_w \, z_c^2 \qquad (12.33)$$

The expression for the factor of safety then becomes:

$$F = \frac{R \, \Sigma c_u b \sec \alpha}{\Sigma RW \sin \alpha + P_w y_c} \qquad (12.34)$$

Note that the width of the slices b need not be the same for all slices.

Undrained analysis stability charts – Taylor's method (Figure 12.10)

Taylor (1937) proposed the use of a stability number, N_s given by:

$$N_s = \frac{c_u}{F\gamma H} \qquad (12.35)$$

where F is the lowest factor of safety obtained from a circular arc analysis of a homogeneous slope with undrained shear strength c_u, bulk unit weight γ and height H. The angle of internal friction for the undrained condition is taken as $\phi_u = 0°$.

The stability number depends on the slope angle β, and the depth factor D where DH is the depth to a rigid stratum. Values of N_s can be obtained from Figure 12.10.

SEE WORKED EXAMPLE 12.4.

A feature of the analysis of a homogeneous slope is that when the slope angle is less than 53° the factor of safety decreases with deeper circles so more critical (lower) values are obtained for deeper-seated 'base' circles.

When the slope angle is greater than 53° all critical circles are toe circles irrespective of the depth to the rigid stratum. Shallower slip circles are likely to occur for non-homogeneous conditions such as when the shear strength increases with depth.

Effective stress analysis

Stability analyses in terms of effective stress can only be carried out successfully when there is a reasonably accurate assessment of pore water pressures within the slope.

For a natural slope and a cutting the more critical long-term condition can be analysed using the effective stress analysis. The pore pressures can be represented by the steady seepage state when a flow net can be drawn or when a known equilibrium water table level exists.

For an embankment the more critical short-term condition can also be analysed using an effective stress analysis but this will be limited by the accuracy with which the excess pore pressures can be estimated.

Effective stress analysis – method of slices (Figure 12.11)

A circular arc slip surface is presumed with radius R. The soil segment above this circular arc is of unit thickness and is divided into a number of vertical slices usually of equal width b and varying height h

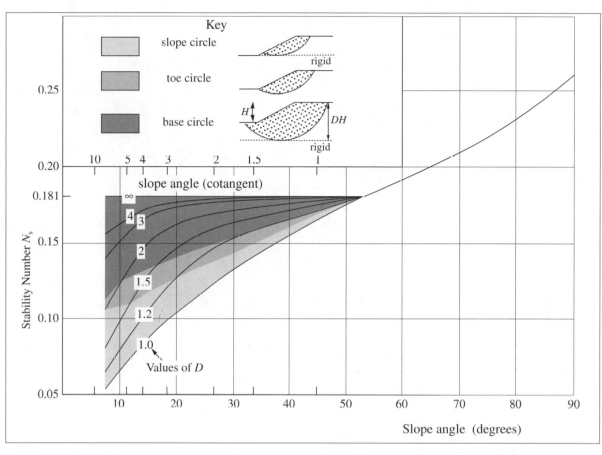

FIGURE 12.10   *Taylor's curves*

as shown in Figure 12.11. With a sufficiently large number of slices the base of each slice can be assumed to be a straight line inclined at an angle α to the horizontal and with a length $l = b \sec\alpha$.

The segment is separated into slices only for convenience of analysis. It is assumed that the segment (all of the slices) rotates around the circle centre O as a rigid body so that the factor of safety is the same for each slice, i.e. there is one factor of safety for the trial surface chosen. This implies that forces must act between the slices, known as inter-slice forces and these are usually taken as normal and tangential to the sides of the slices.

The factor of safety is defined as the ratio of the maximum available shear strength around the slip surface, τ_f to the shear strength, τ_m, mobilised by the overturning moment:

$$F = \frac{\tau_f}{\tau_m} \qquad (12.36)$$

The forces acting on a slice are:

1. The total weight of the slice, $W = \gamma b\, h$ where γ is the bulk unit weight of the soil.
2. The weight of each slice will induce a shear force parallel to its base $S = W \sin\alpha$.
3. The total normal force on the base, $N = \sigma l$.
4. The total normal force is obtained from the total normal stress which has two components, the effective normal force $N' = \sigma'l$ and the water force $U = ul$ where u is the pore water pressure at the centre of the base of the slice.
5. The shearing resistance of the soil will provide a shear force $T = \tau_m\, l$.

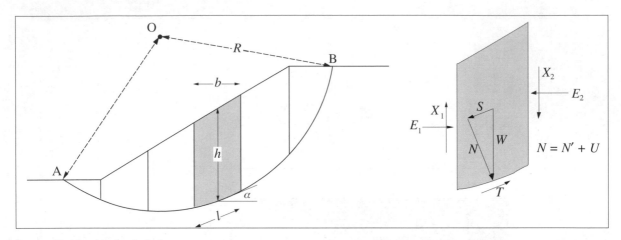

FIGURE 12.11   *Method of slices*

6. The inter-slice forces can be represented as total normal forces E_1 and E_2 and tangential shear forces X_1 and X_2.

Other forces acting on the segment such as a foundation or surcharge at the toe or crest of the slope, or a water-filled tension crack may be included in the analysis.

For moment equilibrium, ignoring inter-slice forces, the overturning moment produced by the forces S must be balanced by the resisting moment of the mobilised shear strength forces T:

$$\Sigma S R = \Sigma T R \qquad (12.37)$$

$$\Sigma W \sin \alpha R = \Sigma \tau_m l R = \Sigma \frac{\tau_f}{F} l R \qquad (12.38)$$

Since F is the same for all slices:

$$F = \frac{\Sigma \tau_f l}{\Sigma W \sin \alpha} \qquad (12.39)$$

The shear strength of the soil in terms of effective stress is given by equation 12.8 and:

$$\tau_f l = c'l + N' \tan\phi' \qquad (12.40)$$

so that, from equation 12.39:

$$F = \frac{\Sigma (c'l + N' \tan \phi')}{\Sigma W \sin \alpha} \qquad (12.41)$$

For a homogeneous slope where c' and ϕ' are constant along the slip surface:

$$F = \frac{c' L_{AB} + \tan \phi' \, \Sigma N'}{\Sigma W \sin \alpha} \qquad (12.42)$$

where L_{AB} is the total length of the arc between A and B.

The solution of this expression requires assumptions to be made about the interslice forces which affects the values of the forces N'. Two methods are given below.

Fellenius method

Also known as the Swedish method, in this analysis it is assumed that the interslice forces are equal and opposite so their resultants are zero:

$$E_1 = E_2 \text{ and } X_1 = X_2. \qquad (12.43)$$

Resolving the forces acting normal to the base of each slice:

$$N' = W \cos\alpha - ul \qquad (12.44)$$

where $u = \gamma_w \, h_w$ and h_w is the height to the water table above the base of the slice or related to the nearest equipotential if a flow net is drawn. From equation 12.42 the factor of safety is then:

$$F = \frac{c' L_{AB} + \tan \phi' \, \Sigma \, (W \cos \alpha - ul)}{\Sigma W \sin \alpha} \qquad (12.45)$$

Values of W, α and u can be determined for each slice and presented in a table, the summations are obtained by addition in the columns.

This solution underestimates the factor of safety compared with more accurate methods of analysis with an error of up to 20%, and will, therefore, be conservative.

Bishop simplified method

In this method (Bishop, 1955) it is assumed that the tangential interslice forces are equal and opposite, i.e.

$$X_1 = X_2 \qquad (12.46)$$

but the normal interslice forces are not equal, $E_1 \neq E_2$. However, resolving forces in the vertical direction (when E_1 and E_2 can be ignored) gives:

$$W = N' \cos \alpha + ul \cos \alpha + \frac{c'l}{F} \sin \alpha + \frac{N'}{F} \tan \phi' \sin \alpha$$

$$\therefore N' = \frac{W - \dfrac{c'l}{F} \sin \alpha + ul \cos \alpha}{\cos \alpha + \dfrac{\tan \phi' \sin \alpha}{F}} \qquad (12.47)$$

Inserting this expression into equation 12.42 and rearranging gives:

$$F = \frac{1}{\sum W \sin \alpha} \sum \frac{[c'b + (W - ub) \tan \phi'] \sec \alpha}{1 + \dfrac{\tan \alpha \tan \phi'}{F}} \qquad (12.48)$$

Since the value of F occurs on both sides of the expression a trial value for F must be chosen on the right-hand side to obtain a value of F on the left-hand side. By successive iteration convergence on the true value of F is obtained.

SEE WORKED EXAMPLE 12.5.

The method is obviously better solved by using a computer program which can obtain the factor of safety of a trial circle in a matter of seconds compared to many minutes when done manually. This speed of calculation is also beneficial in searching for the circle giving the lowest factor of safety where the computer programs will analyse circles over a grid of circle centres and with the circles passing through chosen points on the slope, over a range of radius values or tangential to particular levels. The programs also permit more complex soil profiles, groundwater conditions, external loading and seismic effects.

SEE WORKED EXAMPLE 12.6.

Pore pressure ratio r_u

The pore pressure ratio, r_u can be used to represent overall or local pore pressure conditions in a slope. It is given by the ratio of the pore water pressure to the total stress:

$$r_u = \frac{u}{\gamma h} \qquad (12.49)$$

Equation 12.48 can then be written as:

$$F = \frac{1}{\sum W \sin \alpha} \sum \frac{[c'b + W(1 - r_u) \tan \phi'] \sec \alpha}{1 + \dfrac{\tan \alpha \tan \phi'}{F}} \qquad (12.50)$$

Effective stress analysis – stability coefficients (Figures 12.12 – 12.14)

Bishop and Morgenstern (1960) found a relationship between the factor of safety of a slope and the pore pressure ratio r_u using the Bishop simplified method of analysis with effective stress conditions as:

$$F = m - n \, r_u \qquad (12.51)$$

where m and n are termed stability coefficients. These are related to a number of variables, the slope angle, the soil properties ϕ' and a combined value $c'/\gamma H$ and a depth factor D, similar to Taylor's approach. Values of m and n are plotted and tabulated in their paper and have been extended by O'Connor and Mitchell (1977) and Chandler and Peiris (1989).

This method has been used for many years as a rapid means of assessing the stability of a slope. However, the method of estimating an average value of r_u for a slope can be laborious and inaccurate and representing the pore pressure condition beneath the whole of a slope by a single value can be inappropriate.

This author (Barnes, 1992) has published a method which gives the critical (minimum) factor of safety for long-term effective stress stability of a homogeneous slope in the form:

$$F = a + b \tan\phi' \qquad (12.52)$$

Pore pressures are represented as a steady-state con-

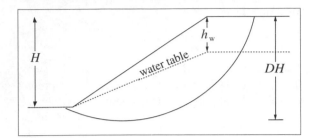

FIGURE 12.12 *Notation for stability coefficients*

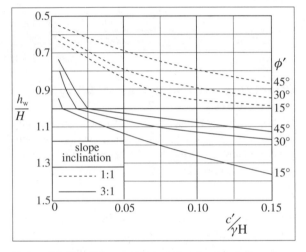

FIGURE 12.13 *Water table location for 'dry' slopes*

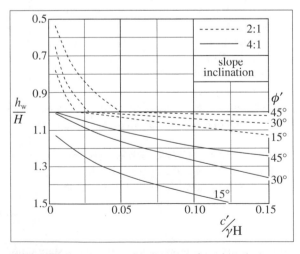

FIGURE 12.14 *Water table location for 'dry' slopes*

dition by a water table at toe level beyond the toe and inclined at various angles within the slope given by the depth below crest level, h_w, see Figure 12.12. This allows the groundwater conditions to be represented directly by an appropriate water table level and enables the effect of water table fluctuations on the factor of safety to be readily determined.

The stability coefficients a and b have been found to be related to the slope angle, the cohesion soil parameter, $c'/\gamma H$, and the water table parameter, h_w/H and are given in Tables 12.1 to 12.4 for slope inclinations of 1:1, 2:1, 3:1 and 4:1, respectively. Values of the coefficients for intermediate values of all of the above parameters can be obtained from the tables with sufficient accuracy by linear interpolation.

SEE WORKED EXAMPLES 12.7 AND 12.8.

When the critical circle cuts below the water table the slope is described as 'wet' and when the critical circle lies entirely above the water table the slope can be considered as a 'dry' slope.

The water table parameter, h_w/H, at which the 'dry' slope condition is obtained has been determined and is plotted in Figures 12.13 and 12.14. If the actual water table lies below this level then the slope can be considered 'dry' with no effect from the water table. The appropriate values of a and b are then obtained from the shaded areas in the Tables.

For steeper slopes, higher values of h_w/H and higher values of ϕ' a lower factor of safety may be obtained for the 'dry' condition. These values of ϕ', above which the slope should be considered as 'dry' are given in brackets in Tables 12.1 and 12.2.

Submerged slopes (Figure 12.15)
When a slope is submerged fully or partially the weight of each slice should be calculated using the bulk unit weight, γ_b, above the external water level, A–B and the submerged or buoyant unit weight γ_{sub} below the line A–B, the shaded area in Figure 12.15. This is because the water in the soil pores in the slope below A–B and its moment about O is balanced by the water outside the slope. Thus, the overturning moment is reduced so the factor of safety increases. As the external water level falls (drawdown) the factor of safety will decrease.

The shearing resistance along the slip surface is determined from effective stresses which are

TABLE 12.1 *Stability coefficients a and b for slope 1:1 (Barnes, 1992)*

$\dfrac{h_w}{H}$	$\dfrac{c'}{\gamma H}=0.005$		$\dfrac{c'}{\gamma H}=0.025$		$\dfrac{c'}{\gamma H}=0.050$		$\dfrac{c'}{\gamma H}=0.100$		$\dfrac{c'}{\gamma H}=0.150$	
	a	b	a	b	a	b	a	b	a	b
0		0.16		0.27		0.36		0.45		0.52
0.10		0.51		0.54		0.60		0.67		0.74
0.20		0.71		0.73		0.78		0.85		0.91
0.25	0.06	0.79	0.22	0.82	0.38	0.87	0.68	0.93	0.97	1.00
0.30		0.87		0.90		0.95		1.01		1.08
0.40		1.01		1.06		1.11		1.17		1.24
0.50		1.15(35)*		1.21		1.27		1.33		1.40
0.60				1.29(35)*		1.36		1.42		1.50
0.70						1.41(40)*		1.50		1.60
0.75						1.43(35)*		1.54(45)*		1.63(45)*
0.80								1.56(40)*		1.65(40)*
0.90										1.70(30)*
1.00										
Dry	0.08	1.12	0.27	1.21	0.44	1.32	0.75	1.46	1.04	1.56

*When ϕ' is greater than the value shown in brackets treat the slope as 'dry'. The shaded area represents the 'dry' condition.

TABLE 12.2 *Stability coefficients a and b for slope 2:1*

$\dfrac{h_w}{H}$	$\dfrac{c'}{\gamma H}=0.005$		$\dfrac{c'}{\gamma H}=0.025$		$\dfrac{c'}{\gamma H}=0.050$		$\dfrac{c'}{\gamma H}=0.100$		$\dfrac{c'}{\gamma H}=0.150$	
	a	b	a	b	a	b	a	b	a	b
0		0.88		1.01		1.11		1.27		1.37
0.10		1.24		1.31		1.38		1.50		1.60
0.20		1.46		1.51		1.57		1.69		1.78
0.25	0.06	1.56	0.24	1.60	0.42	1.66	0.75	1.78	1.07	1.87
0.30		1.64		1.69		1.75		1.86		1.95
0.40		1.81		1.85		1.91		2.02		2.11
0.50		1.96		2.00		2.06		2.17		2.26
0.60		2.07		2.13		2.22		2.33		2.41
0.70		2.17(45)*		2.22		2.31		2.43		2.52
0.75		2.20(30)*		2.27		2.35		2.48		2.57
0.80		2.22(25)*		2.30(45)*		2.38		2.52		2.62
0.90				2.38(30)*		2.44		2.60		2.70
1.00				2.39(25)*		2.50(40)*		2.67		2.79
Dry	0.10	2.08	0.30	2.24	0.50	2.37	0.86	2.56	1.20	2.69

*When ϕ' is greater than the value shown in brackets treat the slope as 'dry'. The shaded area represents the 'dry' condition.

TABLE 12.3 *Stability coefficients a and b for slope 3:1*

h_w/H	$c'/\gamma H = 0.005$		$c'/\gamma H = 0.025$		$c'/\gamma H = 0.050$		$c'/\gamma H = 0.100$		$c'/\gamma H = 0.150$	
	a	b	a	b	a	b	a	b	a	b
0		1.48		1.65		1.77		1.95		2.09
0.10		1.87		1.95		2.03		2.18		2.30
0.20		2.11		2.16		2.24		2.37		2.48
0.25		2.21		2.26		2.33		2.45		2.56
0.30		2.31		2.35		2.42		2.54		2.65
0.40		2.49		2.53		2.59		2.71		2.81
0.50	0.07	2.66	0.25	2.69	0.44	2.75	0.79	2.87	1.12	2.97
0.60		2.79		2.85		2.92		3.02		3.13
0.70		2.91		2.96		3.03		3.14		3.25
0.75		2.97		3.01		3.09		3.20		3.31
0.80		3.02		3.06		3.14		3.25		3.36
0.90		3.11		3.16		3.23		3.36		3.47
1.00				3.25		3.32		3.45		3.57
Dry	0.11	3.10	0.33	3.27	0.56	3.41	0.95	3.63	1.31	3.80

TABLE 12.4 *Stability coefficients a and b for slope 4:1*

$\dfrac{h_w}{H}$	$\dfrac{c'}{\gamma H} = 0.005$		$\dfrac{c'}{\gamma H} = 0.025$		$\dfrac{c'}{\gamma H} = 0.050$		$\dfrac{c'}{\gamma H} = 0.100$		$\dfrac{c'}{\gamma H} = 0.150$	
	a	b	a	b	a	b	a	b	a	b
0		2.05		2.22		2.35		2.53		2.67
0.10		2.46		2.54		2.63		2.78		2.90
0.20		2.72		2.77		2.84		2.98		3.09
0.25		2.83		2.88		2.94		3.07		3.18
0.30		2.93		2.98		3.04		3.16		3.27
0.40		3.13		3.17		3.22		3.34		3.44
0.50	0.07	3.32	0.26	3.35	0.47	3.40	0.84	3.51	1.18	3.61
0.60		3.49		3.52		3.57		3.67		3.77
0.70		3.62		3.66		3.70		3.81		3.90
0.75		3.69		3.72		3.76		3.87		3.97
0.80		3.74		3.78		3.82		3.93		4.03
0.90		3.86		3.89		3.94		4.05		4.15
1.00		3.97		4.00		4.05		4.16		4.26
Dry	0.12	4.11	0.37	4.29	0.60	4.45	1.01	4.70	1.38	4.90

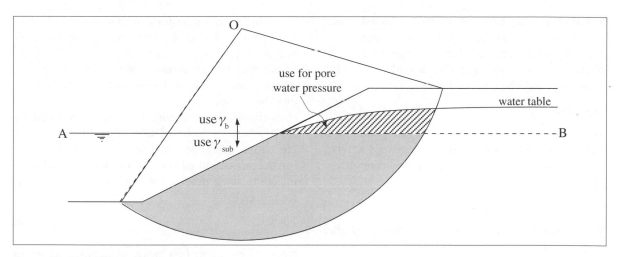

FIGURE 12.15 *Effect of submergence*

obtained partly by assuming submerged unit weights for the slice weights W below the line A–B. If the water table within the slope lies above the line A–B then the remaining pore pressure is determined from the difference between the piezometric head on each slice and the external water level A–B, the hatched area in Figure 12.15.

Rapid drawdown (Figure 12.16)

Figure 12.16 shows the upstream face of an earth dam where the steady seepage condition has become established with flow towards the downstream face. The phreatic surface and an equipotential passing through the point P on a trial slip surface are also shown. The initial pore water pressure u_0 at P in this condition is:

$$u_0 = \gamma_w(h + h_w - h') \qquad (12.53)$$

When the reservoir level is lowered seepage will then commence towards the upstream surface but if the permeability of the soil is low this re-adjustment will not be given sufficient time to establish and a high water level condition and hence high pore pressures will remain in the earth dam when support from the external water has been removed. It is found that because of the very slow adjustment a drawdown

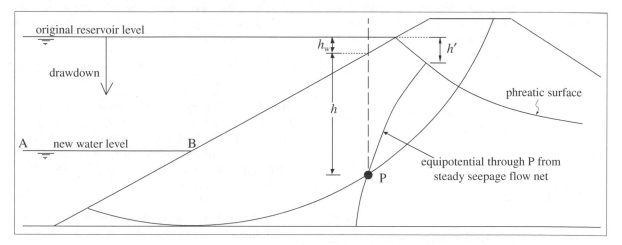

FIGURE 12.16 *Rapid drawdown*

period of several weeks can still be considered 'rapid' with the soil behaving in an undrained manner.

Bishop and Bjerrum (1960) suggested that the change in pore water pressure u under undrained conditions could be represented by:

$$\Delta u = \bar{B}\Delta\sigma_1 \tag{12.54}$$

where $\Delta\sigma_1$ is the change in total stress and represents a combined pore pressure parameter for a partially saturated compacted fill. In this case $\Delta\sigma_1 = -\gamma_w h_w$ so:

$$\Delta u = -\bar{B}\gamma_w h_w \tag{12.55}$$

Therefore, the pore water pressure immediately after drawdown is:

$$u = u_0 + \Delta u \tag{12.56}$$

$$u = \gamma_w [h + h_w (1 - \bar{B}) - h'] \tag{12.57}$$

A conservative approach is obtained by assuming $\bar{B} = 1$ and h' can be neglected since it is generally small, in which case pore pressures on the slip surface are given by:

$$u = \gamma_w h \tag{12.58}$$

The stability can then be analysed in the manner described above assuming a partially submerged slope with an external water level A–B, as shown in Figure 12.15.

Non-circular slip surfaces – Janbu method

Where non-homogeneous soil profiles exist, such as with layered strata, a non-circular slip surface may be more appropriate. Using the method of slices, see Figure 12.11, and considering overall horizontal equilibrium as the stability criterion, Janbu (1973) obtained the following expression for the average factor of safety along the slip surface:

$$F = \frac{\sum [c'b + (W + dX - ub) \tan \phi'] \, m_\alpha}{\sum (W + dX) \tan \alpha} \tag{12.59}$$

where:

$$m_\alpha = \frac{\sec^2 \alpha}{1 + \dfrac{\tan \phi' \tan \alpha}{F}} \tag{12.60}$$

and $dX = X_1 - X_2$, the resultant vertical interslice force.

A similar procedure for the solution of the Bishop simplified method is required using an iterative procedure to converge towards the value of F and an assumption for the position of the interslice forces on each slice (a line of thrust) and their magnitudes must be included.

A simplified procedure has been suggested by Janbu where the factor of safety F_0 is obtained using the above expression and assuming the interslice forces can be ignored, $dX = 0$. The factor of safety, F including the influence of the interslice forces is then given by:

$$F = f_0 F_0 \tag{12.61}$$

where f_0 is a correction factor related to the depth of the slip mass and the soil type, see Figure 12.17.

Morgenstern and Price (1965) published a rigorous method of analysis for slip surfaces of any shape by considering force and moment equilibrium for each slice and assuming a relationship between the normal and tangential interslice forces, usually in the form of $X = \lambda E$. The complex iterative calculations required to obtain values of F and λ for a slip surface mean that use of a computer program is essential.

Wedge method – single plane
(Figure 12.18)
Wedge analysis is a useful technique when assessing the stability of a cross-section with distinct planar boundaries such as backfilled slopes and zoned embankments.

For a single plane wedge, see Figure 12.18, the forces acting along the slip surface are resolved to obtain the factor of safety:

$$F = \frac{c'L + (W \cos \alpha - U) \tan \phi'}{W \sin \alpha} \tag{12.62}$$

where L is the length of the slip plane between A and D. W is the total force obtained by multiplying the area ABD by the bulk unit weight of the soil. U is the pore pressure force obtained from the area ACD multiplied by the unit weight of water, γ_w. The effective force $N' = W\cos\alpha - U$.

Wedge method – multi-plane (Figure 12.19)
When there is more than one plane surface the sliding mass can be separated into wedges with ver-

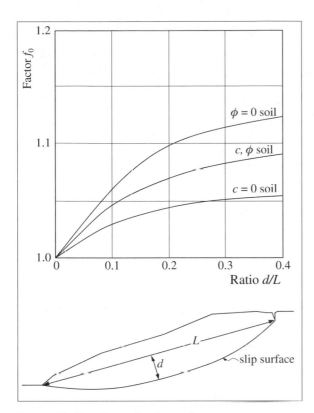

FIGURE 12.17 *Correction factor f_0*

$$\theta = \phi_m \text{ or } \tan \theta = \frac{\tan \phi'}{F} \qquad (12.63)$$

where ϕ_m represents the mobilised angle of shearing resistance. Since F is not known initially a trial approach is required by adjusting the value for F until convergence is achieved.

From Figure 12.19 each wedge is considered separately. The magnitude of the total force, W_1, is obtained from the area of the wedge multiplied by the unit weight of the soil. A polygon of forces can now be drawn for wedge 1. The magnitude and direction of $c'L_1/F$ (the cohesion part of T_1) are known. The direction of N_1' is known. The resultant R_1 of the frictional component is assumed to act at the angle ϕ_m' from the direction of N_1'. The direction of P_1 is assumed so the polygon of forces can be closed and a value of P_1 measured.

The same procedure is adopted for wedge 2 to obtain a value of P_2. If $P_2 = P_1$ the correct value of F has been chosen. Otherwise the procedure must be repeated until $P_2 = P_1$. Alternatively, the value of $P_2 - P_1$ could be plotted versus the factor of safety as shown in Figure 12.19. The correct value of F will occur where $P_2 - P_1 = 0$. This approach will also give some indication of the adjustment required to the value of F.

Factor of safety

The choice of an acceptable factor of safety requires sound engineering judgement due to the multitude of factors which must be considered. It must also be remembered that a factor of safety can only be determined when there is an appropriate method of analysis; some modes of failure such as flow slides or erosion cannot be readily analysed.

The factors to be considered generally fall under two headings: the consequences of a failure occurring and the confidence in the information available.

A higher factor of safety would be chosen where there is a risk to life and adjacent structures and would be related to an ultimate limit state. A serviceability limit state would exist where there is a possibility that the slope could be prone to deformations which could affect adjacent structures even though it may have adequate stability otherwise. Lateral

tical interfaces with an inter-wedge force P. The magnitude of P is not known but can be found from a polygon of forces providing an assumption about its inclination θ is made. The value of the factor of safety obtained can be sensitive to the value of θ chosen. A reasonable assumption is that:

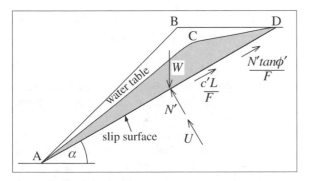

FIGURE 12.18 *Wedge method – single plane*

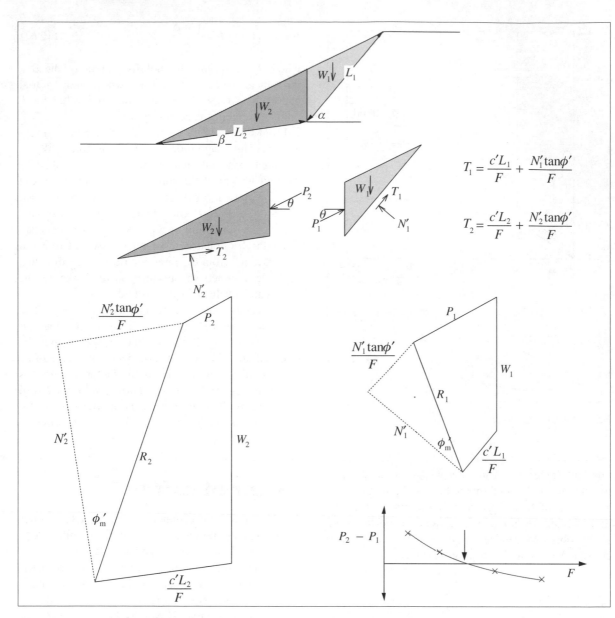

$$T_1 = \frac{c'L_1}{F} + \frac{N_1' \tan\phi'}{F}$$

$$T_2 = \frac{c'L_2}{F} + \frac{N_2' \tan\phi'}{F}$$

FIGURE 12.19 *Wedge method – two planes*

movements and heave at the toe of a slope, lateral movements within the slope and lateral movements and settlements at the crest of the slope could occur, having effects on buried pipes, drains, road surfaces etc.

A lower factor of safety could be chosen where the period of exposure is small such as for temporary works and for economical reasons such as where

instabilities do not affect lives or structures and may be localised requiring simple remedial measures, only when required.

The complexity of the ground conditions, the adequacy of the information obtained from the site investigation and the certainty of the design parameters, such as shear strength and pore pressures, all affect the confidence in the chosen factor of safety.

Previous local experience can be invaluable, especially if there is any knowledge about the presence of pre-existing slip surfaces. Future changes such as to water table levels, surface profiles must also be considered.

The Code of Practice for Earthworks (BS6031:1981) suggests that for cuttings and natural slopes, provided a good standard of site investigation is obtained and considering the factors mentioned above with no more than average importance a factor of safety between 1.3 and 1.4 should be used in the design for first-time slides with a factor of safety of about 1.2 for slides where pre-existing slip surfaces exist. Similar considerations should be applied to embankments although their design and construction may be controlled more by the need to limit pore pressure increases during construction and to minimise the effects of settlements.

For earth dams, with similar considerations to above, factors of safety greater than 1.3, 1.5 and 1.2 could be acceptable at the end of construction, after establishment of the steady seepage condition on the downstream face and after rapid drawdown on the upstream face, respectively. However, such important structures must be monitored during and after construction to compare with predicted behaviour using piezometer systems to measure pore water pressures, surveying techniques, vertical extensometers, and inclinometers to measure deformations.

An approach to the adoption of partial factors and limit state design is given for the plane translational slide method of analysis and can be applied to the other methods.

SEE WORKED EXAMPLES 12.2 AND 12.8.

SUMMARY

Mass movements or landslides on slopes can take several forms depending on the geological and hydrogeological conditions and the types of soil present. For some of these modes there is a method of analysis available and their degree of stability can be assessed but those that cannot be analysed should not be ignored.

As well as the conditions described above the stability of natural slopes and cuttings depends on whether the slope has already failed at some time in the past. A first-time slide will mobilise the peak strength initially, reducing to the critical state strength. The residual strength is to be expected where there is a pre-existing slip surface.

The stability of artificial earth structures, such as embankments and earth dams, is related to the failure of the slopes within the structure and failure within the foundation soils beneath. Placing and compacting fill materials means that slope failures are likely to be first-time slides with the critical state strength mobilised. Foundation failures depend on the natural geological and groundwater conditions beneath and if soil with pre-existing slip surfaces is left in place then re-activation of slipping is likely to occur.

Cuttings and the slopes of embankments tend to fail in the long term after pore pressure redistribution has achieved a steady-state condition. Foundation failure beneath an embankment is most likely near the end of construction when pore pressures generated in the underlying soils are at their highest. Both the upstream and downstream faces of an earth dam can be subject to failure due to slope instability. Several other factors such as erosion, piping, hydraulic fracture are equally important.

Three methods of analysis are presented, the plane translational mode, circular arc analysis using the method of slices and wedge analysis. The degree of stability, represented by the factor of safety, is considered for several conditions.

The significant effect of pore pressures on the stability of a slope cannot be overstated.

First-time slide – Lodalen, Oslo, Norway

Case Objectives:

This case illustrates:

- slopes that have not failed previously can undergo first-time slides
- that peak strength is operative on the slip surface
- that slips are usually instigated by changes to the slope angle and/or the pore water pressure condition
- when unloading occurs, as in this case, a new pore water pressure equilibrium needs to establish itself. In low permeability soils this can take many years
- that stability is determined by the effective stress conditions in the slope

This slide took place in the railway yard at Lodalen on 6 October 1954. The slope that failed had been originally at an inclination of about 1 in 2.5 but in the 1920s as part of the expansion of the rail facilities the lower part of the slope was excavated 5–6 m and the inclination increased to 1 in 2. In 1949 a further 2.5 m was removed, keeping the slope at 1 in 2. Three sheds at the top of the slope did not contribute to the slide but one was left in a precarious condition.

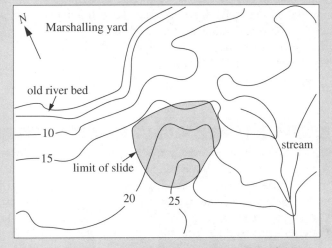

The failure was described as a rotational movement of a monolithic mass of soil producing a scarp at the rear of the slide of about 5 m and a heaved toe that was pushed about 10 m forward, carrying a rail line and three carriages which were badly damaged. The width of the slide was about 50 m with a length of 40 m and a total volume of about 10,000 m³. A ground investigation was carried out involving boreholes, vane tests, piston samples and pore pressure measurements. The slip surface location was detected at two boreholes by careful examination of the samples. A thin remoulded horizon was found with a higher moisture content (by about 5%) than elsewhere. The soil comprised marine clays of late-glacial and post-glacial origin with a drier crust. The clay was soft and firm lightly overconsolidated low plasticity and non-fissured.

Micro-fossil investigations and salt concentrations suggested that the clays had been quick clays and the present state had originated from ancient slides. After remoulding the high sensitivity would have been lost

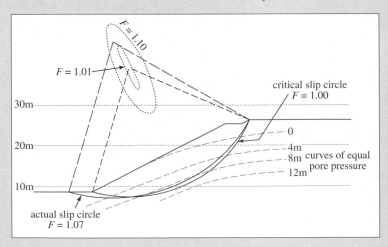

and following consolidation the strength would increase and return to normal sensitivity. The sensitivity varied between 3 and 15. The clay had average soil properties of moisture content = 29% liquid limit = 35% plastic limit = 19% clay fraction = 46% (Skempton, 1964) with peak shear strength parameters of c' = 10 kN/m^2 and ϕ' = 27°. Using these parameters and the Bishop (1955) method of analysis with assumed pore pressure conditions a factor of safety of 1.00 was obtained for a cross-section through the middle of the slope. The critical slip circle also corresponded closely with the actual slip surface.

If a cohesion intercept of zero had been used a factor of safety of 0.70 would have been obtained which is obviously incorrect. This case shows that an analysis based on peak strengths for a first-time slide is appropriate. When the slope was first cut the pore pressure conditions illustrated in Figure 12.2 would have pertained. The pore pressure regime for the long-term case was modelled from the measured values as contours of equal pore pressure relating to a steady seepage condition.

Analyses carried out for cross-sections near the sides of the slipped mass gave factors of safety which were 10–20% higher than in the middle. Slope stability analysis usually just considers a simple circular cylindrical sliding surface with no end effects. In this case the bowl-shaped nature of the slipped mass suggests that a three-dimensional approach would be more appropriate such as proposed by Gens *et al.*, 1988. The fundamental importance of shear strength is discussed in Chapter 7.

CASE STUDY

Reactivation of Old Landslip – Jackfield, Shropshire, UK

Case Objectives:

This case illustrates:

- that slopes that have failed previously can remain stable for many years although their factor of safety remains close to unity
- that slips can be instigated by removal of toe support such as by river erosion and/or worsening pore water pressure conditions
- when slipping is reactivated only the residual strength is available on the slip surface
- that stability is determined by the effective stress conditions on the slip surface
- that thorough investigations must be made when it is suspected that ancient slip surfaces may exist otherwise engineering structures can be seriously affected. A recent example is at Carsington Dam, see below
- that ancient landslipping can be detected by geological, geomorphological and aerial photographic techniques

The village of Jackfield lies on the south side of the River Severn about 1 mile downstream from Ironbridge. The valley sides are marked by the scars of several landslides originating from the retreat of an ice sheet and an ice-dammed lake in late-glacial times. A similar slip occurred in the valley just upstream of the village in 1925.

Mineworkings for coal and clay had taken place in the area but these were fairly deep and not likely to be the cause of the slip which was considered to be part of the normal geological process of valley formation (Henkel and Skempton, 1954). The slope had been stable for a long time until a broken water main serving cottages near the river bank gave warnings of instability in 1950. Towards the end of 1951 at a time of exceptional heavy rainfall, fairly rapid ground movements took place continuing through the winter and by

February 1952 the road was in a dangerous condition. During the next few months the landslide developed dramatically with six of the houses demolished, closure of the road, daily adjustments to the railway track and relaying of mains supplies on the surface. A horizontal movement of about 18 m had occurred on the road in the centre of the moving mass with a smaller movement of 9 m on the river bank. This meant that considerable bulging and cracking occurred between the road and the river causing severe damage to the houses in this area. The rear of the slip comprised a vertical scarp face 1.2 m high.

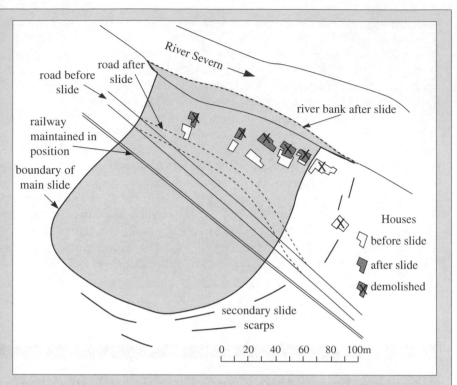

During the summer of 1952 movements virtually ceased but restarted in the autumn. The investigation carried out at this time was able to detect the depth to the slip surface in some of the boreholes. By April 1953 the slide had extended further to the south and east affecting more property. The movement on the road had increased to 24 m and at the river to 14 m. The width of the river had decreased from 38 m to 24 m resulting in erosion of the river bank. It was estimated that the slide mass was about 150,000m³. The hillside is formed of stiff clays and mudstones of the Halesowen Formation (formerly known as the

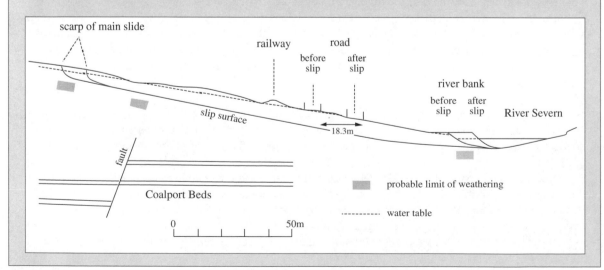

Coalport Beds) of Upper Carboniferous age. These beds dip gently to the south-east with a strike roughly at right angles to the river so it is unlikely that the slip is associated with the bedding. A weathered mantle, 6–8 m thick overlaid these beds and comprised firm and stiff fissured clays.

The slope is inclined at about 10° to the horizontal. The slip surface was found at an average depth of 5.5 m and with a length of slipped mass of 170–180 m giving a ratio of about 1 to 30 this is a good example of a slab or plane translational slide.

During winter groundwater was observed at the surface at several locations with an average depth of 0.6m. The pore pressure at the level of the slip surface was found to be non-artesian.

The average properties of the clay were liquid limit = 45%, plastic limit = 20%, clay fraction = 42%, moisture content = 25% but with a moisture content in the failure zone of 30%. In two of the boreholes the failure zone was found to comprise a layer of soft clay no more than 50 mm thick with a moisture content higher than the surrounding clay.

Effective stress shear strength parameters were determined from triaxial and direct shear box tests:

	c' kN/m^2	ϕ'
Peak strength, outside failure zone	10.5	25
Peak strength, inside failure zone	7	21
Residual strength*	0	19

*The shear box test was taken to its full travel but ϕ_r' was not determined directly and could be lower. Given the large scale of the slab slide edge effects could be ignored and the plane translational mode of failure used in the analysis. Using peak strength values a factor of safety of 2.06 would be obtained so it is clear that much lower values were operative on the slip surface. With a ϕ_r' value of 19° the factor of safety was 1.12. The more likely value of the residual angle of shearing resistance for the clay is 17° since the factor of safety of the failed mass was 1.0.

At Carsington Dam (Skempton and Coats, 1985) a layer of yellow clay had been left intact beneath the upstream slope of the dam. Towards the end of construction the whole upstream slope failed with most of the slip surface passing through the yellow clay. Geological interpretation of the yellow clay was that it was part of a head deposit formed during late-glacial times by downslope movement and associated shearing, a process known as periglacial solifluction. Geotechnical examination of the clay showed it to contain pre-existing shear surfaces making the residual strength the operative value for the clay.

Reference should be made to Chapter 7, Shear Strength.

Worked Example 12.1   Translational slide

A long slope comprising fine sand exists at an inclination of 4:1 (horizontal:vertical) with a water table parallel to the slope. The unit weight of the sand is 20 kN/m³ and the angle of internal friction is 26.6°. Assume the unit weight of water to be 10 kN/m³. Determine the factor of safety of the slope assuming:
i) the water table exists at ground level (worst case)
ii) the slope is dry (best case)
iii) the water table exists at 2.0 m below ground level

$$\tan \beta = \frac{1}{4} = 0.25$$

a) *Waterlogged slope*
assuming $c' = 0$ Equation 12.12 gives:

$$F = \frac{(20 - 10) \times 0.5}{20 \times 0.25} = 1.00$$

i.e. the slope is very close to failure in this condition

b) *dry slope*
Using Equation 12.11

$$F = \frac{0.50}{0.25} = 2.0$$

the slope has more than adequate stability in this condition

c) *water table at 2.0 m below ground level*
with $c' = 0$ Equation 12.10 simplifies to:

$$F = \frac{\tan \phi'}{\tan \beta} \frac{(\gamma z - \gamma_w z + \gamma_w h_w)}{\gamma z}$$

This expression can be used for slip surfaces below the water table:

$z = 2.0$ m $\quad F = \dfrac{(40 - 20 + 20)}{40} \times 2.0 = 2.0$ (dry slope case)

$\qquad\qquad\qquad\qquad\qquad\qquad\qquad\qquad\quad z = 3.0$ m $\quad F = \dfrac{(60 - 30 + 20)}{60} \times 2.0 = 1.67$

$z = 4.0$ m $\quad F = \dfrac{(80 - 40 + 20)}{80} \times 2.0 = 1.50$

$\qquad\qquad\qquad\qquad\qquad\qquad\qquad\qquad\quad z = 6.0$ m $\quad F = \dfrac{(120 - 60 + 20)}{120} \times 2.0 = 1.33$

$z = 10.0$ m $\quad F = \dfrac{(200 - 100 + 20)}{200} \times 2.0 = 1.20$

$\qquad\qquad\qquad\qquad\qquad\qquad\qquad\qquad\quad z = 50$ m $\quad F = \dfrac{(1000 - 500 + 20)}{1000} \times 2.0 = 1.04$

With a constant strength (homogeneous slope) the factor of safety decreases with depth. To assess a likely depth of slip surface any layering of strata should be included or an increase in strength (increasing ϕ') with depth should be considered.
Assuming suction above the water table Equation 12.18 could be used:

$z = 0.5$ m $\quad F = \left(1 + \dfrac{10 \times 1.5}{20 \times 0.5}\right) \times 2.0 = 5.00$

$$z = 1.0 \text{ m} \quad F = \left(1 + \frac{10 \times 1.0}{20 \times 1.0}\right) \times 2.0 = 3.00$$

$$z = 1.5 \text{ m} \quad F = \left(1 + \frac{10 \times 0.5}{20 \times 1.5}\right) \times 2.0 = 2.33$$

Worked Example 12.2   Translational slide – limit state design

A slope comprises a weathered upper layer of residual soil, 3.5 m thick overlying less weathered rock. The properties of the residual soil are:
$\gamma = 19.5 \text{ kN/m}^3$
$c' = 12 \text{ kN/m}^2$
$\phi' = 35°$
With a worst credible water table at 1 m below ground level determine the maximum slope angle that could be sustained by the residual soil for:
i) an overall factor of safety of 1.3
ii) partial factors of $\gamma_c = 1.6$ and $\gamma_\phi = 1.25$

Assume that a plane translational slide occurs within the residual soil parallel to its sloping surface with a maximum depth to a potential slip surface at the top of the rock. The general expression, Equation 12.10, can be used.

$$\text{i)} \quad 1.3 = \frac{12 + \tan 35° \cos^2 \beta(19.5 \times 3.5 - 9.8 \times 3.5 + 9.8 \times 1.0)}{19.5 \times 3.5 \times \sin \beta \cos \beta}$$

giving $\beta = 27.3°$

ii) In this case the properties given are assumed to be characteristic values. The unit weight of the soil is not factored, $\gamma_m = 1.0$. The partial factor on the permanent actions is, $\gamma_G = 1.0$.
Using Equations 12. 26 to 12.28:

$$\frac{12}{1.6} + \frac{\tan 35°}{1.25} \cos^2 \beta(19.5 \times 3.5 - 9.8 \times 3.5 + 9.8 \times 1.0) \geq 19.5 \times 3.5 \times \sin \beta \cos \beta$$

giving $\beta = 26.4°$

Worked Example 12.3 $\phi_u = 0°$ analysis – method of slices

A slope is to be cut into a soft clay with undrained shear strength of 30 kN/m² and unit weight of 18 kN/m³. The slope is 8.0 m high and its inclination is 2:1 (horizontal:vertical). Determine the factor of safety for the trial circle shown on Figure 12.20.

Equation 12.31 is used with values of b, α and mid-slice height h determined for each slice. The weight of each slice is obtained from $W = \gamma bh$.

slice no.	b	h	W	α°	$W\sin\alpha$	$b\sec\alpha$
1	0.65	0.15	1.8	25.7	−0.8	0.72
2	2.0	1.23	43.2	−20.0	−14.8	2.13
3	2.0	2.82	100.8	−1.8	−20.6	2.04
4	2.0	4.06	146.2	−3.9	−9.9	2.01
5	2.0	5.08	182.9	3.9	12.4	2.01
6	2.0	5.82	209.5	11.8	42.8	2.04
7	2.0	6.26	225.4	20.0	77.1	2.13
8	2.0	6.36	229.0	28.6	109.6	2.28
9	2.0	6.02	216.7	38.0	133.0	2.54
10	2.0	0.60	165.6	48.9	124.8	3.04
11	1.7	1.94	58.4	61.7	51.4	3.59
				$\Sigma =$	505.0	24.53

$$F = \frac{30 \times 24.53}{505.0} = 1.46$$

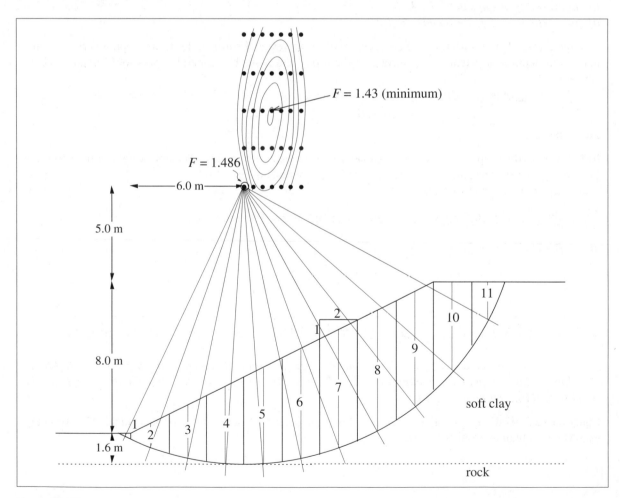

FIGURE 12.20

The included angle θ is 97° and the radius of the circle is 14.6 m so the length of the circular arc is

$$L = R\theta = 14.6 \times 97 \times \frac{\pi}{180} = 24.72 \text{ m}$$

Since this is a more accurate measure of $\Sigma b \sec\alpha$ a more accurate value of the factor of safety is obtained as

$$F = \frac{30 \times 24.72}{505.0} = 1.47$$

The accuracy of the factor of safety F is affected by the number of slices adopted.
A computer analysis (SLOPE ©) of the same slope using 25 slices gave a value of $F = 1.486$.
This circle does not give the lowest factor of safety so it is not the critical slip circle. A computer run (SLOPE©) obtained the F values for circles with their centres on a grid pattern and all tangential to the depth of 1.6 m below the toe. Contours of factor of safety are plotted on this grid and the circle with the lowest value of F lies at the centre of the contour plot, as shown on Figure 12.20. This gives the minimum factor of safety as 1.430.

Worked Example 12.4 $\phi_u = 0°$ analysis – Taylor's method

For the slope in Example 12.3 determine the minimum factor of safety using Taylor's stability number approach.

Depth factor $D = \dfrac{9.6}{8.0} - 1.2$

From the stability chart, Figure 12.10
$N_s = 0.146$

$$\therefore \quad F = \frac{30}{0.146 \times 18 \times 8} = 1.43$$

Worked Example 12.5   Effective stress analysis – Bishop simplified method

A 2:1 slope 10 m high, has been constructed in a stiff clay with effective stress parameters $c' = 5$ kN/m^2 and $\phi' = 30°$ and bulk unit weight 20 kN/m^3. For the circular arc shown in Figure 12.21 determine the factor of safety using the Bishop simplified method for the long-term condition when the pore pressures are related to the water table at 5.0 m below crest level.

Splitting the cross-section into 12 vertical slices values of α, the width and mid-height of each slice, b and h, respectively, are determined. The weight W of each slice $= \gamma bh$. The pore pressure u is determined at the mid-point of each slice as the height above the base of the slice to the water table h_w and $u = \gamma_w h_w$ where γ_w is the unit weight of water $= 9.81$ kN/m^3. Since F appears on both sides of Equation 12.48 a trial value of 1.0 has been assumed for the expression on the right hand side giving a calculated value F of 1.31. With a revised value of 1.4 a calculated value of $F = 1.394$ is obtained.

$$A = c'b + (W - ub) \tan \phi'$$

$$B = \frac{\sec \alpha}{1 + \dfrac{\tan \alpha \, \tan \phi'}{F}}$$

slice no.	b	h	W	α	$W\sin\alpha$	u	A	$F = 1.00$		$F = 1.40$	
								B	$A \times B$	B	$A \times B$
1	2.50	0.40	20.0	−15.6	−5.4	3.9	18.4	1.24	22.8	1.17	21.5
2	2.00	1.38	55.2	−8.8	−8.4	11.1	29.1	1.11	32.3	1.01	29.4
3	2.00	2.58	103.2	−2.9	−5.2	18.0	48.8	1.03	50.3	1.00	48.8
4	2.00	3.58	143.2	2.9	7.2	22.9	66.2	0.97	64.2	0.98	64.9
5	2.00	4.40	176.0	8.8	26.9	26.0	81.6	0.93	75.9	1.01	82.4
6	2.00	5.02	200.8	14.8	51.3	27.2	94.5	0.90	85.1	0.93	87.9
7	2.00	5.40	216.0	20.9	77.1	26.0	104.7	0.88	92.1	0.92	96.3
8	2.00	5.48	219.2	27.3	100.5	21.9	111.3	0.87	96.8	0.93	103.5
9	2.00	5.26	210.4	34.1	118.0	14.8	114.4	0.87	99.5	0.94	107.5
10	2.00	4.68	187.2	41.5	124.0	4.2	113.2	0.88	99.6	0.98	110.9
11	2.00	3.64	145.6	49.9	111.4	0	94.1	0.92	86.6	1.04	97.9
12	1.67	1.57	52.4	59.1	45.0	0	38.6	0.99	38.2	1.15	44.4
					$\Sigma = 642.4$				$\Sigma = 843.4$	$\Sigma = 895.4$	

$$F = \frac{843.4}{642.4} = 1.313 \qquad F = \frac{895.4}{642.4} = 1.394$$

Worked Example 12.6   Effective stress analysis – computer application

As shown in Example 12.3 a search must be made for the critical slip surface which gives the lowest factor of safety.

For a homogeneous slope this can be achieved by specifying a grid of circle centres and some points for the circles to pass through or to be tangential to a line. Using a commercial software program called SLOPE© marketed by Geosolve the slope in Example 12.4 was analysed using a grid of 25 (5 × 5) circle centres and a number of points beyond the toe for the circles to pass through. The contours of equal factor of safety plotted on the grid of centres gives a series of elliptical loops which lead to the critical circle centre. The lowest factor of safety for each point is plotted on Figure 12.21 and shows that the critical circle cuts at 2.5 m beyond the toe of the slope. The lowest factor of safety was determined as 1.40 with 20 slices and 5 iterations (varying F to find F).

Example 12.5 is a hand calculation for the critical slip circle to check the value of F obtained. It will be appreciated that to search for this critical slip circle can entail much trial and error and the benefits of using a computer program are obvious.

Worked Example 12.7   Effective stress analysis – stability coefficients

For the slope in Example 12.5 determine the minimum factor of safety from the stability coefficients method.

Using equation 12.52 and Table 12.2 for a slope of 2:1

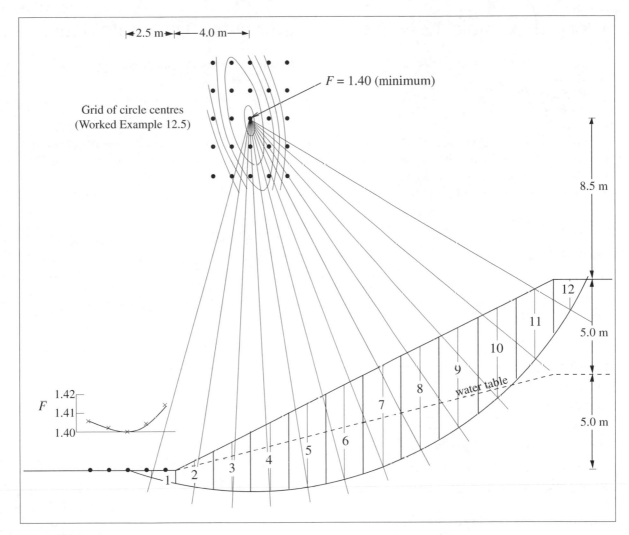

FIGURE 12.21

$$\frac{c'}{\gamma H} = \frac{5}{20 \times 10} = 0.025$$

$$\frac{h_w}{\gamma H} = \frac{5}{10} = 0.05$$

From the Table $a = 0.24$ and $b = 2.00$

$\therefore F = 0.24 + 2.00 \times \tan 30° = 1.395 \qquad (= 1.40)$

Worked Example 12.8 Stability coefficients – limit state design

A slope 10 m high is required to be cut into a stiff clay. The characteristic values of parameters for the clay are:

$\gamma = 21.5 \ kN/m^3$
$c_k' = 8 \ kN/m^2$
$\phi_k' = 36°$

Assuming a worst credible water table level of 2.5 m below the crest of the slope and coincident with the toe level of the slope determine the steepest slope that can be cut, adopting partial factors of $\gamma_c = 1.6$ and $\gamma_\phi = 1.25$. The design values for the clay are:

$\gamma = 21.5 \ kN/m^3$ (assume unfactored)

$$c_d' = \frac{8}{1.6} = 5 \ kN/m^2$$

$$\tan \phi_d = \frac{\tan 36°}{1.25} = 0.581$$

$\phi_d = 30.2°$

$$\frac{c_d'}{\gamma H} = \frac{5}{21.5 \times 10} = 0.0233$$

$$\frac{h_w}{H} = \frac{2.5}{10} = 0.25$$

For a 2:1 slope
Interpolating, from Table 12.2,

$$\frac{(0.0233 - 0.005)}{(0.025 - 0.005)} = 0.915$$

$a = (0.24 - 0.06) \times 0.915 + 0.06 = 0.225$

$b = (1.60 - 1.56) \times 0.915 + 1.56 = 1.597$

$F = 0.225 + 1.597 \times \tan 30.2° = 1.153$

For a 1:1 slope
Interpolating, from Table 12.1,
$a = (0.22 - 0.06) \times 0.915 + 0.06 = 0.206$
$b = (0.82 - 0.79) \times 0.915 + 0.79 = 0.817$
$F = 0.206 + 0.817 \times \tan 30.2° = 0.681$
Interpolating between these slopes

$$\text{slope} = (2 - 1) \frac{(1 - 0.681)}{(1.153 - 0.681)} + 1 = 1.676$$

A slope of 1.68 to 1 or 30.8° is the maximum design slope.

EXERCISES

12.1 A long slope exists at an angle of 14° to the horizontal and comprises overconsolidated clay with the following properties:

$c' = 5$ kN/m^2 $\phi' = 26°$ $\gamma = 19$ kN/m^3

With the water table at ground level and steady seepage parallel to the surface determine the factor of safety against shear failure assuming potential slip surfaces at 2.0 m, 4.0 m and 6.0 m below ground level.

12.2 For the slope described in Exercise 12.1 determine the factors of safety when the water table lies at 4.0 m below ground level with steady seepage parallel to the surface.

12.3 A long slope at an angle of 10° has failed along a slip surface at 2.5 m below ground level parallel to the ground surface. The water table lies at 0.5 m below ground level with seepage parallel to the ground surface. Assuming $c' = 0$ and $\gamma = 21$ kN/m^3 determine the operative value of the residual angle of friction ϕ_r'.

12.4 A slope excavated in soft clay has failed immediately after excavation along a circular slip surface as shown in Figure 12.22. Assuming the strength of the soft clay to be constant determine the mean value of shear strength acting at failure along the slip surface shown on the figure. The bulk unit weight of the clay is 18.5 kN/m^3.

Note: the approach gives the lowest shear strength value applicable only if the circle analysed is the critical one.

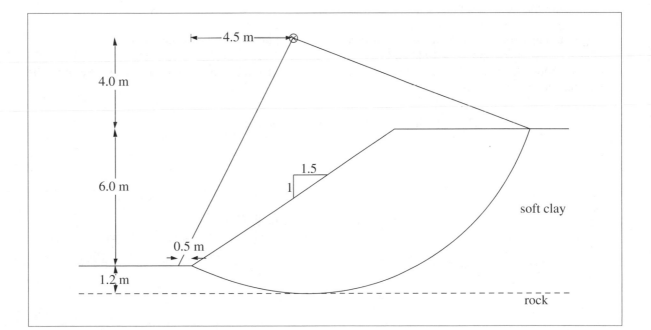

FIGURE 12.22

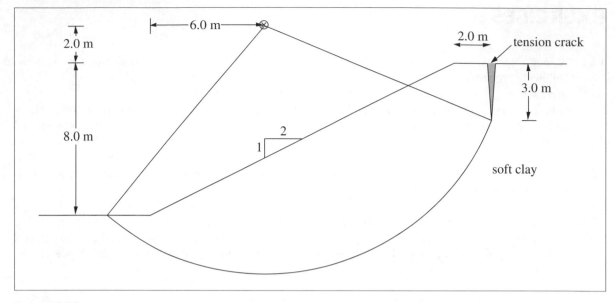

FIGURE 12.23

12.5   Determine the factor of safety for the slip surface shown in Figure 12.23, with a water-filled tension crack, 3.0 m deep.
$c_u = 35$ kN/m^2 $\gamma = 20$ kN/m^3

12.6   For the slope profile in Exercise 12.4 use Taylor's curves to determine the lowest shear strength which could support the slope.

12.7   A cutting slope 7 m high has been constructed at an angle of 2.5:1 in a firm clay with:
$c_u = 55$ kN/m^2 and $\gamma = 19$ kN/m^3.
The bedrock surface lies at 14 m below original ground level. Using Taylor's curves determine the short-term factor of safety.

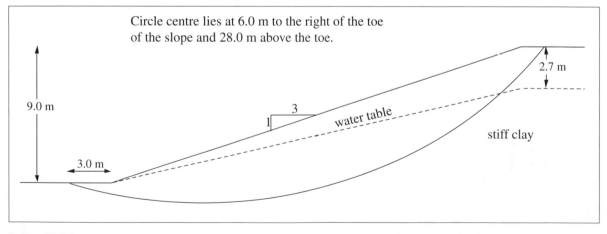

FIGURE 12.24

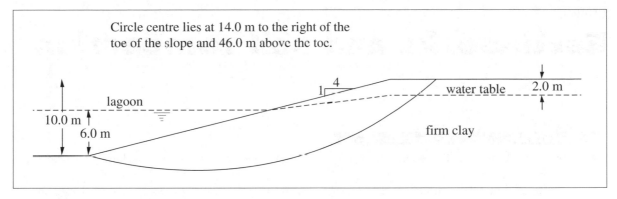

Circle centre lies at 14.0 m to the right of the
toe of the slope and 46.0 m above the toc.

lagoon

10.0 m

6.0 m

water table   2.0 m

firm clay

FIGURE 12.25

12.8   It is required to construct an excavation with cutting slopes 8 m high in a firm clay with $c_u = 40$ kN/m^2 and $\gamma = 20.5$ kN/m^3.
The bedrock surface lies at 12 m below ground level. Using Taylor's curves determine the slope angle which could be adopted to ensure short-term stability with a factor of safety of 1.5.

12.9   For the slope profile in Figure 12.24 the groundwater regime is represented by steady seepage with pore pressures given by the water table level shown. Determine the factor of safety for long-term conditions on the slip surface shown using the Bishop simplified method.
   $c' = 3$ kN/m^2 $\phi' = 26°$ $\gamma = 22.2$ kN/m^3

12.10 For the slope and groundwater conditions in Exercise 12.9 determine the minimum factor of safety using the stability coefficients method, Equation 12.52.

12.11 The slope shown in Figure 12.25 supports a lagoon with water 6 m deep and with a water table inside the slope at 2 m below ground level. Determine the factor of safety on the given slip surface using the Bishop simplified method.
   $c' = 5$ kN/m^2 $\phi' = 23°$ $\gamma = 19$ kN/m^3

12.12 The water level in the lagoon in Exercise 12.11 has been drawn down rapidly to the base level. The groundwater regime is now represented by the water table within the slope as in Exercise 12.11 but with a water table at ground level in the lagoon area. For these slope and groundwater conditions determine the factor of safety on the slip surface shown.

CHAPTER 13

Earthworks and soil compaction

OBJECTIVES

- To appreciate that an earth structure can comprise a variety of material types each performing a different function.
- To understand the factors affecting the acceptability of fill materials and the choice of criteria based on laboratory tests.
- To be aware of the factors that can affect the efficiency of an earthmoving operation including trafficability, softening and bulking.
- To understand the factors affecting the field compaction of soils and to be able to interpret the results of laboratory compaction tests.
- To understand the benefits of lime and cement stabilisation.

Earthworks

Earthworks or earth structures are constructed where it is required to alter the existing topography. They comprise excavation and filling and are most commonly formed for highway works, such as cuttings and embankments but they may also consist of site levelling, such as for industrial or housing estates and excavation and backfilling of quarries, trenches and foundations.

The stability of cutting and embankment slopes is discussed in Chapter 12. In this chapter the construction and compaction of the fill material to form an embankment structure is considered.

Construction plant (Figure 13.1)
The basic functions of construction plant are:

- *Excavation*
  To break up and remove soil or rock from the cut areas. Back-acters, rippers, pneumatic breakers and explosives are used for this purpose.
- *Loading*
  Some items of plant such as dump trucks require an additional item, a back-acter, to load soil into them whereas scrapers excavate and load themselves.

- *Transporting*
  For backfilling pipes, services and foundations the excavated spoil is usually heaped close by so transport is not necessary. On some highway schemes the haul distances between the cut and fill areas can be several kilometres. The overall cost and efficiency of the earthmoving operations will be determined by the ability of the plant to run in high gears (with faster speeds and more fuel economy) along a haul road comprising the soil surface. This haul road will simply be the surface of the excavation or the fill areas so it must have sufficient strength or trafficability to support the plant. Factors such as gradients, rolling resistance and rutting beneath tyres and tracks should be considered.

  Towed scrapers (tractor-pulled) are most efficient for short hauls, say less than 200–300 m, motorised scrapers are required for intermediate haul distances while dump trucks are most efficient over long haul distances, say greater than 1–2 km.
- *Depositing and spreading*
  Some items of plants such as dump trucks, can only deposit their loads by tipping in a heap. The soil must then be spread in a thin layer by a

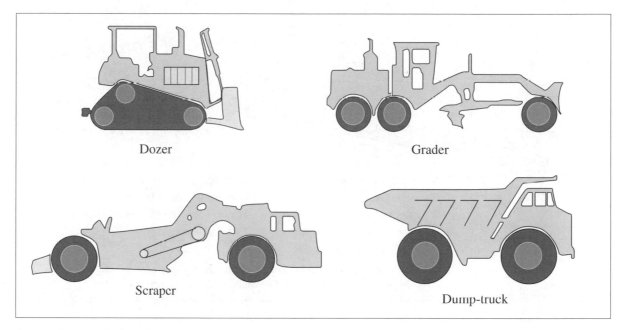

FIGURE 13.1 *Typical earthmoving plant*

dozer or grader in preparation for compaction. Scrapers spread the soil as they unload and provide better mixing of the soils.

- *Compaction*
  On large-scale bulk earthworks this will be carried out using rollers running over the surface of the previously spread layers with sufficient compactive effort to provide a stable fill material. These are described later. In more confined excavations, such as for backfilling services and foundations, plate compactors, tampers, rammers and dropping-weight compactors may be used.

Fill material used for landscaping purposes may be given little or no compaction since high strengths and low compressibility are not so important. Nominal compaction obtained from the tracks of a dozer is often considered sufficient.

There is a wide variety of earthmoving plant for different purposes and scales of operation. Four basic items are illustrated in Figure 13.1.

Purpose and types of materials

(Figure 13.2)
Simple earth structures such as flood banks and levees often comprise a homogeneous mass of one soil type. For more important structures different types of materials are required for different purposes, for example see Figure 13.2. This shows that an earth structure can be more complex than is apparent from the surface.

Table 13.1 lists the types of materials which can form the various parts of a highway structure, summarised from the Specification for Highway Works, 1992.

Material requirements

To ensure stability of an earth fill and to minimise volume changes after construction (swelling and shrinkage) all of the materials used must be of low compressibility, have adequate shear strength after compaction and contain minerals which are not prone to volume and moisture content changes.

In addition they must be inert (unreactive, insoluble), unfrozen (thawing releases excess water), non-degradable (wood tissues, perishable materials), non-hazardous (chemically and physically harmful) and not susceptible to spontaneous combustion, such as some unburnt colliery wastes.

For certain specific purposes they may be required to be:

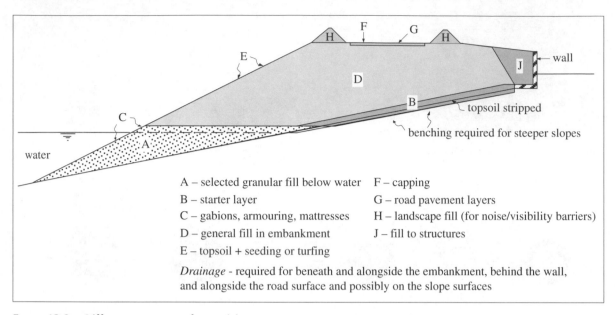

A – selected granular fill below water   F – capping

B – starter layer   G – road pavement layers

C – gabions, armouring, mattresses   H – landscape fill (for noise/visibility barriers)

D – general fill in embankment   J – fill to structures

E – topsoil + seeding or turfing

Drainage - required for beneath and alongside the embankment, behind the wall, and alongside the road surface and possibly on the slope surfaces

Figure 13.2   *Different purposes of materials*

- Free-draining – such as for starter layers and drainage backfill, therefore the fines content must be limited to ensure adequate permeability.
- Non-crushable – a minimum 10% fines test value is normally specified to ensure that the individual particles (of rock) will not break down, produce more fines and reduce the permeability both under the initial compaction stresses and for long-term durability.
- Impermeable – for lining canals, ponds or landfill sites a clay layer can be provided which must have low permeability to contain water, be flexible to allow for movements and be plastic to prevent cracking.

Some typical properties required of the various highway construction materials are given in Table 13.1.

Acceptability of fill material

A material may be deemed acceptable if it has the required properties to enable it to be incorporated into the permanent works (Specification for Highway Works, 1992).

The general fills comprise the bulk of an earthworks structure. For economic and environmental reasons it is important to use as much on-site materials as possible, i.e. use the materials from the cuttings to form the embankments. It is also important to obtain a balanced earthworks when the volumes removed from the cut areas equal the volumes required for the fill areas. Otherwise, material may have to be transported to tips off or on site or obtained from borrow pits close to the site, both providing environmental problems. Useful information and discussion on this subject is contained in the *Proceedings of the Conference on Clay Fills*, 1979.

The stricter criteria applied to the selected fills usually means that it is unlikely they will be won from the site and they must be imported from a nearby suitable source, such as a rock or gravel quarry.

The criteria for the acceptability of a general fill material depend on three factors:

1. *The nature of the works*
   For a highway embankment sufficient strength is required to prevent slope instability and provide an adequate subgrade stiffness (such as the CBR value) for road pavement support. Self-weight settlements within the fill must also be limited, therefore soils with low compressibility must be used. This is usually achieved by using materials which are not excessively moist or excessively

TABLE 13.1 *Acceptable earthworks materials (Abridged version from Specification for Highway Works, 1992)*

CLASS		Material description	Typical use	Typical properties				Compaction requirements*
GENERAL GRANULAR FILL	1A	well graded	General fill	< 125 mm	< 15% fines		$U > 10$	Method 2
	1B	Uniformly graded					$U < 10$	Method 3
	1C	Coarse granular		< 500 mm	< 15% fines		$U > 5$	Method 5
GENERAL COHESIVE FILL	2A	Wet cohesive	General fill	< 125 mm	> 15% fines		$w > PL - 4$	Method 1
	2B	Dry cohesive					$w < PL - 4$	Method 2
	2C	Stony cohesive		< 125 mm	15% > 2 mm			Method 2
	2D	Silty cohesive		< 125 mm	> 80% fines			Method 3
	2E	Reclaimed PFA		< 20% FBA				End product 95% ρ_{dmax} (2.5kg)
GENERAL CHALK FILL	3	Chalk	General fill					Method 4
LANDSCAPE FILL	4	Various	Fill to landscape areas					Nominal
TOPSOIL	5A	On site topsoil or turf	Topsoiling					
	5B	Imported topsoil		Complies with B.S. 3882				
	5C	Imported turf	Turfing	Complies with B.S. 3969				
SELECTED GRANULAR FILL	6A	Well graded	Below water	< 500 mm	< 5% fines		$U > 10$	No compaction
	6B	Coarse granular	Starter layer	< 500 mm	90% > 125 mm		$U > 5$	Method 5
	6C	Uniformly graded		< 125 mm	90% > 2 mm		$U < 10$	Method 3
	6D		Starter layer below PFA	< 10 mm			$U < 10$	Method 4
	6E		For stabilisation with cement to form capping	< 125 mm	LL < 45 PI < 20 org. < 2% fines < 15%	SO4 < 1%		Not applicable
	6F1	Fine grading	Capping	OMC–2 < w < OMC			< 75 mm	Method 6
	6F2	Coarse grading					< 125 mm	
	6G		Gabion filling	< 200 mm	min. size > mesh opening			None
	6H		Drainage layer to reinforced earth structure	< 20 mm	chemically stable			Method 3
	6I	Well graded	Fill to reinforced earth	< 125 mm < 15% fines chemically stable, frictional			$U > 10$	Method 2
	6J	Uniformly graded					$5 < U < 10$	Method 3
	6K		Lower bedding	< 20 mm PI < 6 < 10% fines OMC–2 < w < OMC+1 $U > 5$			chemically stable	End product 90% ρ_{dmax} (Vib)
	6L	Uniformly graded	Upper bedding	Corrugated steel structures	< 10 mm			None
	6M		Surround		< 75 mm	As 6K		As 6K
	6Q	Well graded, uniformly or coarse	Overlying fill	As Class 1A, 1B or 1C				
	6N	Well graded	Fill to structures	< 75 mm minimum strength and permeability			$U > 10$	End product 95% ρ_{dmax} (Vib)
	6P						$U > 5$	
SELECTED COHESIVE FILL	7A	Cohesive	As 6N	15 – 100% fines LL < 45 PI < 25 minimum strength				End product 100% ρ_{dmax} (2.5kg)
	7B	Conditioned PFA	As 6N and to reinforced earth	w controlled minimum strength and permeability				End product 95% ρ_{dmax} (2.5kg)
	7C	Wet cohesive	Fill to reinforced earth	Minimum strength and chemically stable				Method 1
	7D	Stony cohesive						Method 2
	7E		For stabilisation with lime to form capping	> 15% fines PI > 10 org < 2% SO4 low				Not applicable
	7F	Silty cohesive	For stabilisation with cement to form capping	> 15% fines 80% < 2 mm				
	7G	Conditioned PFA		SO4 < 1%				
	7H	Wet, dry, stony or silty cohesive and chalk	Overlying fill for corrugated structures	As Class 2A, 2B, 2C, 2D or 3 chemically stable				
MISC. FILL	8	Class 1, 2 or 3	Lower trench fill	Stones, clay lumps must be < 40 mm				
STABILISED MATERIALS	9A	Cement stabilised well graded	Capping	Class 6E + min. 2% cement				Method 6
	9B	Cement stabilised silty cohesive		Class 7F + min. 2% cement MCV < 12				Method 7
	9C	Cement stabilised conditioned PFA		Class 7G + min. 2% cement				As 7B
	9D	Lime stabilised cohesive		Class 7E + min. 2.5% lime				Method 7

*Methods of compaction are given in Table 13.7

dry. These requirements become less important for landscaping works. For water-retaining structures the criteria must provide long-term permeability, therefore not only are clay materials required but they must also retain flexibility without cracking, such as for a clay core in an earth dam. It will be necessary to place them in a soft consistency and with a higher than normal moisture content.

2. *Earthmoving efficiency*

In order to maintain adequate trafficability for the earthmoving plant and compactability beneath a roller the soils traversed must have sufficient strength otherwise the job will grind to a halt with excessive rutting and plant bogging down.

The Transport Research Laboratory has conducted research into this problem and has shown that the overall efficiency is related to the depth of a rut produced after a single pass of the machine. If the single pass rut depth is less than 50 mm then no difficulties with scraper movement are likely although the rolling resistance will be increased and the maximum speed of travel will be reduced. A rut depth of 100 mm or more will represent severe damage to the formation and considerable loss of productivity.

The economics of the use of inferior quality on-site materials must be balanced with the need for feasible operation of plant and its efficient use.

If a material is deemed to be of particularly poor quality then it may be used for an inferior type of job, such as landscaping, or it may be improved such as by drying or lime modification, otherwise it must be removed from the site.

3. *Compactability*

Soils should be compacted at a moisture content near to their optimum moisture content in order to achieve a high dry density. This is described in more detail in the section on soil compaction, below.

If a soil is compacted at moisture contents too dry of its optimum then it is likely that a high air voids content will be left within the soil making it compressible and initially brittle. With a high permeability this soil will increase in moisture content following infiltration from surface water

leading to a loss of strength, particularly if the soil is clayey. All of these are undesirable properties.

A lower limit to moisture content is specified to eliminate these problems. Since the introduction of the moisture condition test a maximum *MCV* may be specified in the range 12–14 depending on the soil type and degree of compaction required.

If soils are compacted too wet of their optimum then there is a risk of inducing high pore water pressures due to overcompaction of a layer by traversing too often with the compaction plant and within the body of the fill as overlying fill layers are placed. High pore pressures could cause instability of a slope during construction and excessive consolidation settlements following construction. The upper limit to moisture content (or minimum *MCV*) is also specified to restrict these problems.

Acceptability of granular soils

Granular soils are generally considered to be in an acceptable condition provided their natural moisture content lies close to the optimum moisture content (*OMC*) obtained from a laboratory compaction test. A range of values is usually quoted as:

$$OMC \pm x \% \qquad (13.1)$$

where x may be 0 to 2%. Different optimum moisture contents will be obtained from the three laboratory compaction tests available so the test type must be specified. For many granular soils the dry density achievable is not too sensitive to the placement moisture content.

For soils wetter than this range and, therefore, unacceptable it may be possible, depending on weather conditions, to loosely spread out a layer to allow evaporation until the moisture content reduces to the required value. The effectiveness of this approach will depend on factors such as the climate and the fines content of the soil. For soils drier than this range it is feasible to increase their moisture content using sprinklers. The soils should be thoroughly remixed before compaction.

Acceptability of cohesive soils

For cohesive soils there is a requirement to specify a

lower limit of acceptability, usually strength-related, so that the earth structure will not be subjected to slope instability or self-weight settlements, and the construction plant can traffic over the soil surface. An upper limit is also required to ensure that the soil lumps are not too stiff to be remoulded during compaction.

A number of methods have been adopted for specifying a lower limit of acceptability, the most common being:

1. *A maximum moisture content*

   The fill must have a moisture content no greater than a given value. On the basis that undrained shear strength increases as moisture content decreases and the test is easy to perform this approach seems attractive. For example, fill could be specified with a maximum moisture content of, say, 20%. During the contract this would provide simple site control. However, on most large, linear sites different geological conditions occur so the plasticity limits (liquid and plastic) and the particle size distributions vary. If one value of moisture content is specified for the whole site a wide range of shear strengths can be obtained, as shown in Figures 13.6 and 13.7, below. The standard test also requires 24 hours for oven-drying so an assessment cannot be made immediately.

2. *Moisture content related to plastic limit*

   A maximum moisture content w determined from:

   $$w = PL \times factor \qquad (13.2)$$

   has been frequently used as the acceptability criterion in the UK. The factor usually given for scrapers is 1.2, although a factor of 1.3 has been adopted for 'wet' clay fill and a lower factor of 1.0 has been used for Scottish stoney clays. For tracked vehicles this factor could lie between 1.40 and 1.65 (Farrar and Darley, 1975).

   These authors have shown that a single factor is not appropriate. Instead, the factor can depend on the size of scraper used, the degree of damage that can be tolerated and the efficiency of plant movement required. Higher factors could also be obtained for soils containing more than 50% fines.

Although a simple test, the poor repeatability of the plastic limit test does not lend itself to use as an acceptability criterion. Sherwood (1971) has shown that if one operator carries out this test a number of times on one soil, up to one-third of all the results can be more than three units above or below the actual value, due largely to operator variability.

The plastic limit test is also carried out on material finer than 425 µm (sand size) so direct comparison with moisture contents from samples containing gravels, especially glacial clays, without corrections for the stone content, see later, may lead to false conclusions.

3. *Undrained shear strength*

   Rather than using an indirect means such as moisture content to assess strength, it is preferable to determine the strength of a clay fill directly from a re-compacted specimen tested either in the unconfined compression or triaxial compression (quick or unconsolidated undrained) mode.

   A result can be obtained promptly, and for highly cohesive, stone-free soil quite simply using either a hand vane of small diameter or the unconfined compression apparatus either on site or in a laboratory. The main disadvantages are when the soil has poor cohesion to form a cylindrical specimen and when stones are present, large diameter specimens are necessary. There can be difficulty in achieving complete removal of the air voids to achieve full saturation after compaction, especially for stiff fissured clays.

   Farrar and Darley (1975) and Arrowsmith (1979) have shown the operation of earthmoving plant to be related to the strength of the clay, see Table 13.2.

4. *Moisture condition value (MCV) (Figures 13.3 and 13.4)*

   Due to the difficulties associated with the above methods for assessing suitability of earthworks material the Transport Research Laboratory developed the moisture condition test (Parsons, 1976, Parsons and Boden, 1979). The objectives of this test were to be able to provide an immediate result for site use, to be applicable to a wide range of soil types, to eliminate the effect of operator error and to use a large sample for more representative behaviour.

TABLE 13.2 *Operation of earthmoving plant on clay soil*

Plant type	Minimum shear strength kN/m²		
	Farrar and Darley (1975)		Arrowsmith (1979)[3]
	feasible operation[4]	efficient operation[5]	
Small dozer, wide tracks	20		
Small dozer, standard tracks	30		
Large dozer, wide tracks	30		
Large dozer, standard tracks	35		
Towed and small scrapers (less than 15m³)	60[1,2]	140[1,2]	34
Medium and large scrapers (over 15m³)	100[1,2]	170[1,2]	57

1 – Values are vane shear strength which is likely to be higher than triaxial shear strength.
2 – Values were obtained in the fill area which may have 'dried' during dry weather.
3 – Values were obtained from the cut area and relate to 'feasible' operation.
   These soils may be stronger in the fill area during dry weather.
4 – Represents deepest rut of 200 mm after a single pass of machine
5 – Represents depth of rut not greater than 50 mm after a single pass

The test (BS1377:1990, Part 4) consists of determining the compactive effort required to almost fully compact a given mass of soil. 1.5 kg of moist soil (with particles or lumps less than 20 mm) is placed loose in a 100 mm internal diameter steel mould and a lightweight disc, 99.2 mm diameter is placed on top of the soil. Compactive effort is provided by the blows from a free falling cylindrical rammer of 7 kg mass and 97 mm diameter falling onto the disc with a height of drop of 250 mm. As the number of blows increases the particles and lumps of soil move and remould to expel air between them.

The effect of each blow will be related to the density of soil achieved which is assessed by measuring the penetration of the rammer into the mould. On the basis that the change in density with the penetration of the rammer is directly related to the logarithm of the number of blows, the change in penetration between one number of blows, n and four times that number, $4n$ is plotted against the logarithm of the number of blows n, see Figure 13.3.

As a very large number of blows may be required to remove the last remaining air voids a change of penetration of 5 mm was selected to represent the point beyond which no significant change in density occurs. The moisture condition value (MCV) is defined as:

$$MCV = 10 \log_{10} B \qquad (13.3)$$

where B is the number of blows corresponding to a change of penetration of 5 mm on the steepest straight line through the points, see Figure 13.3.

SEE WORKED EXAMPLE 13.1.

The test should only be carried out on cohesive soils and granular soils with cohesive fines, otherwise variable and misleading results may be obtained. However, it is less important to assess the 'clean' sands and gravels for acceptability as they are often considered to be 'all weather' materials and less affected by their moisture condition.

A moisture condition 'calibration' comprises the relationship between moisture content and *MCV*. This is obtained by adjusting the moisture content of the soil by drying or wetting before carrying out the moisture condition test. Parsons and Boden (1979) showed that the moisture content w versus *MCV* calibration usually produces a straight line with an equation of the form:

$$w = a - b \, MCV \qquad (13.4)$$

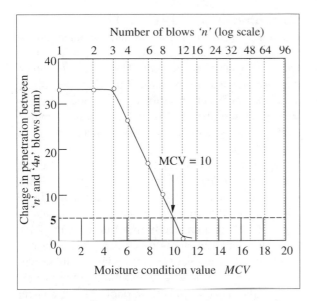

FIGURE 13.3 *Determination of the moisture condition value of a sample of heavy clay (from Parsons and Boden, 1979)*

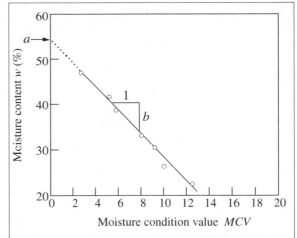

FIGURE 13.4 *Moisture condition calibration of a heavy clay soil (from Parsons and Boden, 1979)*

where a is the intercept or moisture content w when $MCV = 0$ and b is the slope, see Figure 13.4.

Care must be exercised when dealing with soils which are prone to alteration on drying (see Table 2.13 in Chapter 2) and will crush during compaction. It is then preferable to wet up or dry gradually from the natural condition and for crushable soils, to use separate specimens for each determination.

For cohesive soils a relationship between the undrained shear strength, determined by the hand vane and MCV has been given (Parsons and Boden, 1979) in the form:

$$\log c_u = c + d\,MCV \qquad (13.5)$$

with values of the coefficients c and d given in Table 13.3.

As an acceptability criterion it has been found that for UK conditions a minimum MCV of about 8 relates to the lower limits of strength for trafficability purposes as well as for the stability of the earth structure.

Efficiency of earthmoving (Figure 13.5)

The suitability of a soil as it affects the operation and efficiency of earthmoving plant has been assessed by TRRL (Parsons and Darley, 1982). A reasonable relationship was found to exist between:

● the soil condition given by the MCV
● the type of plant including factors such as the number of driven wheels, tyre width, maximum available engine power and total mass, and

TABLE 13.3 *Values of the coefficients c and d (from Parsons and Boden, 1979)*

Soil type	c	d	Number of results	Correlation coefficient
Clay – high plasticity	0.74	0.111	40	0.94
Clay – intermediate plasticity	0.77	0.107	44	0.96
Clay – low plasticity	0.91	0.112	14	0.89
Silt – high plasticity	0.70	0.105	15	0.97

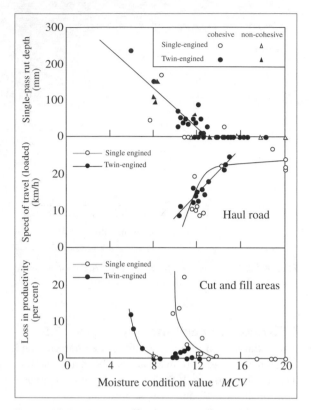

FIGURE 13.5 *Factors affecting operations with medium scrapers (from Parsons and Boden, 1979)*

● the efficiency of operation as represented by the speed of travel which could be achieved on the haul road.

Some of these factors are illustrated in Figure 13.5.

Single-pass rut depth is a measure of the degree of damage likely to be caused in a fill area. The speed of travel on the haul road will determine the cycle time, the time required to travel from the cut area to the fill area and back again and, therefore, the overall volume of fill moved each day.

Material problems

The effects of weather are all too obvious on an earthworks project with softening during wet weather, dust causing visibility and environmental problems during hot, dry weather and frost damage during freezing and thawing conditions. Two factors are considered here, softening and bulking.

Softening (Figures 13.6 and 13.7)

For cohesive soils in particular, and granular soils to a lesser extent their condition will be affected by changes in moisture content. The degree to which they are affected can be termed moisture sensitivity which is a different phenomenon to remoulding sensitivity, as described in Chapter 7. A soil exhibiting both types of sensitivity is likely to deteriorate even more rapidly.

Moisture sensitivity is the reduction in strength produced by an increase in moisture content. This is illustrated for clays of different plasticity in Figure 13.6 and for different gravel (or sand) contents in Figure 13.7. For a change in moisture content of 1% the change in shear strength is much greater for low plasticity clays and very gravelly clays so these will be more sensitive to wetting and more prone to softening. Glacial clays have a variable gravel content and are typically of low plasticity so they can be easily weakened by wetting.

Similar to the plots in Figures 13.6 and 13.7 the value of b from the *MCV* calibration, Equation 13.4,

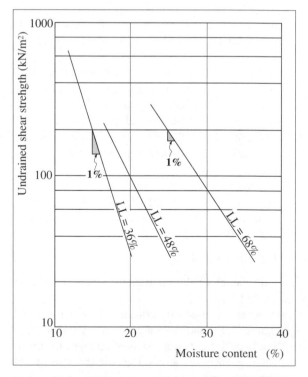

FIGURE 13.6 *Moisture sensitivity – effect of plasticity (from Clayton, 1979)*

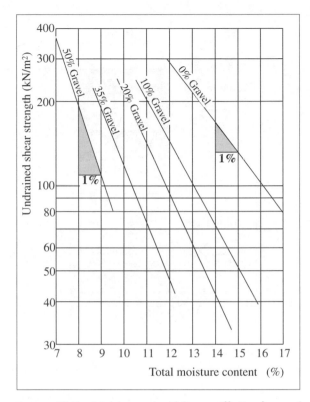

FIGURE 13.7 *Moisture sensitivity – effect of gravel (from Barnes and Staples, 1988)*

TABLE 13.4 *Moisture sensitivity related to MCV calibration slope b (after Matheson, 1983, 1988)*

b	Moisture sensitivity
> 1.0	low
0.5 – 1.0	moderate
0.33 – 0.5	high
< 0.33	very high

is an indicator of moisture sensitivity. Matheson (1988) has suggested levels of moisture sensitivity related to the b value, given in Table 13.4.

Bulking (Figure 13.8)

In estimating the cost of an earthmoving contract it is desirable to achieve a balance between the volumes of soil removed from the cuttings and the volume of soil required to form the embankments in order to avoid the expensive disposal of surplus materials or importation of materials from off-site. For this purpose, the estimator will use a mass haul diagram, the principles of which are illustrated in many textbooks on surveying. However, it would be incorrect to compare the geometric volumes from the cuttings and the embankments directly, due to the phenomenon of bulking.

Soil in its natural *in situ* location in the cutting is referred to as bank volume, measured in bank m³. When it is excavated loosening and breaking up into lumps increases the volume occupied when it is referred to as loose volume, measured in loose m³. Following compaction, at its final location, the soil may occupy a smaller or larger compacted volume, compacted m³, depending on the soil type and level of compaction. This is illustrated in Figure 13.8. To allow for these effects factors must be applied to the bank volumes.

To assess how much bank volume can be removed by one scraper or dump truck each trip the following expression is used:

$$\text{bank m}^3 \text{ removed by each load} = \text{machine payload (m}^3) \times LF \quad (13.6)$$

where LF = load factor.

The load factor is related to the % swell of the material:

$$LF = \frac{100}{100 + \% \text{ swell}} \quad (13.7)$$

Thus a machine cannot carry its own geometrical volume in soil. Some typical values of load factor for soils and rocks are given in Table 13.5.

To assess the compacted volume (in the fill area) which could be formed from a given bank volume (in the cut area) the following expression is used:

$$\text{compacted volume, m}^3 = \text{bank volume, m}^3 \times SF \quad (13.8)$$

where SF = shrinkage factor.

The shrinkage factor is related to the % shrinkage of the material:

$$SF = \frac{100 - \% \text{ shrinkage}}{100} \quad (13.9)$$

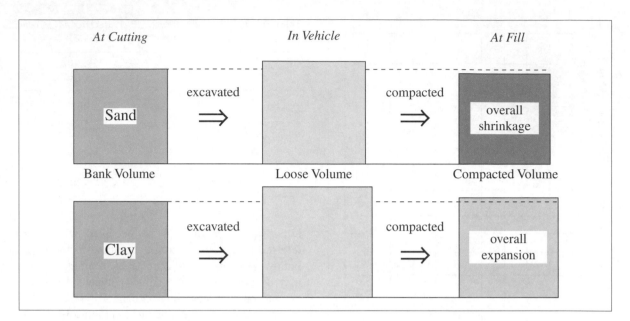

FIGURE 13.8 *Bulking*

TABLE 13.5 *Typical values of swell and load factor*

Material type	% swell	Load factor LF
Soils		
Sand	10–15	0.87–0.91
Gravel	12–18	0.85–0.89
Clay	20–30	0.77–0.83
Peat	30–40	0.71–0.77
Topsoil	30–40	0.71–0.77
Rocks		
Coal	30–40	0.71–0.77
'soft' rocks e.g. mudstone, shale, chalk	30–50	0.67–0.77
Sandstone	40–70	0.59–0.71
Limestone	50–70	0.59–0.67
igneous rocks e.g. basalt, granite, gneiss	50–80	0.56–0.67

SEE WORKED EXAMPLES 13.2 AND 13.3.

An estimate of the shrinkage factor can be obtained from:

$$SF = \frac{\gamma_b}{\gamma_c} \qquad (13.10)$$

where γ_b is the bulk unit weight of the material *in situ* in the cutting obtained from field or laboratory density tests and γ_c is the compacted bulk unit weight obtained from a laboratory compaction test or directly from the compacted soil in the fill area.

Typical values for some soils and rocks are given in Table 13.6. Note that some materials, such as sand and chalk, reduce in volume overall with net shrinkage whereas most other soils and rocks increase in volume with negative shrinkage or net bulk-up.

Further loss of material is to be expected overall on a job due to the creation and maintenance of haul roads, overfilling and trimming of embankments and removal of material following weather deterioration. This can be as much as 10–15% on some sites although 5% is considered more typical.

TABLE 13.6 *Typical values of % shrinkage and shrinkage factor*

Material type	% shrinkage	Shrinkage factor SF
Sand	0–10	0.90–1.00
Chalk	0–15	0.85–1.00
Most other soils, weak rocks	0–10 (–ve)	1.00–1.10
Harder rocks	5–20 (–ve)	1.05–1.20

Soil compaction

After a layer of soil has been deposited and spread in the fill area it must be rendered as strong and stiff as possible to ensure stability and minimise settlements. Energy is applied to the soil to remould lumps of clay, move granular particles together and essentially remove as much air as possible. This process is known as compaction.

For nearly all soils the extent to which air can be removed depends on the strength of the clay lumps or the friction between the granular particles which in turn depend on the moisture content of the soil

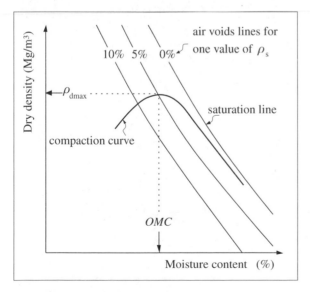

FIGURE 13.9 *Typical compaction curve*

during compaction. The degree of compaction achieved is measured by dry density, ρ_d:

$$\rho_d = \frac{\text{mass of soil particles}}{\text{volume occupied}} \qquad (13.11)$$

This represents the amount of solid soil particles in a given volume. Compaction should aim to achieve as high a dry density as possible.

Factors affecting compaction

The main factors affecting compaction are:

1. *Moisture content (Figures 13.9 and 13.10)*

   At low moisture contents the strength of clay lumps and friction between granular particles is high so a given compactive effort will not be able to remove all air voids leaving the soil in an overall compressible state when it is subjected to stresses from further layers of fill or a structure, and in a potentially collapsible state.

   At higher moisture contents clay lumps become weaker and friction between granular particles reduces so the air voids are more easily removed during compaction. The dry density increases until a maximum value ρ_{dmax} is reached at the optimum moisture content (*OMC*). This is shown on the typical compaction curve, Figure 13.9.

   At moisture contents above the optimum value the soil particles cannot move any closer together because even though most of the air has been expelled there is more water present in the voids, see Figure 13.10.

2. *Compactive effort (Figure 13.11)*

   Applying more energy to a soil will reduce the air voids content further and increase the dry density so more compaction energy can be beneficial especially for soils dry of the optimum value. However, if the soil is already moist, weaker and above the optimum moisture content then applying more energy is wasteful since the air can quickly be removed.

   Applying large amounts of energy to a very moist soil may be damaging since no more air can be expelled but high pore water pressures can build up which could cause slope instability during construction and consolidation settlements as they dissipate after construction.

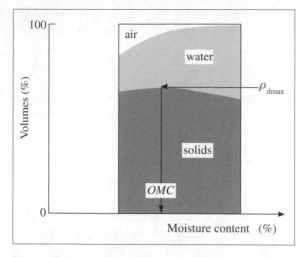

FIGURE 13.10 *Compaction curve – volumes of solids, water and air*

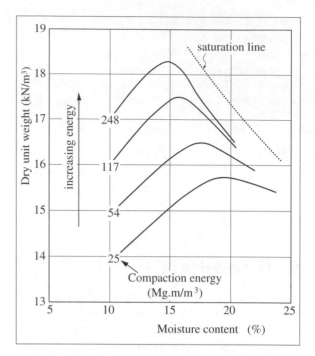

FIGURE 13.11 *Effect of compaction energy (from Lambe and Whitman, 1969)*

3. *Soil type (Figure 13.12)*

The strength – moisture content relationship differs for different soils, as illustrated in Figures 13.6 and 13.7, so the compactability or ease with which soils can be compacted will depend on the soil type. From Figure 13.6, at a given moisture content a clay with low plasticity will be weaker than a heavy or high plasticity clay so it will be easier to compact. For a given compactive effort the air voids can be removed more easily for a low plasticity clay and because it will have a lower moisture content anyway, a higher dry density can be obtained. Compare the sandy clay and heavy clay curves in Figure 13.12.

Difficulties have been experienced in adequately compacting stiff fissured high plasticity clays on some projects due to the strength of the clay lumps.

Field compaction

Compaction is achieved in the field by traversing a fairly thin layer of soil with an item of compaction plant a sufficient number of times (passes) until a required density is achieved.

The layer thickness and number of passes must be chosen to ensure that the required density is produced throughout the layer with no undesirable

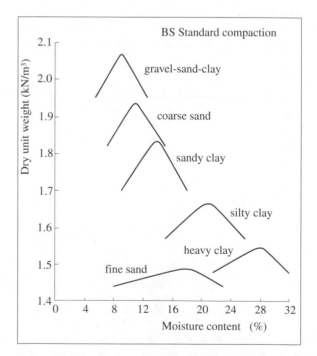

FIGURE 13.12 *Compaction curve – effect of soil type*

density gradients, such as a poorly compacted lower level. Generally, plant of greater weight can transmit compaction energy to lower levels so layer thickness can be increased and the number of passes can be decreased. However, plant which is too heavy can damage a compacted soil surface by applying too much pressure and causing rutting or degradation and some plant may just be too light to remould stiff clay lumps or move granular particles closer.

Compaction plant

To some extent construction traffic provides compactive effort, particularly crawler tracked vehicles, and for some purposes, such as landscaping, this may be sufficient. For the main earthworks compaction must be of known amount and applied more uniformly.

Compaction is achieved by specialist items of plant which are designed to apply energy to the soil by means of pressure and where suited this is assisted by a kneading or remoulding action, vibration or impact.

The main types of compaction plant are decribed below.

Smooth-wheeled rollers

These comprise smooth steel drum rollers which are either towed by a crawler tractor as single or tandem rollers or they are self-propelled with tandem, two- or three-rollers. They travel at a typical speed of about 2.5 to 5 km/hour. The mass can be increased by water or sand ballast. They are suited to firm cohesive and well-graded granular fills providing small layer thicknesses of 125 to 150 mm are adopted. They may become unstable in uniform sands due to the roller pushing itself into the soil and they are generally unsuited for coarse, granular soils without the assistance of vibrations.

They produce a smooth surface which is useful for encouraging rainfall run-off at the end of a work period and for proof-rolling subgrade surfaces. However, the smooth surface provides a poor bond between the layers in a general fill, leaving the earthworks with a laminated type of structure which would be very undesirable for water-retaining earthworks. With some clay soils and plant types, plant-induced shear surfaces can be produced which are polished or slickensided and could promote instabil-

ity. If these surfaces are detected then it will be necessary to adjust the compaction procedure by applying less effort or scarifying each surface before placing another layer.

Vibratory rollers

Vibrational energy assists compaction considerably by shaking the soil and reducing inter-particle friction while the pressure applied moves the particles together. A vibratory roller can be any item of plant with a vibratory attachment, even a sheepsfoot roller, although the smooth drum is the most common.

They can be towed, self-propelled or manually guided with speeds no greater than about 1.5 to 2.5 km/hour. Better compaction is achieved with lower speeds and with a frequency giving the maximum amplitude. Vibrations are applied either by separate engines mounted on the towing frame or by the rotation of eccentric weights within the drum.

They are suited to most soil types although they may be less efficient with moist clays and will become unstable in uniform fine sands due to them pushing in. Towed rollers, rather than self-propelled rollers should then be used.

Tamping rollers

Various shapes of projections or 'feet' have been fitted to a smooth steel drum roller to penetrate the soil layer and produce lateral compaction as well as vertical compaction, high localised pressure and a mixing and kneading action. They provide good interlock between successive layers so they are particularly suited to water-retaining structures such as earth dams.

They may be towed or self-propelled with one or more drums mounted on one or more axles travelling at speeds of between about 4 and 10 km/hour. The feet are circular, square or rectangular of varying lengths and tapers and must be designed to ensure adequate coverage across the drum and penetration of the soil layer but without trapping soil between the projections and clogging up.

For pad-type tamping rollers the end area of each foot is greater than 100 mm square and the sum of the areas of these feet may occupy up to 25% of the surface of an area swept around the ends of the feet.

For sheepsfoot rollers the projections are larger than the usual pad-type tamping roller. They are typically between about 180 and 240 mm long, 70–80 mm square or 75–90 mm diameter with spacings of between 200 and 280 mm. The total end area of the projections is typically about 5–10% of the surface area at the ends of the feet.

Tamping rollers are most suited to cohesive soils, particularly soils of low moisture content and fine granular soils although sheepsfoot rollers may be less efficient in the latter due to disturbance of the soil as the projections retract.

Grid rollers

These consist of an open steel mesh drum ballasted with concrete blocks attached to the frame and towed by a crawler tractor. They provide high localised pressure and are most suited to soft rocks and stiff clays where breaking lumps is beneficial. They are less suited to moist clays and uniform sands where they may become bogged down.

Pneumatic-tyred rollers

A number of rubber-tyred wheels mounted on one or two axles provide a useful kneading effect. On two axle versions the wheels are off-set to provide complete coverage of the soil layer and with the wobbly-wheel type the wheels are mounted so that they move from side to side, increasing the kneading action. The wheels are usually mounted independently or in pairs so that a more uniform pressure can be applied over an uneven surface. The tyres are either small with no tread or large and with tread. The applied pressure and hence the compactive effort is increased by ballast and by adjusting the tyre inflation pressures.

They are most suited to moist cohesive soils and well-graded granular soils. They produce a relatively smooth compacted surface with poor bonding between layers and may be prone to leaving laminations and plant-induced shear surfaces.

Vibrating plate compactors

These comprise a vibrating unit mounted on a steel plate which is manually operated. Weights vary up to about 2 tonnes with plate areas up to 1.6 m². They operate fairly slowly, less than 1 km/hour and are mostly used in small, confined areas, such as for backfilling around structures. They are most efficient when compacting granular soils.

Vibro-tampers

These weigh between 50 and 100 kg and apply compaction by vibrations. They are manually guided, useful in confined spaces and can be used for cohesive and granular soils.

Power rammers

These are machines actuated by explosions in an internal combustion engine causing the machine to impact on the soil layer. They are manually operated, typically weigh about 100 kg and are only suited to compaction in small, confined areas, such as backfilling narrow trenches.

Dropping weight compactors

These consist of a mass of between 200–500 kg lifted by a hoist mechanism and dropped through a height of 1–3 m. They are, therefore, useful for compaction in small, confined areas using cohesive or well-graded granular fill.

Specification of compaction requirements

End-product specification

A logical way of specifying how much compaction effort should be applied to a soil would be to stipulate a required property of the compacted soil, such as a minimum density or a maximum air content. Then the state of the finished earthworks and its properties are known. This approach is referred to as an end-product specification and requires *in situ* density tests and laboratory moisture content tests to confirm that the specified property is achieved.

The two most common approaches are:

1. *Relative compaction (Figures 13.13 and 13.14)*
   The required dry density of the earth fill after field compaction must be greater than a certain proportion of the dry density obtained from a laboratory test, usually the BS light or vibrating hammer compaction test:

$$\text{relative compaction} = \frac{\gamma_d \text{ in the field}}{\text{maximum } \gamma_d \text{ from lab. test}}$$

   (13.12)

   This provides some certainty that the fill will achieve the desired properties and is the

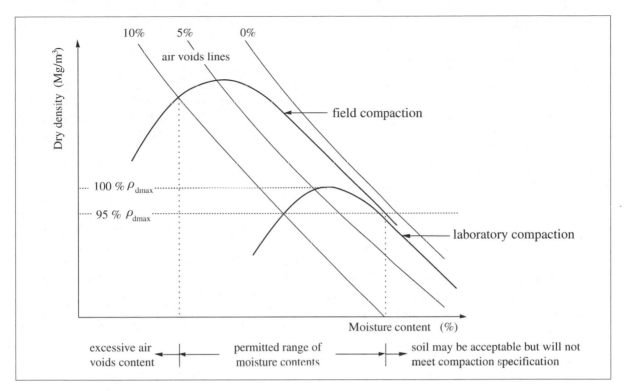

FIGURE 13.13 *Relative compaction*

adopted method for a number of fill types, see Table 13.1.

SEE WORKED EXAMPLE 13.7.

There are several disadvantages with this method, including:

- The cost and time required for testing. Waiting a day for an oven-drying moisture content result, before deciding to place another layer of fill is unacceptable.
- Variability of the soil types. A laboratory compaction test should be carried out on the same type of material each time the field density is determined, otherwise direct comparisons cannot be made.
- Particle size. The fill material may contain coarse gravel, even cobble sizes but the compaction test can only realistically be carried out on material less than 20 mm size. Corrections for the effect of stone content may not be appropriate.

- Mode of compaction. Soil is compacted in a much different way in the field compared to the laboratory test.
- The permitted range of moisture content for field placement must be specified. Soils compacted in the field on the dry side of the laboratory optimum moisture content may achieve the required 95% of the maximum dry density but could still contain an excessive air void content, as shown in Figure 13.13. To ensure a maximum air voids content of say 10% a minimum moisture content must be specified. Soils compacted in the field wet of the laboratory optimum may be considered acceptable for use with adequate strength, but may never achieve 95% of the maximum dry density.
- Density gradients within the soil layer, see Figure 13.14.

2. *Air voids content*
A maximum air voids content, A_v, of 10% has

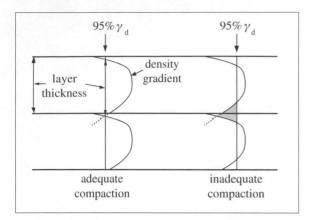

FIGURE 13.14 *Density gradients and layer thickness*

often been quoted as acceptable for the bulk of an earth fill with a value of 5% for the top of an embankment to provide a good foundation support or the subgrade for road pavement layers. The *in situ* bulk density must be determined, together with the moisture content and average particle density so that the air voids content can be calculated from a formula, see Table 2.15.

SEE WORKED EXAMPLE 13.6.

The main disadvantages with this approach are:

- Statistical variation. It has been estimated that for a permitted probability of 90% (9 out of 10 results give A_v less than 10%) the mean value of A_v obtained from the tests must be 7%. To ensure a representative statistical sample a large number of tests are required and this can be prohibitive and time-consuming.
- The range of moisture content must be specified for a particular plant item. If the soil is above its optimum value an upper limit of moisture content is required, otherwise it will be easy to achieve less than 10% air voids with just one pass of the compaction plant, with the risk of inadequate coverage and poor compaction in the bottom half of a layer. A lower limit of moisture content is also required, otherwise an excessive number of passes will be necessary to achieve the desired air voids content. These

effects are illustrated in Figures 13.9 and 13.11.

Method specification

In view of the major difficulties associated with the end-product specification and following extensive research by the Transport Research Laboratory, the Department of Transport requires that fill placed for highway construction be compacted using a Method Specification (Anon, 1992) except for a few selected fill materials.

For a given soil type the DTp Specification states a method of compaction to be adopted. For example, for a well-graded granular general fill material (Class 1A) Method 2 compaction must be used, see Table 13.1. The methods of compaction are given in Table 6/4 of the DTp Specification and an abridged version is included in Table 13.7.

For each method of compaction the types and masses of compaction plant which are unsuitable are stated so the choice becomes limited to those which are suitable. The method of compaction is given as a maximum depth of compacted layer and a minimum number of passes for each type and mass of suitable plant.

For the top 600 mm of an embankment composed of general granular or cohesive fill the number of passes is doubled to provide a stiffer formation to receive the pavement layers.

The main advantages of this approach are:

- It removes most of the disadvantages of the end-product approach.
- The requirements for compaction are more precise.
- The contractor's estimate of costs involved should be more accurate.
- Fill quality control is minimised with much less testing. Some control or compliance testing is advisable.

However, adequate supervision to ensure that the work is carried out as specified is essential, with control of the maximum layer thicknesses and minimum number of passes, see Figure 13.14. This can be time-consuming on large-scale projects. With some soil types there may still be doubt concerning the quality of the compacted fill so control testing could be needed.

TABLE 13.7   Method compaction for earthworks materials: plant and methods (Abridged version from Specification for Highway Works, 1992)

Type of compaction plant	Category	Ref No	Method 1 D	Method 1 N#	Method 2 D	Method 2 N#	Method 3 D	Method 3 N#	Method 4 D	Method 4 N	Method 5 D	Method 5 N	Method 6 N for D=110 mm	Method 6 N for D=150 mm	Method 6 N for D=250 mm	Method 7 N for D=150 mm	Method 7 N for D=250 mm
Smooth wheeled roller	Mass per metre width of roll: over 2100 kg up to 2700 kg	1	125	8	125	10	125	10*	175	4	unsuitable		unsuitable	unsuitable	unsuitable	unsuitable	unsuitable
	over 2700 kg up to 5400 kg	2	125	6	125	8	125	8*	200	4	unsuitable		16	16	unsuitable	unsuitable	unsuitable
	over 5400 kg	3	150	4	150	8	unsuitable		300	4	unsuitable		8	8	unsuitable	12	unsuitable
Grid roller	Mass per metre width of roll: over 2700 kg up to 5400 kg	1	150	10	unsuitable		150	10	250	4	unsuitable		unsuitable	unsuitable	unsuitable	unsuitable	unsuitable
	over 5400 kg up to 8000 kg	2	150	8	125	12	unsuitable		325	4	unsuitable		20	20	unsuitable	16	16
	over 8000 kg	3	150	4	150	12	unsuitable		400	4	unsuitable		12	12	unsuitable	8	8
Tamping roller	Mass per metre width of roll: over 4000 kg	1	225	4	150	12	250	4	350	4	unsuitable		12	20	unsuitable	4	8
Pneumatic-tyred roller	Mass per wheel: over 1000 kg up to 1500 kg	1	125	6	unsuitable		150	10*	240	4	unsuitable		unsuitable	unsuitable	unsuitable	unsuitable	unsuitable
	over 1500 kg up to 2000 kg	2	150	5	unsuitable		unsuitable		300	4	unsuitable		unsuitable	unsuitable	unsuitable	12	unsuitable
	over 2000 kg up to 2500 kg	3	175	4	125	12	unsuitable		350	4	unsuitable		unsuitable	unsuitable	unsuitable	6	unsuitable
	over 2500 kg up to 4000 kg	4	225	4	125	10	unsuitable		400	4	unsuitable		12	unsuitable	unsuitable	5	16
	over 4000 kg up to 6000 kg	5	300	4	150	10	unsuitable		unsuitable		unsuitable		12	unsuitable	unsuitable	4	16
	over 6000 kg up to 8000 kg	6	350	4	150	8	unsuitable		unsuitable		unsuitable		10	16	unsuitable	unsuitable	8
	over 8000 kg up to 12000 kg	7	400	4	175	8	unsuitable		unsuitable		unsuitable		8	12	unsuitable	unsuitable	4
	over 12000 kg	8	450	4	175	6	unsuitable		unsuitable		unsuitable		unsuitable	unsuitable	unsuitable	unsuitable	4
Vibratory roller	Mass per metre width of a vibratory roll: over 270 kg up to 450 kg	1	unsuitable		75	16	150	16	unsuitable		unsuitable		unsuitable	unsuitable	unsuitable	unsuitable	unsuitable
	over 450 kg up to 700 kg	2	unsuitable		75	12	150	12	unsuitable		unsuitable		unsuitable	unsuitable	unsuitable	unsuitable	unsuitable
	over 700 kg up to 1300 kg	3	100	12	125	10	125	6	125	10	unsuitable		16	16	unsuitable	12	unsuitable
	over 1300 kg up to 1800 kg	4	125	8	150	8	200	10*	175	4	unsuitable		6	6	12	10	unsuitable
	over 1800 kg up to 2300 kg	5	150	4	150	4	225	12*	unsuitable		unsuitable		4	5	11	10	12
	over 2300 kg up to 2900 kg	6	175	4	175	4	250	10*	unsuitable		400	5	3	5	10	8	11
	over 2900 kg up to 3600 kg	7	200	4	200	4	275	8*	unsuitable		500	5	3	4	8	8	10
	over 3600 kg up to 4300 kg	8	225	4	225	4	300	8*	unsuitable		600	5	2	4	8	8	8
	over 4300 kg up to 5000 kg	9	250	4	250	4	300	6*	unsuitable		700	5	2	3	7	8	7
	over 5000 kg	10	275	4	275	4	300	4	unsuitable		800	5	2	3	6	6	6

Refer to Table 13.1 for Material Classes related to each method of compaction.

\# – For Material Classes 1A, 1B, 2A, 2B, 2C or 2D within the top 600 mm of the embankment the number of passes is doubled.

* – The roller must be towed by a track-laying tractor. Self-propelled rollers are not suitable.

D – Maximum depth of compacted layer

N – Minimum number of passes

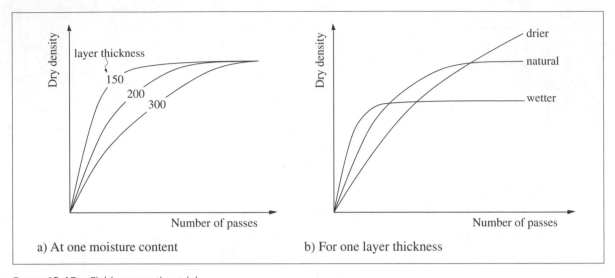

FIGURE 13.15 *Field compaction trials*

Control of compaction in the field
(Figure 13.15)

Where it is necessary to obtain the dry density or air voids content of the compacted fill the *in situ* bulk density must be determined. Suitable methods include the core-cutter method for cohesive soils, the sand-replacement method for granular soils and the nuclear moisture/density gauge for a wide range of materials. In each case the results will have variable accuracy and the latter methods require a calibration. Where thick layers of fill are placed the density should be determined throughout the layer.

On some projects a field trial may be justified to determine the most efficient type of plant and method of compaction. This could comprise spreading layers of different thicknesses and running the compaction plant over, measuring the dry density of the fill at given numbers of passes. This should determine the most economical method to adopt by providing the maximum layer thickness and minimum number of passes which will give a required dry density, see Figure 13.15a.

The soil should be spread initially at its natural moisture content. The exercise could then be repeated with the moisture content of the soil adjusted either by adding or removing moisture, depending on the optimum moisture content and the climate, see Figure 13.15b.

Laboratory compaction

The compaction characteristics of soils can be assessed using standard laboratory tests. The soil is compacted by dropping a mass or vibrating a weight onto thin layers in a cylindrical mould using an amount of compaction energy per unit volume given by:

$$\text{energy} = \frac{m \times h \times b \times n}{V} \tag{13.13}$$

where:

m = weight or mass of hammer
h = height of drop of hammer
b = number of blows per layer
n = number of layers
V = volume of mould

The effect of compaction energy is illustrated in Figure 13.11.

Laboratory tests

Three tests are described in BS 1377:1990, Part 4.

Light compaction

This test may be referred to as BS light compaction, the Proctor method or the 2.5 kg rammer method. The test is carried out on soil with particles larger than 20 mm removed and prepared by drying and then adding water so that its moisture content is fairly low and sufficiently mixed to ensure uniform moisture content throughout. Cohesive soils should be chopped into pieces smaller than 20 mm. Preparation procedures for soils which contain particles greater than 20 mm for the one-litre mould or 37.5 mm for the CBR mould are given in BS 1377:1990 and summarised in Table 13.8.

For soils which may be affected by oven-drying or even air-drying it is preferable to commence the test with the soil at its natural moisture content and then to adjust the moisture content by air-drying or blow-drying for compaction at lower moisture contents and mist-spraying for compaction at higher moisture contents.

The soil is compacted (in three layers of equal thickness) into a metal mould of 105 mm diameter and of 1 litre or 1000 cm^3 capacity or into a CBR mould of 152 mm diameter and 2305 cm^3 capacity. Each layer receives blows from a 2.5 kg mass falling freely through a height of 300 mm, with 27 blows in the one-litre mould and 62 blows in the CBR mould.

The test should be carried out to ensure that the final compacted surface lies just above the top of the mould, but no more than about 5 mm. Otherwise, the test must be discarded. After trimming the soil surface flush with the top of the mould so that its volume can be taken as one litre, the mould and soil are weighed and by subtracting the weight of the mould the bulk density or unit weight of the soil can be determined. The soil is removed from the mould and a smaller specimen taken for moisture content determination. The moisture content of the soil is then adjusted, up or down, and the test is repeated to give at least five density values.

The dry density of the soil is calculated (see Table 2.15) and plotted versus moisture content and providing a reasonable 'inverted parabola' shape can be drawn through the points the maximum dry density, ρ_{dmax} and optimum moisture content, *OMC* are read off, see Figure 13.9.

SEE WORKED EXAMPLE 13.4.

Heavy compaction

This test may be referred to as BS heavy compaction or the 4.5 kg rammer method. The test procedure is the same as for the light compaction test but with five equal layers of soil in the one-litre and CBR moulds and compaction for each layer provided by blows from a 4.5 kg mass falling freely though 450

TABLE **13.8** *Sample preparation methods for compaction tests*

Grading zone	Minimum % passing test sieves			Type of mould	Preparation method
	20 mm	37.5 mm	63 mm		
(1)	100	100	100	1 litre	Test whole sample
(2)	95	100	100		Remove particles > 20 mm, test remainder
(3)	70	100	100	CBR mould	Test whole sample
(4)	70	95	100		Remove particles >37.5 mm, test remainder
(5)	70	90	100		1) Remove particles > 37.5 mm. Replace with same quantity of between 20 and 37.5 mm and test this modified sample OR 2) Remove particles > 37.5 mm, test remainder of sample. Apply correction for stone content to ρ_{dmax} and *OMC*
(X)	less than 70	less than 90	less than 100	Test not applicable	Too coarse to be tested

mm. 27 blows are applied in the one-litre mould and 62 blows are applied in the CBR mould.

Vibrating hammer

This test is more appropriate for granular soils and is not suitable for cohesive soils. The soil for the test can be from grading zones 1 to 5 in Table 13.8. The soil is compacted into a CBR mould, 152 mm diameter and approximately 2305 cm³ capacity, in three equal layers. Each layer is compacted by placing a circular tamper, 145 mm diameter, on top and vertically vibrating it using a vibrating hammer operating at a frequency of between 25 and 45 Hz for a period of 60 seconds. The dry density and moisture content are determined in the same manner as for the light compaction test.

Some particles, especially weak rocks, such as chalks and shale, will break down during compaction so this material must be discarded after compaction and fresh samples used each time the moisture content is adjusted.

Air voids lines

On the plot of dry density versus moisture content the air voids lines for 0, 5 and 10% air voids content, A_v must be plotted using the expression:

$$\rho_d = \frac{1 - \dfrac{A_v}{100}}{\dfrac{1}{\rho_s} + \dfrac{w}{100\rho_w}} \qquad (13.14)$$

where:
ρ_d = dry density (Mg/m³)
ρ_s = particle density (Mg/m³)
ρ_w = water density, assumed to be 1.0 Mg/m³
A_v = air voids content, %
w = moisture content %

SEE WORKED EXAMPLES 13.5 AND 13.6.

Some soils exhibit more than one 'peak' on the compaction curve so the air voids lines are a means of confirming that the most relevant part of the curve has been obtained. The right-hand side of the compaction curve should be adjacent to the saturation or zero air voids line as in Figure 13.9.

The air voids lines are sensitive to the value of particle density so it is preferable to determine ρ_s by

a test, particularly where light or heavy minerals are suspected, rather than assuming a typical value.

Correction for stone content

For *in situ* soils which contain particles larger than 20 mm a 'correction' to the maximum dry density and optimum moisture content values may be required so that direct comparison of the laboratory test result can be made with the *in situ* soil properties.

This is because gravel particles (>20 mm in this case) usually contain little moisture within themselves and exist within a matrix of more moist soil. A good example is a 'boulder' clay or glacial clay which comprises a moist cohesive matrix containing relatively 'dry' coarse gravel or cobble size particles.

The correction is based on the displacement of the soil matrix which has been used in the tests by coarse gravel of particle density ρ_g and assuming the coarse gravel contains no moisture of its own. In BS 1377:1990 the procedure is only suggested for grading zone 5 in Table 13.8. The 'corrected' dry density can be obtained from:

$$\text{'corrected' } \rho_{dmax} = \frac{\rho_{dmax}}{\dfrac{B}{100} + \dfrac{\rho_{dmax}}{\rho_g}\left(1 - \dfrac{B}{100}\right)} \qquad (13.15)$$

where:
ρ_g = particle density of the gravel particles
ρ_{dmax} = maximum dry density obtained from the compaction test with the coarser particles removed
B = percentage of the particles passing 37.5 mm.

The 'corrected' optimum moisture content is given by:

$$\text{'corrected' } OMC = \frac{B}{100} \times \text{test } OMC \qquad (13.16)$$

The effect of this correction is to move the compaction curve upwards and to the left.

SEE WORKED EXAMPLE 13.8.

California Bearing Ratio (CBR) test

Although not a very satisfactory test the parameter from this test is used in the design of road pavements (LR 1132). The test is a quasi-bearing capacity test so the strength condition of the soil will determine the value obtained. The equilibrium or long-term CBR beneath a road pavement depends on several factors, such as the basic soil type, the location of the water table, the efficiency of the sub-surface drains, the protection afforded by the pavement, and the subgrade conditions, particularly the moisture condition during construction. For typical clay soils the range of equilibrium CBR values is small, between 1 and 8% (LR 1132) and with poor reproducibility from the test it is difficult to accurately assess the CBR from tests alone.

Capping materials are designed to provide a stiffness and strength equivalent to a CBR of at least 15% so if the CBR of the subgrade has this value then no capping is required. If the CBR is less than 2% then 600 mm of capping is required with lesser thicknesses for CBR's between 2 and 15% (HA 44/91).

The *in situ* test (BS 1377:Part 9) can be carried out on a natural 'undisturbed' soil surface, such as in a cutting or on the remoulded but compacted soil surface of an embankment. The laboratory test (BS 1377:Part 4) is carried out on a soil specimen obtained either *in situ* by driving or pushing a CBR mould with internal diameter of 152 mm and height of 127 mm into the soil or compacting remoulded soil into the mould. For the latter, the soil is prepared at the required moisture content and compacted to either a given dry density, a required air voids content or with a specified compactive effort.

The laboratory sample may be soaked under water in order to obtain the lowest possible value although this may be too severe for most cases. Measurement of swell movements during the soaking period can detect whether the soil is sensitive to swelling when wetted.

A solid steel plunger 49.65 mm diameter (to give a standard cross-sectional area of 3 square inches) is pushed into the soil surface at a constant rate of penetration of 1 mm/min and the force applied and the penetration are measured until a penetration of at least 5 mm is reached. Steel discs are placed on the soil surface around the plunger to model the surcharge from the pavement layers above. It is often worthwhile testing on both the top and bottom surfaces. The force–penetration curve is plotted and compared with a standard force–penetration curve

which corresponds to a 'standard' compacted and confined crushed rock which would give a CBR value of 100%. On the test curve the forces at 2.5 mm and 5 mm are read off and these are expressed as a percentage of the standard forces at these penetrations , i.e. 13.2 kN and 20.0 kN, respectively. The CBR value is taken as the higher result.

Lime stabilisation

The main uses of lime stabilisation is to provide an improved formation to support road construction in the form of a capping layer (HA44/91), see Figure 13.2, and to improve the quality of soils for general earthworks purposes by producing acceptable soils from wet unacceptable soils. It has also been used to reduce the effects of contaminants in soils by 'locking up' the contaminants.

There are two types of lime used in civil engineering. Quicklime is calcium oxide (CaO) and comes in a granular state with lumps less than 5 mm. It requires more water to hydrate so it has a greater effect on wet soils with a faster drying action. Hydrated lime (or slaked lime) is calcium hydroxide ($Ca(OH)_2$) and is in the form of a fine powder. It has a lower available lime content than quicklime so it is more suited to drier soils but it may be less economical and in urban areas there may be a dust problem. Calcium carbonate ($CaCO_3$) is also referred to as lime (carbonate of lime) but has no benefit in stabilising soil although it is used to adjust soil pH for agricultural purposes. Due to its granular nature quicklime is less easy to distribute homogeneously through the soil and because it is more reactive there are greater risks of skin and eye burns. The addition of lime to a soil can have two effects: modification and stabilisation.

Lime modification

Mixing quicklime to a soil causes the calcium oxide to hydrate, removing some water from the soil. The reaction is exothermic producing large amounts of heat so moisture will be driven off with the heat produced. The result is a soil with a lower moisture content. Hydrated lime will not produce the same effect. For both limes the presence of calcium ions causes the clay minerals to flocculate and produces an immediate increase in the plastic limit. This reaction is little affected by temperature. Thus a soil in a wet and very soft state with a moisture content well above its plastic limit can be quickly modified to a soil with a moisture content below its plastic limit when it will be stiffer, more brittle and crumbly and easier to spread and compact. This is a useful way of making unacceptable material into acceptable material.

Pozzolana

Most clay minerals are pozzolanic in that their siliceous and aluminous constituents can chemically react with lime in the presence of water, to form cementitious compounds.

Lime stabilisation

In the highly alkaline environment (pH > 12) produced by the addition of lime, clay particles react with the calcium ions to produce calcium silicates and aluminates, particularly on the edges of the particles, and these cementitious compounds bind the soil together. The process is temperature and time dependent so better results are obtained in warmer climates and in the long term. In the UK lime stabilisation should only be carried out between March and September and when the shade temperature is above 7°C (HA 74/95).

Soil constituents that could inhibit the stabilisation process are organic matter, sulphates, sulphides and carbon dioxide. Some organic compounds can react with the calcium ions present and the resulting decrease in pH will reduce the reaction between the lime and the clay minerals (Sherwood, 1993). Sulphates are found naturally in many soils and may be found in solution in groundwater. They may also be found in waste materials such as colliery spoil, PFA and contaminated soils. Sulphides are commonly found in unburnt colliery spoil and blastfurnace slag in the form of iron pyrites (FeS_2) which will oxidise to sulphates in the presence of air. Sulphates have no detrimental effects on the lime modification process but they can react, in the presence of water, with the cementitious products of the lime–clay reactions and produce calcium sulphoaluminate (ettringite) with considerable volume expansion resulting in heaving, disruption and loss of strength of the stabilised soil. A maximum total sulphate content of 1% is normally specified (HA44/91) although values as low as 0.25% may cause swelling (Sherwood, 1993).

Carbon dioxide in the air or produced by decomposition of organic matter in the soil can react with both types of lime to form calcium carbonate, a process known as carbonation.

The lime is spread over the layer to be stabilised using a specialist machine to ensure uniform application. A minimum of 2.5% available lime is usually specified, more than about 6% is uneconomical. Layers greater than 250 mm thick require greater control to ensure adequate mixing (SHW, 1992) and pulverisation to break down the soil into sufficiently small lumps. Water may be added through a spray-bar on the machine to ensure adequate slaking of quicklime and to prepare the stabilised soil for compaction. Inadequate compaction is likely if the soil is too dry so a maximum MCV of between 12 and 14 is aimed for.

The surface is sealed with one pass of a lightweight roller and left for a mellowing period of between 1 and 3 days to allow the reactions to take place. The layer is then compacted according to the method specification in Table 13.7, see also Table 13.1.

Lime and cement stabilisation may be combined to produce a stronger product than just lime on its own. This is appropriate for clay soils that do not develop sufficient long-term strength with lime alone and are difficult to stabilise with cement alone, such as high plasticity clays. With this method the soil is processed by adding the lime first, say 3–5%. Following pulverisation and a mellowing period, 4–7% of cement is mixed in and the layer is compacted. Greater long-term strengths can be achieved but laboratory trials should be undertaken to determine the optimum proportions.

Cement stabilisation

Cement stabilised soils can be used for sub-bases to support road pavements and bases within a road pavement, see Figure 13.2. The products would be referred to as cement-bound materials, CBM1 to CBM4 (SHW,1992). The process can also be used to provide capping material to support the subbase. The differences between the materials are in their long-term strength, which will be determined by the type of soil to be stabilised and the cement content applied (Sherwood, 1993). For the higher strength CBM the soil must be a well-grade sand and gravel with very little fines and sufficient cement added to achieve the required strength.

The construction process is similar to lime stabilisation except that no mellowing period is allowed and the final compaction should be completed within two hours following mixing with the soil, otherwise some of the hardening effects will be lost.

The hydration of cement is affected by the presence of organic matter so a maximum value of 2% organics is usually the limit. A maximum total sulphate content of 1% is usually specified although if clay minerals are present sulphates may react with them causing swelling and cracking. In this case a maximum total sulphate content of 0.25% should be adopted.

A mixture of lime and PFA which react to form cementitious compounds can be used to stabilise granular soils and this may be more economical than cement stabilisation. However, if clay minerals are present they may react preferentially with the lime and have a deleterious effect on the lime:PFA reaction.

SUMMARY

An earth structure can be formed from several different soil or fill types, each satisfying a different purpose. The feasibility, economics and efficiency of an earthmoving operation are dependent on the acceptability of the fill materials to be used.

In the UK the moisture condition test is usually specified to assess whether a soil is acceptable for use as a fill material.

In a temperate climate an earthworks project is heavily dependent on the weather conditions, due to the problem of the soil softening when wetted by rainfall and the knock-on effects on earthmoving efficiency. Winter shut-downs are not uncommon.

Compaction of fill materials in the field, usually carried out by rollers, is effective in minimising the air voids content and producing a high dry density providing the appropriate type of roller is chosen for the soil type, the maximum layer thickness is not exceeded, the minimum number of passes of the plant is adhered to, and the moisture content of the soil is within the permitted limits.

Laboratory compaction tests provide the maximum dry density and optimum moisture content of a soil for a given compaction energy. They provide information on the suitable range of moisture contents for placement in the field and are used to provide control of field compaction.

The properties of wet clay soils can be improved by the addition of lime and strong capping layers to support road pavements can be formed from lime stabilisation. Cement-bound materials to form sub-bases and bases can be made from granular soils with the addition of sufficient cement.

CASE STUDY

Use of wet fill – M6 Motorway, UK

Case Objectives:

This case illustrates:

- the economic importance of utilising poor quality materials
- the application of field trials in geotechnical engineering
- the moisture sensitivity of low plasticity low clay content soils
- the development during construction and subsequent dissipation after construction of pore water pressures
- the effectiveness of horizontal drainage layers in accelerating the dissipation of pore water pressures

The construction of a section of the M6 Motorway between Lancaster and Penrith required embankments 5–10 m high in an area where high rainfall and highly moisture-sensitive glacial clays (boulder clays) would result in a fill material in a 'wet' and unacceptable condition. This case study illustrates the successful performance of a very poor quality fill incorporated into the lower part of an embankment, due to the provision of drainage blankets.

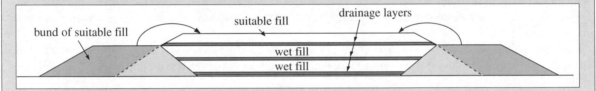

The glacial clay was a well-graded mixture of sand and gravel with a fines content of about 30% and a small clay fraction of typically 5%, making a soil of low permeability. At the correct moisture content it would provide a very suitable fill material when adequately compacted. Its optimum moisture content from the light compaction test was 9% but it had a much higher placement moisture content of up to 18%, hence the description 'wet fill'. In this state it would have provided negligible bearing capacity and would normally have been removed from the site and discarded as unacceptable. Its consistency in this condition was likened to that of wet concrete (Grace and Green, 1979).

The liquid limit was typically 25% with a plastic limit of 18%. Note that these tests were carried out on the less than 425 μm material which was only about 35% of the total. As the material greater than 425 μm was unlikely to contain any moisture a small change in the bulk moisture content would have a large effect on the soil condition. For example, from Equation 2.7 with these values a soil at its liquid limit would have a bulk moisture content of about 9–10%.

Efficiency of horizontal drainage layers

It has been shown (Sills, 1974) that a drainage blanket arrangement will be more effective for thinner clay layers, shorter lengths of drain and fill with a higher vertical permeability. The drainage layers were found both theoretically and from piezometer monitoring to be highly efficient in providing a free-draining horizon and a drainage boundary with zero pore pressure to allow dissipation of the excess pore pressures set up in the clay fill.

The pore pressures set up in the fill during construction virtually mimicked the increase in the height of the fill, with a pore pressure ratio of unity, and little dissipation occurred until after construction. The drains were not particularly effective during the construction period and would not have been of benefit in improving the stability of the embankment. From pore water pressure measurements it was confirmed that dissipation occurred faster with closer drain spacings.

Due to the highly granular and well-graded nature of the glacial clay and its low clay content it was considered that providing it could be placed satisfactorily and consolidated during the construction period under the overlying stresses, its fairly low compressibility would provide small settlements.

It was decided to construct a trial embankment 6 m high at Killington in Cumbria (McLaren, 1968) with the aim of investigating the effectiveness of drainage layers. A control section was built with a single drainage layer at the base and in two adjacent sections a second drainage layer was placed, one with 1.75 m of wet fill and the other with 3 m of wet fill above the drainage layer.

Normal rubber-tyred plant such as scrapers was unable to operate on this material. Excavation in the cut areas was carried out using back-acters or front loaders and loading into dump trucks which transported the material along specially constructed haul roads. The material was tipped in the fill areas and spread by the blades of bulldozers. Due to its high moisture content normal compaction plant could not be employed but compaction was readily achieved by the passage of the tracked dozers, with rapid removal of air voids. Any further compaction would have been counter-productive with the inducement of excess pore water pressures and over-compaction. The drainage layers were crushed rock, 0.45 m thick.

Settlements after construction of only 30 mm were recorded and in the sections including the drains these settlements were complete after 3 months with drains at 1.75 m apart and after 9 months with drains 3 m apart. An average coefficient of consolidation of 3 m^2/yr was estimated for the fill.

The wet fill was confined laterally by bunds of suitable fill which was subsequently excavated and placed as the upper part of the embankment. The embankment has performed satisfactorily since the motorway was opened.

Other relevant chapters are 6, Consolidation and 12, Slope Stability.

Worked Example 13.1   Moisture condition value

The results of a moisture condition test on a sample of silty clay are given in Figure 13.16. Determine the moisture condition value.

The change in penetration from:
1 blow to 4 blows = 81 – 58 = 23 mm
2 blows to 8 blows = 92.5 – 69 = 23.5 mm and so on
From the plot of change in penetration (natural scale) to number of blows (logarithm scale) the steepest straight line through the points cuts the 5 mm change in penetration line at 14.9 blows
∴ $MCV = 10 \log_{10} 14.9 = 11.7$

Worked Example 13.2   Bulking – sand

An embankment with a volume of 20,000 m^3 is to be constructed of sand taken from a cutting. Assuming a shrinkage factor for the sand of 0.90 and a load factor of 0.85 determine:
i) the volume of cutting required (bank m^3)
ii) the production for 3 scrapers each with a heaped capacity of 20 m^3 and a total cycle time of 6 minutes

i) the compacted volume = 20,000 m^3

∴ bank volume required $= \dfrac{20000}{0.90} = 22222 \ m^3$

ii) machine load in bank $m^3 = 20 \times 0.85 = 17.0 \ m^3$

∴ total number of loads required $= \dfrac{22222}{17.0} = 1307$

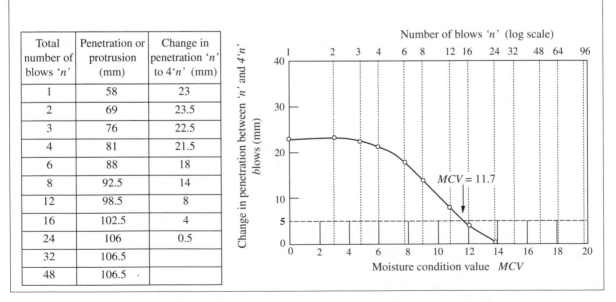

Total number of blows 'n'	Penetration or protrusion (mm)	Change in penetration 'n' to 4'n' (mm)
1	58	23
2	69	23.5
3	76	22.5
4	81	21.5
6	88	18
8	92.5	14
12	98.5	8
16	102.5	4
24	106	0.5
32	106.5	
48	106.5	

FIGURE 13.16   *Worked Example 13.1*

(Note: if bulking was not considered this figure would be 1000)

For average working conditions an efficiency rating given by 50 working minutes per hour is assumed

$$\text{Hourly fleet production} = 17.0 \times \frac{60}{6} \times \frac{50}{6} \times 3 = 425 \text{ bank m}^2/\text{hour}$$

$$\text{Total time required} = \frac{22222}{425} = 52.3 \text{ hours}$$

Worked Example 13.3   Bulking – clay

For the same project as in Example 13.2 determine the total time required if the embankment is to be constructed of clay with a shrinkage factor of 1.1 and a load factor of 0.8.

$$\text{bank volume required} = \frac{20,000}{1.10} = 18,182 \text{ m}^3$$

$$\text{machine load} = 20 \times 0.8 = 16.0 \text{ m}^3$$

$$\text{total number of loads required} = \frac{18182}{16.0} = 1136$$

$$\text{Hourly fleet production} = 16.0 \times \frac{60}{6} \times \frac{50}{6} \times 3 = 400 \text{ bank m}^3/\text{hour}$$

$$\text{Total time required} = \frac{18182}{400} = 45.5 \text{ hours}$$

Worked Example 13.4   Compaction test

The results of a 2.5 kg rammer compaction test on a sample of silty clay are given below. Plot the compaction curve and determine the maximum dry density and optimum moisture content.

Volume of mould used = 1 litre
Weight of mould + base = 1.368 kg

	1	2	3	4	5	6
Wt. of mould + compacted soil (kg)	3.467	3.482	3.497	3.462	3.436	3.370
Wt. of tin (g)	29.31	29.44	29.50	29.43	29.38	29.59
Wt. of tin + wet soil (g)	156.64	152.65	130.18	159.88	132.71	153.08
Wt. of tin + dry soil (g)	140.70	138.12	119.17	147.11	123.06	143.09

For point 1

$$\text{Bulk density} = \frac{3.467 - 1.368}{1.0} \times \frac{1000}{1000} = 2.099 \text{ Mg/m}^3$$

$$\text{Moisture content} = \frac{156.64 - 140.70}{140.70 - 29.31} \times 100 = 14.3\%$$

$$\text{Dry density} = \frac{2.099}{1 + 0.143} = 1.836 \text{ Mg/m}^3$$

The remainder of the results are tabulated below

	1	2	3	4	5	6
Bulk density (Mg/m^3)	2.099	2.114	2.129	2.094	2.068	2.002
Moisture content (%)	14.3	13.4	12.3	10.9	10.3	8.8
Dry density (Mg/m^3)	1.836	1.864	1.896	1.888	1.875	1.840

From the compaction curve, Figure 13.17:
Maximum dry density ≈ 1.902 Mg/m^3
Optimum moisture content ≈ 11.9 %

Worked Example 13.5 Air voids lines

Determine the values of dry density at selected moisture content values for air voids contents of 0, 5 and 10% assuming the particle density = 2.67 Mg/m³.

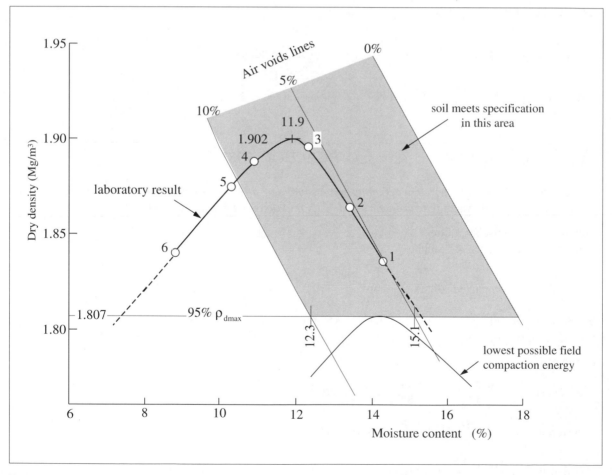

FIGURE 13.17 *Worked Examples 13.4 to 13.7*

Using equation 13.14 the values are tabulated below:

moisture content w (%)	dry density (Mg/m³)		
	$A_v = 0\%$	$A_v = 5\%$	$A_v = 10\%$
8	2.200	2.090	1.980
10	2.107	2.002	1.896
12	2.022	1.921	1.820
14	1.944	1.847	1.750
16	1.871	1.777	1.684
18	1.803	1.713	1.623

These values are plotted in Figure 13.17.

Worked Example 13.6   Void ratio, degree of saturation and air voids content at the optimum value

From the results of Example 13.4 determine the above values.

Expressions to derive these parameters are given in Table 2.15.
Assuming particle density = 2.67 Mg/m³
Bulk density $\rho_b = 1.902 \times (1 + 0.119) = 2.128$ Mg/m³

$$2.128 = \frac{2.67(1 + 0.119)}{1 + e} \quad \text{giving void ratio } e = 0.404$$

$$\text{Porosity } n = \frac{0.404}{1 + 0.404} = 0.288$$

$$\text{Degree of saturation } S_r = \frac{wG_s}{e} \times 100 = \frac{0.119 \times 2.67}{0.404} \times 100 = 78.7\%$$

Air voids content $A_v = 0.288 \, (1 - 0.787) \times 100 = 6.1\%$

Worked Example 13.7   Relative compaction

The minimum dry density to be achieved by field compaction has been specified as 95% of the maximum dry density obtained from the 2.5 kg rammer compaction test. From the results of Examples 13.4 and 13.5 determine the minimum moisture content which must be specified to ensure that no more than 10% air voids may be present. Determine the maximum moisture content w that can be permitted.

The minimum dry density $= \dfrac{95}{100} \times 1.902 = 1.807$ Mg/m³

Using equation 13.14:

$$1.807 = \frac{1 - \dfrac{10}{100}}{\dfrac{1}{2.67} + \dfrac{w}{100}} \quad \text{giving} \quad w = 12.3\%$$

Note that this minimum moisture content would only apply if the field compaction energy is less than the laboratory compaction energy. If the field compaction energy is greater then soils of lower moisture contents can be compacted to the specified requirements.

The maximum moisture content can be obtained assuming that wet of optimum compaction achieves 5% air voids. Then

$$1.807 = \frac{1 - \dfrac{5}{100}}{\dfrac{1}{2.67} + \dfrac{w}{100}} \quad \text{giving} \quad w = 15.1\%$$

Worked Example 13.8   Correction for stone content

A gravelly clay contains 25% and 8% gravel retained on 20 mm and 37.5 mm sieves, respectively. A compaction test has been carried out using the CBR mould on material remaining after the gravel greater than 37.5 mm has been removed. The maximum dry density of the material from the compaction curve is 1.965Mg/m³ and the optimum moisture content is 10.8%. Determine the maximum dry density and optimum moisture content of the whole sample including the gravel greater than 37.5 mm. The particle density of the gravel particles is 2.65 Mg/m³.

According to BS 1377:1990 this material would lie within Grading zone 5 so a correction for the effect of stone content would be permissible. Assuming the coarse gravel merely displaces the remaining matrix and would have no effect on the compaction test result and that it contains no moisture of its own equations 13.15 and 13.16 can be used.

$$\text{'corrected' } \rho_{dmax} = \frac{1.965}{\dfrac{92}{100} + \dfrac{1.965}{2.65}\left(1 - \dfrac{92}{100}\right)} = 2.006 \text{ Mg/m}^3$$

$$\text{'corrected' } OMC = \frac{92}{100} \times 10.8 = 9.9\%$$

The effect of this 'correction' is to move the compaction curve upwards and to the left.

EXERCISES

13.1 The results of a moisture condition test on a sample of stiff sandy clay are given below. Determine the moisture condition value of this clay.

Total number of blows (*n*)	Penetration (mm)
1	43.0
2	56.0
3	63.0
4	68.0
6	75.0
8	80.5
12	87.0
16	91.0
24	95.5
32	98.5
48	100.5
64	101.5
96	102.0
128	102.0

13.2 The results of a moisture condition value calibration are given below. Determine the values of *a* and *b* and assess the moisture sensitivity.

MCV	2.0	3.4	5.2	7.8	10.4	12.9	14.9
water content (%)	30.0	28.9	27.5	25.3	23.6	21.3	19.9

13.3 An embankment of total volume 15450 m^3 is to be constructed using sand taken from a cutting. The bulk density of the sand *in situ* is 1.86 Mg/m^3 and in the compacted state in the embankment it is 1.98 Mg/m^3. Four dump trucks each with a heaped capacity of 15 m^3 are to be used operating on an average cycle time of 8 minutes with an efficiency rating of 0.75. The bulk density of the sand loaded in the dump trucks is estimated to be 1.67 Mg/m^3. Determine the:
a) volume of cutting required to make the embankment
b) total number of loads
c) hourly production rate and total time required

13.4 The results of a BS light compaction test are given below. Plot the compaction curve and obtain the optimum moisture content and maximum dry density for the soil.

water content (%)	14.3	15.8	17.7	19.2	20.9	22.6
bulk density (Mg/m^3)	1.967	2.008	2.065	2.088	2.077	2.072

13.5 For the result obtained in Exercise 13.4 determine the void ratio, porosity, degree of saturation and air voids content for the soil at its optimum moisture content. Assume the specific gravity of the particles to be 2.72.

Site investigation

Site investigation

Inadequate information on the expected ground conditions can lead to faulty design assumptions and mistakes. Problems associated with the ground encountered can lead to cost over-runs, longer construction periods and possibly expensive litigation. It is essential, therefore, to find out as much as possible about the site and its ground conditions by carrying out a thorough site investigation.

Site investigation consists of collecting available information about the site and its environment and carrying out a ground investigation. This information is then used to assess the suitability of the site for designing and constructing the proposed works with regard to stability, serviceability, ease of construc-

tion, and acceptable performance balanced by concern for safety, economics and the environment.

In addition to carrying out site investigations for new works they may also be used for:

● exploring sources of construction materials
● quarrying and mineral prospecting
● hydrogeology and groundwater abstraction
● selecting sites for waste disposal
● assessing the degree of contamination of soils and groundwater
● checking the safety of existing works when constructing new works nearby
● forensic investigations to establish the causes of failures such as with slopes, or defects such as with structures affected by excessive ground movements, and

- to design remedial measures such as for slope stabilisation or underpinning of buildings.

Geotechnical design for new schemes using methods of analysis given in other chapters of this book (such as bearing capacity and settlements of shallow and pile foundations, stability of embankment and cutting slopes, earth pressures on retaining structures) cannot be carried out until the appropriate model of the site is determined and the relevant parameters for each soil type obtained.

A geotechnical engineer should aim to achieve the objectives stated at the beginning of this chapter. However, this subject warrants more than a single chapter and although most of the aspects of a site investigation are described, more detailed reading is recommended. Further coverage of this subject is given in the UK Code of Practice BS 5930:1981 and its draft revision (1996) and books by Clayton, Simons and Mathews (1995), Cottington and Akenhead (1984), Joyce (1982) and Weltman and Head (1983).

Other areas of ground investigation are described in previous chapters, such as Chapter 2, soil description and classification, Chapter 3, permeability testing and Chapters 6 and 7, testing for consolidation and shear strength properties.

Stages of investigation

A degree of discipline must be exercised in carrying out a site investigation, otherwise some information may be overlooked or simply not gathered. To reduce this risk a staged procedure is usually adopted comprising desk study, site reconnaissance, topographic surveys and hydrographic surveys, if necessary, followed by a detailed ground investigation with sampling, laboratory and *in situ* testing and groundwater observations.

These stages will overlap and complement each other and may not be considered complete until all of the stages are concluded and inter-related. The planning of the ground investigation will proceed much more efficiently if the previous stages have been carried out thoroughly.

Geotechnical design is often as much an art as a science, relying on experience and empiricism for its effectiveness. Due to the uncertainties concerning methods of analysis, material properties and behav-

iour, a major element of geotechnical works consists of performance monitoring, particularly where savings in costs and construction periods may be gained and where the consequences of the works not achieving their predicted performance may be serious. Re-design at intermediate stages or remedial works are then a viable solution to progressing the scheme.

Desk study

A desk study is the collection of as much existing information about the site as possible. Comprehensive lists are given in BS 5930:1981 and TRL Report 192 (1996) including land surveys, boundaries, site features, topography, natural and artificial drainage, access, flood risk, utilities such as water, drainage, sewerage, electricity, gas, telephone both for adequate supply to the site and as obstructions below ground. Guidance on carrying out a desk study is given in *BRE Digest* 318:1987.

Much useful information can be readily obtained from maps, both recent and old, coastal charts, aerial photographs, memoirs, local authority records and local libraries. The British Geological Survey publishes geological maps and memoirs for most of Great Britain and has available various other geological records including borehole and well records.

Previous uses of the site must be determined even in 'untouched' rural areas where changes in topography, erosion, deposition, diversions of streams, rivers and drainage conditions could affect the project. In more developed areas records of underground mining, mineral extraction, quarrying operations, waste tipping, and demolished properties will give clues to potential hazards and obstructions in the ground. Local residents and former workers may be helpful in this respect.

Industrial areas must be examined for underground obstructions and the effects of the industrial processes such as removal of groundwater, changes in temperature, ingress of harmful liquid chemicals and incorporation of harmful solids.

Site reconnaissance

A thorough examination of the site should be made by visiting the site and its surroundings as early as possible and preferably in conjunction with the desk study when the information already obtained can be checked and omissions or uncertainties can be further investigated.

A walk-over survey observing the features within and around the proposed works with checks on access, adjacent properties, surface topography, surface water, drainage, present site use, evidence of ground conditions from geomorphology, quarries, cuttings, exposures, type and condition of vegetation, condition of existing structures. A detailed record should be made on site including notes, sketches and surveyed plans, sections or elevations.

Guidance on carrying out a walk-over survey is given in BS 5930:1981 and *BRE Digest* 348:1987.

Ground investigation

Extent of the ground investigation

With the information obtained from the desk study and site reconnaissance, the amount of ground investigation can be assessed by considering the variability expected in the ground conditions and the type and scale of the project.

Skimping on site investigation is very risky and rarely cost-effective since for nearly all projects most risk lies in the ground conditions.

Typical costs lie in the region of 0.2 to 2 % of the total cost of the project (Anon, 1991) whereas contractual claims involving ground conditions which had not been foreseen can cost the client much more than this figure and involve lengthy litigation. *BRE Digest* 322:1987 recommends that a minimum of 0.2% of the project cost should be spent on ground investigation for low-rise buildings and that developers should actively participate in the investigation process to ensure that they appreciate the risks involved with the ground conditions.

The purpose of the ground investigation is to obtain a three-dimensional cross-section of the site and exploratory holes, such as boreholes and trial pits, only establish the ground conditions at a point (one-dimensionally) so the cross-sections must be completed by inferring the intermediate conditions. For uniform homogeneous conditions and simple geology this could be achieved fairly confidently with widely spaced exploratory holes but for variable conditions an accurate impression can only be gained from closely spaced boreholes.

For many sites, the degree of variability will not be known so a constant monitoring and review of the investigation is necessary to ensure that the ground conditions are adequately investigated. To ensure this, a useful procedure is to carry out a preliminary and limited investigation to obtain basic, general information which can be used to plan a more intensive investigation. In some instances, double drilling is carried out by sinking a borehole without undisturbed sampling or testing to determine the soil layering, followed by an adjacent borehole with samples or tests taken at pre-determined depths.

The actual spacing and locations of the holes will depend on the nature of the project, its size and its shape. For housing estates, holes may be sunk at 30–100 m apart and not necessarily related to building layout since this may change. Conversely, the properties may be located to suit the ground conditions encountered. For structures, holes at 10–30 m apart may be appropriate with locations at the corners, some intermediate points and locations of heavy loading or special construction. It should be remembered that more confidence can be placed in interpolation between exploratory holes than extrapolation away from them so a wide coverage is preferable.

For linear structures, such as pipelines and sewers, holes at special crossings or manholes are usually sunk with less investigation between. If these works are constructed by tunnelling then vertical boreholes will provide limited and inefficient information. If the ground conditions require, horizontal drilling from the face of the tunnel using pilot boreholes, can be cost-effective and reduce the risk of unforeseen conditions.

Highway construction comprises major and minor structures and earthworks of cuttings and embankments, each requiring different consideration so the investigation must be designed and carried out to cater for these.

Exploratory holes should be sunk at locations away from proposed foundations, tunnels or shafts if possible. Backfilling trial pits will leave a disturbed and softened zone and boreholes may provide connections between different water-bearing layers producing groundwater contamination or water ingress into the works. It is preferable to backfill boreholes with an impermeable grout.

Depth of exploration

On the cross-section through the site will be super-imposed the proposed construction so the depth of exploration must include the depth of ground which may affect or be affected by the works. In the case of the investigation of mining areas, workings at deep levels may affect the stability of structures which themselves do not stress the mineworkings.

Providing mineworkings are not present a general guide is to take boreholes:

1. in thick compressible strata, to a depth where the changes in stress applied by the works are minimal. For this purpose, a depth of 1.5 times the width of the loaded area is often suggested, where the loaded area may be an individual foundation if the foundations are spaced widely or the whole of the structure if the foundations are closely spaced or the structure is supported on a raft foundation, see Figure 5.6. In the case of a pile group the loaded area would tend to be the equivalent raft, see Chapter 10, part way down the pile length. However, since the length of the pile, particularly a floating pile, would not be known at the investigation stage, if piles are likely to be required then the depth of investigation should be as in (2) below.

2. into relatively incompressible strata, or strata which for the type of construction envisaged will not contribute to settlements or other movements. This is considered as the 'rigid' stratum in the methods for stress distribution (see Chapter 5) and settlements (see Chapter 9).

3. into sound, unweathered bedrock. Beneath the weathered rock horizons a penetration of 2–5 m should be obtained depending on the hardness of the rock. It is also important to make certain that the material is bedrock and not a boulder. Obtaining rock cores by rotary coring is preferable to chiselling in a cable percussion borehole.

Choice of method of investigation

The main factors which affect the choice of method of investigation are:

a) *Access*

Cable percussion rigs, rotary drilling rigs and mechanical excavators require reasonably smooth, unhindered access to a location with space for erection of the apparatus and sufficient headroom. Otherwise, access must be provided by breaking out or removing obstructions, cutting roadways in hilly or hummocky terrain or forming temporary roads on waterlogged or boggy sites. In some instances, especially on steep slopes the drilling rigs are dismantled and re-erected on scaffold staging or other supports. Some companies specialise in 'mini-site investigation' for very limited access.

b) *Equipment limitations*

Mechanical excavators are typically limited to excavation depths of 3–4 m and may damage unforeseen services. Side support or trench shoring must be provided if personnel are to enter the pit and this may obscure some of the exposure.

Hand augering is carried out without support to the sides of the hole so it is limited to self-supporting strata without obstructions present, such as soft or firm clays. A maximum depth of 5–6 m is then possible.

The cable percussion rig has a winch capacity of 1 to 2 tons so the amount of casing it can pull out of the ground is limited. For deep boreholes it is usual to commence with larger diameter casing and then reduce in diameter to extend the borehole. In this way, maximum depths of about 60 m can be achieved, although deeper boreholes, up to 90 m, can be sunk if the lower ground strata are self-supporting and casing is not needed over this depth.

In bedrock and in some heavily overconsolidated glacial soils, such as gravelly clays, rotary drilling is the most effective method with core samples obtained.

c) *Types of ground*

Made ground is best investigated with trial pits since it is at shallow depths and its variability can be assessed on a large scale. Hand dug pits may be necessary if underground services or buried structures are to be exposed without damage.

Silts, sands and gravels are best investigated with cable percussion boring which can support the soil with casing during drilling. They can provide the access for carrying out *in situ* SPTs, permeability tests or installing piezometers. As well as sinking boreholes in these strata the Dutch cone penetration test rig is also used to provide a continuous, undisturbed plot of the layering and densities of the strata. For soils which contain cobbles and boulders, cable percussion boring can be time-consuming because of the chiselling required and not truly representative due to the relatively small casing diameters used. If the nature of the scheme warrants, large-scale pits, with dewatering if necessary may be the most appropriate method but will be limited to shallow depths.

Cable percussion boreholes are most commonly used in clays to obtain U100 samples and perform *in situ* vane tests although difficulties are often experienced when the clays contain cobbles or boulders and care must be exercised with soft, sensitive and laminated clays.

Methods of ground investigation

Trial pits

Trial pits are excavated with a hydraulic back-hoe excavator forming a trench about 3–5 m long with a width equal to the back-acter bucket, usually 0.6 or 0.9 m and a depth determined by the reach of the hydraulic arms, between 3 and 6 m depending on the type of machine.

The sides must never be assumed to be stable. Trench collapse is still unfortunately a major cause of death on construction sites. The sides must be supported at all times if the pits are to be entered. If they are not supported then collapse often occurs due to the surcharge on the side of the pit where the spoil is temporarily stock-piled or due to groundwater inflows. Access is usually preferred when the bucket smears the sides of the pit and obscures the *in situ* structure.

A trial pit record should describe all four faces if they are different and dip and direction measurements of bedding, joints, fissures, unconformities should be taken, if present. 'Undisturbed' samples can be taken using hand-auger rods with 38-mm diameter sampling tubes or preferably 100-mm diameter sampling tubes if the clay is not too stiff and *in situ* vane tests can be carried out. Bulk and smaller disturbed samples are taken from the pit, the bucket or the spoil heap. Samples of the groundwater must not be forgotten.

Headings or adits

Headings are excavated using hand-tools and timber supports in a near horizontal direction from the bottom of shafts or from the surface of sloping ground. They are not commonly used for investigation purposes alone due to their cost and slow progress although they are used occasionally for construction of short sections of tunnels.

Hand auger (Figure 14.1)

A hole, usually 100-mm diameter, is formed by manually rotating a cross-piece above ground level attached by rods to an auger bucket below ground level. Soil is collected in the auger, removed from the ground and inspected to obtain a record of the strata. A cylindrical or barrel-auger (post-hole or Iwan-type auger) is used to remove clayey soils and a flat circular auger with a flap-valve is used to remove gravel although variable success can be expected with the latter due to the obstructions.

Disturbed samples are taken from the cuttings and 38-mm diameter sample tubes can be pushed or driven in for 'undisturbed' samples. Vane tests

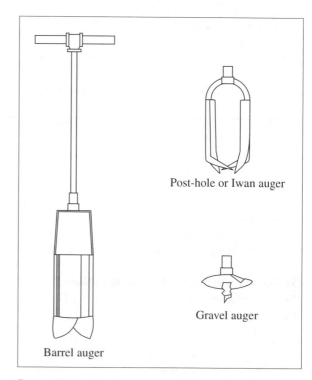

Post-hole or Iwan auger

Gravel auger

Barrel auger

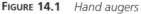

FIGURE 14.1 *Hand augers*

may also be carried out *in situ*. Depths of 5–6 m are achievable if the soil is self-supporting, e.g. soft or firm clay but progress can be halted by one gravel particle or when water-bearing strata are entered.

Cable percussion boring (Figures 14.2 and 14.3)

Often referred to as shell and auger boring an auger is rarely used with this method in the UK. The rig consists of a four-leg 'A-frame' or derrick with a diesel-powered winch lifting and releasing a cable running over a pulley wheel so that tools attached to the end of the cable are lifted vertically and dropped free-fall.

Once the rig is erected and stabilised the hole should be commenced by hand excavation to about 1 m depth to ensure that shallow underground services are avoided. Then a heavy steel tube (clay-cutter or shell) is dropped into this hole a number of times to collect soil inside it and extend the depth of the hole. The tube is removed, the soil is cleaned out and the process repeated.

As the hole progresses the soils may not be self-supporting and soil at the top of the hole may fall in so steel lining tubes called casing are driven into the hole to support the soil, to provide a guide for the boring tools and to ensure a 'clean' hole. Casing consists of short sections of steel tube connected together by their threaded ends with a protective steel cutting shoe at the base and a driving head at the top. Casing diameters of 150 and 200 mm are most common.

The clay-cutter (Figure 14.3) is an open-ended tube to allow clay to enter and with slots in the sides to enable removal of the clay. An alternative is the cross-blade clay cutter which is lighter and requires attached weights called sinker bars to enable penetration but allows easier removal of the clay.

The shell (Figure 14.3) is used for removing granular soils. It comprises a steel tube without slots but with a flap-valve or 'clack' at its lower end. Sands and gravels are drawn into the shell by a pumping action with water added to the borehole to 'fluidise' the soils. With short up and down movements the soil enters the shell and is retained in the shell by the clack.

If cobbles or boulders are encountered then they must be broken up by dropping a heavy chisel on top of them so they can be removed by the shell or displaced sideways.

Where soils are in a loose sensitive state the percussive action of the tools can cause disturbance and excessive piston and suction action may loosen a dense sand so samples or tests must be taken from at least one diameter below the bottom of the borehole to ensure a more representative result. Care is required in taking disturbed samples of granular soils from the shell to prevent loss of the finer particles. Addition of water to a borehole precludes the testing of samples for moisture content.

Mechanical augers (Figure 14.4)

Two basic types exist – solid rod bottom augers or hollow stem continuous flight augers.

The bottom augers may comprise either a flight-

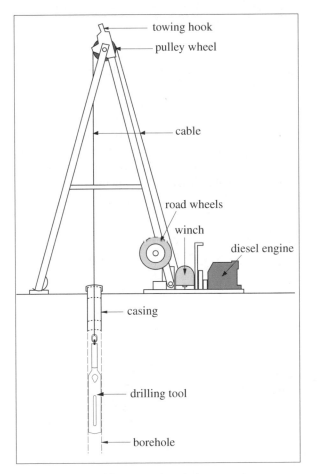

FIGURE 14.2 *Cable percussion rig in operation*

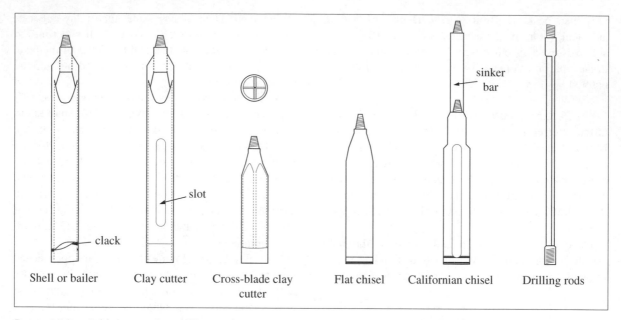

FIGURE 14.3 *Cable percussion drilling tools*

auger with short helical flights up to about 600 mm diameter or a bucket auger with cutters on an angled base-plate with diameters between 300 and 1800 mm. Both are rotated at the end of drill rods to cut into the soil and retain it on the flights or in the bucket. They are then raised from the borehole and emptied for identification of the strata and taking disturbed samples. Addition of water should not be necessary. To progress the borehole this procedure is repeated.

Undisturbed samples or tests can be taken at intervals from the bottom of the borehole. Significant lengths of casing cannot be inserted, therefore these boreholes are only suited to self-supporting ground, such as clayey soils and soils above the water table.

The continuous flight auger method comprises a

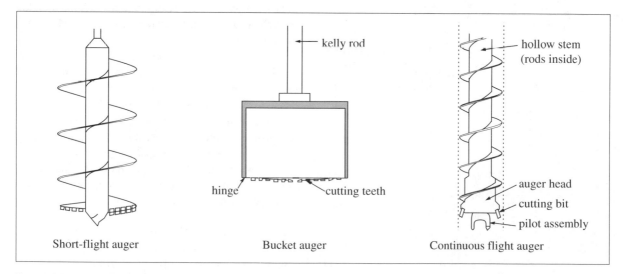

FIGURE 14.4 *Mechanical augers*

full-length helical flight wrapped around a hollow tube with a central rod and pilot assembly at the bottom of the auger to prevent soil entering the hollow stem. As the auger rotates soil rises up the spiral for identification and disturbed sampling. The augers are typically 150–250 mm outside diameter with hollow stem diameters of 75–125 mm and can reach to depths of 30–50 m in suitable soils.

At intervals, the central rod and pilot assembly can be removed to allow sampling tubes to be pushed or driven in at the bottom of the borehole or Standard Penetration Tests may be carried out. Water balance is not usually maintained so disturbance due to piping of the ground below a water table may occur.

The torque required to turn the auger can become considerable so a heavy duty rotary drilling rig is necessary. These are lorry mounted and will, therefore, be more expensive and require better access conditions than the light cable percussion rig. As the soil is brought to the surface on the flights it is diffi-

cult to assess its depth in the ground, so changes of strata will not be determined accurately unless frequent undisturbed sampling is carried out.

Penetration into the ground can be severely hampered in coarse cohesionless soils. Finer silts and sands can flow up the hollow stem when drilling below the water table, causing excessive disturbance of the ground at the bottom and around the auger.

Rotary open hole and core drilling (Figure 14.5)

These techniques are used in harder rock exploration when cable percussion or auger methods can no longer penetrate.

Open hole drilling consists of rotating a rock roller bit at the end of hollow drill rods to cut a cylindrical hole. The drilling fluid which is usually compressed air or pumped water passes down the hollow rods to cool the bit and remove cuttings by flushing them back up the outside of the rods. These cuttings and the rate of penetration are the only means of recording information so the accuracy of

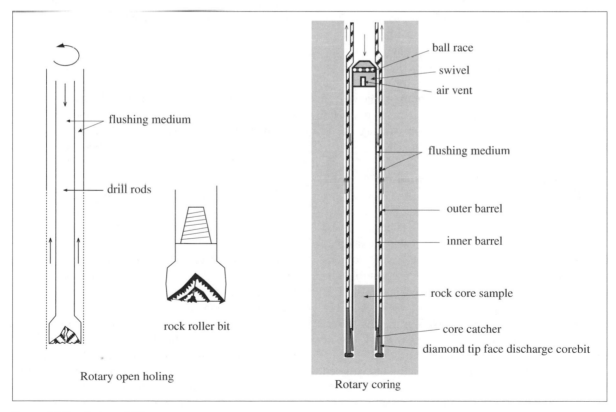

Rotary open holing

flushing medium

drill rods

rock roller bit

ball race
swivel
air vent

flushing medium

outer barrel

inner barrel

rock core sample

core catcher
diamond tip face discharge corebit

Rotary coring

FIGURE 14.5 *Rotary drilling*

rock descriptions and identification of changes of strata will be limited.

Rotary coring consists of cutting a cylinder of rock using a rotating double-tube core barrel with a coring bit attached to the bottom of the outer barrel. A swivel mechanism at the top of the inner barrel allows it to remain stationary as the outer barrel and coring bit cut an annular hole and produce the rock sample.

The flushing medium which may be compressed air, pumped water, mud or foam passes down the hollow drill rods, between the inner and outer barrels to cool the drill bit, remove the cuttings and flush them up the borehole outside the core barrel. As the inner barrel does not rotate and the flushing medium passes outside, the rock core sample is not disturbed, smeared or eroded. This is especially important with friable, weathered rocks.

A plastic liner inside the inner barrel enables the rock core to be removed from the core barrel with minimal disturbance, keeping the rock pieces in their correct positions for inspection.

It is essential to obtain 100% core recovery or as near as possible. Core loss means that the strata are not identified and since these are likely to be the weakest materials they are the most critical. The type of core bit used must be matched to the nature of the rock with larger diameter coring adopted in weaker rocks. Core sizes between 18 and 165 mm diameter may be obtained, with H size (76 mm) usually specified as a minimum requirement in the UK.

Undisturbed sampling – sampling quality

Due to the relief of total stresses, pore pressure changes and the cutting action while taking samples, it is not possible to obtain truly undisturbed samples from the ground. The aim must be, therefore, to minimise disturbance as much as possible. A useful classification of sample quality is given in Table 14.1, based on the soil properties that can be reliably determined from a sample. Soil properties determined from laboratory tests on a lower class of sample should be treated with some caution.

Difficulties in obtaining good quality undisturbed samples can be experienced with soft clays, sensitive clays, laminated clays, gravelly clays, partially saturated soils, collapsing soils, organic soils, coarse soils and cohesionless soils with even poorer quality likely below a water table. For further discussion on sampling quality, see Clayton *et al.*, 1995.

Types of samples
Block samples

These are cut by hand from the soil at the base of an excavation. They can be quite large and carefully selected. However, they can be heavy, difficult to remove, handle and transport so they are time-consuming and liable to be expensive. They are not confined by sampling equipment so the soil must have some self-supporting capability and they must not be allowed to change in moisture content.

General purpose open-tube sample (U100) (Figure 14.6)

This is the most commonly adopted method of sampling when using cable percussion boring in all cohesive soils.

The sampler comprises a steel or aluminium alloy tube 450 mm long with a cutting shoe threaded on to its lower end. The sampling tube is threaded into a driving assembly with an overdrive space and a non-return valve above the tube.

The borehole is cleaned of loose debris and the sample tube is driven or pushed into the soil at the bottom of the borehole a sufficient distance so that it is full of 'undisturbed' soil with air and any water passing through the non-return valve. The remaining debris or softened soil entering the overdrive space is discarded. The sample must not be over-driven as this would compress the soil.

The sampler and tube are then removed from the borehole with the non-return valve closed, providing a suction effect to assist in the retention of the soil. Sometimes the soil is not recovered in the tube so a core-catcher device is frequently used. The samples are trimmed at the ends, sealed with wax, labelled top and bottom and retained with screw caps for transport to the laboratory.

The cutting shoe is larger than the inside and outside diameters of the tube to reduce friction or adhesion on the inside for minimum disturbance to the sample and on the outside to facilitate withdrawal of the sampler from the borehole. The internal diameter of the cutting shoe D_c is about 1% less than the internal diameter of the tube

TABLE 14.1 *Sampling quality (from Rowe, 1972)*

Quality class	Properties	Purpose	Typical sampling procedure
1	Remoulded properties Fabric Water content Density and porosity Compressibility Effective stress parameters Total stress parameters Permeability Coefficient of consolidation	Laboratory data on *in situ* soils	Piston thin walled sampler with water balance
2	Remoulded properties Fabric Water content Density and porosity Compressibility Effective stress parameters Total stress parameters	Laboratory data on *in situ* insensitive soils	Pressed or driven thin or thick walled sampler with water balance
3	Remoulded properties Fabric A 100% recovery 　　　　Continuous 　　　B 90% recovery 　　　　Consecutive	Fabric examination and laboratory data on remoulded soils	Pressed or driven thin or thick walled samplers. Water balance in highly permeable soils
4	Remoulded properties	Laboratory data on remoulded soils. Sequence of strata	Bulk and jar samples
5	None	Approximate sequence of strata only	Washings

D_s to provide clearance. This must not be exceeded since the sample would be allowed to expand, open up macro-fabric and possibly soften if the sample was taken from a wet borehole. The outside diameter of the cutting shoe D_w should be slightly greater than the outside diameter of the tube, D_T.

The area ratio (Equation 14.1), defined in Figure 14.6, represents the volume of soil displaced during sampling compared to the volume of the sample. It is related to disturbance effects so it should be as low as possible.

Thin-walled tube sampler (Figure 14.6)
Using a thin-walled tube (Shelby tube) with a sharp integral cutting edge rolled slightly inwards gives an area ratio of about 10% and an inside clearance so these tubes are more suited to softer, sensitive soils. They are usually 38, 75 or 100 mm diameter. They are not suited to stiff soils or soils containing gravel particles when they are liable to damage and cohesionless soils may not be retained in the tube. The length of sample recovered will often be less than the length of sampler pushed into the soil due to fric-

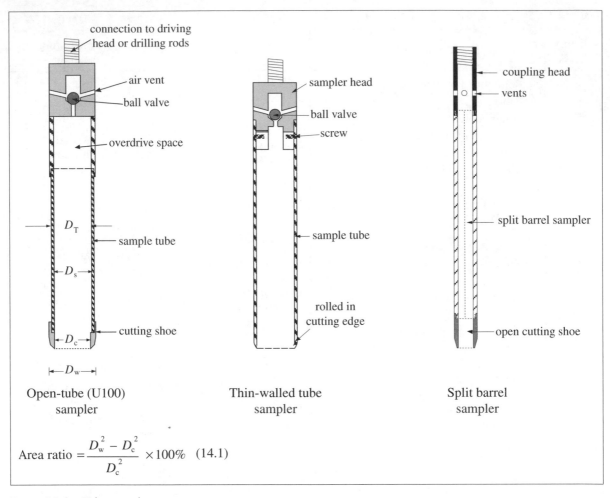

$$\text{Area ratio} = \frac{D_w^2 - D_c^2}{D_c^2} \times 100\% \quad (14.1)$$

FIGURE 14.6 *Tube samplers*

tion on the inside of the tube exceeding the bearing capacity of the soil and preventing soil entering the tube.

Split-barrel SPT sampler (Figure 14.6)

This sampler is used in conjunction with the standard penetration test, see below. Two half-cylinders are joined together by a threaded open steel cutting shoe at the bottom and threaded onto drilling rods at the top. Following removal from the borehole a sample is recovered by undoing the two halves. Due to inside friction a full-length sample is seldom obtained and because the cutting shoe has a thick wall with an outside diameter of 51 mm and an inside diameter of 35 mm giving an area ratio of more than 100% the samples are highly disturbed.

Piston sampler (Figure 14.7)

For soils which are particularly sensitive or may not be recovered on removal of the sampler, such as very soft clays, silts, muds and sludges, the piston sampler should be used. The purpose of the piston is to retain soil in the sampling tube during removal by means of suction.

The piston is attached to an inner solid rod and the sampling tube and sampler head is attached to hollow sleeve rods. With the piston at the bottom of the tube the piston and tube are locked together. The whole assembly is lowered down the borehole using

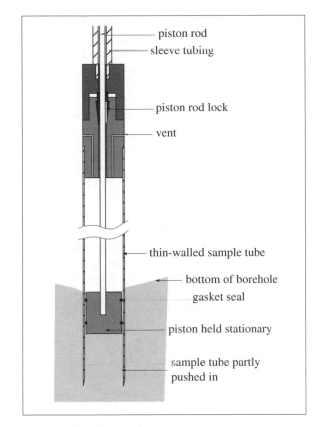

piston rod

sleeve tubing

piston rod lock

vent

thin-walled sample tube

bottom of borehole

gasket seal

piston held stationary

sample tube partly pushed in

FIGURE 14.7 *Piston sampler*

extension pieces of solid and hollow rods and then pushed into the bottom of the borehole and beneath the disturbed zone. The rods are then unclamped and with the piston held stationary the sampling tube is pushed downwards, not driven.

Short length tubes may be pushed in manually but for longer length tubes and firmer soils a block and tackle arrangement attached to the drilling rig is required. When the sampling tube has been pushed in for its full length (and no more) the piston rods and sampling tube rods are clamped together and twisted to shear the soil and break the suction at the bottom of the tube. The complete assembly is then slowly withdrawn from the borehole.

The sampler is available in diameters between 20 and 125 mm with lengths between 0.15 and 1.5 m but the 100-mm diameter 600-mm long sampler is the most common. When the importance of the project has justified their use, 250-mm diameter samples have been taken for testing in the large

diameter hydraulic cell apparatus for consolidation and permeability properties.

The sampler produces good quality, undisturbed specimens because it is pushed beneath any disturbed zone, and it has a low area ratio. It may not withstand pushing into stiff soils and may be damaged if gravel particles are present. Good recovery is achieved due to the piston suction although this may be eliminated when sampling partially saturated soils and cohesionless soils.

Delft continuous sampler

This sampler was developed by the Laboratorium Voor Grond Mechanica, Delft, Holland for obtaining continuous core samples either 29, 66 or 100 mm diameter to depths of 18 m in soft normally consolidated clays, silts, sands and peat.

The 29-mm sample is used for detailed examination of the soil layering and for density and classification tests. The continuous sample means that more confidence in the soil profile is gained, compared to other methods of investigation which may miss some layers or fabric of significance.

The 66- and 100-mm samples may be used for other tests, although some compression during sampling may occur which will affect the results. The sampler provides a good indication of where to take subsequent piston samples, perform vane tests or install piezometers etc.

Disturbance is limited by using a sharp, longer cutting edge, by pushing the sampler into the ground and eliminating side friction inside the sampling tube with an impervious nylon sleeve. The Dutch deep-sounding machines are used to push in the sampler so a preliminary sounding using a cone penetrometer is made to obtain an indication of the soil profile and ensure that the sampler will not be damaged by resistant or gravelly strata.

The sampler comprises two tubes about 1.5 m long. The liner tube is a thin-walled plastic tube filled with a bentonite based fluid of similar density to the surrounding ground. This fluid will support the sample and lubricate the space between the sample and the inner tube. Enclosed inside the outer tube is a magazine containing a rolled up tube of nylon stockinette which is treated with a rubberising fluid to make it impervious and flexible. The end of this nylon sleeve is fixed to the top cap.

As the sampler is pushed into the ground the cutting shoe cuts a continuous sample which is fed into the stockinette and supported in the plastic inner tube by the bentonite fluid. As the sampler is advanced extension tubes are added until the final depth is reached. The soil is then locked in the sampler by a sample-retaining clamp above the magazine and the whole sampler is withdrawn from the ground. Samples are cut into 1 m lengths and placed in purpose-made core boxes. For inspection the stockinette is cut, the samples are split in half, examined, described and photographed as a continuous record of the soil profile.

Window sampler

This system comprises a set of sampling tubes each about 1 m long and in a range of diameters, with a longitudinal slot or 'window' on one side. They are coupled together and with a screw-on cutting shoe are driven into the ground by percussion hammer methods. The largest diameter tube, say 80 mm is driven as far as possible, up to about 2 m, extracted and the retained sample is examined through the slot, described and specimens taken for laboratory testing. A smaller diameter tube coupled to extension rods is then driven through the hole formed, to sample the ground beneath in the same way. The process is repeated with successive reductions in diameter until no further penetration is possible. The samples recovered are disturbed and can be compressed with less soil recovery than length driven. The system is often used for investigations in contaminated ground.

Methods of *in situ* testing

Standard penetration test

This test was developed in the USA in the 1920s to assess the density of sands. It is now the most commonly used *in situ* test in cable percussion boring.

The test measures the penetration resistance of the split-barrel sampler, see above, when driven into the soil at the bottom of a borehole in a standard manner. The N value is the number of blows required to achieve 300 mm penetration and indicates the relative density of a sand or gravel. It has also been found useful as an indication of the consistency of other soils such as silts or clays and the strength of weak rocks such as chalk, shale and mudstone.

Over the years, many empirical correlations with other parameters have been obtained and methods of analysis such as for settlements of foundations have been developed empirically using the N value. Skempton (1986) has examined various factors that affect the results.

The test is described in BS 1377:1990, Part 9 and ASTM D1586. The split-barrel sampler (Figure 14.6) is attached to stiff drill rods and lowered to the bottom of the borehole. A standard blow consists of dropping a mass of 64 kg free fall through 760 mm onto an anvil at the top of the rods and ensuring that this amount of dynamic energy is transferred to the sampler as much as possible.

The number of blows required to achieve each 75 mm penetration is recorded for a full penetration of 450 mm. The initial 150 mm penetration is referred to as the seating drive and the blows required for this penetration are not included as this zone is considered to be in disturbed soil. The next 300 mm of penetration is referred to as the test drive and the number of blows required to achieve this fully is termed the penetration resistance or N value.

In dense or hard soils the seating drive may not achieve 150 mm penetration so the seating drive is considered as the penetration for the first 25 blows. If the full 300 mm penetration is not achieved with a further 50 blows the test is terminated and the amount of penetration achieved is recorded.

In very loose soils it can be worthwhile continuing the penetration beyond 450 mm to confirm that the soil is loose and has not been disturbed by drilling.

Water balance should be maintained for tests carried out below the water table, to prevent loosening of the soil due to piping. Often very low results are obtained in sands immediately beneath a clay layer when sub-artesian groundwater flow up the borehole causes extensive disturbance and these results should be discarded. In gravelly soils or weak rocks a solid shoe in the form of a 60° cone is sometimes used to reduce damage. However, no sample is recovered and higher N values may be obtained.

The test is simple to carry out, quick and inexpensive so it is frequently used and may be the sole means of assessing the strength of the ground when difficult conditions prevent other samples and tests being carried out. However, the results are sensitive to variations in the test equipment used and the procedure adopted (Thorburn, 1984) and several factors affect the results even when the test is carried out in a standard manner (Skempton, 1986). For these reasons, judgement is required in the assessment of the results and they should always be considered as approximate. If any doubt exists then the results should be supplemented by and compared with other tests, such as the Dutch cone penetration test.

Dutch cone penetrometer (Figure 14.8)

A review of this type of test is given in Meigh (1987). The test was originally devised to provide parameters for the determination of settlements of foundations on sands. Experience with the test has enabled the soil types and layers to be assessed so a continuous strata record with undisturbed testing is available, although no samples are obtained. The results should be compared with boreholes and visual identification of samples.

With a high rate of penetration and rapid data acquisition and analysis, several tests can be carried out in a day to depths of 15–40 m depending on the soil conditions.

The electric friction cone penetrometer test is preferred to the mechanical types of cone penetrometer. A piezometric sensor and filter (piezocone) may be incorporated in the penetrometer tip to record equilibrium pore water pressures at intervals and excess pore pressures during penetration and some devices include a temperature measuring sensor.

The tip shown in Figure 14.8 comprises a solid $60°$ cone with a diameter of 35.7 mm and a base area of 1000 mm^2 (10 cm^2), a cylindrical friction sleeve above the cone with a surface area of 150 cm^2, internal sensing devices for measuring the axial thrust on the cone and the frictional force on the sleeve and an inclinometer to check vertical alignment and drift.

The electrical signals are transmitted by an umbilical cable within the hollow push rods to recorders/ analysers/plotters at ground level as the tip is pushed into the ground at a constant rate of penetration of 20 mm/second.

The point or cone resistance q_c (axial force/cross-sectional area), the side or sleeve friction resistance f_s (frictional force/sleeve surface area) and the friction ratio R_f:

$$R_f = f_s / q_c \qquad (14.2)$$

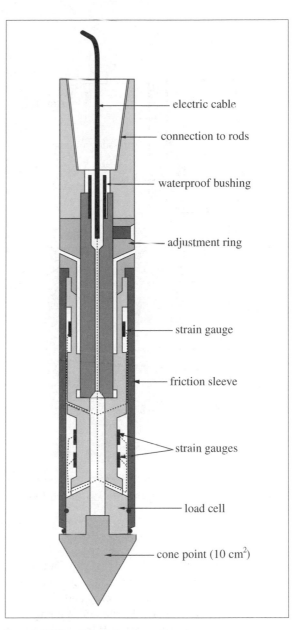

FIGURE 14.8 *Cone penetrometer*

are plotted continuously with depth.

The maximum thrust that can be applied depends on the type and weight of truck or rig used to mount the penetrometer. Tests usually have to be terminated when dense sands, gravelly soils or rock is encountered or when the vertical mis-alignment is excessive.

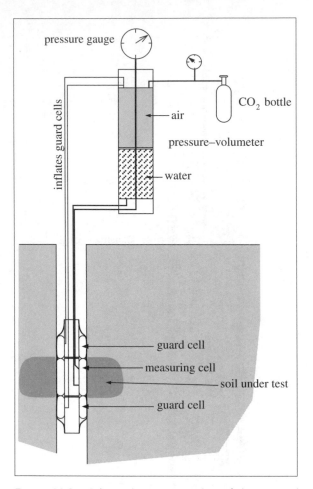

FIGURE 14.9 *Schematic representation of the Menard type pressuremeter*

Vane test

This test is described in BS 1377:1990 Part 9 and ASTM D2573. It consists of rotating cruciform-shape thin steel plate blades in a clay and shearing a cylinder of the soil relative to the surrounding soil. The test is appropriate for very soft to firm saturated clays and is assumed to obtain the undisturbed unconsolidated undrained shear strength.

The height H of the vane blades should be twice the vane width (or diameter D) and the blades must be sharpened and thin enough to give an area ratio of less than 12%. Vane widths of 50 mm and 75 mm are usually adopted for firm and soft clays, respectively.

The vane is attached to rods and pushed about 0.5 m below the bottom of the borehole into undis-

turbed soil. A torque head is then attached to the top of the rods and the rods rotated at a speed of between 0.1 and 0.2 degrees/second until the soil is sheared. The maximum torque reading T is taken and the vane shear strength is obtained as illustrated in Figure 7.30. The torque required to rotate the rods on their own should be measured and this value deducted.

The vane may then be rotated rapidly at least 10 times to remould the soil on the blade edges. Repeating the test will provide the remoulded shear strength and a measure of sensitivity of the soil.

The test is quick and simple and plots of vane shear strength with depth can be obtained to determine the rate of increase of strength. This is particularly important when laboratory tests may indicate no increase of strength due to the greater stress relief and disturbance which can occur with conventional sampling from boreholes at depth.

Bjerrum (1972) showed that the vane shear strength is usually larger than the *in situ* shear strength. He proposed the relationship:

$$in\ situ\ c_u = \mu\ vane\ c_u \tag{14.3}$$

and gave μ values as shown on Figure 7.30. However, the μ value is dependent on more than the mineralogy of the clay or plasticity index and a wide range of values has been obtained, as shown by the results of Ladd *et al.* (1977).

Pressuremeter test (Figures 14.9 and 14.10)

This test was developed in the 1950s by Louis Menard. The test is not included in the British Standard but is described in ASTM D4719 and pressuremeter testing is discussed in general in Mair and Wood (1987).

The pressuremeter consists of three parts, top and bottom cylindrical guard cells and an intermediate water-filled measuring cell. The apparatus is lowered to a test level in a borehole and the top and bottom guard cells are inflated to minimise end-effects on the measuring cell which is, in turn, inflated to apply radial stresses to the sides of the borehole. The inflation is provided by water pressurised by nitrogen or carbon dioxide.

The apparatus must be calibrated so that corrections can be applied for expansion resistance of the

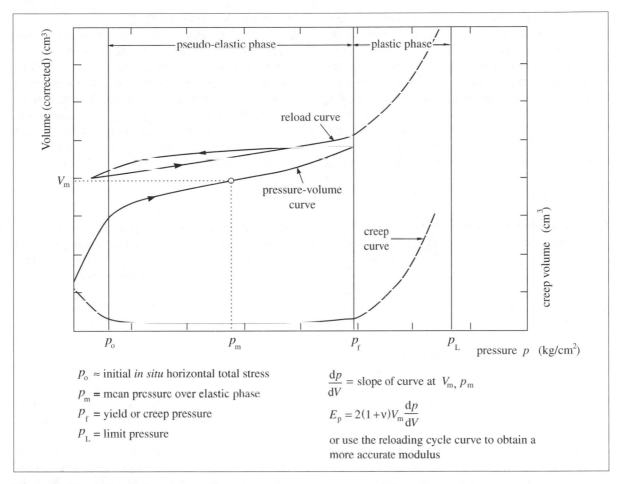

$p_o \approx$ initial *in situ* horizontal total stress

p_m = mean pressure over elastic phase

p_f = yield or creep pressure

p_L = limit pressure

$\dfrac{dp}{dV}$ = slope of curve at V_m, p_m

$E_p = 2(1+v)V_m\dfrac{dp}{dV}$

or use the reloading cycle curve to obtain a more accurate modulus

FIGURE 14.10 *Pressuremeter test results*

measuring cell and volume increase of the connecting tubes. The volume change of the central test cell is measured as the pressure is increased in stages and a pressure–volume change plot is obtained. At an intermediate stage the pressure can be reduced and re-applied to obtain a reloading modulus.

Each pressure increment is maintained for 1 or 2 minutes, recording the volume change at 15, 30, 60 and 120 seconds. The difference in volume change readings between 30 and 60 seconds is referred to as 'creep' and a creep curve can be useful in distinguishing the pseudo-elastic and plastic phases.

A typical result is shown in Figure 14.10. The value p_o gives a measure of the *in situ* horizontal total stress, the shear modulus G and elastic modulus E_p can be estimated from the pressure–volume gradient in the elastic region and the limit pressure p_L can be used to obtain an estimate of the undrained shear strength c_u.

The pressuremeter tests a fairly large volume of soil in its *in situ* condition and may be the most suitable method for ground which is difficult to sample, such as glacial clays and weak rocks. However, the test is carried out in a pre-formed borehole where some self-supporting capacity is required. The apparatus tests an annulus of soil around the borehole so the method of forming the hole and its preparation, with varying degrees of disturbance, will have an effect on the results obtained. The test will provide parameters related to the horizontal direction which may not be relevant for vertical loading by structures, especially if the ground is anisotropic.

To reduce the effects of soil disturbance self-boring pressuremeters have been developed (Wroth *et al.,* 1973, Baguelin *et al.,* 1973). As this pressuremeter is pushed into the soil a rotating cutter and a flushing medium remove the spoil. At required depths the pressuremeter is expanded using gas pressure and its expansion is measured by three spring feelers which sense the deformation of the membrane and enable measurement of radial deformation. The device requires skilled operators to insert without causing excessive disturbance, and will not penetrate soils containing obstructions.

Groundwater observations

The application of soil mechanics is fundamentally dependent on the effective stresses present. Since these cannot be measured directly they are obtained from an estimate of the total stresses and measurements of the pore water pressures. Thus it is imperative that the latter be carried out. Groundwater movements and seepage forces may also affect civil engineering works and the quality or chemistry of the groundwater is important if it is to be used for abstraction or where it may affect material durability.

Careful observations and recording of the groundwater encountered during drilling cannot be overemphasised. The depths of entries, rates of seepage, any change in level after 15 to 20 minutes, when water is sealed off by casing, when water is added, depths of casing, morning and evening water levels should all be recorded. From the impression of groundwater conditions observed during boring and the stratification of the soils, the depths for the installation of the piezometers, their response zones and seals can be more appropriately specified. Very slow seepages into a borehole should not be ignored because in low permeability soils high pore pressure conditions may be masked by small inflows.

Groundwater levels should never be considered to be static. They may vary with seasonal, tidal or weather conditions and they could be higher or lower than the level obtained during the period of on-site work. Monitoring water levels over appropriate periods of time should, therefore, be carried out to establish equilibrium water levels and their variations. Some instruments for measuring water levels are described below.

Standpipe (Figure 14.11)

This is the simplest method of measuring water levels. It comprises plastic tubing, perforated along its whole length or a lower section, and backfilled with a gravel or sand filter material. Considerable volumes of groundwater will be required to saturate the backfill and to fill the tube up to the equilibrium level. Depending on the permeability of the soil this could take a long time, known as the response time.

Water level recordings are made using an electrical device called a dipmeter probe. This responds to the completion of an electrical circuit causing a light to illuminate or a sound to be heard when the probe lowered down the standpipe enters the water.

The main disadvantages of the standpipe are its slow response time and no distinction is made between groundwater from different layers.

Standpipe piezometer (Figure 14.11)

The Casagrande-type piezometer consists of a porous plastic or ceramic cylindrical element or tip attached to the bottom of unperforated plastic tubing and placed at a pre-determined depth to monitor water levels at that horizon only. This is achieved by providing a short response zone of sand backfill around the piezometer tip with a seal above and below.

The seal may comprise bentonite or a bentonite cement mixture in the form of granules, pellets, balls or pumpable grout. It is preferable to backfill the whole of the borehole above and below the sand response zone with grout.

If more than one piezometer is installed in a borehole there may be doubt concerning the separation of the two response zones by adequate seals so it is preferable to install piezometers at different levels in separate boreholes. To speed up the response time the sand backfill should be placed in a near saturated condition and water can be added to the standpipe tube up to the anticipated water level. The piezometer tip should be fully saturated by soaking in water. This ensures that the pores contain no air which may cause faulty readings.

The installation can be used for the measurement of the *in situ* permeability of the soil surrounding the response zone, as described in Chapter 3.

In soft soils a driven type of piezometer can be installed. The piezometer tip is fitted with a point

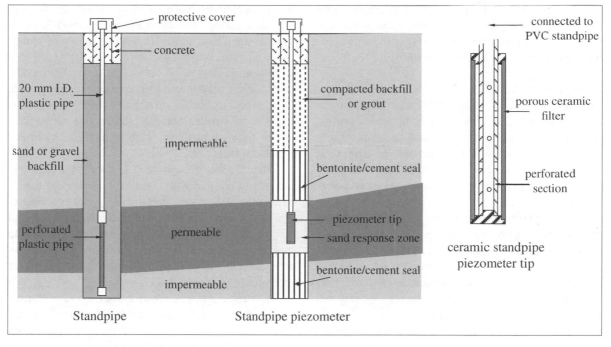

FIGURE 14.11 *Standpipe and standpipe piezometer installations*

and attached to steel tubes for driving to depths of up to 10 m. The tip is protected by a sleeve during driving to avoid smearing and clogging. At the required depth the tip is driven beyond the sleeve. A period of time will be required to allow dissipation of any pore pressures set up around the tip during driving. The tip may be used for the measurement of *in situ* permeability although the results can be affected by disturbance and smearing.

Hydraulic piezometer (Figure 14.12)

These comprise porous ceramic tips about 100 to 200 mm long and 40 to 50 mm outside diameter. They are installed in a compacted fill or earth structure for monitoring construction and post-construction pore pressures. The tips are connected to a mercury–water manometer, a Bourdon gauge or a pressure transducer at a location remote from the piezometer tip for direct readings of pore water pressure.

The tubes connecting the tip to the readout must be completely full of water with no air present so twin tubing is used to allow periodic flushing of the system by circulating de-aired water. Pressure read-

ings from each tube can be taken as a check. The tubing comprises nylon covered with polythene so that it is impervious to both air and water. Pressures in the range of –5 to +200 m head of water can be measured with this device for long-term monitoring. For most applications it is preferable to place the tips and the measuring devices below the anticipated water level to give a positive pressure head and avoid the problem of air cavitation.

If the ground is fully saturated then a ceramic with a large pore size (mean diameter of 60 μm) and low air entry value can be used. However, if air or gas is present this ceramic will not be suitable as the gases will enter the tip and then only pore air pressure will be measured.

For partially saturated ground such as compacted fill or when negative pore pressures are anticipated a finer ceramic (mean pore size of 1 μm) with a high air entry value should be used. This type will permit water to pass through but a high air pressure (air entry value) will be required to cause air to penetrate the ceramic so the air should be excluded.

With small diameter bores and the minimum length of tubing possible the response time of this piezometer is short so it is useful for measuring pore pressure changes due to fill placement, superimposed loads or tidal variation. The tip can be used for the measurement of *in situ* permeability.

Pneumatic piezometer (Figure 14.12)

This piezometer is used in similar applications as the hydraulic piezometer. It comprises a porous ceramic tip with one tube attached to a gas supply and pressure measurement device and the other tube is attached to a flow indicator. The measuring devices are at a remote location from the piezometer.

Compressed air or nitrogen is passed down one tube (the pressure line) until the pressure is equal to the pore water pressure inside the tip. At this stage a flexible rubber diaphragm-type valve is activated and the compressed air can pass up the return line to the flow indicator. These piezometers have a short response time but there is no means for checking the readings, they cannot normally measure negative pore pressures, and cannot be de-aired following installation.

Vibrating wire piezometer

This piezometer contains a tensioned stainless steel wire attached to a diaphragm. One side of the diaphragm is in contact with the groundwater pressure inside a porous ceramic tip with the other side of the diaphragm connected to atmospheric pressure by an air line. As the pore water pressure deflects the diaphragm the tension in the wire and, hence, its frequency of vibration changes. This frequency is recorded and calibrated to pressure readings.

The piezometer gives a rapid response time but requires calibration which cannot be checked after installation for the standard type and may give misleading results if gas bubbles can enter the tip because it cannot be de-aired after installation.

Investigation of contaminated land

In many countries the demise of certain industrial activities and poor control of pollution from chemical processes has left the ground in a contaminated state. Hazards to human health and construction materials must be assessed so that appropriate remedial measures can be determined. The draft BS 5930 (1996) defines contamination as 'land in which the ground or groundwater has concentrations of one or more potentially harmful substances significantly elevated above "normal background levels", or in which significant concentrations of toxic or explosive soil gases occur'.

Contamination may be in the form of metals, salts, acids, alkalis, explosives, organic compounds and vapours, fibrous materials such as asbestos, pathogens, viruses, bacteria, radioactive substances, gases such as methane, carbon dioxide, hydrogen sulphide, radon, etc.

There is, therefore, a need to conduct a chemical site investigation involving a desk study, site reconnaissance and ground investigation. For economical reasons it is preferable to undertake the chemical and geotechnical investigations together but to assess the health and safety risks it may be advisable to conduct a chemical investigation first, especially for potentially high risk sites.

The desk study, site reconnaissance and walk-over survey must be undertaken to identify potential risks, particularly for sites with a present or past use such as industrial or urban development, but also for some agricultural uses. Sources of contamination include:

● waste disposal sites
● quarries, pits, brickworks, docklands
● heavy industrial sites such as steelworks, shipbuilding
● metal processing plants
● gas works sites
● paint manufacturing
● pharmaceutical plants
● chemical manufacturing plants
● tanneries, textile and dyeing processes
● power stations, boiler houses
● pulp and paper manufacturing
● sewage farms and sewage treatment works
● scrapyards, timber treatment works, railway sidings
● fuel storage facilities, garages, petrol station forecourts
● areas of made ground
● former mining sites and their waste disposal areas, for both coal and metalliferous mines

Types, numbers and locations of sampling points must be considered carefully to ensure adequate coverage. Solids, liquids and gases should be sampled and analysed independently.

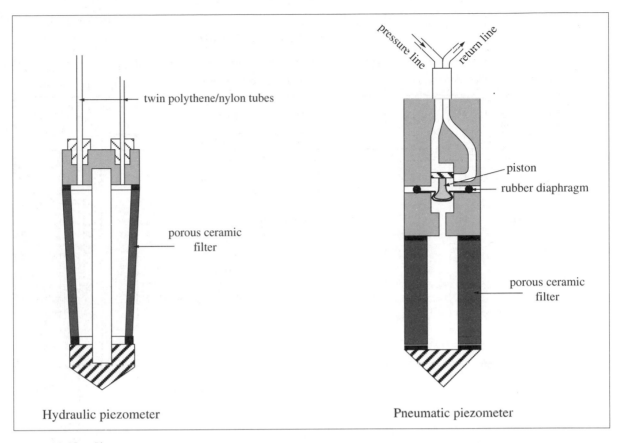

twin polythene/nylon tubes

porous ceramic filter

Hydraulic piezometer

pressure line   return line

piston
rubber diaphragm

porous ceramic filter

Pneumatic piezometer

FIGURE **14.12**   *Piezometers*

Guidance on risk assessments for the contaminated ground investigation, health and safety requirements, phasing of the investigation, exploration techniques, sampling practices and analytical techniques is given in DD175:1988, Draft BS 5930 (1996), Anon (1995).

The assessment of contaminant concentrations is conducted by comparing with trigger concentrations such as those given in ICRCL (1987). If the concentrations are below the trigger values then no remedial treatment of the site is necessary. The need for treatment is not dependent on the general level of contamination across a site but is determined by the maximum concentrations found in 'hot spots' and their distribution.

More detailed advice can be obtained from ICRCL (1983), Leach and Goodger (1991) and Cairney (1993).

Site investigation reports

All of the information is collected and collated and assessments of the ground conditions are made so that the works can be designed, methods of construction can be determined and the overall costs and risks can be estimated. Thus various parties, the designer, the contractor and the client will require information and its interpretation but their needs, responsibilities and concerns will differ.

It is unlikely that these parties will have seen the samples taken or have observed the tests carried out, especially the contractors. In many cases several months, if not years, elapse between the site investigation period and construction period when the samples have either been thrown away or are unrepresentative. It is essential, therefore, that the site investigation report provides an accurate, clear,

Name of Site Investigation Company						**BOREHOLE RECORD**		BOREHOLE 3		
Client						Site		Sheet 1 of 1		
Department of Transport						Proposed Southerly By-pass		Job Ref. 2000/56		

Casing depth / Daily progress	Water level	Samples in situ tests and coring runs			N value or core recov.	Depth (m)	Description of strata	Reduced level	Legend	287.043 N 654.529 E
		Depth (m) From	To	Type			Ground level (metres OD)	9.50		
12.7.00						0.30	Topsoil	9.20		bentonite/cement
	17.7 ▼	1.00 1.50	1.45	U D			Soft becoming firm with depth grey organic silty CLAY with thin bands or partings of silt			
		2.00 2.50	2.45	U D		(4.70)	Vertical root holes in upper horizon			
							Layer of peat at 3.50-4.00 m			
		4.00 4.50	4.45	U D						
	▲	5.00	5.45	SD	N=5	5.00	(Estuarine Clay)	4.50		sand
(6.00) 13.7.00	17.7 ▼	6.00	6.45	SD	N=15	(3.00)	Medium dense brown fine and medium SAND and fine to coarse GRAVEL			
		7.50	7.95	SD	N=25		Gravel mostly subrounded to rounded (Flood Plain Gravel)			
		8.00 8.50	8.45	U D		8.00	Very stiff fissured brown silty CLAY with much fine to coarse subangular to subrounded gravel and occasional cobbles	1.50		bentonite/cement
		10.00 10.50	10.45	U D		(6.00)	Lenses of sand at 10.2 and 12.5 m			
		12.00 12.50	12.45	U D						
	▲	14.00	14.45	SD	N=76	14.00	(Glacial Clay)	–4.50		
(10.00) 15.7.00		15.00	16.50	H	T=78 S=47 R=32	(>4.0)	Greyish brown highly becoming moderately weathered weak becoming moderately strong thinly bedded fine to medium grained SANDSTONE			sand
		16.50	18.00	H	T=95 S=75 R=45	18.00	Bedding subhorizontal Frequent joints at 45° (Coal Measures) (Westphalian B)	–8.50		
						END				

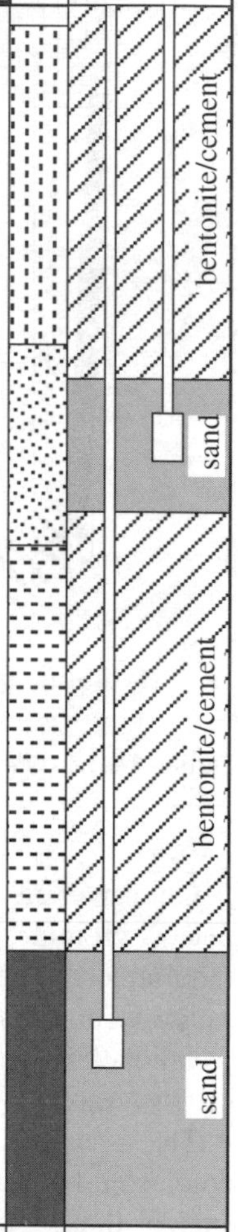

T – Total core recovery S – Solid core recovery R – Rock quality designation	Piezometer sealed at 6.00 m, water level at 2.00 m 17.7.00 Piezometer sealed at 15.00 m, water level at 7.00 m 17.7.00

Boring equipment and methods	Remarks
Cable percussion, 150 mm, GL – 15.00 m Rotary coring, water flush 15.00 – 18.00 m	Chiselling sandstone 1 hour Very fast seepage at 5.00 m, level rose to 2.20 m after 20 mins. Sealed off by casing at 10.00 m Moderate seepage at 14.00 m, level rose to 12.00 m after 20 mins Water level at 2.00 m on morning of 13.7.00 and at 8.00 m on 15.7.00
Logged by: GEB	Scale 1:100	

FIGURE 14.13 *Typical borehole record*

unambiguous and complete picture of the site conditions at and below ground level.

Site investigation should be treated as an on-going process, with information collected affecting decisions about how to proceed. Thus a fixed lump sum contract approach can be counter-productive although persuading many clients that this is the case can prove difficult. To assist in this process for larger schemes, a preliminary smaller scale site investigation can prove invaluable in assessing overall viability for the scheme, the likely types and methods of construction and identifying problem areas before more detailed investigation is carried out.

Many reports are separated into two parts or separate volumes:

- factual report
- interpretative report

Factual report

This comprises all of the factual information collected during the ground investigation and is the document provided by the site investigation contractor as the concluding part of his contract. Most of the factual information is included in Appendices. A brief written section should be included at the beginning of the report consisting of:

Report
- *Introduction*
  The site name, its location and a brief description. The name of the client, architect, consulting engineer or other interested parties. The purpose of the investigation and the party responsible for planning the investigation and issuing instructions.

- *Scope of the work*
  The types of field investigation carried out, the depths of investigation, numbers and locations of exploratory holes, frequency and type of *in situ* testing, sampling, groundwater observations and monitoring devices. The period of the investigation, delays incurred and problems encountered such as access, obstructions, equipment limitations, sample recovery, inappropriate testing, weather conditions. The laboratory tests carried out with a standard reference for the test proce-

dures adopted and with an explanation of any deviations from the standard procedures.

- *Anticipated geology*
  From available geological records a summary of the geology and other relevant conditions of the site are described since these could affect the scope of the work.

- *Ground conditions encountered*
  A brief summary of the sequence of deposits, their nature, thicknesses and variability. The groundwater conditions should also be described although this may be more open to interpretation.

- *Signatures*
  The persons within the site investigation company responsible for the organisation and preparation of the report.

Appendices
These will form the bulk of the report and will include the following:

- *Site location plan*
  This is usually an extract from an OS map.

- *Site plans*
  These identify existing and proposed features and give locations of the exploratory holes. On some jobs, it can be useful to give dimensions in relation to existing features or preferably surveyed co-ordinates since some features may be removed after the investigation.

- *Records of boreholes, trial pits, hand augers, probing, window sampling, hand excavations*
  A typical borehole record is given in Figure 14.13. Descriptions of the strata on these records must be complete, clear and unambiguous. Guidance is given on standardised descriptive terms in BS 5930:1981 and in Chapter 2 of this book.

  The borehole record or log is often the least complete piece of information yet it is the most important. It is compiled from the strata boundaries and simple descriptions given by the driller on his daily log, descriptions of the jar and bulk samples and the results of *in situ* and laboratory tests. The borehole record will be enhanced if a geologist or geotechnical engineer can include

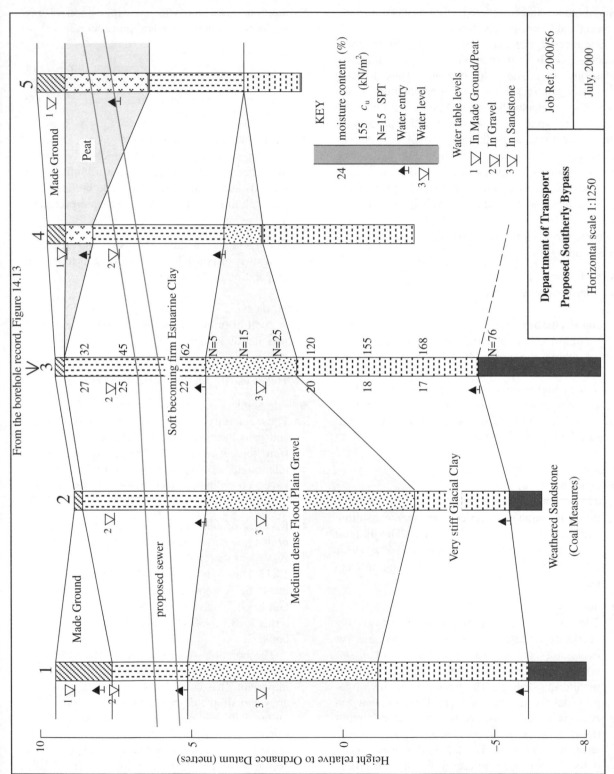

FIGURE 14.14 *Typical soil section*

descriptions and details obtained from his own field records.

The samples will have been taken at specified intervals, with gaps in knowledge (apart from the driller's observations) and often the U100s which have not been extruded for testing are discarded without inspection. The client pays for obtaining these samples so a little more expense incurred to obtain descriptions of these better quality samples will be cost-effective.

- *Records of in situ test results*
  These comprise both the information collected on site and the calculated test result or parameter. They could include the SPT, cone penetrometer plots, vane tests, pressuremeter tests, permeability (open borehole, packer or piezometer) tests. This information is best presented in a standard form and to the same scale so that information can be readily compared.

- *Groundwater observations*
  The observations made in the exploratory hole during the drilling period are reported on the records. The readings obtained subsequently in standpipes or piezometers should be tabulated or plotted on a time base.

- *Instrumentation and monitoring*
  As well as the groundwater observations there may be gas monitoring and for specialised investigations there may be inclinometers for measuring horizontal displacements or extensometers for recording settlements. There may also be other types of investigation such as geophysical methods.

- *Laboratory test results*
  The schedule of laboratory testing is usually specified by the client or his engineer or proposed by the contractor for approval.

  The results are presented in a standard form and could include classification tests (moisture content, bulk density, particle density, liquid and plastic limits, particle size distribution), shear strength tests (undrained total stress and effective stress strengths), consolidation tests (m_v and c_v), compaction tests (optimum moisture content, maximum dry density, *MCV*) and chemical tests (sulphates, chlorides, pH, chemical testing).

Interpretative report

This report is often prepared separately from the factual report for ease of reading and assimilation of information. It may be written by the site investigation contractor, or it is more often prepared by the geotechnical adviser acting for the client, architect or engineer.

The geotechnical adviser has better and more up-to-date information about the project, is better placed to incorporate changes which inevitably occur and is able to provide continued advice to other disciplines during the development of a project. When the client does not appoint a geotechnical adviser the site investigation contractor will prepare the interpretative report either on a lump sum or fee basis.

Separate consideration should be given to the contaminated land issues affecting the site, either in the same report or as a separate report. The contamination assessment must take account of the geotechnical conditions at the site.

This report is mostly a written document comprising:

- *Introduction*
  Apart from the information given in the factual report the introduction should state the terms of reference and engagement, the brief from the client with a fuller explanation of the purpose of the project and a clear understanding of his requirements and any limitations or gaps in the information gathered.

- *Walk-over survey, desk study, anticipated geology*
  These could be combined or separate sections and would describe the information obtained, its consequences and how it has affected and been addressed by the site investigation.

- *Site model (Figure 14.14)*
  This is essential for design and construction purposes. The geometry of the ground conditions is best displayed on cross-sections through the site with the ground investigation information superimposed, such as borehole records or CPT plots.

  An interpolation of the ground conditions should be carried out between these locations based on an understanding of the overall geology of the site. The engineer/geologist should demonstrate enough confidence in his/her knowledge and the adequacy of the investigation to interpolate the boundaries between deposits and strata. Alternative interpolations may be included to illustrate possible complexities. Lines showing water table levels may also be interpreted.

A typical soil section using the information obtained from Borehole 3 (in Figure 14.13) is plotted in Figure 14.14 showing how borehole information can be interpreted to produce a site model.

Once the layers have been determined, parameters for design purposes can be allocated to each layer. Test results plotted on the sections can be useful in showing variations with depth or across the site. Plots of soil properties with depth or reduced level should be made with each soil type identified separately. These plots should be treated with caution if there is no consideration for the different soil types, the deposits are dipping or the ground surface is sloping.

An assessment of the groundwater conditions in relation to the geological boundaries is required. Aquifers, aquicludes and aquitards should be identified with estimations of their permeabilities, pressure heads and seepage flows. Their relationships with any surface water should be determined such as in river and coastal areas where seasonal and tidal fluctuations are likely.

● *Advice and recommendations*
This section condenses all of the information gathered and the calculations and analyses carried out using methods described in other chapters in this book, into advice and recommendations concerning the design, construction, monitoring and potential risks involved with reference to the purpose of the proposed scheme.

Firstly it is necessary to identify all of the problems of the site in relation to the project needs. Then various options should be discussed, followed by recommendations for the design and construction of earthworks, foundations, walls, basements, pavements, tunnels, dams and other types of structures. It should include comments and advice on construction problems and expedients such as groundwater control, drainage, grouting, ground improvement, excavation stability and support, sources of materials, chemical or contamination hazards.

Any general hazards should be identified such as land subsidence, mining subsidence, landslides, drainage problems, swelling and shrinking soils, collapsible soils, erosion, seismic effects.

SUMMARY

Unlike other engineering disciplines the geotechnical engineer cannot undertake analysis and design by choosing the required proportions and properties of any construction material available. The ground is part of nature and man has had no control over its character. It must be investigated.

An accurate, complete and unambiguous picture of the site conditions must be obtained. Skimping on site investigation is for fools.

A site investigation should be phased into a desk study, site reconnaissance and then a geotechnical investigation. In view of our industrial heritage it is increasingly important to undertake a chemical investigation either prior to or alongside this investigation.

It should be emphasised that the methods of exploration, procuring samples, conducting *in situ* and laboratory testing are not perfect. They cause disturbance to the soil structure, they change the stress state and do not always model appropriately the conditions applied by the engineering works. Their limitations should be appreciated. Development in site investigation techniques is to be encouraged.

To prepare a thorough interpretative report the geotechnical engineer not only needs a good knowledge of geology, soil mechanics and geotechnical engineering (with some chemistry as well!) but must have well-developed communication and presentation skills.

Answers to exercises

Chapter 2

2.1 2.65 Mg/m^3

2.2   100, 99.1, 95.9, 93.8, 90.2, 84.4, 75.2, 63.8, 50.4, 38.9, 27.8, 15.2, 3.8 %

2.3   12.2 %, 2.22 Mg/m^3, 21.8 kN/m^3, 1.98 Mg/m^3, 19.4 kN/m^3

2.4   0.38, 0.27, 88 %, 3.3 %

2.5 2.25 Mg/m^3, 13.8%

2.6 2.12 Mg/m^3

2.7   55 %, 27 %, 28 %, CH

2.8   0.29, 0.71

Chapter 3

3.1 $8.6 \times 10^{-5} \text{ m/s}$

3.2 $2.7 \times 10^{-8} \text{ m/s}$

3.3 $2.4 \times 10^{-7} \text{ m/s}$

3.4 $60.5 \text{ m}^3\text{/day}$

3.5 $2.9 \times 10^{-5} \text{ m/s}$

3.6 $8.6 \times 10^{-4} \text{ m/s}$

3.7   5 litres/min/m

3.8   23 litres/min/m

3.9   4.0, 2.3

3.10   22.5 litres/hour

Chapter 4

4.1   139, 98, 41 kN/m^2

4.2   90, 49, 41 kN/m^2

4.3   a) 105, 29.4, 75.6 kN/m^2
       b) 195, 78.4, 116.6 kN/m^2

4.4   42.8, 67.4 kN/m^2

4.5   1) 41.1, 69.1, 97.1 kN/m^2
       2) 57.7, 69.1, 97.1 kN/m^2
       3) 57.7, 85.7, 113.7 kN/m^2

4.6 19.6 kN/m^2

4.7   1.0, 1.50, 1.18

4.8   0.96, overconsolidated

4.9   0.34, 0.58, 0.74, 0.88, 0.95, 0.98

4.10   0, 0.22, 0.16, 0.08, 0.01, −0.07, −0.13

Chapter 5

5.1 62 kN/m^2

5.2 172 kN/m^2

5.3   59.2, 37.2, 9.4 kN/m^2

5.4   50, 21 kN/m^2

5.5   48, 20 kN/m^2

5.6   57, 23 kN/m^2

5.7 45 kN/m^2

Chapter 6

6.1   0.856, 0.852, 0.848, 0.840, 0.803, 0.753, 0.704, 0.712, 0.726, 0.737

6.2 125 kN/m^2, 2.2

6.3   0.086, 0.087, 0.087, 0.201, 0.138, $0.070 \text{ m}^2\text{/MN}$
       0.09, $0.15 \text{ m}^2\text{/MN}$

6.4   0.16

6.5 $3.3 \text{ m}^2\text{/year}$, $0.20 \text{ m}^2\text{/MN}$, $2.0 \times 10^{-10} \text{ m/s}$

6.6   70 mm

6.7   a) 12 kN/m^2 b) 42 kN/m^2

6.8   46 mm

6.9   53 weeks

6.10   83 %

6.11   1.35 m

Chapter 7

7.1   57.7, 20.2 kN/m^2

7.3 24 kN/m^2

7.4   21.5°, 0.822

7.5 15 kN/m^2, 39°

7.6   1.5

7.7   248.1, 238.1, 244.3 kN/m^2
       124, 119, 122 kN/m^2

7.8 $a = 16 \text{ kN/m}^2$ $\alpha = 30.5°$
       $c' = 20 \text{ kN/m}^2$ $\phi' = 36°$

7.9 $a = 10 \text{ kN/m}^2$ $\alpha = 26.6°$
       $c' = 11.5 \text{ kN/m}^2$ $\phi' = 30°$

7.10 $\lambda = 0.14$ $N = 2.47$

7.11   0.87

7.12   184 kN/m^2

7.13   0.60, 0.50

7.14   a) 65, 97 kN/m^2 b) 129, 195 kN/m^2

Chapter 8

8.1   a) 586   b) 686 kN/m^2

8.2   2393 kN

8.3   3.1 m

8.4   9.3

8.5   a) 2380 kN/m^2 b) 9.2

8.6   a) 4065 kN/m^2 b) 15.7

8.7   a) 4.2   b) 8.0

8.8   a) 3.1   b) 4.9

Chapter 9

9.1   28 mm, 11 mm

9.2   1) 14 mm   2) 9 mm

9.3   1) 21 mm   2) 23 mm

9.4   42 mm, 12 mm

9.5   1) 19 mm   2) 29 mm

9.6   164 mm

9.7   165 mm

9.8   41 mm, 14 mm

9.9   84 mm, 177 mm

9.10   19 mm, 28 mm

9.11   22 mm, 33 mm

Chapter 10

10.1   1086 kN

10.2   16.4 m

10.3   1477 kN

10.4   9498 kN

10.5   a) 4570 kN b) 5890 kN

10.6   1206 kN

10.7   729 kN

Chapter 11

11.1   96 kN, 1.43 m

11.2   806 kN, 1.76 m

11.3   82 kN, 1.49 m

11.4   1) 86 kN, 1.39 m   2) 18.5 kN

11.5   1) 6.4   2) 2.4   3) 172, 54 kN/m^2

11.6   1) 5.6 m   2) 395 kN

11.7   7.7 m

Chapter 12

12.1   1.51, 1.23, 1.13

12.2   3.53, 2.24, 1.81

12.3   15.7°

12.4   17 kN/m^2

12.5   1.38

12.6   17 kN/m^2

12.7   2.5

12.8   30°

12.9   1.34

12.10   1.30

12.11   1.90

12.12   1.30

Chapter 13

13.1   14.4

13.2   32, 0.8

13.3   1) 16447 m^3

          2) 1221

          3) 303 bankm^3/hour, 54.3 hours

13.4   18.5%, 1.76 Mg/m^3

13.5   0.55, 0.35, 92%, 2.7%

Glossary

accretion   accumulation of sediments

alluvium   loose deposits of materials transported by rivers and streams

aquiclude   a relatively impermeable layer that prevents the flow of water

aquitard   a fairly low permeability layer that retards but does not prevent the flow of water

arenaceous   material composed mainly of sand sized particles

argillaceous   material composed mainly of clay sized particles

bentonite   mudrock composed of montmorillonite

boulder clay   non-stratified, poorly sorted material deposited directly by ice usually with a wide range of particle sizes, see till.

comminution   breaking down into smaller pieces

diatomaceous   comprising microscopic marine or freshwater algae with siliceous skeletons

earthflow   downslope movement of rock and soil

emergence   exposure of land as a result of land uplift or lowering water level

erratics   gravel to boulder sized pieces of rock transported by glaciers

escarpment   sharp change in topography produced at the rear of a landslip

esker   ridge of sand and gravel deposited by glacial melt waters bemeath the ice

fen peat   formed from grasses, sedges and reeds in low lying areas

fluvial   associated with rivers

gleyed   reduction of iron oxides to a grey-blue colour in high water table and poorly drained areas

granulometry   study of the nature of grains

H-section piles   steel pile formed in the shape of a universal column or H shape

holocene   the most recent epoch of geological time, last 10,000 years

humification   process of decomposition

humus   decomposed animal and vegetable matter

hydraulic fracture   high fluid pressures applied within a soil above the value of the minor principal stress cause the formation and opening of fractures

infiltration   the gradual movement of water into a soil

kame   mound of sand and gravel formed at the edges of ice masses

lacustrine   associated with lakes

laminated clay   clays with a macroscopic small scale (<10 mm) bedded structure as opposed to a homogeneous mass

leaching   removal of mineral salts and soluble substances as a result of percolation

levee   an embankment alongside a river formed naturally or artificially

lithorelic   a corestone of less weathered material surrounded by a matrix of weathered material

mudflow   flow of water-laden fine-grained sediment

ooze   very fine skeletal remains forming deposits on the bottom of deep oceans

PCB   polychlorinated biphenyl

perched water table   a water table lying above an aquiclude usually of localised extent and not necessarliy associated with the general water table

percolation   movement of water through a soil

periglacial   associated with the region at the edge of a glacier

phreatic water   the water in the zone of saturation

PFA   pulverised fuel ash, the residue from coal-fired power stations

pleistocene the epoch between 10,000 and 2 million years ago distinguished by ice sheet advances with warmer interglacial periods

rudaceous material composed mainly of gravel or cobble sized particles

saprolite rock decomposed to a soil *in situ* by chemical weathering

slickensided polished and striated surface on a slip plane

solifluction slow movement downslope of water-laden soils, comon in permafrost regions

solum rock totally decomposed to a residual soil

sphagnum peat peat formed from mosses

spring the location where the water table intersects the ground and emanates as a flow

thixotropy a process causing a soil to become stiffer with time if left undisturbed but when the soil is remoulded it softens or even liquefies. The process is reversible

till see boulder clay

tremie a pipe or tube inserted to the bottom of a hole beneath a fluid to allow grout or concrete to be inserted

under-reamed piles the base of a large diameter bored pile is enlarged using a cutting tool to provide a larger base area. The method relies on the soil being self-supporting

vadose water water that occurs above the water table

varved clay thin layers of fine and dark soil interlaminated with coarser and lighter soil formed in glacial lakes with cyclic sedimentation, usually annually

References

Alpan, I. (1967). The empirical evaluation of the coefficient K_o and K_{or}. *Soils and Foundations*, **7**, No. 1, p. 31.

Andersland, O.B. (1987). *Frozen ground engineering, in Ground Engineer's Reference Book*, F. G. Bell, (ed.) Butterworth and co (Publishers) Ltd

Anon (1977). The description of rock masses for engineering purposes. *Report by the Geological Society Engineering Group Working Party*, vol. 10, pp. 355–388.

Anon (1983). Review of granular and geotextile filters. *Hong Kong Geotechnical Engineering Office, Hong Kong Government.*

Anon (1990). Tropical residual soils. *Geological Society Engineering Group Working Party Report*, vol. 23, No. 1, pp. 1–111.

Anon (1991). *Inadequate site investigation*. Report by the Ground Board of the Inst. Civ. Engrs., Thomas Telford, London.

Anon (1995). *Site investigation for highway works on contaminated land*. Design Manual for Roads and Bridges, Vol. 4. Geotechnics and drainage, Section 1, Earthworks, Part 7, HA 73/95. The Highways Agency.

Anon (1999). Escape Guide: Venetian bars. *The Observer*, 9 May.

API (1993) *RP2A-WSD recommended practice for planning, designing and constructing fixed offshore platforms-working stress design*, 20th edition, American Petroleum Institute, Washington, DC, USA.

Arrowsmith, E.J. (1979). Roadwork fills – a materials engineer's viewpoint. *Proc Conf. Clay Fills, ICE,* London, pp. 25–36.

ASTM, (1992). *Annual book of ASTM standards*, vol. 04.08. American Society for Testing and Materials, Philadelphia, PA.

Atkinson, J.H., Bransby, P.L. (1978). *The mechanics of soils – an introduction to critical state soil mechanics*. McGraw-Hill Book Company (UK) Limited.

Baguelin, F., Jezequel, J.F., Le Mehaute, A. (1974). Self-boring placement of soil characteristics measurement. *Proc. Conf. on subsurface exploration for underground excavation and heavy construction*, ASCE, Henniker, New Hampshire, pp. 312–332.

Bandarin, F. (1994). The Venice Project: a challenge for modern engineering. *Proc. Inst. Civ. Engrs., Civ. Engng*, **102**, Nov., pp. 163–174.

Barnes, G.E. (1992). Stability coefficients for highway cutting slope design. *Ground Engineering*, **2**, No. 4, pp. 26–31.

Barnes, G.E. and Staples, S.G. (1988). The acceptability of clay fill as affected by stone content. *Ground Engineering*, 21, No. 1, pp. 22–30.

Barron, R.A. (1948). Consolidation of fine-grained soils by drain wells. *Trans. ASCE*, **113**, Paper 2346, pp. 718–54.

Berezantzev, V.G., Khristoforov, V. and Golubkov, V. (1961). Load bearing capacity and deformation of piled foundations. *Proc. 5th Int. Conf. Soil Mechanics and Foundation Engineering, Paris*, **2**, pp. 11–12.

Bishop, A.W. (1955). The use of the slip circle in the stability analysis of slopes. *Geotechnique*, **5**, No. 1, pp. 7–17.

Bishop, A.W. and Al-Dhahir, Z.A. (1970). Some comparisons between laboratory tests, in situ tests and full scale performance, with special reference to permeability and coefficient of consolidation. *Proc. Conf. In Situ Investigations in Soil and Rocks, ICE, London*, pp. 251–264.

Bishop, A.W. and Bjerrum, L. (1960). The relevance of the triaxial test to the solution of stability problems. *Proc. Research Conf. Shear Strength of cohesive soils, ASCE, Boulder, Colorado*, pp. 437–501.

Bishop, A.W., Green, G.E., Garga, V.K., Andresen, A. and Brown, J.D. (1971). A new ring shear apparatus and its application to the measurement of residual strength. *Geotechnique*, **21**, No. 4, pp. 273–328.

Bishop, A.W. and Henkel, D.J. (1957). *The measurement of soil properties in the triaxial test*. Edward Arnold (Publishers) Ltd. London.

Bishop, A.W. and Henkel, D.J. (1962). *The measurement of soil properties in the triaxial test*. Edward Arnold (Publishers) Ltd. London. 2nd Edition.

Bishop, A.W. and Morgenstern, N. (1960). Stability coefficients for earth slopes. *Geotechnique*, **10**, No. 4, pp. 129–150.

Bjerrum (1972). Embankments on soft ground. *Proc. ASCE Conf. Performance of Earth and earth-supported structures*, **2**, pp. 1–54.

Bolton, M.D. (1986). The strength and dilatancy of sands. *Geotechnique*, **36**, No.1, pp. 65–78

Boussinesq, J. (1885). *Application des potentials à l'étude de l'équilibre et du mouvement des solids elastiques*, Gauthier-Villars, Paris.

BRE Digest 240 (1980). *Low-rise buildings on shrinkable clay soils: Part 1.* Building Research Establishment Digest, HMSO, London.

BRE Digest 241 (1990). *Low-rise buildings on shrinkable clay soils: Part 2.* Building Research Establishment Digest, HMSO, London.

BRE Digest 242 (1980). *Low-rise buildings on shrinkable clay soils: Part 3.* Building Research Establishment Digest, HMSO, London.

BRE Digest 251 (1990). *Assessment of damage in low-rise buildings.* Building Research Establishment Digest, HMSO, London.

BRE Digest 274 (1991). *Fill Part 1: Classification and load carrying characteristics.* Building Research Establishment, HMSO, London.

BRE Digest 298 (1987). *The influence of trees on house foundations in clay soils.* Building Research Establishment Digest, HMSO, London.

BRE Digest 318 (1987). *Site investigation for low-rise buildings: desk studies.* Building Research Establishment Digest, HMSO, London.

BRE Digest 322 (1987). *Site investigation for low-rise buildings: procurement.* Building Research Establishment Digest, HMSO, London.

BRE Digest 343 (1989). *Simple measuring and monitoring of movement in low-rise buildings. Part 1: cracks.* Building Research Establishment Digest, HMSO, London.

BRE Digest 344 (1989). *Simple measuring and monitoring of movement in low-rise buildings. Part 2: settlement, heave, out-of-plumb.* Building Research Establishment Digest, HMSO, London.

BRE Digest 348 (1989). *Site investigation for low-rise buildings: the walk-over survey.* Building Research Establishment Digest, HMSO, London.

BRE Digest 361 (1991). *Why do buildings crack?* Building Research Establishment Digest, HMSO, London.

BRE Digest 412 (1996). *Desiccation in clay soils.* Building Research Establishment Digest, HMSO, London.

Brinch Hansen, J. (1970). A revised and extended formula for bearing capacity. *Bulletin No. 28, Danish Geotechnical Institute, Copenhagen*, pp. 5–11.

British Standards Institution (1981). BS 5930: *Code of practice for site investigations*, B.S.I., London.

British Standards Institution (1981). BS 6031: *Code of practice for earthworks*, B.S.I., London.

British Standards Institution (1986). BS 8004: *Code of practice for foundations*, B.S.I., London.

British Standards Institution (1987). BS 8002: *Draft code of practice for earth retaining structures*, B.S.I., London.

British Standards Institution (1988). DD 175: 1988. *Draft for development: Code of practice for the identification of potentially contaminated land and its investigation.* B.S.I., London.

British Standards Institution (1989). BS 8081: *Ground anchorages.* B.S.I., London.

British Standards Institution (1989). BS 812: *Testing aggregates*: Part 124: Method for determination of frost-heave. B.S.I., London.

British Standards Institution (1990). BS 1377: *British Standard Methods of test for soils for civil engineering purposes*, B.S.I., London.

Part 1: General requirements and sample preparation.
Part 2: Classification tests.
Part 3: Chemical and electro-chemical tests.
Part 4: Compaction – related tests.
Part 5: Compressibility, permeability and durability tests.
Part 6: Consolidation and permeability tests in hydraulic cells and with pore pressure measurement.
Part 7: Shear strength tests (total stress).
Part 8: Shear strength tests (effective stress).
Part 9: In situ tests.

British Standards Institution (1994). BS 8002: *Code of practice for earth retaining structures.* B.S.I., London.

British Standards Institution (1995). BS 8006: *Code of practice for strengthened/reinforced soils and other fills*, B.S.I., London.

British Standards Institution (1995). Draft revision of BS 5930: 1981. *Code of Practice for site investigations.* B.S.I., London.

Bortolami, G., Carbognin, L. and Gatto, P. (1986). The natural subsidence in the lagoon of Venice, Italy. *Int. Association Hydrological Sciences* Publication 151, pp. 777–784.

Bromhead, E.N., (1978). A simple ring-shear apparatus. *Ground Engineering* **12,** 5, pp. 40–44.

Brooker, E.H. and Ireland, H.O. (1965). Earth pressures at rest related to stress history. *Canadian Geotechnical Journal,* **2**, no. 1, pp. 1–15.

Burland, J.B. (1970). Discussion on session A. *Proc. Conf. In Situ Investigations in Soils and Rocks. British Geotechnical Society, London*, pp. 61–62.

Burland, J.B. (1973). Shaft friction of piles in clay – a simple fundamental approach. *Ground Engineering*, **6**, no. 3, May, pp. 30–32.

Burland, J.B. (1997). *Propping up Pisa*. Royal Academy of Engineering, London.

Burland, J.B. (1999). Pisa goes critical. Ingenia, *Quarterly Journal of the Royal Academy of Engineering*, July.

Burland, J.B., Broms, B.B. and de Mello, V.F.B. (1978). Behaviour of foundations and structures. *Proc. 9th Int. Conf. Soil Mechanics and Foundation Engineering, Tokyo*, Session 2, pp. 495–546. Also published as BRE Current Paper CP51/78.

Burland, J.B. and Burbridge, M.C., (1985). Settlement of foundations on sand and gravel, *Proc. I.C.E.*, **78** (Dec.), pp. 1325–81.

Burland, J.B., Butler, F.A. and Dunican, P., (1966). The behaviour and design of large diameter bored piles in stiff clay. *Proc. Symp. Large Bored Piles, I.C.E., London*, pp. 51–71.

Burland, J.B. and Potts, D.M. (1995). Development and application of a numerical model for the Leaning Tower of Pisa. *Int. Symposium on Pre-failure deformation characteristics of geo-materials*. IS-Hokkaido '94, Japan, vol.. 2, pp. 715–738.

Burland, J.B., Potts, D.M. and Walsh, N.M. (1981). The overall stability of free and propped embedded cantilever retaining walls. *Ground Engineering*, **14**(5), pp. 28–38.

Burland, J.B. and Wroth, C.P., (1974). Settlement of buildings and associated damage. State of the art review, *Proc. Conf. on Settlement of Structures, Cambridge*, pp. 611–654, Pentech Press, London.

Butler, F.G. (1974). Heavily overconsolidated clays, *Proc. Conf. on Settlement of Structures, Cambridge*, pp. 531–578, Pentech Press, London.

Butterfield, R. and Banerjee, P.K. (1971). A rigid disc embedded in elastic half space. *Journal S.E. Asian Soc. Soil Engineering*, **2**, No. 1, pp. 35–52.

Cairney, T. (1993). *Contaminated land: problems and solutions*. Blackie Academic and Professional, Glasgow.

Caquot, A. and Kerisel, J. (1948). *Tables for the calculation of passive pressure, active pressure and bearing capacity of foundations*. Gautheir-Villars, Paris.

Carbognin, L. and Gatto, P. (1984). An overview of the subsidence of Venice. *Int. Association Hydrological Sciences* Publication 151, pp 321–328.

Casagrande, A. (1936). The determination of the preconsolidation load and its practical significance. *Proc. 1st Int. Conf. Soil Mechanics and Foundation Engineering, Cambridge, Mass. USA*, **3**, pp. 60–64.

Cedergren, H. (1989). *Seepage, drainage and flow nets*. 3rd ed., Wiley, New York.

Chandler, R.J. (1972). Lias clay : weathering processes and their effect on shear strength. *Geotechnique*, **22**, No. 3, pp. 403–431.

Chandler, R.J. and Apted, J. (1988). The effect of weathering on the strength of London Clay. *Q.J.E.G*, **21**, pp. 59–68.

Chandler, R.J., Crilly, M.S. and Montgomery-Smith, G. (1992). A low-cost method of assessing clay desiccation for low-rise buildings. *Proceedings Institution of Civil Engineers, Civil Engineering*, **92** (2), pp. 82–89.

Chandler, R.J. and Davis, A.G. (1973). Further work on the engineering properties of Keuper Marl. C.I.R.I.A. Report 47, October 1973.

Chandler, R.J. and Peiris, R.A. (1989). Further extensions to the Bishop and Morgenstern slope stability charts. *Ground Engineering*, **22**, No. 4, pp. 33–38.

Christian, J.T. and Carrier, W.D. (1978). Janbu, Bjerrum and Kjaernsli's chart reinterpreted. *Canadian Geotechnical Journal*, **15**, pp. 123–128.

Civil Engineering Code of Practice No. 2 (1951). *Earth retaining structures*. Institution of Structural Engineers.

Clay Fills. *Proceedings of the Conference held at the Institution of Civil Engineers*, November 1978, published 1979, London.

Clayton, C.R.I. (1979). Two aspects of the use of the moisture condition apparatus. *Ground Engineering*, May, pp. 44–48.

Clayton, C.R.I., Simons, N.E. and Matthews, M.C. (1995). *Site Investigation: a handbook for engineers*. Second edition. Blackwell Scientific Ltd, Oxford.

Cole, K.W. (1972). Uplift of piles due to driving displacement. *Civil Engineering and Public Works Review*, March, pp. 263–269.

Collins, K. and McGown, A. (1974). The form and function of microfabric features in a variety of natural soils. *Geotechnique*, **24**, No. 2, pp. 223–254.

Cooke, R.W. (1974). Settlement of friction pile foundations. *Proc. Conf. on Tall Buildings, Kuala Lumpur*, No. 3, p. 7–19. Also BRE Current Paper C.P. 12/75.

Cooke, R.W. and Price, G. (1973). Strains and displacements around friction piles. *Proc. 8th Int. Conf Soil Mechanics and Foundation Engineering, Moscow*, **2–1**, pp. 53–60. Also BRE Current Paper C.P. 28/73.

Cooke, R.W., Price, G. and Tarr, K. (1980). Jacked piles in London Clay : interactional group behaviour under working conditions. *Geotechnique*, **30**, No. 2, pp. 97–136.

Coulomb, C.A. (1776). *Essai sur une application des*

règles des maximis et minimis à quelque problèmes de statique relatif à l'architecure. Memoirs Divers Savants, Academie Science, vol. 7. Paris.

Crilly, M.S. and Chandler, R.J. (1993). *A method for determining the state of desiccation in clay soils.* BRE Information Paper, IP 4/93. Building Research Establishment, Watford, England

Davis, E.H. and Poulos, H.G. (1972). Rate of settlement under two and three-dimensional conditions. *Geotechnique,* **22**, No.1, pp. 95–114

D'Appolonia, D.J., Poulos, H.G. and Ladd, C.C. (1971). Initial settlement of structures on clay. *Proc. ASCE,* **97**, No. SM10, pp. 1359–77.

Department of Transport (1978). *Reinforced earth retaining walls and bridge abutments for embankments.* Technical Memorandum (Bridges) B.E. 3/78, London.

Department of Transport, Scottish Office, Welsh Office, Department of the Environment for Northern Ireland (1992). *Specification for Highway Works.* Four volumes, H.M.S.O., London.

Driscoll, R. (1983). The influence of vegetation on the swelling and shrinking of clay soils in Britain. *Geotechnique,* **33**, No. 2, pp. 93–105

EC7 (1995). Eurocode 7: Geotechnical design–Part 1: General rules. (together with the United Kingdom National Application Document). *DD ENV 1997–1:1995.* London, British Standards Institution.

Eyles, N. and Sladen, J.A. (1981). Stratigraphy and geotechnical properties of weathered lodgement till in Northumberland, England. *Q.J.E.G.,* **14**, pp. 129–141.

Fadum, R.E. (1948). Influence values for estimating stresses in elastic foundations. *Proc. 2nd. Int. Conf. Soil Mechanics and Foundation Engineering, Rotterdam,* **3**, pp. 77–84.

Farrar, D.M. and Darley, P. (1975). The operation of earth-moving plant on wet fill. Department of the Environment, TRRL Report L.R. 688, Crowthorne, Berks.

Figueroa Vega, G.E. (1984). Case History No 9–8, Mexico. In, Poland, J.F. (1984) *op. cit.,* pp. 217–232.

Fox, E.N. (1948). The mean elastic settlement of a uniformly loaded area at a depth below the ground surface. *Proc. 2nd Int. Conf. Soil Mechanics and Foundation Engineering, Rotterdam,* **1**, p. 129.

Fox, L. (1948). Computations of traffic stresses in a simple road structure. *Proc. 2nd Int. Conf. Soil Mechanics and Foundation Engineering, Rotterdam,* **2**, pp. 236–246.

Fleming, W.G.K. and Sliwinski, Z.J. (1977). *The use and influence of bentonite in bored pile construction.* C.I.R.I.A./D.O.E. Report P.G.3.

Fleming, W.G.K., Weltman, A.J., Randolph, M.F. and

Elson, W.R. (1985). *Piling Engineering.* Surrey University Press/Halstead Press.

Fraser, R.A. and Wardle, I.J. (1976). Numerical analysis of rectangular rafts on layered foundations. *Geotechnique,* **26**, No. 4, pp. 613–630.

Gens, A., Hutchinson, J.N. and Cavounidis, S. (1988). Three-dimensional analysis of slides in cohesive soils. *Geotechnique,* **38**, No. 1, pp. 1–23.

Gibson, R.E. (1963). An analysis of system flexibility and its effect on time-lag in pore water pressure measurements. *Geotechnique,* **13**, pp. 1–11.

Gibson, R.E. (1966). A note on the constant head test to measure soil permeability in situ. *Geotechnique,* **16**, No. 3, pp. 256–259.

Gillott, J.E. (1979). Fabric, composition and properties of sensitive soils from Canada, Alaska and Norway. *Eng. Geol.,* **14**, pp. 149–172.

Giroud, J.P. (1970). Stresses under linearly loaded rectangular area. *Journal Soil Mech. Found. Div., A.S.C.E.,* **96**, No. S.M.1, pp. 263–268.

Grace, H. and Green, P.A. (1979). The use of wet fill for the construction of embankments for motorways. *Proceedings of the Conference on Clay Fills, Institution of Civil Engineers, London,* 1979, pp. 113–118.

HA44/91 (1991). *Earthworks: Design and Preparation of Contract Documents.* Highways Agency, Design Manual for Roads and Bridges, 4.1, Part 1.

HA74/95 (1995). *Design and Construction of Lime Stabilised Capping.* Highways Agency, Design Manual for Roads and Bridges, 4.1, Part 6.

Hambly, E.C. (1979). *Bridge foundations and substructures.* Building Research Establishment.

Head, K.H. (1982). *Manual of soil laboratory testing. Volume 2: permeability, shear strength and compressibility.* Pentech Press.

Head, K.H. (1986). *Manual of soil laboratory testing. Volume 3: effective stress testing.* Pentech Press.

Head, K.H. (1989). *Soil technician's handbook.* Pentech Press, London.

Head, K.H. (1992). *Manual of soil laboratory testing. Volume 1: soil classification and compaction tests,* Second Edition. Pentech Press, London.

Henkel, D.J. and Skempton, A.W. (1955). A landslide at Jackfield, Shropshire in an overconsolidated clay. *Geotechnique,* **5,** 2, pp. 131–137.

Holtz, R.D., Jamiolkowski, M.B., Lancelotta, R. and Pedroni, R. (1991). *Prefabricated vertical drains: Design and Performance.* C.I.R.I.A. Ground Engineering Report, Butterworth-Heinemann, Oxford.

Holtz, W.G. (1983). The influence of vegetation on the swelling and shrinking of clays in the United States of America. *Geotechnique,* **33**, No.2, pp. 159–163.

Hooper, J.A. (1979). *Review of behaviour of piled raft foundations*. C.I.R.I.A. Report R83.

Huber, F. (1991). Update: Bridge Scour. *Civil Engineering,* **61**, 9, pp. 62–63.

Huntington, W.C. (1957). *Earth pressures and retaining walls.* John Wiley, New York.

Hvorslev, M.J. (1951). *Time lag and soil permeability in ground water observations.* U.S. Waterways Experiment Station Bulletin 36, Vicksburg.

Ikuta, Y., Maruoka, M., Aoki, M. and Sato, E. (1994). Application of the observational method to a deep basement excavated using the top-down method. *Geotechnique,* **44**, 4, pp. 655–664.

Ingold, T.S. (1979). The effects of compaction on retaining walls. *Geotechnique,* **29**, 3, pp. 265–283.

Institution of Civil Engineers (1979). *Proceedings of the Conference on Clay Fills.*

Institution of Civil Engineers (1991). *Inadequate site investigation,* Ground Board, I.C.E., Thomas Telford Ltd., London.

Institution of Structural Engineers (1975). *Design and constrution of deep basements.* I.Struct.E., London.

Institution of Structural Engineers (1989). *Soil-structure interaction : The real behaviour of structures.* I.Struct.E., London

I.C.R.C.L. (1987). *Guidance on the assessment and redevelopment of contaminated land.* Interdepartmental Committee on the Redevelopment of Contaminated Land. Guidance Note 59/83, 2nd edition, 1987.

Irvine, D.J. and Smith, R.J.H. (1983). *Trenching practice.* C.I.R.I.A. Report 97, London.

Jaky, J. (1944). The coefficient of earth pressure at rest. *Journal for Society of Hungarian Architects and Engineers, Budapest, Hungary,* October 1944, pp. 355–358.

Jaky, J. (1948). Pressure in silos. *Proc. 2nd Int. Conf. Soil Mechanics and Foundation Engineering,* **1**, pp. 103–107.

Jamiolkowski, M., Lancellotta, R., Pasqualini, E., Marchetti, S. and Nova, R. (1979). Design parameters for soft clays. General Report, *Proc. 7th European Conf. on Soil Mechanics and Foundation Engineering,* **5**, pp. 27–57.

Janbu, N. (1973). Slope stability computations. In *Hirschfield and Poulos (eds.). Embankment Dam Engineering, Casagrande Memorial Volume,* John Wiley and Sons, New York, pp. 47–86.

Janbu, N., Bjerrum, L. and Kjaernsli, B. (1956). *Veiledning ved løsning av fundamenteringsoppgaver.* Norwegian Geotechnical Institute, Oslo, Publication 16, pp. 30–32.

Jewell, R.A. and Wroth, C.P. (1987). Direct shear tests on reinforced sand. *Geotechnique,* **37**, No. 1, pp. 53–68.

Johannessen, I.J. and Bjerrum, L. (1965). Measurement of the compression of a steel pile to rock due to settlement of the surrounding clay. *Proc. 6th ICSMFE,* Montreal, Canada, vol. 2, pp. 261–264.

Johnson, S.J. (1970). Precompression for improving foundation soils. *Proc. ASCE Journal Soil Mech. and Found. Div.,* **96**, SM1, pp. 111–144.

Kerisel, J. and Absi, E. (1990). *Active and passive earth pressure tables.* Third edition, A. A. Balkema, Rotterdam.

Ladd, C.C., Foott, R., Ishihara, K., Schlosser, F. and Poulos, H.G. (1977). Stress-deformation and strength characteristics: state-of-the-art report. *Proc. 9th Int. Conf. Soil Mechanics and Foundation Engineering, Tokyo,* **2**, pp. 421–494.

Landva, A.O., Korpijaakko, E.O. and Pheeney, P.E. (1983). Geotechnical classification of Peats and Organic Soils. *A.S.T.M. S.T.P. 820 Testing of Peats and Organic Soils.* American Society for Testing and Materials, pp. 37–51.

Landva, A.O. and Pheeney, P.E. (1980). Peat fabric and Structure. *Canadian Geotechnical Journal,* **17**, pp. 416–435.

Leach, B.A. and Goodger, H.K. (1991). *Building on derelict land.* C.I.R.I.A. Report S.P. 78.

Leonards, G.A. (1962). *Foundation Engineering.* McGraw-Hill, New York.

Leonards, G.A. and Davidson, L.W. (1984). Reconsideration of failure initiating mechanisms for Teton Dam. *Proc. Int. Conf. on Case Histories in Geot. Eng.* Univ. of Missouri-Rolla, vol. 3, pp. 1103–1113

Lupini, J.F., Skinner, A.E. and Vaughan, P.R. (1981). The drained residual strength of soils. *Geotechnique,* **31**, No. 2, pp. 181–213.

McLaren, D. (1968). Trial embankment at Killington. *T.R.R.L. Report LR 238,* Department of Transport, Crowthorne, Berks.

Mair, R.J. and Wood, D.M. (1987). *Pressuremeter testing : methods and interpretation.* C.I.R.I.A. Ground Engineering Report: in-situ testing, Butterworths, London.

Marsland, A. (1971). The shear strength of stiff fissured clays. *Proc. Roscoe Memorial Symposium, Cambridge,* March, 1971. Also published as B.R.E. current paper C.P. 21/71.

Matheson, G.D. (1988). *The use and application of the moisture condition apparatus in testing soil suitability for earthworking.* S.D.D. Applications Guide No. 1, T.R.R.L. Scothill Branch.

Mayne, P.W. and Kulhawy, F.H. (1982). K_o – OCR relationships in soil. *A.S.C.E, J.G.E.D.,* **108**, pp. 851–872.

Meigh, A.C. (1976). The Triassic rocks, with particular reference to predicted and observed performance of some major foundations. 16th Rankine Lecture, *Geotechnique*, **26**, No.3, pp. 391–452.

Meigh, A.C. (1987). *Cone penetration testing: methods and interpretation*. C.I.R.I.A. Ground Engineering Report: in-situ testing, Butterworths, London.

Mesri, G. (1973). Coefficient of secondary compression. *Proc. A.S.C.E.*, **99**, S.M.1., pp. 123–137.

Mesri, G., Rokhsar, A. and Bohar, B.F. (1975). Composition and compressibility of typical samples of Mexico City clay. *Geotechnique*, **25**, 3, pp. 527–554.

Meyerhof, G.G. (1953). The bearing capacity of foundations under eccentric and inclined loads. *Proc. 3rd Int. Conf. Soil Mechanics and Foundation Engineering, Zurich*, **1**, pp. 440–445.

Meyerhof, G.G. (1976). Bearing capacity and settlement of pile foundations. *Proc. A.S.C.E., Journal Geot. Eng. Div.*, **102**, G.T.3., pp. 197–228.

Milovic, D.M. and Tournier, J.P. (1971). Stresses and displacements due to rectangular load on a layer of finite thickness. *Soils and Foundations*, **11**, No. 1, March, pp. 1–27.

Mitchell, J.K. (1993). *Fundamentals of soil behaviour.* second edition. John Wiley and sons, New York, USA.

Morgenstern, N.R. and Price, V.E. (1967). The analysis of the stability of general slip surfaces. *Geotechnique*, **15**, No. 1, pp. 79–93.

Muir Wood, D. (1990). *Soil behaviour and critical state soil mechanics.* Cambridge University Press, Cambridge, England.

NAVFAC DM7. (1971). *Design Manual: soil mechanics, foundations and earth structures*, U.S. Department of the Navy, Washington, D.C.

NAVFAC DM7. (1982). *Design Manual: Soil mechanics, foundations and earth structures, soil dynamics, deep stabilisation and special geotechnical construction.* Design Manual 7, U.S. Department of the Navy, Alexandria, Va.

Newmark, N.M. (1942). *Influence charts for computation of stresses in elastic foundations.* Engineering Experiment Station Bulletin No. 338, University of Illinois, Urbana, Ill.

N.H.B.C. (1985). *Building near trees.* National House-Building Council, Practice Note 3.

Norbury, D.R., Child, G.H. and Spinks, T.W. (1986). A critical review of Section 8 (B.S. 5930) – soil and rock description. *Geological Society, Engineering Geology Special Publication*, No. 2, pp. 331–342.

Nordlund, R.L. and Deere, D.U. (1970). Collapse of Fargo Grain Elevator. *Proc. ASCE*, **96**, No. SM2, pp. 585–607.

O'Connor, M.J. and Mitchell, R.J. (1977). An extension of the Bishop and Morgenstern slope stability charts. *Can. Geot. Journal*, **14**, pp. 144–151.

O'Neill, M.W. and Poormoayed, N. (1980). Methodology for foundations on expansive clays. *Proc. ASCE, Journal Geot. Eng. Div.*, 106, GT12, pp. 1345–1367.

Padfield, C.J. and Mair, R.J. (1984). *Design of retaining walls embedded in stiff clays.* C.I.R.I.A. Report 104, London.

Parsons, A.W. (1976). The rapid measurement of the moisture condition of earthwork material. *T.R.R.L. Report LR 750*, Department of the Environment, Crowthorne, Berks.

Parsons, A.W. and Boden, J.B. (1979). The moisture condition test and its potential applications in earthworks. *T.R.R.L. Supplementary Report 522*, Department of Transport, Crowthorne, Berks.

Parsons, A.W. and Darley, P. (1982). The effect of soil conditions on the operation of earthmoving plant. *T.R.R.L. Report LR 1034*, Department of Transport, Crowthorne, Berks.

Peck, R.B. (1969). Advantages and limitations of the observational method in applied soil mechanics. *Geotechnique*, **19**, 2, pp. 171–187.

Peck, R.B. and Bryant, F.G. (1953). The bearing capacity failure of the Transcona Elevator. *Geo- technique*, 3, pp. 201–208.

Peck, R.B., Hanson, W.E. and Thornburn, T.H. (1974). *Foundation engineering.* 2nd edition, John Wiley and Sons, New York.

Penman, A.D.M. (1977). The failure of Teton Dam. *Ground Engineering*, 10, No.6, pp. 18–27.

Penman, A.D.M. (1979). Construction pore pressures in two earth dams. *Proc. Conf. Clay Fills, Institution of Civil Engineers*, London, pp. 177–187.

Penman, A.D.M. (1986). On the embankment dam. *Geotechnique*, **36**, No.3, pp. 303–348.

Perry, J. and West, G. (1996). Sources of information for site investigations in Britain (Revision of TRL Report LR403) *TRL Report 192*, Transport Research Laboratory, Crowthorne, Berkshire.

Poland, J.F. ed. (1984). *Guidebook to studies of land subsidence due to groundwater withdrawal.* Studies and Reports in Hydrology, UNESCO, 40.

Poulos, H.G. (1968). Analysis of the settlement of pile groups. *Geotechnique*, **18**, No. 4, pp. 449–471.

Poulos, H.G. and Davis, E.H. (1974). *Elastic solutions for soil and rock mechanics.* John Wiley and Sons, New York.

Poulos, H.G. and Davis, E.H. (1980). *Pile foundation analysis and design.* John Wiley and Sons, New York.

Poulos, H.G. and Mattes (1971). Settlement and load distribution of pile groups. *Aust. Geomechanics Journal,* **G1**, No. 1, pp. 18–28.

Preene, M., Roberts, T.O.L., Powrie, W., Dyer, M.R. (1997) *Groundwater control:design and practice.* Funders Report FR/CP/50, CIRIA, London.

Randolph, M.F., Dolwin, J. and Beck, R. (1994). Design of driven piles in sand. *Geotechnique,* **44**, No.3, pp. 427–448.

Randolph, M.F. and Wroth, C.P. (1982). Recent developments in understanding the axial capacity of piles in clay. *Ground Engineering,* **15**, No. 7, pp. 17–25.

Ricceri, G. and Butterfield, R. (1974). An analysis of compressibility data from a deep borehole in Venice. *Geotechnique,* **24**, 2, pp. 175–192.

Richards, B.G., Peter, P. and Emerson, W.W. (1983). The effects of vegetation on the swelling and shrinking of soils in Australia. *Geotechnique,* **33**, No.2, pp. 127–139.

Roe, P.G. and Webster, D.C. (1984). Specification for the TRRL frost-heave test. *TRRL Supplementary Report 829,* Department of Transport, Crowthorne, Berkshire.

Roscoe, K.H., Schofield, A.N. and Wroth, C.P. (1958). On the yielding of soils. *Geotechnique,* **8**, No. 1, pp. 22–53.

Rowe, P.W. (1962). The stress-dilatancy relation for static equilibrium of an assembly of particles in contact, *Proc. Royal Society of London, Series A,* pp. 500–527.

Rowe, P.W. (1968). The influence of geological features of clay deposits on the design and performance of sand drains. *Proc. Inst. Civil Engineers, Suppl. 1,* pp. 1–72.

Rowe, P.W. and Barden, L. (1966). A new consolidation cell, *Geotechnique,* **16**, No. 2, pp. 162–170.

Schmertmann, J.H. (1955). The undisturbed consolidation behaviour of clay. *Trans. A.S.C.E.,* **120**, pp. 1201–1227.

Schmertmann, J.H. (1970). Static cone to compute static settlement over sand. *Proc. A.S.C.E.,* **96**, No. S.M.3, Paper 7302, pp. 1011–1043.

Schmertmann, J.H., Hartmann, J.P. and Brown, P.R. (1978). Improved strain influence factor diagrams. *Proc. A.S.C.E.,* **104**, No. G.T.8, pp. 1131–1135.

Schofield, A.N. and Wroth, C.P. (1968). *Critical state soil mechanics.* McGraw-Hill, London.

Semple, R.M. and Rigden, J. (1986). Shaft capacity of driven pipe piles in clay. *Ground Engineering,* January, pp. 11–19.

Serota, S. and Lowther, G. (1973). SPT practice meets critical review. *Ground Engineering,* **6**, No. 1, pp. 20–25.

Sevaldson, R.A. (1956). The slide in Lodalen, October 6th, 1954. *Geotechnique,* **6**, 4, pp. 167–182.

Sheppard, G.A.R. and Aylen, L.B. (1957). The Usk scheme for the water supply of Swansea. *Proc. ICE,* vol. 7, pp. 246 265.

Sherard, J.L. Dunnigan, L.P. and Talbot, J.R. (1984a). Basic properties of sand and gravel filters. *Proc. ASCE, Journal Geot. Eng. Div.* 110, No.6, pp. 684–700.

Sherard, J.L. Dunnigan, L.P. and Talbot, J.R. (1984b). Filters for silts and clays. *Proc. ASCE, Journal Geot. Eng. Div.* 110, No.6, pp. 701–718.

Sherwood, P.T. (1971). The reproducibility of the results of soil classification and compaction tests. *T.R.R.L., Report L.R. 339,* Department of Transport, Crowthorne, Berkshire.

Sherwood, P.T. (1993). *Soil stabilisation with cement and lime.* TRL state of the art review. Transport Research Laboratory, HMSO.

Sills, G.C. (1974). An assessment, using three field studies, of the theoretical concept of the efficiency of drainage layers in an embankment. *Geotechnique,* **24**, No. 4, pp. 467–474.

Simpson, B. and Driscoll, R. (1998). *Eurocode 7: a commentary.* Building Research Establishment. Construction Research Communications Ltd, Watford.

Site Investigation Steering Group (1993). *Site investigation in construction.* Thomas Telford Limited, London.

Part 1 – Without site investigation ground is a hazard.

Part 2 – Planning, procurement and quality management.

Part 3 – Specification for Ground Investigation.

Part 4 – Guidelines for the safe investigation by drilling of landfills and contaminated land.

Skempton, A.W. (1953). The post glacial clays of the Thames Estuary. *Proc. 3rd Int. Conf. Soil Mechanics and Foundation Engineering, Zurich,* **1**, p. 302.

Skempton, A.W. (1953). The colloidal "activity" of clays. *Proc. 3rd Int. Conf. Soil Mechanics and Foundation Engineering, Zurich,* **1**, pp. 57–61.

Skempton, A.W. (1954). The pore pressure coefficients A and B. *Geotechnique,* **4**, No. 4.

Skempton, A.W. (1959). Cast in-situ bored piles in London Clay. *Geotechnique,* **9**, pp. 153–178.

Skempton, A.W. (1961). Horizontal stresses in an overconsolidated eocene clay. *Proc. 5th I.C.S.M.F.E., Paris,* **1**, pp. 352–357.

Skempton, A.W. (1964). Long-term stability of slopes. *Geotechnique,* **14**, No. 2, pp. 77–101.

Skempton, A.W. (1966). Summing up. *Proc. Symp. on Large Bored Piles, London.*

Skempton, A.W. (1986). Standard penetration test procedures and the effects in sands of overburden pressure, relative density, particle size, ageing and overconsolidation. *Geotechnique,* **36**, No. 3, pp. 425–447.

Skempton, A.W. and Bjerrum, L. (1957). A contribution to the settlement of foundations on clay. *Geotechnique,* **7**, No. 4, pp. 168–178.

Skempton, A.W. and Coats, D.J. (1985). Carsington dam

failure. *Failures in Earthworks,* pp. 203–220, Thomas Telford, London.

Skempton, A.W. and Hutchinson, J.N. (1969). Stability of natural slopes and embankment foundations. *Proc. 7th Int. Conf. Soil Mechanics and Foundation Engineering, Mexico City,* State-of-the-Art Volume, pp. 291–340.

Skempton, A.W. and MacDonald, D.H. (1956). Allowable settlements of buildings. *Proc. I.C.E., part 3,* **5**, pp. 727–768.

Somerville, S.H. (1986). *Control of groundwater for temporary works.* CIRIA Report 113, London.

Stroud, M.A. and Butler, F.G. (1975). The standard penetration test and the engineering properties of glacial materials. *Proc. Symp. Engineering Properties of Glacial Materials.* Midlands Soil Mechanics and Foundations Society.

Taylor, D.W. (1937). Stability of earth slopes. *Journal Boston Soc. Civ. Engrs.,* **24**, pp. 197–246.

Taylor, D.W. (1948). *Fundamentals of soil mechanics.* John Wiley and Sons, New York.

Terzaghi, K. (1943). *Theoretical soil mechanics,* John Wiley and Sons, New York.

Terzaghi, K. (1954). Anchored bulkheads. *Trans. A.S.C.E.,* **119**, paper 2720, pp. 1243–1281.

Terzaghi, K. and Peck, R.B. (1967). *Soil mechanics in engineering practice.* 2nd edition, John Wiley and Sons, New York, first edition published 1948.

Terzaghi, K., Peck, R.B. and Mesri, G. (1996). *Soil mechanics in engineering practice.* 3rd edition, John Wiley and Sons, New York.

Thorburn, S. (1984). Field Tesitng: the standard penetration test. *Proc. 20th Reg. Meeting, Eng. Group of Geol. Soc.,* Surrey, Sept, 1984 on Site Investigation in Practice:Assessing BS5930, vol II, pp. 37–48.

Tomlinson, M.J. (1970). *The adhesion of piles driven in stiff clay.* C.I.R.I.A. Research Report No. 26.

Tomlinson, M.J. (1971). Some effects of pile driving on skin friction. *Proc. Conf. on Behaviour of Piles, I.C.E., London,* pp. 107–114.

Tomlinson, M.J. (1986). *Foundation design and construction.* Fifth edition. Longman Scientific and Technical.

Tomlinson, M.J. (1987). *Pile design and construction practice.* Third edition, Viewpoint Publications, Palladian Publications Limited.

Ueshita, K. and Meyerhof, G.G. (1968). Surface displacement of an elastic layer under uniformly distributed loads. *Highway Research Board Record,* No. 228, pp. 1–10.

US Army Corps of Engineers (1965). *Soils and geology-pavement design for frost conditions.* Tech. Manual TM 5-818-2, Dept. of the Army, Washington, D.C.

Vesic, A.S. (1967). *A study of bearing capacity of deep foundations.* Final Report, Project B-189, Georgia Inst. Tech., Atlanta, Georgia, pp. 231–6.

Vesic, A.S. (1969). Experiments with instrumented pile groups in sand. *A.S.T.M., S.T.P. 444,* pp. 177–222.

Vesic, A.S. (1975). Bearing capacity of shallow foundations. In *Foundation Engineering Handbook,* Edited by Winterkorn, H.F. and Fang, H-Y., Van Nostrand Reinhold Company.

Vesic, A.S. (1977). *Design of piled foundations.* N.C.H.R.P. Synthesis of Highway Practice, No. 42, Transportation Research Board, Washington, D.C.

Vijayvergiya, V.N. and Focht, J.A.J. (1972). A new way to predict the capacity of piles in clay. *4th Annual Offshore Tech. Conf., Houston,* **2**, pp. 865–874.

Weltman, A.J. (1980). *Pile load testing procedures.* C.I.R.I.A./D.O.E. Report P.67.

Weltman, A.J. and Healy, P.R. (1978). *Piling in 'boulder clay' and other glacial tills.* C.I.R.I.A./D.O.E. Piling Development Group Report P.G.5, Nov. 1978.

Whitaker, T. (1957). Experiments with model piles in groups. *Geotechnique,* **7**, No. 1, pp. 147–167.

Whitaker, T. (1970). *The design of piled foundations.* Oxford : Pergamon.

Whitaker, T. and Cooke, R.W. (1966). An investigation of the shaft and base resistances of large bored piles in London Clay. *Proc. Symp. on Large Bored Piles, London,* pp. 7–49.

White, L.S. (1953). Transcona Elevator failure: Eyewitness account. *Geotechnique,* 3, pp. 209–214.

Williams, B.P. and Waite, D. (1993). *The design and construction of sheet-piled cofferdams.* CIRIA Special Publication 95, London.

Wilkinson, W.B. (1968). Constant head in situ permeability tests in clay strata. *Geotechnique,* **18**, pp. 172–194.

Wroth, C.P. (1979). Correlations of some engineering properties of soils. *Proc. 2nd. Int. Conf. on Behaviour of Offshore Structures, London,* **1**, pp. 121–132.

Wroth, C.P. and Hughes, J.M.O. (1973). An instrument for the in-situ measurement of the properties of soft clays. *Proc. 8th Int. Conf. Soil Mechanics and Foundation Engineering, Moscow,* **102**, pp. 487–494.

Wynne, C.P. (1988). *A review of bearing pile types.* C.I.R.I.A./P.S.A. Report PG1, second edition.

Index